AF835657

State of the Environment

ZAMBEZI BASIN 2000

The World Conservation Union

SARDC Southern African Research and Documentation Centre

Sida

The State of the Environment in the ZAMBEZI BASIN 2000

A report by Southern African Development Community (SADC-ELMS and WSCU),
The World Conservation Union (IUCN)
Zambezi River Authority (ZRA) and
Southern African Research and Documentation Centre-Musokotwane Environment Resource Centre for Southern Africa (SARDC-IMERCSA)

SADC Environment and Land Management Sector (ELMS)
Private Bag A284, Maseru 100, Lesotho
E-mail sadcelms@lesoff.co.za
Fax (266) 310190 Tel (266) 312158

SADC Water Resources Sector Coordination Unit (WSCU)
Private Bag A440, Maseru 100, Lesotho
E-mail sadcwscu@lesoff.co.za
Fax (266) 310465 Tel (266) 320722

IUCN-The World Conservation Union
Regional Office for Southern Africa
Box 745, Harare, Zimbabwe
E-mail root@iucnrosa.org.zw
Fax (263-4) 720738 Tel (263-4) 728266/7

Zambezi River Authority (ZRA)
Box 30233, Lusaka, Zambia
E-mail zaraho@zamnet.zm
Fax (260-1) 227498 Tel (260-1) 236601

Southern African Research and Documentation Centre-Musokotwane Environment Resource Centre for Southern Africa (SARDC-IMERCSA)
Box 5690, Harare, Zimbabwe
E-mail cep@imercsa.sardc.net
Fax (263-4) 737301 Tel (263-4) 720814

ISBN: 1-77910-009-4

Citation: Chenje, M., (ed.), *State of the Environment Zambezi Basin 2000,* SADC/IUCN/ZRA/SARDC, Maseru/Lusaka/Harare, 2000

Cover and text design: Inkspots Design Studio
Cover photograph: N. Greaves/APG
Origination: CREDA Communications (Pty) Ltd
Printed by: CREDA Communications (Pty) Ltd

State of the Environment ZAMBEZI BASIN 2000

Editor
Munyaradzi Chenje

Writers
Munyaradzi Chenje
Mats Kullberg
Tendayi Kureya
Elton Laisi
Clever Mafuta
Cathrine Mutambirwa
Munyaradzi Saruchera
Lovemore Sola

This report was prepared as part of the Communicating the Environment Programme (CEP), a southern African partnership of SADC-ELMS, IUCN and SARDC-IMERCSA. This partnership was expanded to include SADC-WSCU and ZRA. The two-year initiative was funded by the Swedish International Development Cooperation Agency (Sida).

ZAMBEZI BASIN

THE STATE OF THE ENVIRONMENT REPORTING PROGRAMME IN THE ZAMBEZI BASIN (SOEPROZ)

Steering Committee

C. Obol, Programme Officer, SADC Environment and Land Management Sector (ELMS), Lesotho
P. Ramoeli, Sector Coordinator, SADC Water Sector Coordination Unit (WSCU), Lesotho
T. Matiza-Chiuta, Regional Programme Coordinator, IUCN ROSA, Zimbabwe
A.S. Murty, Senior Engineer, Zambezi River Authority (ZRA), Zambia
M. Chenje, Director, SARDC-IMERCSA, Zimbabwe

Scientific Advisory Committee

C. Obol, Programme Officer, SADC ELMS, Lesotho
P. Ramoeli, Sector Coordinator, SADC WSCU, Lesotho
E.H. Zulu, Programme Manager, SADC Mining Sector Coordination Unit (MSCU), Zambia
T. Matiza-Chiuta, Regional Programme Coordinator, IUCN ROSA, Zimbabwe
A.S. Murty, Senior Engineer, ZRA, Zambia
K. Kalaote, Senior Hydrological Engineer, Department of Water Affairs, Botswana
P.W.R. Kaluwa, Principal Water Officer, Ministry of Water Development, Malawi
S. Mukuni, Senior Chief, Traditional Authority, Zambia
J. Timberlake, Technical Coordinator, Biodiversity Foundation for Africa, Zimbabwe

SARDC-IMERCSA Staff

Munyaradzi Chenje – Director
Lovemore Sola – Head of Programme
Clever Mafuta – CEP Programme Coordinator
Elton Laisi – SOEPROZ Coordinator
Mats Kullberg – Environmental Communications Specialist
Tendayi Kureya – Senior Researcher/Writer
Munyaradzi Saruchera – Senior Researcher/Writer
Wilson Yule – Resource Centre Manager
Cathrine Mutambirwa – Researcher/Writer
Maria Mutama – Research Assistant
Joshua Chigodora – Senior Documentalist
Natural Nyaganzi – Assistant Documentalist
Sekai Mubako – Administrative Assistant

Production and Design

Paul Wade, Inkspots Design Studio

CONTENTS

LIST OF MAPS, TABLES, FIGURES, BOXES, STORIES AND CASE STUDIES

MAPS

TABLES

FIGURES

BOXES

STORIES

CASE STUDIES

ACRONYMS AND ABBREVIATIONS

AAA	Angolan Association for the Environment
ADMADE	Administrative Management Design Programme for Game Management
AIDS	Acquired Immuno-Deficiency Syndrome
Amsl	Above mean sea level
AMCEN	African Ministerial Conference on the Environment
ASIP	Agricultural Sector Investment Programme
BASP	Biodiversity Strategic Action Plan
BOD	Biological Oxygen Demand
BPC	Botswana Power Corporation
CAB	Congo Air Boundary
CAMPFIRE	Communal Areas Management Programme for Indigenous Resources
CAP	Community Action Project
CAPCO	Central African Power Corporation
CBD	Convention on Biological Diversity
CBNRM	Community Based Natural Resources Management
CBR	Crude Birth Rate
CCD	Convention to Combat Desertification
CDC	Centres for Disease Control and Prevention
CECT	Chobe Enclave Conservation Trust
CEDAW	Convention on the Elimination of All Forms of Discrimination Against Women
CEP	Communicating the Environment Programme
CGIAR	Consultative Group on International Agricultural Research
CITES	Convention on the International Trade of Endangered Species of Wild Fauna and Flora
CMS	Convention on the Conservation of Migratory Species of Wild Animals
COMESA	Common Market for Eastern and Southern Africa
CP	Cleaner Production
CPC	Cleaner Production Centres
CPS	Central Power Systems
DAP	Draught Animal Power
DDT	Dichloro-diphenyl-trichloroethane
DEAP	District Environmental Action Plan
DEMS	Department of Electrical and Mechanical Services (Botswana)
DHS	Demographic and Health Survey
DMC	Drought Monitoring Centre
DNAF	National Directorate of Agriculture and Forestry
DNPWLM	Department of National Parks and Wildlife Management
DNPWS	Department of National Parks and Wildlife Services
DNRSE	Department of New and Renewable Sources of Energy
DOE	Department of Energy
DRC	Democratic Republic of Congo
EAD	Environmental Affairs Department
ECZ	Environmental Council of Zambia
EDF	Environmental Development Fund
EDG	Energy Development Group
EDM	Electricidade de Mozambique
EIA	Environmental Impact Assessment
EMS	Environmental Management Systems
ENE	Empresa Nacional de Electricidade (Angola)
ENSO	El-Niño/Southern Oscillation
EPZ	Export Processing Zones
ERP	Economic Reform Programme
ESCOM	Electricity Supply Commission of Malawi
ESMAP	Energy Sector Management Assistance Programme
ESP	Environmental Support Programme
FANR	Food, Agriculture and Natural Resources
FAO	Food and Agriculture Organisation
FCCC	Framework Convention on Climate Change
GCOS	Global Climate Observing System
GDP	Gross Domestic Product
GEF	Global Environment Facility
GHG	Greenhouse Gas
GIS	Geographic Information Systems
GMA	Game Management Area
GMNRMA	Gorongoza-Marromeu Region Natural Resource Management Area
GPZ	Zambezi River Development Authority
GRR	Gross Reproduction Rate
GRZ	Government of the Republic of Zambia
GTZ	German Technical Cooperation
HCB	Cahora Bassa Hydropower Company
HDI	Human Development Index
HDR	Human Development Report
HIV	Human Immunodeficiency Virus
ICDS	Intercensal Demographic Survey
ICF	International Crane Foundation
ICRAF	International Centre for Research in Agroforestry
ICRISAT	International Crops Research Institute for the Semi-Arid Tropics
IDF	Forestry Development Institute
IFFW	Inland Fisheries Forestry and Wildlife
IFS	Inland Fisheries Sector
IKS	Indigenous Knowledge System
ILO	International Labour Organisation
IMERCSA	Musokotwane Environment Resource Centre for Southern Africa
IMR	Infant Mortality Rate
IPCC	Intergovernmental Panel on Climate Change
IPP	Independent Power Producer
IRDNC	Integrated Rural Development and Nature Conservation
IRI	International Research Institute
ISO	International Organisation for Standardisation
ITCZ	Inter-Tropical Convergence Zone

IUCN ROSA	IUCN Regional Office for Southern Africa
IUCN	World Conservation Union
IWRB	Wetland International Waterfowl Research
IWRM	Integrated Water Resource Management
JICA	Japanese Aid
LDC	Less Developed Countries
LIFE	Living In Finite Environment
LIRDP	Luangwa Integrated Resource Development Project
MAFF	Ministry of Agriculture, Food and Fisheries
MASAF	Malawi Social Action Fund
MEAs	Multilateral Environmental Agreements
MEDP	Micro-Enterprise Development Programme
MIASA	Mining Industry Association of Southern Africa
MMR	Number of maternal deaths per 100,000 live births
MOHCW	Ministry of Health and Child Welfare
MoU	Memorandum of Understanding
MVA	Manufacturing Value-Added
NATURE	Nature programme
NCS	National Conservation Strategy
NCZ	Nitrogen Chemicals of Zambia
NEAP	National Environmental Action Plan
NEMC	National Environment Management Council
NETCAB	Networking and Capacity Building Initiative for Southern Africa
NGO	Non-Governmental Organisation
NOCZIM	National Oil Company of Zimbabwe
NP	National Park
NPWS	National Parks and Wildlife Service
NRM	Natural Resources Management
NRMP	Natural Resources Management Programme
NRR	Net Reproduction Rate
NRSE	New and Renewable Sources of Energy
O&M	Operation and maintenance
ODA	Official Development Assistance
OUZIT	Okavango Upper Zambezi International Tourism
PARCS	Protected Areas Conservation Strategy
PE	Potential Evaporation
PET	Potential Evapotranspiration
PFA	Platform for Action
PIC	Prior Informed Consent
PMU	Programme Management Unit
PPP	Polluter pays principle
PRB	Population Reference Bureau
PRSD	Programme Review and Strategy Development
PV	Photovoltaic
RE	Renewable Energy
RET	Renewable Energy Technology
RETOSA	Regional Tourism Organisation for Southern Africa
SABONET	Southern African Botanical Diversity Network
SACCAR	Southern African Centre for Cooperation in Agricultural Research
SADC ELMS	SADC Environment and Land Management Sector
SADC FTCU	SADC Forestry Technical Coordination Unit
SADC IFFW	SADC Inland Fisheries, Forestry and Wildlife
SADC MCU	SADC Mining Sector Coordination Unit
SADC TAU	SADC Technical and Administrative Unit
SADC WSCU	SADC Water Sector Coordination Unit
SADC	Southern African Development Community
SADCSTAN	SADC Standardisation
SAfAIDS	Southern Africa Aids
SAP	Structural Adjustment Programme
SAPES	Southern African Political and Economic Series
SAPP	Southern African Power Pool
SARDC	Southern African Research and Documentation Centre
SATAC	Southern Africa Technical Advisory Committee
SAZ	Standards Association of Zimbabwe
SDI	Spatial Development Initiative
SEA	Strategic Environmental Assessment
SEI	Stockholm Environment Institute
SHER	Revue Hydroelectric Society
Sida	Swedish International Development Cooperation Agency
SLAMU	South Luangwa Area Management Unit
SME	Small and Medium Sized Enterprises
SMIP	Sorghum and Millet Improvement Programme
SO	Southern Oscillation
SOE	State of the Environment
SOER	State of the Environment Reporting
SOS	State of Sustainability
SRAP	Sub-regional Action Programme
SSA	Sub-Saharan Africa
SST	Sea Surface Temperature
SWH	Solar Water Heaters
TANESCO	Tanzania Electric Supply Company
TAU	Technical and Administrative Unit
TAZARA	Tanzania – Zambia Railway
TBCA	Transboundary Conservation Area
TCU	Tourism Coordination Unit
TDS	Total Dissolved Solids
TFR	Total Fertility Rate
TPDC	Tanzania Petroleum Development Corporation
TSS	Total Suspended Solids
UN	United Nations
UNCED	United Nations Conference on Environment and Development
UNDP	United Nations Development Programme
UNEP	United Nations Environment Programme
UNESCO	United Nations Educational, Scientific and Cultural Organisation
UNFPA	United Nations Population Fund
UNIDO	United Nations Industrial Development Organisation
UNIFEM	United Nations Development Fund for Women
USAID	United States Agency for International Development

VOCs	Volatile Organic Compounds
WCC	Wankie Colliery Company
WCRP	World Climate Research Programme (WCP)
WH	World Heritage Convention
WHO	World Health Organisation
WIDSAA	Women in Development Southern Africa Awareness
WMO	World Meteorological Organisation
WSCU	Water Sector Coordination Unit
WSM	Wildlife Society of Malawi
WTO	World Tourism Organisation
WTOTB	World Trade Organisation Technical Barriers to Trade
WUC	Botswana Water Utilities Corporation
WWF	World Wide Fund for Nature
ZACPLAN	Zambezi River Action Plan
ZACPRO	Zambezi River Action Plan Projects
ZAMCOM	Zambezi Basin Commission
ZAWA	Zambia Wildlife Authority
ZBWCRUP	Zambezi Basin Wetlands Conservation and Resource Utilisation Project
ZCCM	Zambia Consolidated Copper Mines
ZERO	Regional Environment Organisation
ZESA	Zimbabwe Electricity Supply Authority
ZESCO	Zambia Electricity Supply Corporation
ZRA	Zambezi River Authority

UNITS OF MEASURE

Quantity	Symbol	Unit name
Concentration	ppm	parts per million
Energy	W	watt
	KW	kilowatt
	MW	megawatt
	KWh	kilowatthour
	MJ	megajoule
	GJ	gigajoule
	GWh	Gigawatthour
	TJ	terajoule
	KV	kilovolt
Flow	cu m/s	cubic metre per second
Length	mm	millimetre
	cm	centimetre
	m	metre
	km	kilometre
Speed	m/s	metre per second
Weight	mg	milligramme
	g	gramme
	kg	kilogramme
	t	tonne
	kt	kilotonne
	Mt	megatonne
	toe	tonnes of equivalent
	tpy	tonnes per year
Area	ha	hectare
	sq km	square kilometre
Time	s	second
	h	hour
	µs	microsecond
Volume	cu m	cubic metre
	cu km	cubic kilometre
	l	litre
	Ml	megalitre
	ml	millilitre
Gases	CH_4	methane
	CO_2	carbon dioxide
	CO	carbon monoxide
	NO_x	nitrous oxide
	SO_2	sulphur dioxide
	N_2O	dinitrogen oxide
	O_3	Ozone

Prefix	Symbol	Multiplying factor
exa	E	10^{18}
peta	P	10^{15}
tera	T	10^{12}
giga	G	10^{9}
mega	M	10^{6}
kilo	K	10^{3}
hecto	h	10^{2}
deca	da	10^{1}
deci	d	10^{-1}
centi	c	10^{-2}
milli	m	10^{-3}
micro	µ	10^{-6}
nano	n	10^{-9}
pico	p	10^{-12}
femto	f	10^{-15}

COMMON AND SCIENTIFIC NAMES

Plants

Common name	Scientific name
	Acacia
	Acacia mellifera
African Ebony	*Dalbergia melanoxylon*
	Afzelia quanzensis
	Albizia antunesiana
	Annona stenophylla
	Berlinia giorgii
Bloodwood, Mukwa or Kiaat	*Pterocarpus angolensis*
	Brachystegia
	Brachystegia spiciformis
	Combretum
	Cryptosepalum exfoliatum
	Cryptosepalum pseudotaxus
	Cryptosepalum pseudotaxus
	Diospyrus mespiliformis
False Mopane or Mchibi	*Guibourtia coleosperma*
	Grewia flavescens
	Isoberlinia
	Julbernardia
	Julbernardia globiflora
	Julbernardia paniculata
Kariba Weed	*Salvinia molesta*
	Lannea antiscorbutica
Leadwood	*Combretum imberbe*
	Lumnitzera racemosa
	Marquesia
Marula	*Hyphaene petersiana*
	Maytenus buchananii
Mopane	*Colophospermum mopane*
Muwanga	*Pericopsis angolensis*
Natal Mahogany	*Trichilia emetica*
Nile lettuce	*Pistia stratiotes*
Olive tree	*Olea europaea*
Panga Panga	*Millettia stuhlmannii*
	Parinari capensis
	Parinari curatellifolia
Pod Mahogany	*Afzelia quanzensis*
	Pteleopsis spp.
	Pterocarpus angolensis
	Pterocarpus antunesii
Red Mahogany	*Khaya anthotheca*
Red Mangrove	*Rhizophora mucronata*
	Sclerocarya birrea
	Syzygium abysinica
	Syzygium guineense
	Syzygium huillense
Tamarind tree	*Tamarindus indica*
	Terminalia
	Thelypteris confluens
	Trichilia quadrivalvis
Velvet bushwillow	*Combretum molle*
	Warburgia salutaris
Water Fern	*Azolla pinnata*
Water Hyacinth	*Eichhornia crassipes*
Water Lettuce	*Pistia stratiotes*
Waterberry	*Syzygium*
White Mangrove	*Avicennia marina*
Winter Thorn	*Faidherbia albida*
	Xylia torreana
Zambezi Teak	*Baikiaea plurijuga*
	Ziziphus mucronata

Mammals

Common name	Scientific name
Baboon	*Papio spp*
Black Lechwe	*Kobus leche*
Black Rhinoceros	*Diceros bicornis*
Blue Duiker	*Cepalophus monticola*
Blue Wildebeeste	*Connocahetes taurinus*
Buffalo	*Syncerus caffer*
Colobus Monkey	*Colobus angolensis*
Elephant	*Loxodonta africana*
Giraffe	*Giraffa camelopardalis*
Hippopotamus	*Hippopotamus ambibius*
Kafue Lechwe	*Kobus leche kafuensis*
Leopard	*Panthera pardus*
Lion	*Panthera leo*
Pangolin	*Hystrix africaeaustralis*
Red Lechwe	*Kobus ellipsiprymnus*
Samango Monkey	*Cercopithecus mitis*
Serval Cats	*Felis serval*
Tssessebe	*Damaliscus lunatus*
White Rhinoceros	*Ceratotherium simum*
Zebra	*Equus burchelli*

Birds

Common name	Scientific name
Black-cheeked Lovebird	*Agapornis nigrigenis*
Slaty Egret	*Egretta vinaceigula*
Swynnerton's Robin	*Swynnertonia swynnertoni*
Chirinda Apalis	*Apalis chirindensis*
Roberts' Prinia	*Prinia robertsi*
Rock Pratincole	*Glareola nuchalis*
African Skimmer	*Rynchops flavirostris*
Carmine Bee-eater	*Merrops nubicoides*
Blue Swallow	*Hirundo atrocaerulea*
White Stork	*Ciconia ciconia*
Harriers	*Circus spp*
Lesser Kestral	*Falco naumanni*
Corncrake	*Crex crex*
Black-winged Pratincole	*Glareola nordmanni*
Wattled Crane	*Grus carunculata*
Red-billed Quelea	*Quelea quelea*
White-winged Flufftail	*Sarothrura ayresi*
Shoebill	*Balaeniceps rex*
Lesser Flamingo	*Phoeniropterus minor*
Great Snipe	*Gallinago media*
Eastern White Pelican	*Pelecanus onocrotalus*
Pinkbacked Pelican	*Pelecanus rufescens*
Goliath Heron	*Ardea goliath*
Purple Heron	*Ardea purpurea*
Blackheaded Heron	*Ardea melanocephala*
Egrets	*Egretta spp*
Cattle Egret	*Bubulcus ibis*
Black Stork	*Ciconia abdimii*
Yellowbilled Stork	*Mycteria ibis*
Marabou Stork	*Leptoptilos crumeniferus*
Saddlebilled Stork	*Ephippiorhynchus senegalensis*
African Openbilled Stork	*Anastomus lamelligerus*
Woolynecked Stork	*Ciconia episcopus*
Hamerkop	*Scopus umbretta*
Glossy Ibis	*Plegadis falcinellus*
Sacred Ibis	*Hreskiornis aethiopicus*
Spurwinged Goose	*Plectropterus gambensis*
Knobbilled Duck	*Sarkidiornis melanotos*
Whitefaced Duck	*Dendrocygna viduata*
Fulvous Duck	*Dendrocygna bicolor*
African Fish Eagle	*Haliaeetus vocifer*
African Marsh Harrier	*Circus ranivorus*
Wattled Crane	*Bugeranus carunculatus*
Grey Crowned Crane	*Balearica crassirostris*
Longtoed Plover	*Vanellus crassirostris*
Greyheaded Gull	*Larus cirrocephalus*

Fish

Common name	Scientific name
Eels	*Anguilla nebulosa labiata*
Lake Tanganyika sardine	*Limnothrissa miodon*
Lungfish	*Protopterus annectens brieni*
Manyame	*Hunyani labeo*
Nile Tilapia fish	*Oreochromis niloticus*
Tiger fish	*Hydrocynus forskalii*

Photo: IUCN

CONTRIBUTORS AND WRITERS BY CHAPTER

1. Regional Overview: People and Environment (by M. Saruchera and L. Sola)
Munyaradzi Chenje, Director, SARDC-IMERCSA
Clever Mafuta, CEP Programme Coordinator, SARDC-IMERCSA
Ravai Marindo, Lecturer, Centre for Population Studies, University of Zimbabwe

2. Physical Features and Climate (by E. Laisi and M. Saruchera)
Maurice R. Muchinda, Chief Meteorologist, Department of Meteorological Services, Zambia
Hillary Dandaula, Former Director, Department of Meteorological Services, Malawi
Tharsis M. Hyera, Directorate of Meteorology, Tanzania
Cathrine Mutambirwa, Researcher/Writer, SARDC-IMERCSA

3. Water and Wetland Resources (by E. Laisi)
Shirley Bethune, Chief Ecologist, Department of Water Affairs, Namibia
Tabeth Matiza-Chiuta, Regional Programme Coordinator, IUCN-ROSA

4. Biological Resources and Diversity (by M. Kullberg and T. Kureya)
Jonathan Timberlake, Technical Coordinator, Biodiversity Foundation for Africa, Zimbabwe
Mats Kullberg, Environmental Communications Specialist, SARDC-IMERCSA

5. Agriculture (by M. Saruchera)
Cathrine Mutambirwa, Researcher/Writer, SARDC-IMERCSA
Munyaradzi Saruchera, Senior Researcher/Writer, SARDC-IMERCSA

6. Industry (M. Kullberg and M. Saruchera)
Elton Laisi, Programme Coordinator, SARDC-IMERCSA

7. Energy (by M. Kullberg and T. Kureya)
Bothwell Batidzirai, Southern Centre for Energy and Environment, Zimbabwe
Oscar S. Kalumiana, Energy Planner, Department of Energy, Zambia

8. Tourism (by M. Kullberg)
Mwape Sichilongo, Executive Director, Wildlife and Environmental Conservation Society, Zambia
Elias P. Nyakunu, Chief Executive Officer, Zimbabwe Council for Tourism

9. Pollution (by C. Mafuta)
Ngoni Moyo, Lecturer, Biology Department, University of Zimbabwe
Paul Banda, Chief Inspector, Environmental Council of Zambia
Hubert Meena, Senior Consultant, Centre for Energy, Environment, Science and Technology (CEEST), Tanzania

10. Poverty (by E. Laisi and Cathrine Mutambirwa)
Elton Laisi, Programme Coordinator, SARDC-IMERCSA
Jean Theu, Assistant Chief Economist, National Economic Council, Malawi

11. Gender and Women's Roles (by M. Kullberg)
Barbara Lopi, Zambia Association for Research and Development
Tabeth Matiza-Chiuta, Regional Programme Coordinator, IUCN-ROSA

12. Environmental Management and Regional Cooperation (by M. Kullberg and M. Saruchera)
Hillary Masundire, Senior Lecturer, Department of Biological Sciences, University of Botswana
Munyaradzi Chenje, Director, SARDC-IMERCSA

13. Trends and Scenarios (by M. Saruchera)
Munyaradzi Chenje, Director, SARDC-IMERCSA

REVIEWERS BY CHAPTER

1. Regional Overview: People and Environment

Jacquie Chenje, Programme Coordinator, IUCN-ROSA
Michael Tumbare, Chief Executive, Zambezi River Authority, Zambia
Takawira Mubvami, Lecturer, Department of Rural and Urban Planning, University of Zimbabwe
Robson Silitshena, Professor, Africa University, Zimbabwe

2. Physical Features and Climate

Marufu C. Zinyowera, Director, Department of Meteorological Services, Zimbabwe

3. Water and Wetland Resources

Harry Chabwela, Department of Biology, University of Zambia

4. Biological Resources and Diversity

Susan Richardson-Kageler, Lecturer, Department of Statistics, University of Zimbabwe
Phillip O. Nkunika, Department of Biological Sciences, University of Zambia

5. Agriculture

SARDC-IMERCSA Staff

6. Industry

SARDC-IMERCSA Staff

7. Energy

Lloyd Thole, Environmental Council of Zambia

8. Tourism

Zivanai Tamangani, Business Studies Department, University of Zimbabwe Graduate School/MBA
Robin Heath, Senior Lecturer, Department of Geography, University of Zimbabwe

9. Pollution

Sibekile Mtetwa, Ministry of Rural Resources and Water Development, Zimbabwe
Joseph Kazombo, Ministry of Water Development, Malawi

10. Poverty

Kenneth Odero, Lecturer, Department of Rural and Urban Planning, University of Zimbabwe
Marvelous Mhloyi, Centre for Population Studies, University of Zimbabwe

11. Gender

Thoko Ruzvidzo, Consultant, Wilthor Consulting Services, Zimbabwe
Patricia Kalindekafe, Chancellor College, University of Malawi
Monica B. Kethusegile, Head of Programme, SARDC-WIDSAA

12. Regional Cooperation

Flora Musonda, Economic and Social Research Foundation, Tanzania
Larry Swatuk, Deparment of Political Studies, University of Botswana
Tom Mpofu, University of Zimbabwe and SARDC Board member, Zimbabwe
Jacquie Chenje, Programme Coordinator, IUCN ROSA, Zimbabwe

13. Trends and Scenarios

Emmanuel Guveya, Programme Officer, IUCN-ROSA
Jacquie Chenje, Programme Coordinator, IUCN-ROSA
Edward H. Zulu, Programme Manager, SADC Mining Sector Coordination Unit (MSCU), Zambia
Jonathan Timberlake, Technical Coordinator, Biodiversity Foundation for Africa, Zimbabwe

All Chapters

Scientific Advisory Committee

ACKNOWLEDGEMENTS

Many hands make light work, is an old adage that is as relevant today as when it was coined. It holds true in terms of the preparation of the State of the Environment in the Zambezi Basin 2000, a comprehensive report borne out of collaboration among many different experts spread across in different countries in the Southern African Development Community (SADC) region in general.

Various Individuals and institutions have over the past two years worked together to define the environmental issues in the Zambezi River Basin, the most shared natural resource in the whole of the SADC region. Experts attended workshops, exchanged ideas and concepts, and contributed and reviewed the chapters of this report. Institutions made many documents available to enable the partners in this initiative make this report as current as possible.

Words alone can never be enough to thank all the expert individuals and organisations which made time and their expertise available to enable us to start public awareness on the natural asset that is the Zambezi River Basin. However, on behalf of the partners in this initiative, we would like to thank the directors of SADC ELMS, SADC WSCU, IUCN-ROSA, ZRA and SARDC for the foresight in supporting this exercise from inception to the publication of this report.

In particular, we at SARDC-IMERCSA would like to thank Tabetha Matiza-Chiuta of IUCN-ROSA for buying into the initiative and sharing her expertise in making this initiative successful. Jacqueline Chenje also of IUCN for laying the foundation with the successful coordination of the inception mission study. Charles Obol of SADC ELMS for his valuable support in making this initiative a reality. Phera Ramaoeli of the SADC WSCU for the invaluable input and professional advice. A.S. Murty of the Zambezi River Authority for his usual dedication and invaluable comments over the past two years. To all of you, we thank you for making the light of institutional and personal collaboration across different fronts anf frontiers in the SADC region burn so bright.

We would also like to thank our Scientific Advisory Committee members (listed elsewhere in this report) for guiding the process and ensuring that most bases are covered. Thank you also for your individual input in the various chapters. We hope that this product comes close to what you had in mind.

A number of workshops were organised in Zimbabwe and were attended by participants from professional background. For example, social scientists, ecologists, geographers, economists, chiefs, hydrologists, environmentalists and politicians participated in these workshops, reinforcing our commitment to integrating state of the environment reporting across sectors and administrative units. The participants are listed elsewhere in this report. We would, however, like to particularly thank Zimbabwe's Mines, Environment and Tourism Minister Simon K. Moyo for officially opening the first regional workshop in July 1998, which launched this initiative. We would also like to thank Zimbabwe's former Transport and Energy Minister Enos

Chikowore for supporting the workshop. Our thanks are also due to the Government of Zimbabwe for providing an abling environment under which this initiative and process has been undertaken.

We would also like to thank all participants to the Trends and Scenarios workshop whose insights enabled us to write the last chapter of the report. Without you, our vision of the future could have been cloudy.

Phyllis Johnson, our Executive Director, we thank you for your words of encouragement and motivating us all to be the best. We owe it to you that we have managed to be where we are today. Monica B. Kethusegile, Head of Programme for WIDSAA, thank you for your invaluable input in the Gender chapter. Sabelo Mapasure, SARDC's Chief Documentalist, we thank you for indexing.

The coordination of all these activities fell on the shoulders of SARDC-IMERCSA, particularly project coordinator Elton Laisi. Thank you for your work and your usual calmness in the eye of the storm.

Among other IMERCSA staff who went beyond the call of duty are Lovemore Sola, and Clever Mafuta. Thank you for rising to the occasion. Munyaradzi Saruchera, Tendayi Kureya, Cathrine Mutambirwa and Maria Murama, thank you for your invaluable input. Godwell Nhamo, we also thank you for your input at the start of the initiative.

Shehnilla Mohamed and Hugh McCullum, we thank you for your editorial input on the human-interest articles, which breathe life into an otherwise abstract subject.

Wilson Yule and his colleagues in the IMERCSA documentation centre, we thank you for keeping track of the background materials which were crucial to our finishing the work on time.

Sekai Mubako thank you for logistical arrangements and ensuring that our workshops went on smoothly. To the rest of the logistics team, including Memory Sachikonye at SARDC, you have certainly made logistics an art form.

Our thanks go to Paul Wade of Ink Spots Design for designing the cover and text of this report as well as many other materials produced during the course of this initiative. Wayne Miller, also of Ink Spots Design, we thank you for doing the maps.

The publishers also acknowledge all sources of information reproduced in this book. Every effort has been made to trace copyright holders, but if any have been overlooked, the necessary arrangements will be made at the first opportunity.

The preparation of this report and other support materials aimed at raising awareness on environmental and water issues in the Zambezi Basin could not have been possible without a substantial funding investment from the Swedish International Development Cooperation Agency (Sida). We are very grateful for your continued support. In particular, we would like to thank Jakob Granit, the Sida representative in Harare who supported the activities of this initiative at so many different levels. Your support went beyond the call of duty and for that we are extremely grateful. We would also like to thank Jakob's successor, Katarina Perrolf, for your invaluable support from the outset.

We would also like to thank Sida for facilitating the attachment with SARDC-IMERCSA of Mats Kullberg, an environmental communications specialist. Your hardwork and professional approach enabled us to achieve what was otherwise an over-ambitious workplan.

The designation of geographical entities, use of any name in this publication, and the presentation of the material do not imply the expression of any opinion whatsoever on the part of the partners of this initiative concerning the legal status of any country or territory, or area of its authority, or concerning the delimitation of its frontiers or boundaries.

All contributors, reviewers and other supporters may not share all or some of the views expressed in this publication. While a process of wide consultation was undertaken, and every effort made to ensure a balanced presentation, we at IMERCSA are entirely responsible for any errors of judgement, fact or balance, omission or commission.

Munyaradzi Chenje (Editor)
March 2000

FOREWORD

Natural resources in member states of the Southern African Development Community (SADC) define our wealth as nations. They are critical to regional growth, economic integration, and collective economic independence in an increasingly globalising world.

Since most of the natural resources to which I refer are shared, achieving sustainable natural resource management requires regional cooperation, an integrated ecosystem approach and a common understanding of the natural resource base. The rationale for sustainable natural resource management is to achieve a balance between human demands on natural resources and the natural environment's ability to meet these demands.

The Zambezi River Basin, which links eight SADC members, including my own country – Mozambique – and is home to close to 40 million people, perhaps best represents the best of what we have in SADC in terms of our natural capital. Within the basin's large expanse, we have our water resources, land and soils, forests, and wildlife. All these resources define our economic activities which range from agriculture and forestry, manufacturing and mining to conservation and tourism, and scientific monitoring and research.

All of us in the basin states depend on the natural environment for energy supplies, water, food, shelter, tourism, rural development, and jobs, for example. As a region, we therefore need to maintain the Zambezi River Basin's healthy productive ecosystems to meet the challenges of not only intra-generational but also inter-generational equity.

As the most shared resource in the SADC region, the Basin provides a litmus test in terms of meeting one of the objectives of the SADC Treaty. Article 5 commits us all to, among other objectives, "achieve sustainable utilisation of natural resources and effective protection of the environment."

This has to be balanced with another of the objectives articulated in the Treaty which is to, "achieve development and economic growth, alleviate poverty, enhance the standard and quality of life of the peoples of southern Africa and support the socially disadvantaged through regional integration."

Achieving both these and other objectives of the Treaty is a balancing act, which we can only pull off with the full participation of stakeholders at different levels. We realise that various factors militate against us achieving this entirely within the region without global influence. However, I would like to emphasise that development in our context should not be at the expense of our people, impoverishing them further by overexploiting our natural resources for short-term benefits. This is why at the international level, SADC countries will continue to seek equitable trade regimes that do not run roughshod over our own approach to sustainable development. We believe that sustainable development will remain a chimera if the international trade playing field is not levelled.

As a transboundary resource, which is subject to management and use by various sectoral and national interests, the Basin may be overexploited for immediate and unsustainable gain rather than for long-term and sustainable development. Already there are many environmental concerns, which are associated with development planning that lacks an integrated ecosystems perspective. Land degradation, poor watershed management, sewage and industrial pollution, drainage of wetlands, water abstraction and general infrastructural development have reached a magnitude which calls for urgent action in terms of environmentally sound economic development and management.

Most of the current projects in the basin are focused on sectoral approaches targeted at strengthening development capacities and potentials while integrated approaches to sustainable natural resource management remain weak. This is primarily the result of a sectoral focus, and weak inter-sectoral and transboundary coordination mechanism and structures. These do not provide for sound management of transboundary resources; show a lack of knowledge and information on the dynamics and functions of ecosystems, are due to the absence of effective legislative frameworks as well as limited human resources sufficiently trained and qualified to study and demonstrate the hydrological and more general environmental impacts of manipulating and using dynamic and complex resources such as ecosystems of the Basin. The issues highlighted above are perpetuated by lack of information and effective mechanisms for reviewing, exchanging and sharing of information in the basin.

With water being the key resource to economic development and survival in this generally arid region, it is important that we don't overlook the importance of the Basin as we try to satisfy these growing demands on water. Ensuring the long-term balance between demands and the resource base's ability to meet these demands requires an integrated, coordinated and long-term management perspective. We have to accept that supplying more and more water is not the only solution. We have to learn to do with what we have. We simply have no choice. Managing demand for water and other resources is, therefore, critical to our long-term planning.

As stated in Agenda 21, integrated management of natural resources is the key to maintaining ecosystems and the essential services that they provide. The acquisition and provision of timely, effective information on the state of our natural resources is an important factor to the attainment of sustainable natural resource management.

We therefore, welcome the preparation of the *State of the Environment in the Zambezi Basin*; an innovative exercise we hope will strengthen collaboration between our policymakers and the public in our collective efforts to effectively manage our heritage.

Our hope is that the production of the report on the Zambezi Basin is not an end in itself, but a long-term process, which should continue for many decades to come. The publication of the SOE report is based on the principle that information is the key to transformation, and that people need knowledge in order to act. Without this popular participation, environmental issues will be solely the domain of government institutions and international donor agencies. We are committed to ensuring that this does not happen.

The preparation of this report is in line with the 1996 *SADC Policy and Strategy for Environment and Sustainable Development* whose aims are to strengthen the analytical, decision-making, legal, institutional and technological capacities for achieving sustainable development in southern Africa; increase public information, education and participation on environment and development issues in southern Africa; and expand regional integration and global cooperation on environmental and natural resources management for sustainable development. The provision and dissemination of accurate information on the state of the basin's natural resources is a very important prerequisite to the maintenance of a healthy productive ecosystem. Sustainable use and equitable access to resources can significantly contribute towards poverty alleviation strategies in the region. Our consumption patterns and poverty will determine whether or not we will be sustainable. It is, therefore, important that we redouble our efforts at poverty alleviation and raise awareness among our people about the excesses of consumption.

We applaud the partnership of two of our sector coordination units – Environment and Land Management Sector (ELMS) and Water Sector Coordination Unit (WSCU) – and the World Conservation Union Regional Office for Southern Africa (IUCN-ROSA), the Zambezi River Authority (ZRA), and the Southern African Research and Documentation Centre-Musokotwane Environment Resource Centre for Southern Africa (SARDC-IMERCSA) for preparing such a timely report. This reflects the spirit of SADC, which encourages cooperation at different levels to foster regional integration. Multi-stakeholder cooperation can only strengthen our resolve to achieve sustainable development and raise the standard of living of the majority of our people.

J. Chissano
President of Mozambique
and Chairman of SADC

INTRODUCTION

The *State of the Environment in the Zambezi Basin 2000* marks the first time that an assessment of a single ecosystem has been undertaken and reported upon in southern Africa. Although state-of-the-environment reporting in the Southern African Development Community (SADC) region is relatively new, the traditional approach has been to focus on national boundaries, natural resources or sectors. This report breaks new ground, showing the region's acknowledgement of the importance of ecosystems management in the efforts of SADC countries, which are aimed at sustainable development and regional integration.

As the largest river basin in the SADC region, encompassing eight nations, the Zambezi River Basin was a natural choice in terms of undertaking an assessment and state-of-the-environment reporting initiative. The human and ecological processes within the basin are so complex that undertaking this exercise was a challenge.

The challenges with which partners in this initiative had to grapple, mainly focused on data, which has never been collected using the ecosystems approach but along national and sectoral priorities. The challenge was then to interpret the limitations of such priorities and project them across the region to get a picture on the state of the environment in the Zambezi Basin.

The *State of the Environment in the Zambezi Basin 2000* looks at the shared natural resources of the basin, taking into account ecological, social and economic issues. These three factors are critical to achieving a state of sustainability in the Zambezi Basin. The rationale for sustainable natural resource management is to achieve a balance between human demands on natural resources and the natural environment's ability to meet these demands.

The Zambezi Basin plays a key role in the development of the region. The effective management and sustainable use of the basin's natural resources are of prime importance. The basin has been described as the focal point in southern Africa "in terms of energy, wildlife, resettlement models, biodiversity, wetlands and water resources, conservation, pollution concerns, tourism etc. Many countries stand to lose a great deal if it is not well managed." These are some of the major issues addressed in this report.

Sustainable management of the basin is therefore, crucial to the development of not only the basin states but the rest of SADC as well. The *State of the Environment in the Zambezi Basin 2000* is therefore, an attempt to highlight the environmental issues in the basin, and flag areas of concern.

FORMAT

The *State of the Environment in the Zambezi Basin 2000*, which is being published simultaneously in English and Portuguese, has 13 chapters divided into three distinct sections – four chapters providing background information on the people and physical aspects of the basin. The second section deals with social and environmental management issues while the last chapter deals with trends and scenarios.

An executive summary, which has been published as a separate document, provides the main issues raised in the 13 chapters of this report. The summary is intended to provide the busy policymaker with the major areas needing immediate attention.

Chapter 1 gives an overview of the people and issues in the Zambezi Basin, highlighting some of the social factors critical to environmental management.

Chapter 2 looks at the physical features and climate of the basin, emphasising the point that both factors influence human settlements and activities, and species distribution.

Chapter 3 looks at the water and wetland resources of the basin, highlighting their importance in terms of human settlements and development, and the survival of habitats and species.

Chapter 4 on biological resources and diversity shows how the basin is rich in biodiversity, both flora and fauna, hosting some of the most attractive tourist destinations in the SADC region. The chapter, emphasises the importance of effective measures to conserve biodiversity while meeting the human needs at the same time.

Chapter 5 covers agriculture, the largest economic sector in the basin both in terms of areal coverage and the large population of people it supports. Virtually all rural people in the basin states depend on agriculture for their livelihood. The threats of uncontrolled agricultural expansion to wildlife conservation are also discussed.

Chapter 6 is on industry in the region. It looks at how industry has evolved and the new challenges it poses in terms of pollution and environmental management. The basin states have experienced significant industrial growth over the last three decades, increased industrial waste generation and new environmental problems, which pose health risks to people and the environment.

Chapter 7 covers energy issues in the basin, showing that the majority of the people still depend on their traditional source of energy, fuelwood. It also looks at other sources of energy generation and their impact on the environment. The message is that traditional sources of energy cannot sustain growing populations and economies of the basin states.

Chapter 8 is on tourism – the fastest growing industry in the basin. The chapter looks at both positive and negative aspects of tourism growth, highlighting the need for effective management to ensure that tourist attractions are not degraded.

Chapter 9 is on increasing pollution and its impacts on the health of people and the environment in the basin.

Chapter 10 covers poverty issues, showing that poverty is both a cause and result of environmental degradation, and that it threatens sustainable development. The chapter shows that the majority of people in the basin live in poverty, and emphasises the need for effective poverty alleviation programmes.

Chapter 11 on gender issues breaks new ground as it is the first time that these issues have been covered in a separate chapter in state-of-environment reporting in the SADC region. Gender issues have been recognised as critical to human development in the basin. The chapter emphasises that gender sensitivity in natural resource management is critical, particularly if the role of women as resource managers is considered in decision-making and all management activities.

Chapter 12 focuses on regional cooperation in various spheres of water and environmental management. The chapter tries to show that while many of the environmental, social and economic problems may appear localised, they often have a basin-wide impact. It adds that it is imperative for governments in the region to cooperate at different levels.

Chapter 13 unveils the trends and scenarios, showing where the basin countries are at and where they ought to be if they are to achieve a state of sustainability.

THE PROCESS

The *State of the Environment in the Zambezi Basin 2000* report was prepared as part of the ongoing Communicating the Environment Programme (CEP). While the original CEP partnership involved the SADC Environment and Land Management Sector (ELMS), the World Conservation Union Regional Office for Southern

Africa (IUCN-ROSA), and the Southern African Research and Documentation Centre-Musokotwane Environment Resource Centre for Southern Africa (SARDC-IMERCSA), two new regional institutions: SADC Water Sector Coordination Unit (WSCU) and the Zambezi River Authority (ZRA) were involved as full partners in the preparation of this report.

The preparation of the report started in early 1998 with an inception mission to various states in the basin. The main objective was to consult experts in the various countries on the major environmental issues in the Zambezi Basin. The inception mission also provided an opportunity to "image" the issues by taking photographs in different countries.

Following the inception mission, a regional workshop on the State of the Environment in Zambezi Basin was convened in Zimbabwe to provide experts with an opportunity to exchange experiences. Zimbabwe's Mines, Environment and Tourism Minister Simon K. Moyo opened the one-week workshop, which was attended by representatives of at least five SADC sector coordination units. In his opening speech, the Minister clearly articulated the objective of the whole initiative:

"The State of the Environment Reporting Programme for the Zambezi Basin aims to communicate the status of the environment in this basin, a task pivotal for instituting sound and smart planning for a sustainable future. Without popular participation, environmental issues will remain solely the domain of policymakers and a selected minority. When people have access to information about the ecological processes that affect their lives, their resources, their future, and the environmental problems they face, they are better able to respond. The provision and dissemination of accurate information on the state of the basin's natural resources is, therefore, a very important requisite to the maintenance of a healthy productive ecosystem."

The workshop laid the foundation for the preparation of this report with the participants finalising the chapter outline of the issues to be covered. A Scientific Advisory Committee (whose members are listed elsewhere in this report) reviewed the chapter outline before it was finalised and circulated for further comment. Once finalised, the partners approached many different experts in different basin states to contribute to the various chapters. Two or more contributors provided the material, which was used by SARDC-IMERCSA to finalise draft chapters.

Building upon the extensive consultative process, SARDC-IMERCSA worked with other experts to review and comment on the first 11 chapters of the report. Following the incorporation of reviewers' comments, a second regional workshop was convened in Harare in August 1999 to critique the draft manuscript. In addition to some members of the Scientific Advisory Committee, new participants who had not been involved with the process before, were invited at the workshop.

In addition to critiquing the draft manuscript, the participants also held a one-day workshop to look at the trends and scenarios in terms of environmental management in the Zambezi Basin. The workshop, which was also attended by senior diplomats representing some of the basin states such as Mozambique and Tanzania, provided the material used in writing the last chapter on trends and scenarios.

Following the workshop, the draft manuscript was completed and circulated for final review. The first draft of the trends and scenarios chapter was also circulated to workshop participants for their comments.

The *State of the Environment in the Zambezi Basin 2000* meets one of the key aims of Agenda 21, the blueprint for sustainable development during this century. Agenda 21 stresses that integrated management of natural resources is the key to maintaining ecosystems and the essential services that they provide. The acquisition and provision of timely, effective information on the state of the natural resources is an important factor to the attainment of sustainable natural resource management. In most developing regions, such as southern Africa, relevant information that is essential for effective planning and sustainable resource management is not readily available.

This report adds on to other SOE materials that have been produced over the past two years as part of this initiative. A total of 12 factsheets, 10 educational posters, seven newsletters, a bibliography of literature materials on the basin, an electronic photographic database and a comprehensive website have been published. Except for the photographic database, which is available on CD-ROM, the published materials have been distributed in virtually all the basin states, with the possible exception of Angola where there is war. The distribution has been possible with the participation of national collaborating centres (NCCs) – both government and non-government organisations which were involved in the initiative.

The partners in this initiative hope that the *State of the Environment in the Zambezi Basin 2000* and all the other materials which have been referred to, will popularise social, economic and environmental issues in the Zambezi Basin. The materials are aimed at a wide target audience, from politicians and policymakers to civil society and communities in the region – both rural and urban.

The production of this SOE report for the Zambezi Basin is not an end in itself, but a long-term process, which should continue long after this initiative. There is a critical need to improve integrated environmental monitoring and assessment using the ecosystems approach. This involves massive investment in terms of funding and institutions but the dividends of taking the Zambezi Basin as an ecosystem cannot be overemphasised.

While this report unveils the areas of concern in the basin, attention should be given to addressing the concerns through popular participation, integrated planning, monitoring and capacity building as well as allocating adequate resources required to undertake these initiatives.

ENVIRONMENTAL INFORMATION

The overall aim of this report is to challenge people, governments, organisations, researchers and the media to strive for the sustainable utilisation of resources.

The initiative, which led to the development of this report, is based on the principle that information is key to transformation. When people have access to information, they are better placed to weigh their options and respond accordingly. Environmental management is about choices and the best choices can only be made on the basis of up to date and reliable information.

The philosophy of information sharing appears in documents of environmental organisations at the grassroots and at the global level. The United Nations Conference on Environment and Development (UNCED), held in Rio de Janeiro in June 1992, set out a series of principles for bringing together environment and development in a sustainable way. The Rio Declaration recognises the pivotal role of environmental information.

Principle 10 of the Declaration states:

"Environmental issues are best handled with the participation of all concerned citizens, at the relevant level. At the national level, each individual shall have appropriate access to information concerning the environment that is held by public authorities. … States shall facilitate and encourage public awareness and participation by making information widely available."

Among the objectives set out by the 103 heads of state and government who attended the Rio Earth Summit, were:

"Countries should cooperate with each other and with various social sectors and population groups to prepare educational tools that include regional environment and development issues and initiatives using learning materials and resources suited to their own requirements.

"Countries and international organisations should review and strengthen information systems and services in sectors related to sustainable development, at … international levels. Special emphasis should be placed on the transformation of existing information into forms more useful for decision-making and on targeting information at different user groups."

In line with the Rio Earth Summit principles, the partners in this initiative believe the Zambezi Basin countries and their people can use their environment for sustainable development if there is a high level of environmental awareness. It is our view that the basin in particular and southern Africa in general, are still in a position to make choices about environmental policy and management, and that an informed population will not only widen and enhance the discussion, but will take positive action to achieve a state of sustainability.

Map 0.1 Southern Africa

Map 0.2 Zambezi Basin

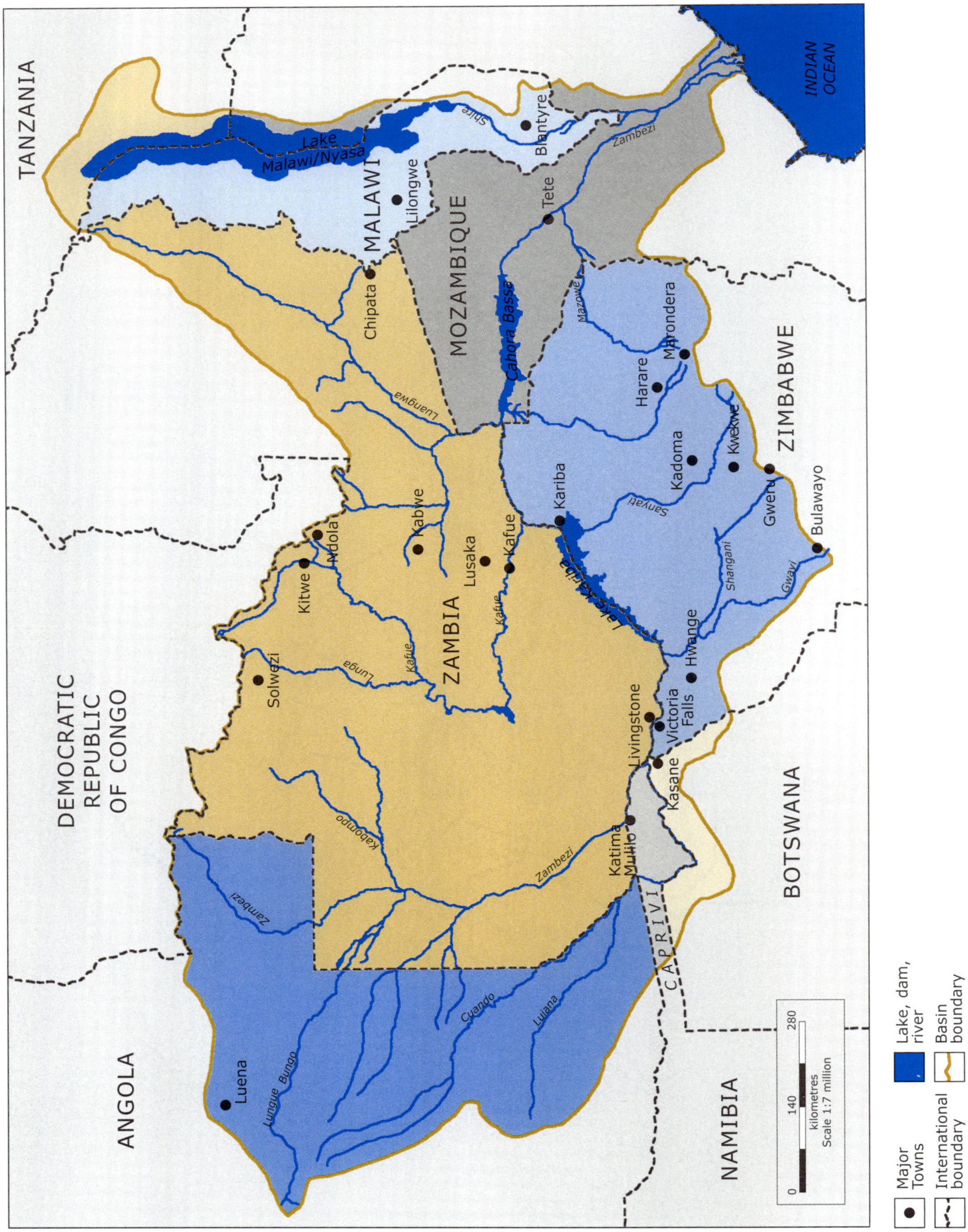

Source: Zambezi River Basin Atlas, SADC/DENCONSULT/ZRA, Lusaka, 1998

1

REGIONAL OVERVIEW: PEOPLE AND ENVIRONMENT

The Zambezi Basin is located between 8° – 20° S latitude, and 16.5° – 36° E longitude, draining an area of about 1.335 million square kilometres (sq km).[1] It covers about 25 percent of the total area of its eight riparian states: Angola, Botswana, Malawi, Mozambique, Namibia, Tanzania, Zambia and Zimbabwe. The Zambezi River rises on the Central African Plateau in the Kalene Hills in northwestern Zambia at 1,585 metres (m) above sea level to the delta in Mozambique before flowing into the Indian Ocean.[2] The river's major tributaries include the Luena and Lungue-Bungo in Angola; Chobe in Botswana; Shire in Malawi; Luiana in Namibia; Kabompo, Kafue and Luangwa in Zambia; and Manyame, Sanyati and Gwayi in Zimbabwe.

The Zambezi River Basin is larger in size than any of the 14 Southern African Development Community (SADC) countries except the Democratic Republic of Congo (DRC). The basin is more than 14 times the size of Malawi, which at 94,080 sq km,[3] is the smallest basin state. Zambia takes up the largest share of the Zambezi Basin with an estimated 41 percent, followed by Zimbabwe with 19 percent. Angola and Mozambique occupy 11 percent each, Malawi eight percent, Botswana six percent, and Namibia and Tanzania, two percent each.[4]

The Zambezi River and its dense network of tributaries and associated ecosystems constitute one of southern Africa's most important natural resources. The basin is a SADC regional centre of endemism due to its vast terrestrial biodiversity in Malawi, Zambia and Zimbabwe, and

Photos: IUCN

Rich in diversity, human and natural, social and political, economic and ecological, the basin faces many challenges in realising the regional goal of integration and sustainable development.

Photos: M Chenje

The flora of the Zambezian unit is reportedly the richest and most diverse in Africa with more than 6,000 species of flowering plants, 650 species of birds and 200 species of mammals.

Figure 1.1: Area details of basin sharing

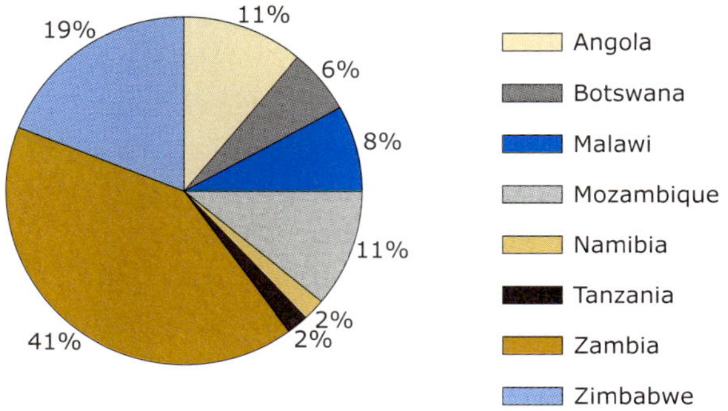

Source: Tumbare M., J.; Zambezi Basin: An Overview. For SARDC, 1999

some parts of Angola, Botswana, Mozambique and Tanzania. The flora of the Zambezian unit is reportedly the richest and most diverse in Africa with more than 6,000 species of flowering plants, 650 species of birds and 200 species of mammals.[5]

The basin provides some of the region's best sub-ecosystems, biomes and landscapes characterised by teak forest reserves in Zambia and Zimbabwe, wetlands and protected areas such as the Chobe National Park in Botswana, Liwonde in Malawi, Gorongoza-Marromeu in Mozambique, Kafue and Luangwa in Zambia and Hwange in Zimbabwe.

The basin's rivers also provide most of the region's power generation, support the subsistence economies of some of the poorest communities of southern Africa and represent a crucial asset in the region's tourism and recreation industry. Outside South Africa, the Zambezi Basin is the second largest industrial and commercial region in SADC, boasting a diversity of conglomerates, including those in mining, manufacturing and agriculture.

Rich in diversity, human and natural, social and political, economic and ecological, the basin faces many challenges in realising the regional goal of integration and sustainable development. The challenges are as much social and political as they are economic and ecological.

PEOPLE OF THE BASIN

About 30 ethnic groups are found in the basin with at least a third located in Zambia. Many of the groups are found along at least two national borders sharing common cultural traits and vocabularies even though dialectical differences exist.

Most of the people speak languages of the Bantu-lineage, in addition to English in six of the countries, and Portuguese in Angola and Mozambique.

Population

The Zambezi Basin is home to about 38.4 million people, mainly concentrated in Malawi, Zambia and Zimbabwe. Table 1.1, shows the area and population of the basin. Of the 38.4 million basin inhabitants, about 22 percent are in Zambia and 29 percent in Zimbabwe, meaning more than half of the basin population is in these two countries. About 90 percent of the people in Malawi live in the basin, representing 31 percent of the total basin population. More than 70 percent of Zambia's population and 72.1 percent of Zimbabwe's population live within the basin. Almost 33 percent of the total population of the basin states lives in the basin.

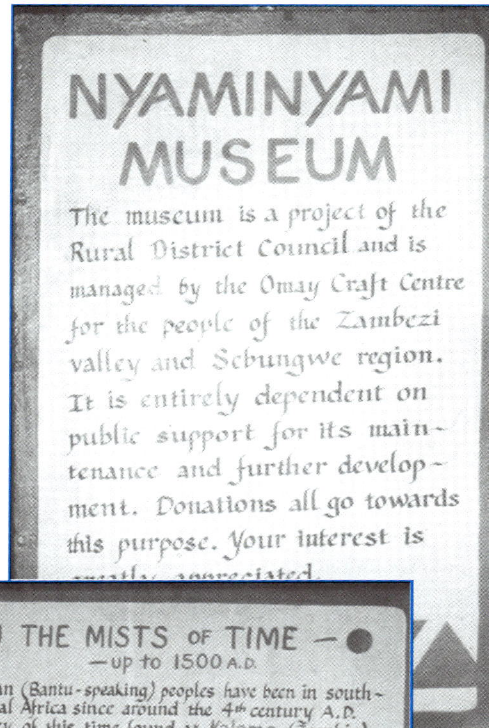

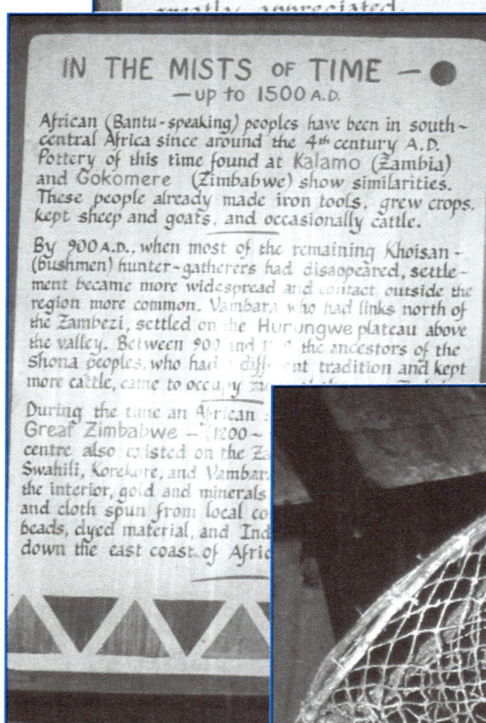

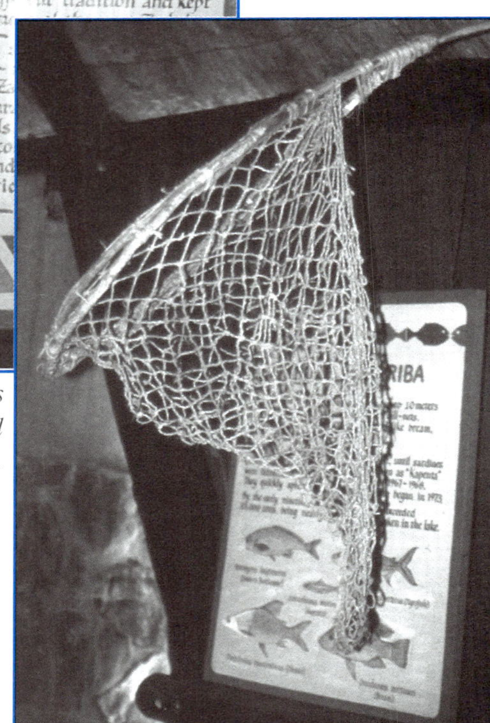

Photos: M Chenje

Many of the ethnic groups in the basin have a well documented history.

Table 1.1: Area and population of the Zambezi Basin

Country	Total area (sq km)	Area in Basin (sq km)	%	1998 Total Population	Population in Basin	%
Angola	1,246,700	145,000	11.6	13,168,000	487,200	3.7
Botswana	582,000	84,000	14.4	1,500,000	12,000	0.8
Malawi	118,484	110,390	93.2	11,407,000	9,821,400	86.1
Mozambique	799,380	140,000	17.5	20,791,000	3,991,870	19.2
Namibia	824,269	24,000	2.9	1,645,700	60,890	3.7
Tanzania	945,087	27,000	2.9	31,798,000	1,271,920	4.0
Zambia	752,614	540,000	71.7	10,037,400	7, 046,250	70.2
Zimbabwe	390,759	251,410	64.3	12,552,000	9,050,000	72.1
Total	5,659,293	1,321,800		102,899,100	31,741,530	30.8

Note: Estimated total population was 38.4 million in 1999.
Source: Denconsult, ZACPRO 6 Sector Study 3, Lusaka,1998.

Population growth

The basin's population is growing rapidly. The average population growth rate for the basin is about 2.9 percent annually, although rates for individual countries vary. In most of the countries, over 40 percent of the population is under 14 years of age, implying a high dependency rate. Population growth rates are quite high for Angola, Botswana, Malawi, Mozambique, Namibia, Tanzania, Zambia and Zimbabwe.[6] If the present growth rates of population are sustained, the population will double within the next generation.

Three major population dynamics are evident in the basin countries:

- the population will expand up to year 2015;
- the demographic structure of societies will place a great burden on national economies because of the youthful population; and
- the greatest urban growth is taking place in small urban centres where people depend on agriculture and other-related industries.[7]

Table 1.2: Population trends in the Zambezi Basin

Country	Population (millions) 1998	Average annual growth rate (%) 1990-95	Population in Urban Areas (%) 1970	Population in Urban Areas (%) 1995	Annual urban growth rate (%) 1990-95
Angola	13.0	3.7	15	32	6.3
Botswana	1.5	3.1	8	28	7.0
Malawi	11.4	3.5	6	14	6.2
Mozambique	20.7	2.4	6	34	7.4
Namibia	1.6	2.7	19	37	5.9
Tanzania	31.7	3.0	7	24	6.1
Zambia	10.0	3.0	30	45*	3.5
Zimbabwe	12.5	2.6	17	32	5.0

Source: Dalal-Clayton B., Southern Africa Beyond the Millennium: Environmental Trends and Scenarios to 2015, London, March 1997; for 1998 Population: Denconsult, ZACPRO 6 Sector Study 3, Lusaka,1998.
* Chenje, M; Reporting the S. African Environment, Harare, 1998.

Fertility is high among the basin countries. Total Fertility Rate (TFR) ranged between 4.4 - 6.7 births per woman for the early 1990s period. Malawi had the highest fertility rate of 6.7 births per woman, followed by Mozambique, Zambia and Zimbabwe.[8] In terms of population/environment issues, a high population growth rate indicates a high possibility of overpopulation, a short doubling time and possibilities of population pressure on land and other natural resources.

Population density and distribution

Population density in the basin states averages 28 people per sq km. Countries in the basin have skewed population distribution patterns. Except for Malawi, population densities in most basin states are relatively low. Malawi is the most densely populated country in the basin, with 105 persons per sq km, followed by Tanzania with 36; Zambia, 13; and Zimbabwe, 28.5-people/sq km.[9]

In Mozambique, the population is very much unevenly distributed, ranging from 3.9 persons/sq km in some areas of low agricultural potential to 20 persons/sq km in fertile areas. In urban areas, the population density is even higher, with 2,590 people per sq km in the capital, Maputo, which is outside the basin.

About 69.4 percent of the total population of the riparian countries lives in the rural areas.[10] The present distribution of land rights in the basin indicates that some areas are highly utilised while others are underutilised in relation to their potential. The evident trends are rural overcrowding in areas with communal land tenure in virtually all the states and higher densities around regional transportation routes, as is the case along Zambia's main railway line.

Mortality

The basin experiences high levels of mortality. The poor health status, lack of well staffed health centres, lack of drugs, poverty and the impact of HIV/AIDS create and support conditions of high mortality.[11] Infant mortality rates (IMR) range between 66 - 148 deaths per 1,000 live births for the 1990s. Mozambique has the highest IMR while Zimbabwe has the lowest. There are indications, however, that the IMR for Zimbabwe may be increasing because of HIV/AIDS.[12] Preliminary current estimates suggest that there may be about 120 deaths per 1,000.[13]

Maternal mortality levels are also high with Mozambique being the highest in the world at 1,092 deaths per 100,000 live births. Indications are that mortality conditions are a lot worse than the data shows because of the impact of AIDS. Estimates of mortality for the early 1990s are likely to underestimate the real mortality levels for countries such as Malawi, Zambia and Zimbabwe because they do not reflect current deaths. The absence of vital registration data on deaths between 1997-98 suggests that mortality estimates may be grossly underestimated.

Settlements

The basin's countries share similar settlement patterns that are characterised by both low and high population densities. The settlements are scattered and sometimes concentrated around service centres. Some of the settlements, especially in Zambia, are more concentrated along the Zambezi riverbanks and its plains. The road and other transport infrastructure significantly influence settlement patterns and distribution in the basin countries.

The countries of the basin experience high levels of rural-urban migration. Large volumes of rural-urban population movements were in the recent past due to both internal and cross-country civil wars that made rural areas unliveable. For example, Malawi experienced a huge influx of Mozambican refugees in the mid-1980s who numbered around one million, representing 10 percent of the total population at that time. In addition, Malawi has and is currently receiving refugees from Burundi, Rwanda and DRC. In early 2000, Zambia was host to about 200,000 refugees from the Angolan war.

Mozambique has experienced a long and protracted civil war, which displaced about 4.5 million people from the rural areas, most of whom moved into the urban areas. Urban population growth rates rose from four percent in the 1960s to 30 percent in the 1990s. The 30 percent rise appears to be an underestimate, because it excludes poor settlements at urban periphery. The rate of urban growth was estimated at 5.5 percent per year in the 1990s.[14] More than half of the Mozambican urban population lives in unsanitary and unhealthy conditions and the levels of urban unemployment are extremely high. The lack of development in the rural areas coupled with unemployment and poverty, force able-bodied people to migrate from rural to urban centres.

In Malawi, the urban population increased from five percent in the 1960s to 13 percent in 1995. Three-quarters of this urban population resides in the major urban centres of Blantyre, Lilongwe, Mzuzu and Zomba. The growth rate for the urban population is currently estimated at 5.6 percent annually.[15]

Zambia has a long history of internal migration from rural areas to the mining areas, which began during the colonial period. Due to labour demand, about 50 percent of the Zambian population is urban-based, making Zambia the most urbanised country in the basin.[16] The urban population is poorly distributed, with 85 percent of all the urban residents living in only two provinces of the Copperbelt and Lusaka. The rate of growth of the urban population is, however, decreasing and has fallen from 8.9 percent annually in the 1960–70 period to 3.7 percent annually between 1980–90.

Zimbabwe has also experienced a large volume of rural-urban migration, with 31 percent of the country's population living in urban areas. About 46 percent of this urban population lives in Harare.[17] The large volume of rural-urban migration has given rise to urban problems of inadequate sanitation, overcrowding and lack of adequate health, schooling and employment opportunities.

Photo: IUCN

Road and other transport infrastructure significantly influence settlements.

Poverty

Poverty covers a wide range of deprivations affecting people and involves lack of knowledge, denial of opportunities for a creative and productive life, social exclusion, lack of freedom and deprivation of income, and health.[18] Within the basin states, income distribution is highly skewed with the richest 20 percent of the peo-

Photo: M Chenje

Poverty forces some rural communities to brew very potent alcohol for either to sell or home consumption.

ple in most countries owning a larger portion of the national income. Botswana, Namibia and Zimbabwe exhibit extreme income inequalities while Tanzania and Zambia have relatively less income inequality although poverty levels are high.

Poverty among the people of the basin states is due to rapid population growth, slow economic growth and a fragile natural resource base. Uneven distribution of resources is also one of the major causes of population pressure that ultimately causes poverty. Malawi, Mozambique, Tanzania, Zambia and Zimbabwe are implementing economic Structural Adjustment Programmes (SAPs) that are resulting in massive job losses, increasing poverty levels and pressure on the environment.

Poverty and environmental degradation are linked in a vicious cycle in which poor people cannot afford to take proper care of the environment since they have no alternative but to use environmental resources unsustainably for their basic survival. Forests are being decimated at unprecedented rates, faster than that of natural regeneration, resulting in environmental problems. People become more vulnerable, both socially and economically, because degraded environments are producing less resources.

Policies that entrench unequal distribution of resources, lack of tenure and low commodity prices drive the poor into marginal environments, causing more poverty. To address poverty and curb environmental damage, more attention should be paid to degraded environments, promoting public expenditure, empowering communities and building their capacity to own and manage initiatives aimed at improving their well being.

Box 1.1: Strategies to address population-environment problems

- To educate people on sustainable use of natural resources as a national strategy and to be educated by elders and other local people on sustainable use of natural resources (Indigenous Knowledge Systems-IKS).
- To listen to the needs of those who live closest to the natural environment and involve them as participants in decision-making about sustainable use of resources. To challenge these current needs with needs of future generations.
- To educate and inform populations about the human population's role in environmental degradation and to be informed and educated by people about their views of population and environment issues.
- To have strong sustainable and implementable policies and legal frameworks to ensure resource use is adequately controlled for this generation and the next. These policies and frameworks have to be developed with the participation and help of the people so that local people can claim ownership.
- To address problems of rural to urban migration and involve both rural and urban populations in suggesting solutions.
- To recognise and respect local customs and traditions regarding natural resource use but to inform traditional leaders and educate them about global changes and global economic systems and explore possibilities for changes of traditional systems under modernisation.
- To recognise and adequately deal with environmental consequences that have a negative impact on population.
- To set up an institutional mechanism for cocoordinating the planning and management of development. This applies to coordination between the different agencies within each country in the basin and across borders. Such mechanisms should include participation of the public and private sectors, and the local communities.

These strategies are not new, but they emphasise the need for a feedback process between the rural people, the urban populace and the decision-makers.

For these strategies to work however, there is need for adequate monetary and human resources and for providing alternative livelihood for populations through adequate education and employment opportunities, recognising the role of poverty and the need for peaceful political situation in the basin.

Source: Marindo, R., Population and Settlement. For SARDC, 1999.

Case Study 1.1: Population growth compatible with sustainable development in Malawian district

It is commonly believed that rapid population growth is incompatible with sustainable management of the environment, but the experience in the eastern part of Mwanza District in southern Malawi clearly demonstrates otherwise.

Chikwekwe, Gobede, George, Kam'mwamba and Manyenje villages in Mwanza District, along the Shire valley, had in the 1970s large areas of land under natural forest due to relatively low population and limited development, with poor road infrastructure.

This however, changed in the mid-1980s with the opening up of the area following the construction of a road network. The roads facilitated the movement of large numbers of people into the area.

The settlers cleared forest areas to develop gardens. Due to the easy access to the commercial city of Blantyre and other towns, there was a boom almost overnight in the demand for charcoal and firewood. The area, which was once covered by thick forests, began to experience deforestation at the rate of 1.6 percent annually.

Concerned about the deforestation, the Wildlife Society of Malawi (WSM) with technical expertise from the SADC Forestry Technical Coordination Unit (SADC FTCU) launched the US$220,000 Mlambe Natural Resource Management project to reduce forest clearance; integrate women and other marginalised groups in sustainable management of natural resources, and encourage the harvesting of non-timber forestry products.

Rankin Mwamadi, WSM forestry programme officer, says the project campaigned to create awareness among the local communities against wanton cutting down of trees through drama, community debates and local media.

The Wildlife Society also mobilised people to form village forest areas, which are communal forests under the care of village members. Communities also demarcated land to conserve their own trees for home needs under the individual forest areas concept.

Mwamadi says conservation of forests cannot work if the people living adjacent to resources are not allowed to benefit from them. To ensure that the communities earn a living from the natural resources while sustaining the indigenous forests, the project has been facilitating the formation of groups to carry out the management, utilisation, and marketing of non-timber forest products.

The major income generating activity introduced by the project to stop people from over-depending on timber forest products is guinea fowl rearing. Communities were trained to tame guinea fowls from the wild. Following training, the communities formed over 40 clubs and received over 1,000 birds. Many of the community members are now raising their own birds.

Emily Ng'omba, one of the farmers, says guinea fowl rearing has proved to be a successful income generating activity. She boasts that one bird fetches the minimum price of US$4.5, which is twice the price of a 50-kilogram bag of charcoal.

In addition to guinea fowl rearing, bee keeping is another income generating activity in which the communities are involved. The project has encouraged communities to stop using box-hives made from natural tree barks and use those made from fast growing exotic trees. This has helped reduce the destruction of natural forests.

Currently, there are 24 bee-keeping clubs with 82 beehives most of which have already been colonised and farmers harvest honey for sale.

Besides bee-keeping and guinea fowl rearing, the establishment of the project has also provided a ready market for vast amounts of wild fruits in the project area. Indigenous fruits currently marketed at the project headquarters include baobab and tamarind. The fruits are processed into juices. The community members are being trained to process the natural fruits to prepare them to take over the project.

Under the project, community members have learnt to process lemons, oranges, tangerines into products like jelly, jam and marmalade. Women are mainly involved in the full production.

Phillip Liwonde, of Kam'mwamba Forest Area, says it is slavery to go into the bush, spend time felling trees and burning charcoal over three nights, to produce 10 bags of charcoal for sale at US$2.5 when it is easy to raise US$30 through the sale of fruits from the forests.

The project has dramatically reduced charcoal and fuelwood exploitation and trafficking. The Mlambe Project is one of the shining examples in Malawi and probably the entire Zambezi Basin, which shows that rapid population growth is not necessarily a passport to poverty, environmental degradation.

By Brian Ligomeka, Blantyre

People and the environment

Rapid changes in the number of people in a particular area may cause environmental change and degradation. People may fail to adapt to the ecological environment to such an extent that they may even cause environmental damage in large new ecological zones.[19] Inappropriate farming methods, for example, and burning of new areas every season, are a constant threat to biodiversity because fire destroys bushes, insects and other animals. The extensive use of mechanical implements on delicate soils weakens the soil, exposing it to wind or water erosion.

Since all natural resources are limited, a high population growth rate and a short doubling time means a higher rate of depletion of resources. In the absence of technological development, which can replace real natural resources with artificial ones, a high population growth rate can only lead to increased destruction and exploitation of natural resources and increased poverty in the basin as a whole.

The relationship between population and environment can be perceived in three ways:

- Population pressure impacts on the environment, leading to environmental change and degradation;
- Environmental change has negative demographic consequences; and
- Population and environment are related in a feedback process in which they impact on each other.

In the Zambezi Basin, growth rates are high, doubling time is short and the level of technological development is low. The impacts of population growth on natural resources are negative.

SOME ENVIRONMENTAL ISSUES IN THE BASIN

Climate

Differences in topography determine the climatic divisions of the basin with much of it comprised of plateaux lying between 1,000 – 1,500 m above sea level. In terms of rainfall and temperature patterns, the basin has three distinct seasons: a cool and dry season (April–August) a hot and dry season (September–October) and a warm and wet season (November–March).[20]

Climate variability in the basin countries is projected to increase due to global warming. Such changes will significantly impact the global ecosystem, and as the climate changes, the basin is bound to face new risks and opportunities. Water resources and food security are likely to be threatened as a result of worsening precipitation and evaporation patterns. Studies carried out in some parts of the basin indicate that climate change will adversely affect agriculture.

Perhaps the most critical weather phenomenon in the basin is drought. The basin states have experienced frequent drought periods in the 1980s and 1990s, which seriously affected food production and security, hydroelectricity generation, and wildlife populations.

It is generally accepted that droughts follow a cyclical pattern of five to 10 years.[21] The 1992 drought, which hit most of the basin states and the rest of the SADC region, has been described as one of the worst in recent memory, adversely affecting the economies of the basin states. Economic recovery for most of the countries has been very slow, and the 1994–95 drought further slackened such recovery. Climatic issues are covered in greater detail in Chapter 2.

Water resources

The distribution, occurrence and availability of water resources are uneven in the individual basin countries and availability depends on rainfall.[22] The variation of physical characteristics and climate as well as population growth, economic development and political, socio-cultural issues determine the state and management of water.

Rainfall remains the primary source of freshwater although it varies in amount, duration and intensity and is critical to the survival of flora, fauna, people and industry. The estimated average annual rainfall for the basin is about 1,200 cubic kilometres of which nearly 1,100 cu km

Photo: IUCN

Irrigation is one of the major water consumers in the basin.

is evapotranspirated from vegetated surfaces, bare soil, lakes and reservoirs. Only 8-10 percent is available for usage.[23]

The demand for water in the basin is of crucial concern because of the increasing human population, growing irrigation and food needs as well as general climatic changes. Water-consuming sectors are broadly divided into urban and rural domestic supply, industrial use, agriculture and fisheries, hydropower generation, navigation, wildlife, recreation and tourism.[24] Each sector has different requirements for water availability and flow conditions, and the demands vary geographically and during seasons.

However, it is estimated that current total evaporation loss from major hydropower reservoirs is approximately 14.3 cubic metres annually, although the figure includes some uncertainty resulting from problems in measuring the evaporation over large water surfaces. Annually, the basin-wide "loss" from hydropower reservoirs is approximately 10 percent of the available water.[25]

Agriculture consumes less water for irrigation than hydropower generation. The total water consumption for irrigation is estimated at 1.6 cu km for the entire basin or 1.5 percent of the annual runoff.[26]

The gap between demand and supply of water among the basin states is fast closing, resulting in water stress or scarcity. Although the demand for groundwater in the basin is increasing, there is insufficient data on supply, quality, and levels of usage and recharge rates. Water issues are covered in greater detail in Chapter 3.

Land

Land is an important resource in the basin, on which most economies depend. However, such dependency is undermined by climatic changes and land tenure arrangements. A common trend in the basin is that the majority of the people have less or no access to arable land. Often access to land is through inheritance, allocation and purchase that follow patriarchal lines. For example in Zambia, 90 percent of the land available for agriculture falls under customary land controlled by chiefs who follow patriarchal principles in terms of allocation.[27] The result is that the majority of the women do not own land.

Land tenure is one of the major constraints to proper land-use and conservation in the basin in which three systems of land ownership are distinct: small-scale communal farms, small to large-scale commercial farms and state land.

Land degradation is particularly severe in the small communal farming areas due to overcrowding and insecure ownership. However, the title deeds issue is often seen as a technical problem below which are other deep-seated problems of poverty and limited access to natural resources. Agricultural practices that are not adapted to fragile soils in the basin have significantly degraded land, contributing to the reduction of its productivity.

Photos: M Chenje

Land tenure is one of the major constraints to proper land-use and conservation in the basin.

One of the greatest impacts of poor land tenure system is reduced food security. In cases of Zambia and Zimbabwe where the land tenure system emphasises on usufruct rights rather than security of tenure, people are often not motivated to improve the land, resulting in yield declines over years. For example, the index of agricultural production on a per capita basis for Zambia and Zimbabwe declined from 91 to 86 and from 109 to 85 respectively, during the period 1984-96.[28]

Another important land-related issue is access to land-based resources. This is especially important if the dominant form of ownership is communal, as is the case in the basin. In Malawi, for example, customary land accounts for 66 percent of the country's total land area.[29] The ten-

Table 1.3: Land Cover/Use figures for the Zambezi Basin

Class	Area (Km^2)	%
Forest		
Forest (not specified)	49,980	3,7
Evergreen	20,710	1,5
Deciduous	5,620	0,4
Total Forest	76,310	5,6
Open Land		
Open Land (not specified)	258,000	19,1
Grass dominated	74,940	5,5
Bush/Open woodland	630,490	46,7
Total Open Land	963,430	71,3
Agriculture		
(Commercial irrigated and rain-fed)	120,590	9,0
Small Scale farming (communal)	86,260	6,4
Total Agriculture	206,850	15,4
Water Spreads		
Lake/reservoir/river	38,190	2,8
Wetlands/swamps	66,000	4,9
Total Water Spreads	104,190	7,7
Non-Vegetated		
Degraded/barren land	79	0,0
Built-up Areas	506	0,0
Total Non-Vegetated	585	0,0
Grand Total All Areas	1,351,365	100

Source: Denconsult, ZACPRO 6, Sector Study 3 base maps, Lusaka, 1998.

dency in such cases is to turn the land-based resources into common property, with little regard to their judicious use and management.

Land degradation

Land degradation refers to the process by which the quality of the natural or existing land surface deteriorates. This loss in quality comes about through soil erosion, reducing potential productivity; vegetation clearance; landscape alterations; the introduction of toxins into the soil; and the accumulation of surface waste materials. Additionally, overstocking of domestic livestock and wildlife results in significant land degradation, especially when combined with poor cultivation techniques and/or excessive cultivation, particularly in marginal areas.

A major challenge in the basin is feeding the growing population. Agricultural production is increasing by, among others, opening up forests, grazing lands and irrigation. Land degradation ruins agricultural land, contributing to the reduction of crop yields. Given the limitation of suitable land, there is increasing pressure to use unsuitable marginal lands. This is likely to add significantly to the degradation problem.

The basin states have various programmes both at national and SADC regional levels aimed at curbing further land degradation. For example, the SADC Sub-Regional Action Programme to Combat Desertification in southern Africa has identified the following issues as priorities for action:

- capacity building and institutional strengthening;
- strengthening of early warning systems;
- cooperation in the sustainable management of shared natural resources and ecosystems;
- information collection, management and exchange;
- development and transfer of appropriate technology to the community level; and
- development of alternative sources of energy.

Deforestation

The basin is witnessing an increasingly high rate of deforestation, mainly due to human activities. For example, between 1972-90 Malawi's total forest cover declined by 41 percent, representing an average loss of 2.3 percent annually.[30]

The increasing demand for building materials and fuelwood for cooking and heating is contributing to deforestation. However, the major cause of deforestation is the clearance of woodlands for agriculture. Deforestation is widespread in many communal areas and some are completely denuded of woodland cover, exposing soils to both water and wind erosion.

The results of deforestation are increased soil erosion, sedimentation and siltation as well as siltation-induced flooding within the basin, at points far away from the deforested areas. Environmental impacts such as soil erosion, siltation and ultimately land degradation are the main problems associated with deforestation. Excessive exploitation of forests reduces the region's capacity to curb climate change since forests are sinks of carbon dioxide.

Biodiversity

Despite the fair species richness, biological resources and diversity in the basin are threatened by many factors, that include overexploitation of species or group of species such as elephant and rhino, extraction of valued timber species such as Mukwa, African Ebony and Zambezi Teak, and fish from lakes Malombe and Malawi/Nyasa and the Lower Shire. Land clearance for agriculture and human settlement expansion is also a major threat to biological resources in the basin.

Dam construction has had probably the greatest effect on biodiversity of wetlands and aquatic species and on wetlands ecological processes. Dams such as the

Kariba, Cahora Bassa, Itezhi-tezhi and Kafue have changed hydrology, greatly modified flooding regimes and habitat and species composition.

The introduction of alien species has seen some of them out-compete native species or modify the ecology of an area. Alien species of concern include the water hyacinth and the Kariba weed. However, the introduction of the Lake Tanganyika sardine into Lake Kariba in the 1960s has been a major commercial success, spawning the *kapenta* inland fishery for the benefit of Zambia and Zimbabwe.

Photo: D Martin

The introduction of the Lake Tanganyika sardine into Lake Kariba in the 1960s has been a major commercial success, spawning the kapenta inland fishery for the benefit of Zambia and Zimbabwe.

Energy resources

About 74 percent of the total energy needs of the eight basin states comes from biomass.[31] Woodfuel takes up a larger share of energy resources in all the basin states with Botswana getting 73 percent of its energy needs from fuelwood; Malawi, 90 percent; Namibia, 91; Tanzania, 90; Zambia, 62; and Zimbabwe, 57 percent.[32]

The woodfuel energy is mostly used for domestic purposes, including cooking and lighting. Other rural industries that consume significant amounts of woodfuel include brick manufacturing, lime production, fish-smoking, beer-brewing, and coffee, tea and salt-drying.

The high costs of alternative energy sources such as electricity in urban areas have been translated into high urban woodfuel demands, leading to severe forest and woodland depletion within the urban periphery. Both renewable energy resources such as biomass, hydroelectric power and non-renewable energy resources such as thermal power, coal, petroleum and natural gas are available in the basin. New technologies, especially solar and wind energy are also gaining prominence.

Pollution

Mining, industrial, agricultural and domestic activities have negative impacts on the basin, including land and water pollution. Coal mining and use impact negatively on the environment in various ways. Fine coal dust, gases and smoke pollute the air while the excavated spoils piled up in dumps are unsightly. Mine dumps are also washed away into watercourses during storms, leading to groundwater pollution from sediments and dissolved harmful substances.

Signs of deterioration of lakes, streams and rivers in the basin are evident in many places. These range from eutrophication of lakes such as Chivero and Kafubu to high levels of conductivity, as well as high presence of dissolved solids, nitrates and suspended solids.[33]

The management of solid waste has become a major environmental problem in urban areas of the basin. Growing numbers of people together with poor or weak disposal systems of collection, landfill and recycling are exacerbating the problem. For example, Lusaka produces 1,400 tonnes of solid waste daily of which only 10 percent is collected by the city council for disposal.[34]

ECONOMIC DEVELOPMENT

While economic development in the basin has generally increased over the years, growth has been slow in the whole of the SADC region, averaging less than two percent in most countries.[35] The problems besetting basin economies range from huge deficits and debt to high interest rates, high inflation and massive unemployment.

Even though economic growth is largely still resource-based, the basin states have achieved significant strides in industrialisation and urbanisation. The major sectors, which have experienced growth in the basin, are tourism, agriculture, mining, forestry and manufacturing.

Photo: M Chenje

The Victoria Falls is one of the major tourists attractions in the basin.

The main component of the economic structure in the basin countries by the end of the 1960s, was the existence of a thriving export enclave economy based on the exploitation and export of natural resources such as copper in Zambia, diamonds in Namibia, beef in Botswana, while tea in Malawi was the main export crop until the mid-1970s when tobacco emerged as the leading export. In the Lusophone countries, Angola and Mozambique, estate farming concentrated on coffee (Angola) and sugar (Mozambique). Zimbabwe's agrarian capitalism concentrated on the production of tobacco and maize.

Tourism is the fastest growing industry in the basin states. The prime tourist areas in the basin are located in Victoria Falls, Livingstone, and Kasane in Botswana, Lake Kariba and the shores of Lake Malawi/Nyasa. Community-based tourism development programmes now exist in countries such as Botswana, Malawi, Mozambique, Zambia and Zimbabwe. Tourism development has contributed to the destruction of the natural habitat and the disturbance of wildlife, for example in Victoria Falls where some wildlife corridors have been fenced off.[36]

Economic development and growing population have combined to bring environmental problems, which did not exist in the basin before. Some of the problems include increased industrial and domestic waste generation, air and water pollution.

MANAGEMENT RESPONSES AND CHALLENGES

The region faces a number of population and environmental challenges that are currently being addressed by different national, regional and international organisations. Under the general SADC framework, the sectors of Food, Agriculture and Natural Resources (FANR), Water Sector Coordination Unit (WSCU) and Environment and Land Management Sector (ELMS) have launched different programmes that are aimed at enhancing the region's natural resource management and human development.

These programmes include:

- the land and water management research;
- sorghum and millet improvement;
- agroforestry research;
- Zambezi River Action Plan (ZACPLAN), integrated water resources management (IWRM) in the Zambezi River Basin;
- fisheries training programme; and
- the SADC Natural Resources Management (NRM) coordinating community-based natural resources management (CBNRM) programmes in five countries of the basin. The programme coordinates Botswana's Natural Resources Management Project (NRMP), Malawi's NATURE project, Namibia's Living in a Finite Environment (LIFE), Zambia's Administrative Management Design Programme for Game Management (ADMADE) and Zimbabwe's Communal Areas Management Programme for Indigenous Resources (CAMPFIRE).

However, outside the SADC NRMP core projects, other community-level projects are also being implemented in other countries, for example Mozambique's Tchuma Tchato project. The CBNRM programme and other projects have remarkably improved wildlife management in the basin and communities are benefiting in the form of revenue realised and directed to other communal development projects.

The basin countries have also responded to general environmental degradation in their respective countries by developing National Conservation Strategies (NCSs), National Environmental Action Plans (NEAPs) that have been decentralised to district levels (DEAP) in the case of Zimbabwe. Such developments have been supported by international development organisations such as the United Nations Development Programme (UNDP) and IUCN–The World Conservation Union.

Further programmes such as cleaner production and the Biodiversity Strategic Action Plan (BASP,) dealing with pollution and biodiversity conservation issues, are also being implemented at both national and regional levels through both private sector and the Global Environment Facility (GEF) support. In the area of energy, GEF support has also been extended to solar projects that offer alternatives to biomass sources of energy for the rural people.

The issue of land reform is very central to the survival and development of the region in the basin. Virtually all basin countries' economies are agro-based and depend on the accessibility of the land resource to the majority of the people for moral, socio-economic, political and environmental reasons. Today in Malawi, Mozambique, Namibia, Zambia and Zimbabwe the debate on land reform is topical while concerted efforts to help previously disadvantaged people access land are underway. In addressing the land question, issues of equity, poverty and environmental degradation will, to some extent, be automatically addressed once the appropriate criteria and approach are used.

In terms of population growth, there has been no common basin-wide or regional approach apart from the encouragement of family planning methods aimed at both men and women. The success stories for the basin countries vary although generally there has been acceptance of family planning. In dealing with the HIV/AIDS pandemic, there is a regional approach over and above the different national level strategies in the countries of the region. Southern Africa AIDS (SAfAIDS), a regional organisation providing an AIDS information disseminating service in SADC, was established to deal with the issue among other strategies. UNAIDS and the International Labour Organisation (ILO) are also playing their part in exploring the issue of HIV/AIDS in the work place. Voluntary testing and counselling organisations have also been established in different countries to deal with HIV/AIDS.

In response to the general state of poverty in the region, poverty alleviation programmes that cut across socio-economic and environmental issues have been initiated alongside economic reform programmes being implemented in most basin countries. The issue of poverty has been strongly linked to environmental management and human development hence specific programmes dealing with poverty.

National level

The Zambezi River Authority Act (1987) was passed in Zambia and Zimbabwe and the subject of the agreement is the utilisation of the Zambezi River as it passes through both countries. Zambia and Zimbabwe have agreed to utilise, operate and maintain the existing Zambezi complex (dam, reservoirs all telemetry stations and other existing installations) as well as any additional dams, reservoirs and installations that may be constructed or installed on the Zambezi River. Other duties of the Zambezi River Authority include collection and processing of hydrological and environmental data, making recommendations to the council on the effective use of the Zambezi River and to submit development plans to the Council.

Photo: M Chenje

The Zambezi Basin offers many opportunities for learning from each other and regional cooperation.

Photo: D Martin

All the riparian countries benefit from the basin's resources, highlighting the need for cooperation in natural resource management.

Most basin countries have an array of environmental standards and regulations that regulate developments that impact on the environment, monitor human impacts and help enforce environmental laws. However, the regulation, monitoring and enforcement of the various well-intended pieces of legislation are hampered by lack of resources and poor coordination, among other factors.

With globalisation fast setting in, there is need for harmonisation of the environmental and related regulations, and standards of the region's countries for meaningful cooperation and integration.

Regional level

All the riparian countries benefit from the basin's resources, highlighting the need for cooperation in natural resources management. Under the SADC Treaty, various protocols have been developed to address specific issues (see Chapter 12: Environmental Management and Regional Cooperation). These protocols provide the political framework that enables and encourages regional cooperation in the use and management of the basin and region's resources, trade and other issues.

Cooperation among basin countries should be cross-sectoral and take into account social and economic policies and programmes over and above environmental resource management. The southern African region at large and basin countries in particular, should harmonise policies on environment, population, human settlements and distribution, human resources development, poverty alleviation, education, urbanisation and industrialisation and other critical areas central to regional cooperation to achieve sustainable development.

Without losing current gains, SADC's regional programmes and projects should strive to become regional in outlook, implementation and benefits. The current sectoral approach used in project implementation has resulted in most projects benefiting the respective countries in which they are coordinated at the expense of the region. More resources need to be committed to the region's programmes by the member countries if they are to be sustainable.

Global level

In response to the various international environmental conventions, the basin countries under the auspices of SADC have and are implementing programmes in line with various agreements. The Convention to Combat Desertification (CCD), Convention on Biological Diversity (CBD), Convention on International Trade of Endangered Species (CITES), Montreal Protocol on Pollution and Climate Change as well as Agenda 21 are some of the common agreements and documents guiding the region's environmental conservation agenda with adaptation of local issues on the ground.

The region is playing its part in implementing some of the above conventions in various ways. Through various regional and international institutions that are operating under the auspices of SADC and its various technical coordination units, projects and programmes such as Networking Capacity Building (NETCAB), Zambezi Basin Wetlands Conservation and Resource Utilisation Project (ZBWCRUP), Communicating the Environment (CEP) and State of the Environment Reporting (SOER), the region is implementing global agreements that are key to sustainable development.

CONCLUSION

The following chapters discuss in greater detail many of the issues that have been raised above. Keeping the pressure-state-response (PSR) concept in mind, the chapters present the problems, how they have affected the environment, and how society is responding in trying to address them.

LINKAGES TO OTHER CHAPTERS

Chapter 2: Physical features and Climate
The physical setting of the Zambezi River Basin and its climate are critical to life, both human and natural, influencing human activities of millions of people in the eight basin countries.

Chapter 3: Water and Wetland resources
Water and wetlands resources have a great influence on human settlements and activities in the basin. Integrated management of both is important for the sustainability of people's well-being and the environment.

Chapter 4: Biological Resources and Diversity
The Zambezi Basin is rich in biodiversity, in terms of culture, and flora and fauna, hosting some of the most attractive tourist destinations in the SADC region. It is, however, important that measures continue to be taken to conserve biodiversity and meet human needs at the same time.

Chapter 5: Agriculture
Increasing population and economic growth continue to place more pressure on ecologically sensitive areas of the basin. Effective management of environmental resources and population growth is necessary to achieve sustainable development.

Chapter 6: Industry
While the basin states have experienced significant industrial growth over the last three decades, increased industrial waste generation and new environmental problems, pose health risks to people in the basin.

Chapter 7: Energy
Traditional sources of energy (thermal and fuelwood) cannot sustain growing populations and economies of the basin states. The growth in population and economies has raised energy demand, resulting in environmental problems such as deforestation, habitat loss and emission of greenhouse gases.

Chapter 8: Tourism
Tourism, which is mainly dependent on wildlife and the basin's watercourses, is the fastest growing industry, and with strategic planning may be crucial to alleviate poverty, particularly in rural areas where community-based ventures are possible.

Chapter 9: Pollution
Pollution has a bearing on the health of people and the environment in the basin. Various sources of pollution, gaseous, liquid and solid, need to be considered in integrated environmental management.

Chapter 10: Poverty
Poverty is both a cause and result of environmental degradation, threatening sustainable development. With the majority of people in the basin living in poverty, concerted action is needed to alleviate poverty.

Chapter 11: Gender and Women's Roles
Gender has generally not been a factor in environmental management programmes. Gender sensitivity in natural resource management is critical, particularly if the role of women as resource managers is considered in decision-making and all management activities.

Chapter 12: Environmental Management and Regional Cooperation
Think basin-wide but act locally is probably the best approach to tackle the numerous environmental problems facing the Zambezi Basin states. While many of these problems may appear localised, they often have a basin-wide impact, making it imperative for governments in the region to cooperate at different levels.

Chapter 13: Trends and Scenarios
Population and economic growth will continue into the foreseeable future, placing increasing pressure on the environment. Effective management of natural resource demand is critical to the basin in order to achieve economic integration, alleviate poverty, and realise sustainable development.

ENDNOTES

1 SARDC-IMERCSA, "Zambezi River Basin Series Factsheet No. 1", SARDC, Harare, 1998
2 ibid
3 WRI/UNEP/UNDP/World Bank, *World Resources 1998-1999,* Oxford University Press, Oxford/New York, 1998
4 ibid
5 Timberlake J., "Biodiversity of the Zambezi Basin", for SARDC, 1999
6 UNDP/SADC/SAPES, SADC *Regional Human Development Report,* Harare, 1998, SARIPS/SAPES Trust, Harare, 1998
7 Dalal-Clayton B., *Southern Africa Beyond the Millennium: Environmental Trends and Scenarios to 2015,* London, March 1997
8 Udjo, E.O., "Is fertility declining in Zimbabwe?" Journal of Biosocial Science, Vol. 28 No.1, 1995
9 Denconsult. ZACPRO 6 *Sector Studies 3, Final Report,* Lusaka, 1998
10 FAO, World Forests Report, Rome, 1997
11 United Nations. *Too Young to Die: Genes or Gender;* Department of Economic and Social Affairs, Population Division, New York, 1998
12 Ministry of Health and Child Welfare, *Zimbabwe National Health Profile,* Government of Zimbabwe, Harare, 1997
13 Marindo, R., "Recent Mortality Estimates for Zimbabwe", unpublished report, University of Zimbabwe, Harare, February 1999
14 UNFPA, "Mozambique Country Report and Strategy Development", Report No. 25, 1995
15 World Bank, *Malawi Population Sector Study,* Washington D.C., 1994
16 UNFPA, "Zambia Country Report and Strategy Development", Report No. 42, 1995
17 Government of Zimbabwe, *The 1994 Zimbabwe Demographic and Health Survey,* Central Statistical Office, Harare, 1995
18 op. cit. 6
19 Zaba, B. and J. Clarke., "Current Directions in Population and Environment Research", In: Zaba, B. and J. Clarke (eds.), *Environment and Population Change,* IUSSP, Ordinary Edition, 1994
20 Dandaula, H., "Climate and Drainage in the Zambezi Basin", For SARDC, April 1999
21 Tumbare, M. J., "Cyclic Hydrological Changes of the Zambezi River Basin: Effects and Mitigatory Measures", Proceedings of the IX Water Congress, Montreal, 1997
22 Chenje M., and P. Johnson (eds.), *Water in Southern Africa,* SADC/IUCN/SARDC, Maseru/Harare, 1996
23 Denconsult, ZACPRO 6, *Sector Studies Chapter 3, Introductory Volume,* Final Report, Lusaka, 1998
24 ibid
25 ibid
26 ibid
27 op. cit. 6
28 op. cit. 3
29 Environmental Affairs Department, *State of Environment Report for Malawi 1998,* Environmental Affairs Department, Lilongwe, 1998
30 ibid
31 SARDC-IMERCSA, "Zambezi River Basin Series Factsheet No. 2", SARDC, Harare, Zimbabwe, 1998
32 ibid
33 Kasonde, J.M., "Aspects of Industrial Production and Pollution in Zambia", National UNCED Seminar, Lusaka, 1991
34 Agyemang, O., et al, *An Environmental Profile of the Greater Lusaka Area: Managing the Sustainable Growth and Development of Lusaka,* Lusaka City Council/Government of Zambia, Lusaka, 1997
35 op. cit. 6
36 Meynell, P.J., Sola, L. and N. Nalumino., *Strategic Environmental Assessment of Developments around Victoria Falls,* IUCN ROSA, Harare, 1996

REFERENCES

Bell-Cross, G., *The Fish Fauna of the Zambezi River System,* National Museums and Monuments, Harare, 1972

David, L.J., Golubev, G.N. and Nakayama, M, *The Environmental Management of Large International Basins,* UNEP, Nairobi, 1988

Derman, W., *Preliminary Reflections on Research Issues and Strategies for Long Term (Five Years of Longer) Study of Common Property and Natural Resources Management with Particular Emphasis upon the Zambezi River Basin,* CASS, Harare, 1987

Hiscock, H.J., Matiza-Chiuta, T. and Lash, T.J.F., *Zambezi Basin Wetlands Conservation and Resource Utilisation Project: Inception Mission Report,* IUCN ROSA, Harare, 1996

Kalapula, E.S., "Zambezi Action Plan: Environmental Management and Sustainable Development of the Common Zambezi River System", University of Zambia, Lusaka, 1989

Marshall, B.E., "The Influence of Man on the Physical Environments of the Zambezi Valley", University of Zimbabwe, Harare, 1992

Matiza, T. and S.A. Crafter (eds.), *Wetlands Ecology and Priorities for Conservation in the Zimbabwe,* IUCN, Gland, 1994

Matiza, T., Crafter, S. and Dale, P. (eds.), *Water Resource Use in the Zambezi Basin,* Proceedings of a Workshop held at Kasame, Botswana, 28 April to 2 May 1993, IUCN, Geneva, 1995

Pallet, J. (ed), *Sharing Water in Southern Africa,* DRF, Windhoek, 1997

SADC ELMS, Proceedings of the Workshop on the "Development of an Integrated Water Resources Management Plan for the Zambezi Basin", Livingstone, 2-6 May 1994

Tumbare, M.J., *A Strategic Action Plan for the Sustainable Development of the Water Resources of the Zambezi River Basin,* University of Strathclyde, Strathclyde, 1996

ULG Consultants, *Zambezi Valley Environmental Impact Study:* Interim Report on Significant Issues for Discussion and Comment, ULG Consultants, Harare, 1990

Vorosmarty, C., *Hydrology, Deforestation and Soil Losses in the Zambezi Basin: Estimating the Impacts of Deforestation and Vegetation Change in Africa,* 1989

Map 2.1 Sub-basins of the basin

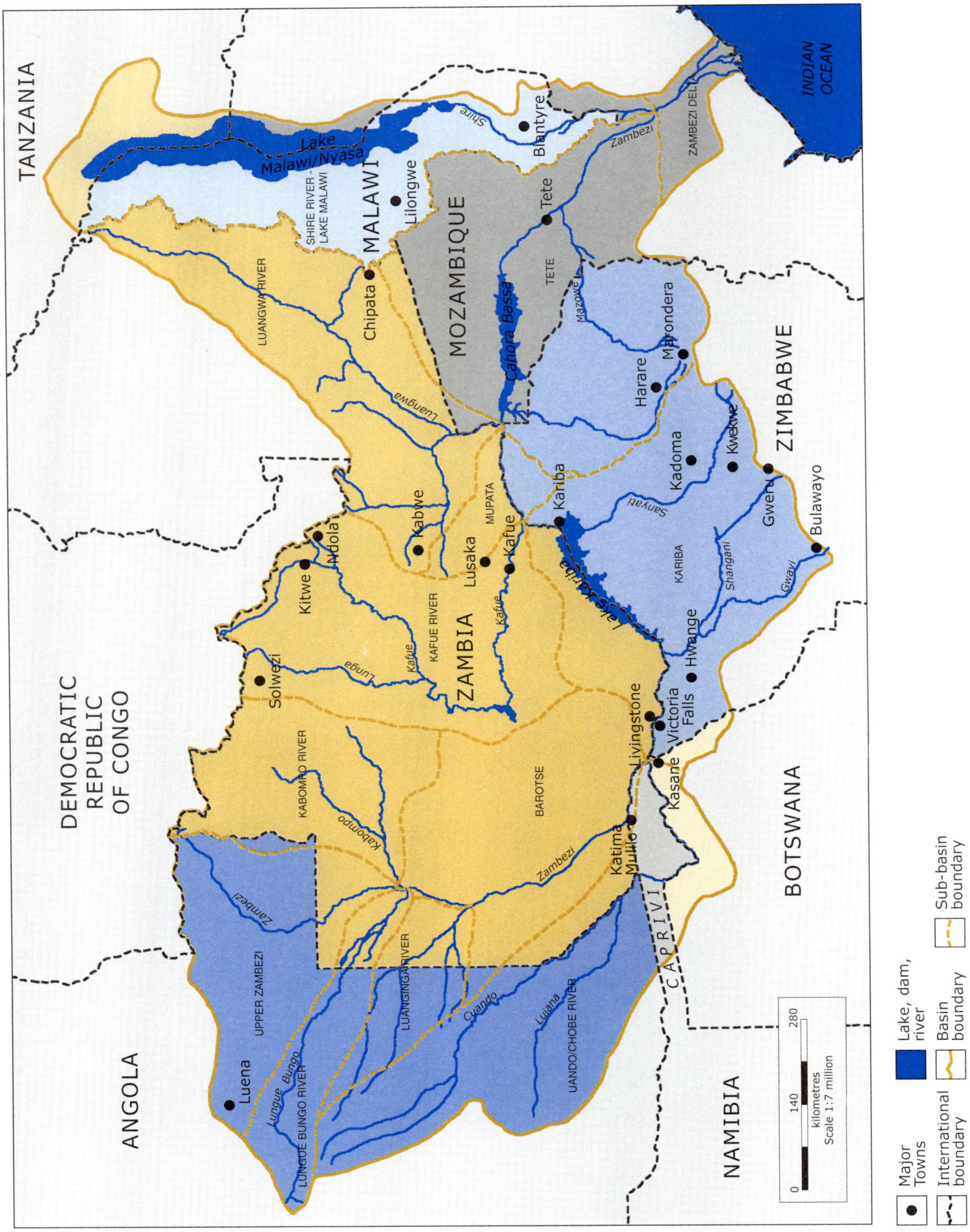

Source: Zambezi River Basin Atlas, SADC/DENCONSULT/ZRA, Lusaka, 1998

2

PHYSICAL FEATURES AND CLIMATE

The Zambezi River Basin is the fourth largest in Africa, flowing eastward for more than 2,800 km from the Kalene Hills in northern Zambia to its mouth at the Indian Ocean in central Mozambique. The Zambezi is the largest river in Africa flowing into the Indian Ocean and the basin is the largest in the Southern African Development Community (SADC) region.

Much of the Zambezi Basin comprises of plateaux that lie at an altitude of between 1,000 – 1,500 m above mean sea level (amsl). The topography of the basin varies in altitude from sea level at its delta on the Indian Ocean to over 1,500 m on the plateau with some mountainous areas lying beyond 2,500 m amsl. These plateaux are deeply dissected by the river valleys that form the tributaries of the Zambezi River.

Photos: M Chenje

The topography of the basin varies from mountain tops to valleys and falls.

Ancient rocks of Triassic and Jurassic age (more than 120 million years old) underlie the Zambezi River catchment, as well as that of its tributaries such as the Kafue, Cuando, Gwayi, Manyame, Mazowe, Luangwa and Shire.

The present-day Zambezi system can be divided into three separate sections, each differing in its landscape characteristics, its geological history and its biodiversity. These are the Upper Zambezi , the Middle Zambezi – that part of the basin between Victoria Falls and the confluence with the Luangwa River, and the Lower Zambezi – from the confluence with the Luangwa River to the coast at Chinde. It also includes the Shire River in Malawi. The Lake Malawi/Nyasa catchment is often considered part of the Lower Zambezi, even though it has its own unique characteristics.

The physical features of the Zambezi Basin impact human-nature interactions, influencing population distribution, settlement patterns, human activities, industrial development, wildlife distribution and movement, and conservation measures.

GEOMORPHOLOGY

The Upper Zambezi

The Upper Zambezi is composed of six sub-basins;

- the Cuando/Chobe river sub-basin,
- the Barotse sub-basin,
- the Luanginga river sub-basin,
- the Lungue Bungo river sub-basin,
- the Upper Zambezi sub-basin, and
- the Kabompo river sub-basin.

All of them except for the Barotse sub-basin originate in the high rainfall Angolan and Zambian highlands associated mainly with the Zambezi/Congo divide.

The Upper Zambezi sub-basin comprises the headwaters of the Zambezi and includes the Kalene Hills of Mwinilunga District of Zambia. In the north along the Congo basin watershed in Angola, the terrain is hilly and rises to 1,500 m amsl.

The northern limits of the Kalahari sands traverse the Upper Zambezi sub-basin in an east-west direction through Angola. To the south of this limit, the terrain is gently undulating and is covered by a deep layer of sand, interspersed with treeless floodplains and seasonally flooded *dambos.*

The Zambezi Floodplain that dominates the Western Province of Zambia begins at Chavuma where the Zambezi re-enters Zambia from Angola. Two important tributaries join the Zambezi near Lukulu, the Lungue Bungo from the north-west and the Kabompo from the north-east (sub-basins Lungue Bungo and Kabompo respectively). Both these catchments are rather similar in character and can be conveniently divided into three zones; upper, middle and lower. The upper reaches of these rivers are characterised by a hilly terrain while the middle zone in the extreme eastern and western areas, wind blown sand is absent. In the lower zone, both rivers enter the Zambezi floodplain characterised by treeless plains and seasonally inundated *dambos.* Bordering on these areas are large expanses of Kalahari sands elevated from the plains.

The Luanginga river sub-basin joins the Zambezi River upstream of the town of Mongu. Within Zambia, the basin is characterised by gently undulating terrain covered by a deep layer of wind-blown Kalahari sands and interspersed with large floodplains and dambos. A significant part of

Floodplains are an important physical feature across the Zambezi Basin, influencing habitats and wildlife distribution.

Photos: M Chenje

the sub-basin is taken up by floodplains and the largest of these, the Liuwa Plain is a national park. In Angola, the sub-basin has characteristics similar to those of the middle zone of the Kabompo and Lungue Bungo sub-basins.

The Barotse sub-basin lies almost entirely in Zambia. The sub-basin is characterised by a gently undulating landscape overlain by a deep layer of wind-blown Kalahari sands which are interspersed with numerous floodplains and *dambos*. A sizeable portion of the entire Barotse sub-basin in Zambia, is occupied by the seasonally flooded Zambezi Floodplain.[1] Despite the general covering of Kalahari sands in the sub-basin, there are areas in the north and west most notably around Kaoma, which are not sand covered. These areas are agriculturally the most productive in the sub-basin.

The Cuando/Chobe river sub-basin joins the Zambezi in the south. Most of this sub-basin lies in Angola where the main tributaries rise in the Central Highlands. In all four countries through which this sub-basin passes (Angola, Botswana, Namibia and Zambia), the rivers flow through flat to gently undulating terrain which is overlain by Kalahari sands.

The Middle Zambezi

The Middle Zambezi, between Victoria Falls and the confluence with the Luangwa River, is much younger and more environmentally heterogeneous than the Upper Zambezi. It includes the Luangwa Valley and most of northern Zimbabwe and the Kafue Flats.

Before the capture of the Upper Zambezi, this section would have been relatively small, with typical narrow floodplains, rapids and large fluctuations in seasonal flow. Today, it is a dissected landscape, ranging from mountains and other highlands down to broad dry valleys. Its biological character has been greatly modified in recent years by the construction of two huge dams, Kariba and Cahora Bassa, and a series of smaller dams on the tributaries, particularly in Zimbabwe.

Two features of over-riding significance, the ancient Central African Plateau and the more recent rift valley faulting, dominate the geomorphology of the Middle Zambezi. The Central African Plateau is about 1,500 m high. The down faulting of the Great East African Rift Valley system is associated with splinter rifts. In their classical form, these rift valleys show a number of common features which are clearly evident in the Middle Zambezi. Tectonic activity has lowered the entire valley floor along linear marginal fault lines, leaving a flat valley bottom between 10–60 km wide with a general elevation of about 500 m. Down faulting of the valley floor has caused displacements that have thrown up a series of scarps rising in steps to elevations that in most instances exceed the original plateau.

Photo: M Chenje

The Victoria Falls Bridge has become one of the major artificial physical features of the Middle Zambezi.

Within this zone, the river flows through one such splinter rift and to the north, the Luangwa, Lunsemfwa and Lukushashi rivers are all similarly confined. These splinter rifts converge at the town of Luangwa where the Luangwa River joins the Zambezi. This point also marks the political frontier between Mozambique, Zambia and Zimbabwe. The splinter rifts occupy about one-third of the entire Middle Zambezi, with their associated scarp rifts. Due mainly to the effects of rain shadows, the rift valleys experience low and erratic rainfall of between 650–750 mm.

There are four sub-basins in the Middle Zambezi which are the:

- Kafue river sub-basin, which is entirely in Zambia;
- Kariba sub-basin, shared by Zambia and Zimbabwe (and a small pocket in Botswana);
- Luangwa river sub-basin, almost entirely in Zambia; and
- Mupata sub-basin shared by Zambia and Zimbabwe.

The headwaters of the Kafue Basin are located along the Congo/Zambezi divide. In the lower reaches, the basin narrows down rapidly and passes over the Zambezi escarpment. The Kafue River is dammed at Itezhi-tezhi.

In Zambia, the Kariba sub-basin is characterised by very steep and dissected country with short, flash-flooding rivers draining into Lake Kariba and the Zambezi. The extent of the plateau at the top of the escarpment is very limited. The major rivers in Zimbabwe, rise along the watershed from Harare to Bulawayo. The 1,200 m high main plateau in the eastern part of the sub-basin is an undulating landscape with broad, fertile interfluves interspersed occasionally with massive granite inselbergs. The plateau slopes gently northward from the watershed. In the north, it is interrupted by the precipitous escarpment of Mavuradonha in Zimbabwe, which drops down into the flat bottom of the ancient rift valley and through which the major rivers have carved deeply incised gorges.[2]

The Luangwa sub-basin includes the highest (Mafinga Mountains) and lowest (Luangwa) geographic points in Zambia. On the eastern side of the rift valleys (Eastern Province of Zambia), the productive plateau is narrowest in the north but widens out to a maximum in Chipata district before narrowing again to disappear completely around Nyimba.

The Mupata sub-basin straddles the Zambezi rift with productive plateau areas found only in the extreme north and south.

The Lower Zambezi

The Lower Zambezi extends from the confluence with the Luangwa river to the Indian Ocean coast. This section also includes the Lake Malawi/Nyasa catchment and the Shire River although they are a biologically separate unit as regards aquatic biodiversity. This section of the Zambezi is typified by a broad floodplain, often with many parallel channels and shifting sandbanks, while the coastal portion includes extensive grasslands and freshwater swamps, dunes and mangroves. Although its geological history is unclear, this part of the basin is biologically similar to the Pungwe/Buzi system, and is linked during floods to the Gorongoza area and Urema Trough.[3]

The geomorphology of the Lower Zambezi, is dominated by the Great East African Rift Valley with the whole of Malawi lying within the rift's trough. The Lower Zambezi Valley in Mozambique which is inclined northwest/southeast, intersects the north/south inclined Rift Valley downstream of Sena. The Shire River enters the Zambezi along the rift from the north and the Zangue River enters from the south along the Urema graben (rift valley).

Photo: M Chenje

Lake Malawi/Nyasa occupies the greater part of the Great East African Rift Valley trough.

Lake Malawi/Nyasa occupies the greater part of the rift valley trough and in many locations the phenomenon of bordering scarps being uplifted beyond the ancient plateau level are evident. Examples are the Livingstone mountains in Tanzania, the Nyika and Viphya plateaux, and Zomba and Mulanje mountains in Malawi. In more northern latitudes, the Rift Valley is tectonically active, whereas in the Lower Zambezi, including Malawi, it is now relatively quiescent. However, in ancient times the trough that Lake Malawi/Nyasa now occupies, was cut off from Lake Tanganyika trough by massive and violent activity which created the Mbeya Highlands which now separate the two lakes and limit the extent of the Lower Zambezi. Downstream of Sena, the Zambezi enters its delta, which is not as extensive as might be expected for such a large river. In its lower reaches, the Zambezi is a relatively youthful river.

In Malawi, altitude varies markedly with the highest 3,011 m and lowest 100 m points being only 100 km apart. Due to rift faulting, areas of extremely steep slopes have been created. This sub-basin of the Zambezi includes parts of the very productive southern highlands of Tanzania such as Tukuyu where rainfall is high. It also includes parts of Mozambique in the central lake region and in the Lower Shire valley region.

The Tete sub-basin covers Lake Cahora Bassa and the Zambezi River downstream as far as Sena. In the east, the sub-basin boundary is exactly along the Malawi/Mozambique border through the western highlands of the rift valley, culminating in the high Angonia plateau. In the north, the boundary passes through the fertile plateau area of the Eastern Province of Zambia. The major rivers in this sub-basin rise along the Zimbabwean watershed from Nyanga to Marondera. The sub-basin traverses a large spectrum of agro-ecological regions, the most productive of which is the main plateau that makes up the bulk of the sub-basin in Zimbabwe.

Uplifting associated with the Rift valley north of the Zambezi along the Morrumbala range, provides the western boundary of the Zambezi Delta sub-basin. South of the Zambezi, all the tributaries of the Zangue are included in the sub-basin which, accounts for a westerly boundary in this sector.

The topography of the sub-basin is determined by the intersection of the rift valley with the Lower Zambezi. Higher elevations occur in a north/south direction through Morrumbala to Inhaminga. The delta proper begins at Mopeia, 120 km from the sea and tidal waves cover the last 80 km. Leveés along the Zambezi and its minor tributaries and old channels, together with relict beaches have created elevated areas. Near the coast, there are higher lying dunes that are most prominent in the northern part of the delta.

SOILS

Soil formation is generally determined by the parent rock material, the climate and the biological activity. The dominant soil types in the basin are acid, leached out tropical soils of low fertility known as tropical red soils (ferralsols) which are derived from geologically old or ancient rock types characteristic of the basin.

Fertile soils are located in areas where younger geological strata form the base or where humic materials are found. The fertility of the basin soils are inherently low.[4] However, soils such as black cotton soils which are of very high quality, are found in the basin. Where sufficient water is available and where reasonable fertilisation is applied with appropriate management, excellent crops may be harvested. Some of the good quality soils are often found in the river valleys, where they are sometimes sacrificed for the benefit of hydropower reservoirs or reserved for wildlife conservation.

Soil classification in the basin is difficult because the basin states use different terminology and classification systems. However, the general soil types of the basin can be established. A large part of the Upper Zambezi is covered by a deep blanket of wind-blown Kalahari sands[5] which severely limit agricultural production. However, these sands are often well timbered with economically important hard-woods.

Photo: M Chenje

Soil classification in the basin is difficult because the basin states use different terminology and classification systems.

The soils in the Middle Zambezi are generally of alluvial and colluvial origin and they are usually of a mixed type. Kalahari wind-blown sands cover significant tracts of land in the western part of the basin in Botswana and

Zimbabwe, and western Zambia where the rainfall is low and erratic. In Zambia, the Kariba sub-basin is characterised by generally fragile, shallow and stony soils with areas of more recent alluvial deposits. In Zimbabwe, the sub-basin consists of soils which vary from whitish-grey sandy loams (depending on the parent materials). Rich alluvial soils are mainly found in the valley bottoms of the main rivers.

The soils found in the Lower Zambezi are very variable. The ferruginous soils which are generally red in colour with a sandy clay loam topsoil over a sandy clay or clay subsoil, are among the most fertile in the sub-basin. They are well represented in the Lilongwe plain and parts of the Southern Province of Malawi.[6] Weathered ferrallitic soils, which are not very fertile, are found in the inland plains north of Kasungu while fertile fragmented alluvial soils are found on the Lake Malawi shore plain and in many parts of the Shire Valley.

The dominant ferralsols found in the basin are part of fersiallitic soils which mainly occur on parent rock materials rich in ferromagnesian minerals (dolomite, calcareous schist, etc), but may even cover old alluvium – 'Kafue basin alluvium'. They have a moderate base status (pH 5-7) and are formed on highlands in the Central Province (Lusaka, Mumbwa, Kabwe), in the Monze-Mazabuka districts, and in Eastern Province (Petauke-Chipata) of Zambia. They are also the most extensive soil group in Zimbabwe and are critical to agriculture in the country. Most of Zimbabwe's maize belt is covered by these soils.

About half of Zambia is covered by ferrallitic soils which derive from various parent rocks, including granite, gneiss, sandstone, and schist. These soils, which have a clay content that increases with depth, are partly suitable for cultivation. Ferrallitic soils require careful management due to their sandy nature. The northern ferrallitic soils in Zambia are leached because of the high rainfall in the area and are widely used for *chitemene* shifting cultivation. The red ferrallitic soils in the basin are found in Mozambique and these are well drained, clayey, deep and are associated with humid and sub-humid climates, covering considerable stretches in the uplands, rainy regions of Niassa, and Zambezia provinces.[7] These soils are very fertile and have a great agricultural potential.

In Malawi, the humic ferrallitic soils are reddish to yellowish and are of moderate depth characterised by a humic topsoil with 5–10 percent organic matter. Other types of ferrallitic soils are paraferrallitic soils which are derived in-situ from granite parent rock and orthoferralitic which result from deep weathering of underlying rock, followed by intense leaching of bases.

Lithosols are derived from granite, gneiss, schist or sandstone. They are shallow escarpment soils which have not developed beyond their present depth either because of climatic conditions or other environmental factors. The characteristic texture of lithosols ranges from sand through loamy sand to sandy clay loam, rock material increasing with depth. These soils are not arable because they are shallow and commonly found in national parks.

Lithoidal soils, poorly developed, coarse-grained stony soils, which occur in areas of transition between humid and semi-arid zones are found in large areas of Tete and Zambezia in Mozambique. These soils are mainly used for extensive natural pastures and are very vulnerable to erosion.

Regosols are mostly developed on Kalahari sands. These are immature soils developed from sands of lacustrine origin and are very sandy, structureless, and with little profile development except for slight humus accumulation in the top soils.[8] Vegetation types found on these soils include fairly closed woodland in some areas whereas in others, it is thick thorny scrub. Regosols are found in western Zimbabwe from around Nyamandhlovu just north of Bulawayo to Victoria Falls in a zone which includes Hwange, Lupane and Nkayi. This belt continues into Botswana where the soils are much more extensive. Their major limitation is the low nutrient reserves and high permeability associated with low water holding capacity.[9]

Among the immature soils, the alluvial soils possess the greatest agricultural potential. They cover considerable areas in the extensive Zambezi Delta and along the banks of numerous rivers.

Vertisols or black cotton soils are dark brown to black in colour and are locally called *makande* soils in Malawi. They are characterised by a strongly developed blocky structure and are very friable when dry but sticky when wet. The major areas of vertisols are associated with basalt and mudstone formations. The Kafue vertisols are found in the floodplains of the Kafue Flats, covering an estimated 7,000 sq km between Itezhi-tezhi and the Kafue Gorge.[10] These soils are also found as small bodies along other rivers and swamps in the basin. Vertisols of the river valleys cover the Luangwa and its tributary valleys, and parts of the Zambezi Valley and they are derived from karoo sediments by colluvial and alluvial processes.

Hydromorphic soils are clays with a strongly developed, very coarse, blocky to prismatic structure, and are seasonally waterlogged. They are locally called *dambo* soils in most of the basin states and are black-grey or mottled in colour. During the rainy season, the water-table is located close to the surface but falls to lower horizons in the dry season. The marsh soils in the basin constitute permanently waterlogged hydromorphic soils.

Other soil types found in the basin include;

- the greyish-red sandy soils of Mozambique which were derived from volcanic rocks,
- calcimorphic soils which occur on nearly level depositional plains, and
- calcimorphic alluvial soils which are grey to dark brown soils formed from alluvium and these have developed mainly on the depositional plains of the Lake Malawi/Nyasa shore and on the Upper and Lower Shire Valley.

Also found in the basin are siallitic soils, arenosols which are found in the eastern Caprivi Strip, Barotse sands which are deep, loose and structureless, and sodic soils whose distribution is determined by the amount of sodium-releasing minerals in the parent material.

MINERAL RESOURCES

The greenstone belts of the central plateau are sources of gold, chrysotile, asbestos, chrome, nickel and iron ores. The mineralisation in the Zambezi belt is mainly stratiform, disseminated copper (and cobalt) sulphides. The mineralisation is found in quiet, near shore marine waters (protected embayments, lagoons, with arid tidal flats).[11] Copper grades are generally 2.5 – 3.5 percent and locally up to 15 percent and rarely 30 percent. The main ore minerals are chalocopyrite, the main source of copper, bornite, the second most important source of copper and chalcocite. Other mineral resources include carrolite which is the main source of cobalt followed by cobaltiferous pyrite. Pyrite is also a main sulphide in the Copperbelt ores. Important secondary copper minerals include malachite, azurite, and cuprite. Minor late veins of uranium-bearing minerals and copper sulphides are found in the Basement and in Footwall and Hangingwall Formations.[12]

In Angola, the following mineral resources are found in the Zambezi Basin: manganese, uranium, gold, iron and diamonds.[13]

Coal fields found in the Tanzanian part of the basin include Mbamba Bay, Liweta, Ngaka, Ketewaka,

Photo: Thompson Publications

Mining is a major factor in terms of changing the state of the physical environment.

Mchuchuma and Songwe-Kiwira. There is also gold mining by small-scale miners in the northeastern part of the basin.

Zimbabwe has a wide variety of minerals, but many occur in superficial deposits. Surface conditions generally give good guidance to mineralisation, except where the Kalahari sands are 100 m or more thick, or the suspected deposits, such as coal and oil in the Zambezi Valley are deeply buried. There are various mining activities in the basin within Zimbabwe. Their locations and minerals mined are, Hwange – coal, Kariba – slate, and Freda Rebecca – gold.

Small-scale mining includes gold panning, red ochre, clays, calcite and industrial minerals. Other mineral resources include those found on the Great Dyke which hosts substantial chrome deposits, platinum and gold as by-product, the Magondi mobile belt and the Zambezi belt. Gold occurs in extensive fields in the basement schists throughout the highveld and in the bordering mid-level areas. Copper and silver are mined, particularly around Chinhoyi and Mhangura. Asbestos, like gold, is found in the Basement schists. Coal deposits occur in the younger rocks at the northern and southern edges of the basement shield. At Hwange, the coal is mostly of a poor quality, with a high content of ash, sulphur and phosphorus.

Kyanite deposits occur at the Kyanite mine in northeastern Zimbabwe in the Mudzi district, mica at Mwami mica field around Karoi and dolomites occur around Rushinga. Some micaceous quartzites "glitterstones" are found in the Makuti Group near Kariba.

A number of mineral deposits exist in Malawi. These include lime, coal in the Northern Region and in the

Lower Shire Valley, phosphates, sulphur, gypsum, glass, high-quality ceramic clay, iron sulphides, rubies and sapphires (small-scale).

In Mozambique, the principle mines and quarries in the basin include the Moatize coal mines in Tete. Heavy minerals are found in Mozambique's coastal sands which are concentrated in dunes and beach sands.

Zambia's principal mineral resource is copper. Further mineral resources include lead, zinc, silver, manganese at Mansa in Luapula Province, coal in the Gwembe Valley and high grade iron ore at Sanje, near Mumbwa. Small quantities of precious and semi-precious stones (except diamond) are obtainable in the Southern Province, including amethysts. Small-scale exploitation of mica has occurred in Eastern Province.[14]

CLIMATE

Climate can broadly be described as the "average" of a series of weather events, or more precisely defined by the statistical measurement of weather variability over a period of time. The earth's climate is affected by factors that cause a change in the redistribution of energy within the atmosphere or between the atmosphere, land and ocean.

Photo: M Chenje

The position of the sun determines how much heat can be received in the basin.

By its location, the Zambezi Basin is much influenced by the prevailing wind systems that include the South-Easterlies, Congo Air, the North-Easterlies and the Inter-Tropical Convergence Zone (ITCZ). These air masses also interact with mid-latitude weather systems. The emergence of tropical cyclones in the Indian Ocean has time and again affected the Zambezi Basin states at varying degrees, sometimes influencing large storms of rain that have devastated life and property. Related to climatic features is the presence of large water bodies in the basin that play a role in modifying weather conditions. These include lakes Malawi/Nyasa, Kariba, Cahora Bassa and others.

Temperature

The sun and its position to the location of the basin determine how much heat can be received at any one time. Temperature is an important factor in plant growth, and its variations are due to cloud cover, wind speed, humidity and other factors. In the dry season, the sun is furthest north and temperatures are at their lowest. Clear skies at night over most of the basin permit rapid loss of heat from the surface.

The highest temperatures are at the end of October or early November and the lowest temperatures in July, particularly in mountain areas. In the low-lying areas, especially the major river valleys such as those of the Luangwa, Shire and the Zambezi, temperatures are appreciably higher though many of these areas experience periodic invasions of cool air, which lower the temperatures.[15]

Temperature inversions frequently develop in valleys and lower areas of the basin. Frost, especially ground frost, may occur locally in some parts of the basin between June-August. As there is little cloud cover, bright sunshine is usually in excess of eight hours per day throughout the dry season.

Mean daily temperatures in July (see Map 2.2) indicate that the coolest (14–16°C) part of the basin is in the north-central region, all of which is entirely within Zambia,[16] as well as the extreme south-eastern area of the basin in Zimbabwe.

Higher mean daily temperatures (22–24°C) are recorded in the lower Zambezi, mainly in the south-eastern part of the basin, in Mozambique. The high temperatures in this region are a reflection of the low elevation above sea level. The rest of the basin has mean daily temperatures of between 16–22°C. The Lower Zambezi and a small part of the Middle Zambezi, have high mean daily temperatures of between 29–32°C in November (see Map 2.3).[17]

Much of the southern part of the basin and northern and central parts have mean daily temperatures of between 25–29°C and 21–25°C respectively. The extreme northeastern region records the lowest mean daily temperature in November of between 19–21°C.

Due to its generally high altitude and with considerable parts of the area more than 1,000 metres above sea level, the tropical heat in the Zambezi Basin states is

Map 2.2: Mean daily temperatures in July

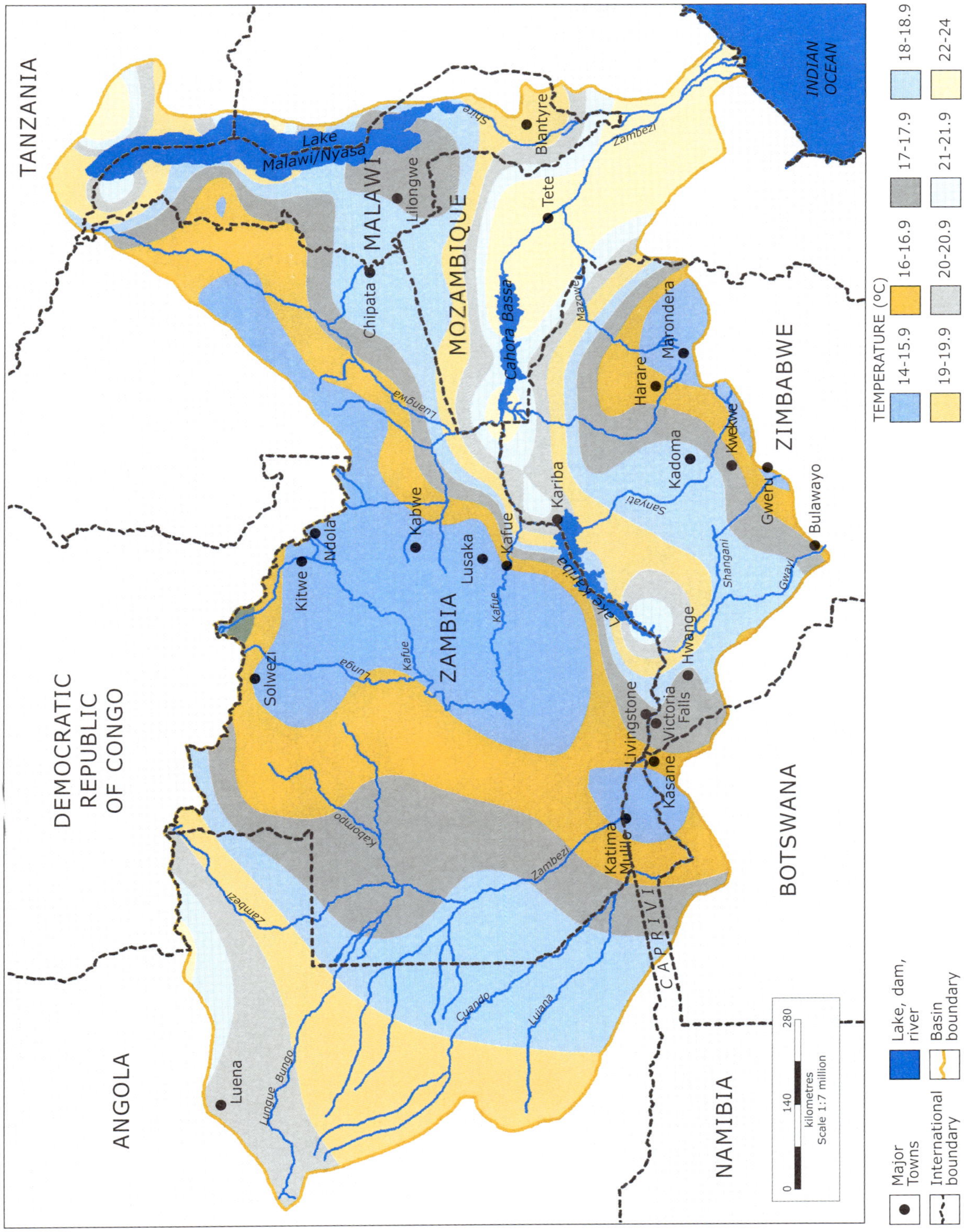

Source: Zambezi River Basin Atlas, SADC/DENCONSULT/ZRA, Lusaka, 1998

Map 2.3: Mean daily temperatures in November

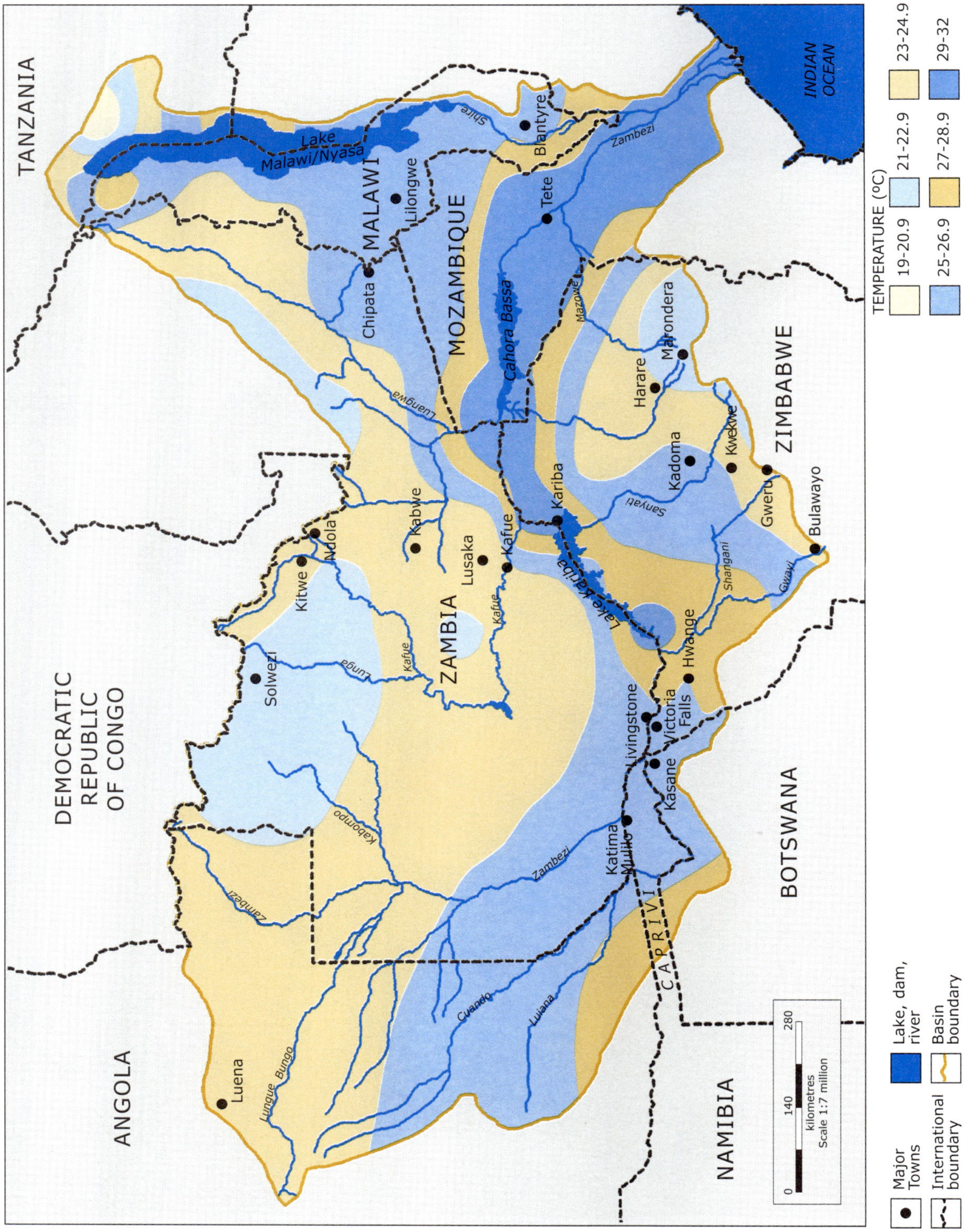

Source: Zambezi River Basin Atlas, SADC/DENCONSULT/ZRA, Lusaka, 1998

reduced close to temperate levels. The rainy season in the summer half of the year and cloudiness, lower the seasonal rise of temperatures before they peak, providing further relief from heat.

The basin experiences a big daily range of temperatures. The average range is about 10°C in the rainy season and as much as 20°C in the dry season in the southern parts of Zambia and in Zimbabwe. Small day-to-day temperature changes between 1–2°C are experienced especially in Central and Northern Malawi and the north of Zambia. These areas are remote from the inversions of cool maritime air that affects Zimbabwe and southern Malawi. The temperature falls sharply over rising ground during the invasions of cool maritime air, which is a cold gusty south-easterly wind accompanied by light rain.

Relative humidity

Relative humidity is a measure of the moisture content in the atmosphere. The measurement and computation of relative humidity in the Zambezi Basin are done only at a few monitoring stations. Humidity data play a significant role in many socio-economic sectors such as agriculture, industry and human health. Humidity data is important in weather forecasting, air pollution monitoring and climate classification.

Generally, the areas with the highest annual relative humidity in the basin are those that receive the highest amount of annual rainfall. Conversely, the lowest annual relative humidity is observed in the areas with the lowest amount of rainfall. In the Zambian part of the basin, mean monthly relative humidity ranges from 30–40 percent in September to about 70–80 percent during the rainy months of December–February.[18]

Rain-bearing systems

Rainfall in the basin is, to a large extent, governed by the ITCZ. With the southward migration of the sun, pressure falls over the interior of the continent, allowing the southward shift of the ITCZ.

The rainy season peaks during January and February when the ITCZ is centred on southern Zambia. Despite the daily control exercised by the pressure systems further south, the rainy season is largely set by the behaviour of the ITCZ, as it moves between latitudes. The ITCZ tends to take up a favoured position each year to which it keeps returning after each displacement. In some years, this position is about 15–17°S, but on occasion the ITCZ moves as far south as the Limpopo River, giving exceptional rains over Zimbabwe.

Figure 2.1: Air movement over southern Africa in summer

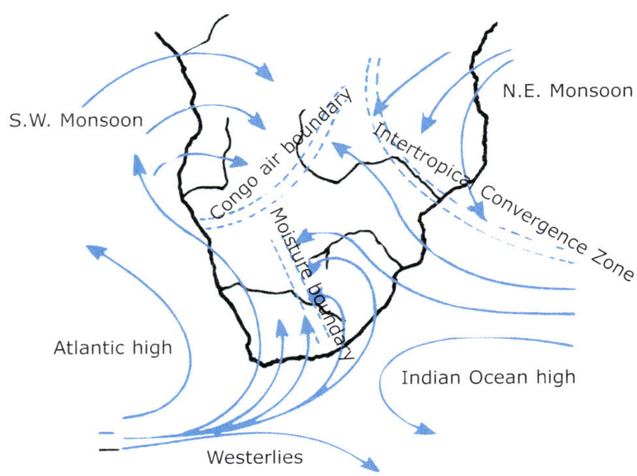

Figure 2.2: Air movement over southern Africa in winter

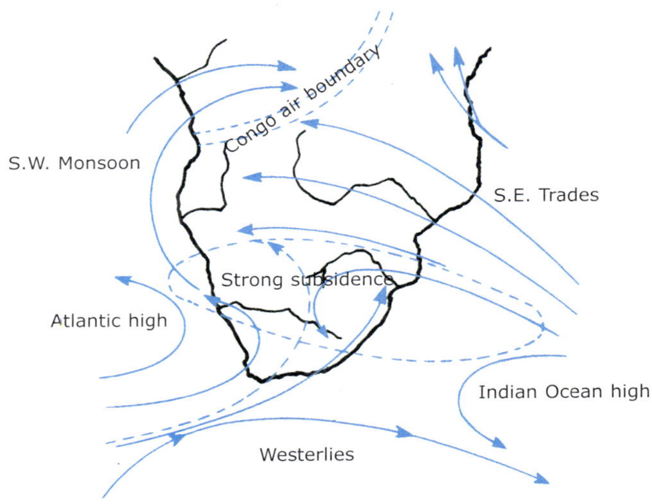

Photo: M Chenje

The rain-bearing systems are governed by the ITZC.

When a tropical cyclone enters the north of the Mozambique Channel, it is likely to draw Congo Air across Zambia into Malawi, resulting in wetter weather. At the same time, the southeast winds over Zimbabwe strengthen and veer, and then displace the ITCZ northwards, influencing a decrease in rainfall over the country. The tropical cyclone may move further south allowing the winds across Zimbabwe to become north-westerly. This generally reverses the dry conditions.

Apart from the northeast monsoons and the southeast trades, two other air masses are involved with the ITCZ in the basin;

- the Congo Air from the northwest, which brings much rain and thunderstorms, and
- the transitional or modified air between the two boundaries of the ITCZ.

Tropical cyclones

The majority of the storms develop over the Indian Ocean in the region bounded by latitude 5–20° South and longitudes 50–90° East. Most of these storms usually move initially westwards towards or across Madagascar and the eastern shores of Mozambique, eventually turning southward into the westerly airstream.

A cyclone following the west coast of Madagascar or the centre of the Mozambique Channel is noted for causing a clearing in the weather, resulting in dry spells in some parts of the Zambezi Basin. Considering the dependence of the economies of the basin states on agriculture, continuous periods of dry weather would have a marked impact on agricultural output. Dry spells caused by cyclones are due to the prevalence of southeasterly winds that become very dry in the upper levels.

In Zambia, cyclones usually induce increased rainfall. Winds on the north side of the trough line build up from the west, and cloudiness and rainfall tend to increase in the east of the country and over Malawi as moist Congo Air spreads eastwards. In March 1994, *Cyclone Nadia,* with sustained winds of up to 185 km/hour claimed a dozen lives as it crossed the northern tip of Madagascar. After crossing the Mozambique Channel, *Nadia* went into northern Mozambique, where more than 200 lives were lost and many more people left homeless.[19] In February 2000, cyclone Eline was more devastating, causing hundreds of deaths in Mozambique and Zimbabwe and massive flooding in more than three decades.

Rainfall

Rainfall is the most crucial climatic element whose abundance or deficiency strongly impacts on the welfare of the people of the basin. The onset, amount, distribution both in time and space, and the end of the rainfall season are critical to agriculture, the economic mainstay of many of the basin states. The amount of rainfall also affects water resources and availability in the basin. A dry spell of a few days or weeks, for example, impacts negatively on the annual agricultural production, as the majority of the people depend on rain. More importantly, intensive and extensive droughts impact severely on water supply sources, including water levels in lakes and rivers and on wildlife, agriculture, industry and the whole economy of the basin states. Too much rain also adversely affects agricultural production, as was the case in

Photo: M Chenje

The heavy rains in the Zambezi Basin carry massive loads of silt which is deposited in reservoirs and floodplains downstream.

Map 2.4: Mean annual rainfall in the Zambezi Basin

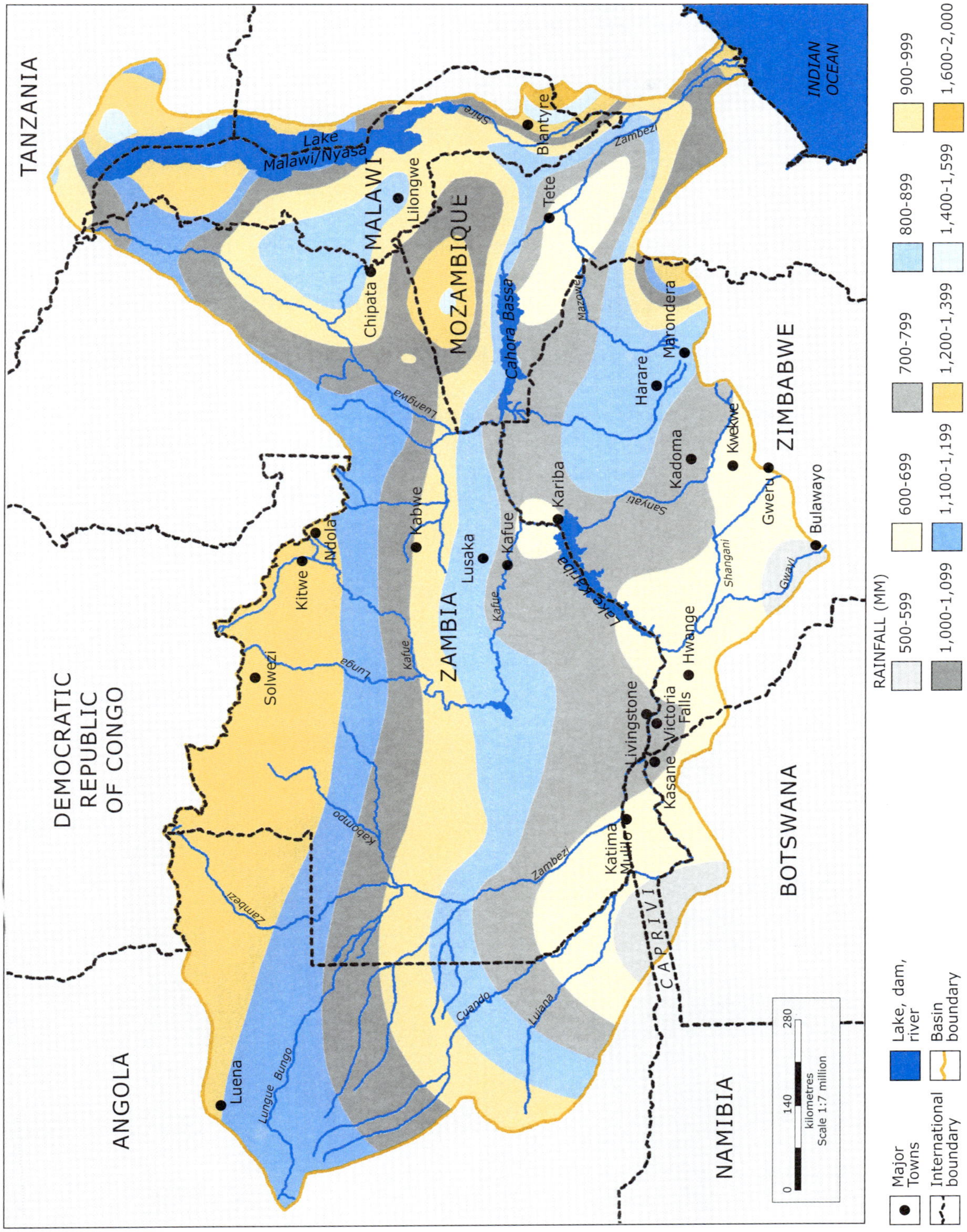

Source: Zambezi River Basin Atlas, SADC/DENCONSULT/ZRA, Lusaka, 1998

Zimbabwe during the 1998–99 season which saw many areas waterlogged.

Floods have also played havoc. They may wipe away a whole field of a good crop and dash the expectations of the farmer. In Malawi, the Ruo, a tributary to the Shire, causes untold misery to the inhabitants of the Lower Shire almost every year. Other rivers that have caused excessive floods in recent years in Malawi, include the Phalombe, Bua, the North Rukuru and Songwe between Malawi and Tanzania.

The highlands of Angola and Zambia on the headwaters of the Zambezi River, and others such as the Misuku Hills and Mount Mulanje in Malawi, and the Mbeya region of Tanzania, are areas of high rainfall and low evaporation rates. Low rainfall and high evaporation dominate the Zambezi River Valley, the Caprivi Strip in Namibia and the Botswana part of the basin. Similarly, lakes Malawi/Nyasa, Kariba and Cahora Bassa, and the wetlands regions of the basin have high evaporation rates. Potential evapotranspiration in the basin is estimated at 1,560 mm, which is much higher than the mean rainfall of about 970 mm, meaning that the Zambezi Basin is a source of water loss into the atmosphere.[20]

Some regional differences are evident. Rainfall is much heavier northwest of Zambia; frost is more frequent in the south-western portions of the basin. Over the greater part of Zambia, relief is relatively uniform and so the rainfall is nearly uniform than elsewhere in the basin.[21] Malawi, on the other hand, with its major lake and rugged topography, has the most variable weather conditions.

The mean annual rainfall in the Zambezi Basin exhibits a strong north-south gradient, with the headwaters receiving between 1,100–1,600mm annually. Much of the southern region of the basin receives low rainfall between 600–700mm. Two pockets, one centred to the west of Bulawayo and the other around the central part of the Caprivi Strip (including adjoining areas to the north and south of this Strip) receive the lowest mean annual rainfall of about 500–600mm. The remainder of the basin receives mean annual rainfall of between 700–1,100mm.

Across most of the basin, the rainy season usually starts in October and ends in April. This applies, with only minor exceptions, to the whole region although the proportion between the lengths of the dry and rainy seasons change from about seven to five months respectively in the south of Zimbabwe, to about five to seven months respectively in the south of Tanzania. The earliest rains in the basin start in eastern Angola and north-western Zambia in September, in the Mwinilunga and Kawambwa areas, progressing steadily towards the southeast. In Zimbabwe, the earliest rains are received in the south of the country and progress with time northwards where the start is as late as mid-November such as in northern Malawi and southern Tanzania. In Tete Province of Mozambique, the rainy season lasts for less than five months, from November to March.

In Zimbabwe, thunder activities start after mid-October. Central Mozambique and southern Malawi start receiving rain in this period. North-eastern Zambia and southern Tanzania usually receive rains last, starting in November.

Rainfall variability and drought

Year to year rainfall variability is high in the basin countries with the 1970s being relatively wet while the period 1980 to 1990 have been relatively dry.[22]

Drought is the single most crucial natural disaster that affects the Zambezi Basin. Research shows that droughts occur every 10–15 years in the basin countries. Since the turn of the 20th century, eight spells of more or less than nine years have alternated with similar periods of rainfall that is below normal. These spells, however, have not always occurred at exactly the same time nor have they affected all countries equally, which make the trends difficult to follow.

Photo: M Chenje

Drought is the single most crucial natural disaster in the basin, often leading to deaths of wildlife.

The droughts of the 1980s and 1990s have had a marked negative impact on the Zambezi Basin states. For example, the lake level at Kariba on the Zambezi River dropped from 487.5m amsl in 1981 (maximum retention level: 488m amsl) to 475.9m amsl (minimum retention level: 475.5m amsl) in 1992 reflecting a drop of 11.6 metres of water within a decade.[23] With regard to water availability and hydroelectric power generation in some of the Zambezi Basin countries, the severe droughts of 1991–92 and 1994–95 had near-disaster consequences.

The period between 1985–95 was disastrous for many rural communities, especially in Mozambique, southern Zambia and Zimbabwe. Boreholes dried up, making it impossible to grow crops or maintain livestock, and forced women and children to walk long distances to collect water for their families' needs. The 1991–92 drought hit particularly hard, putting more than 18 million people in 10 countries at risk of starvation." A WWF report says that, "...urban dwellers were also affected. Water supplies in Zimbabwe's capital, Harare, sank to alarmingly low levels during 1995 and power shortages became commonplace. By October 1995, the Kariba Dam, which produces most of the country's electricity, was running at only 14 percent capacity".

Since the 1991–92 drought that wreaked havoc in most economies of southern Africa, recovery has been rather slow. After a short respite during the 1992–93 rainy season, there has been a virtual persistence of droughts in sympathy with *El Niño*/Southern Oscillation (ENSO). For instance, there was a 17 percent fall in maize harvests in 1993–94 season compared to the 1992–93 season, as most countries in the region suffered a climatically unfavourable growing season. Several rainfall deficits over most of the sub-region resulted in extensive crop failures in southern Zambia and parts of Zimbabwe. Only the abundant rains of 1995–96 averted an energy and agricultural crisis for Zambia and Zimbabwe.

There is marked decrease in flows indicated by long-term trends as evidenced from hydrological information for the Zambezi at Victoria Falls and the Shire River in Malawi. Low Lake Malawi/Nyasa levels have been recorded due to low inflows from its feeder rivers. In Malawi, levels of Lake Malawi/Nyasa dropped to a minimum of 472.89 amsl in 1996, reducing the flows in the Shire River for the first time since the 1930s to less than the required minimum flow of 170 cu m per second for energy generation at Nkula and Tedzani.[24] These occurrences have forced the government to regulate the outflow from the lake at Liwonde to enable power generation at Nkura and Tedzani during the dry season. However, natural lakes such as Lake Malawi/Nyasa are not as vulnerable compared to artificial ones such as Kariba and Kafue that depend on direct runoff. Also, while the global rainfall conversion to runoff is 30 percent, that of the Zambezi Basin is only 10 percent.

Evaporation

Evaporation in the basin is generally greater than precipitation during most of the year. Evaporation is dictated by surface air temperature, relative humidity or saturation deficit of the air, wind speed and solar energy. Direct evaporation measurements and data available in the basin are mainly from pan evaporimeters. Such data is mainly useful at local scales to assess evaporation from land and dams.[25] However, the network of monitoring stations in the basin has, together with systematic observations, declined in recent years.

The high levels of solar radiation are reflected in the high rates of evaporation. The annual total is between 1,800–2,000 m, or averaging about five mm per day. The comparatively low winter temperatures are offset by the largely uninterrupted incidence of sunshine. Rising temperatures in September and October accompanied also by

Box 2.1: Drought impacts on Zimbabwe's economy

Largely as a result of the 1991–92 drought, Zimbabwe's manufacturing output declined by 9.3 percent in 1992. The drought alone led to a minimum 25 percent reduction in volume of manufacturing output and a six percent reduction in foreign currency receipts from manufactured exports or a two percent reduction in total export receipts.

The most severely affected sub-sectors were textiles (including cotton ginning), clothing and footwear, non-metallic mineral products, metal and metal products and transport equipment.

The drought impacted on the manufacturing sector through water and electricity shortages. Most of the municipalities imposed rationing, with severe water shortages affecting urban areas such as Bulawayo, Chegutu (both of which are in the basin) and Mutare in the Eastern Highlands.

Electricity shortages led to load-shedding. Rationing from September 1992 and increased electricity tariffs affected the whole country. Load-shedding caused a reduction of about US$56 million (Z$560 million) in GDP in 1992, US$20 million in exports and 3,000 job losses.

The value of the Zimbabwe Stock Market declined by 62 percent in 1992 partly as a result of the drought, leading the International Finance Corporation to identify the country's stock market as the worst performer of 54 world stock markets in that year.

Source: Benson, C. and Clay, E., "The Impact of Drought on Sub-Saharan African Economies: A Preliminary Examination – Working Paper 77", Overseas Development Institute, London, 1994

the windiest conditions of the year, increase evaporation rates, reaching between 9–10 mm per day. During the rainy season, the evaporation rate drops to about four mm per day in most areas due to the high humidity and cloudiness, and low wind speeds.

An average of about 65 percent of all the rainfall in the basin evaporates soon after it occurs. About 20 percent of the total rainfall is returned to the atmosphere through evapotranspiration. However, the figure is much lower in relatively cooler and more humid areas of the basin, but can be as high as 83 percent in the semi-arid areas such as the Luangwa and Zambezi valleys. An average of about 14 percent of the rain is actually available as surface runoff. Only a very small proportion of the total rainfall, about 10 percent, goes into aquifers.

SOME CHANGES OF THE PHYSICAL ENVIRONMENT

Major activities that affect the geomorphology include mining, agriculture, and infrastructural development. Since the 1970s, the environment and key natural resources in most African countries have been facing escalating and unsustainable pressures from fast growing populations and cities as well as expanding agricultural and industrial activities. Throughout Africa, reducing poverty is the overriding priority for governments. Poverty is a major cause and consequence of the environmental degradation and resources depletion, which threaten present and future economic growth.[26]

Land degradation is a threat to the survival of the economy and the physical features of the basin states. Key issues include escalating soil erosion, declining fertility, increasing salinity, soil compaction, agrochemical pollution and desertification.[27] In the southern African sub-region increased livestock has caused escalating land degradation.

Agriculture in the basin is the most extensive land-based activity, contributing to changes in the land resource base. The use of tractors and other machinery for land preparation, planting, cultivation, harvesting and processing to improve productivity, especially in commercial farming, increases the risk of soil erosion and compaction, deforestation and chemical pollution.[28] Poor agricultural practices and land management also exert pressures on the land resource base.

Soil quality depends on parent material and management while soil distribution patterns are controlled by climate, geology and relief. Poorly developed, sandy soils blanket the basin and in the plateau areas, leached latosols or ferralsols predominate where rainfall is high. Good plateau soils occur in the Lilongwe - Kasungu plains and in the tea-producing areas of Thyolo, Mulanje and Nkhata Bay districts of Malawi.[29]

Sodic soils are very fragile and once disturbed, degradation is difficult to prevent. These soils are not favoured for agriculture or engineering because of their chemical and physical properties. They are largely impenetrable to roots and impermeable to water hence if water penetrates, the sodium in the soil dissolves. The major areas of sodic soils tend to be the low to medium rainfall zones, for example the areas between the Great Dyke, Chegutu and Gweru in Zimbabwe.

Photo: M Chenje

Land degradation is a threat to the survival of the economy and the physical features of the basin states.

The construction of dams has major ecological effects on the margins of the dam and riparian areas upstream and downstream. Settlement and demographic patterns are also affected during and after construction. The construction of the Kariba Dam changed the land-use and affected the ecosystem of the area.

Mining plays a major role in terms of changing the state of the physical environment. However, it is critical to the development of the economies of some of the basin states, such as Zambia and Zimbabwe. The areal expansion of mining activities has had negative effects on the

Photo: M Chenje

The construction of dams such as the Kariba has major ecological effects on the margins of the dam and riparian areas upstream and downstream.

environment in general and on the physical appearance and nature of the land in particular. Mine dumps have created artificial topography and have disastrous effects on the ecosystem if not rehabilitated. The physical appearance and makeup of the Copperbelt area has been changed due to mining. Urban centres have also been established in these mining areas increasing, pressure on the land.

Photo: M Chenje

Urban settlements have added a new dimension to the basin's physical features, contrasting the natural and built environment.

The physical features of the basin have determined settlement patterns (radial or linear) particularly in the rural areas. All major industrial towns and centres in central Zambia are located along the Kafue River, thus the Kafue Basin has become the most important and most highly utilised river basin in the country. It has the potential for irrigated agriculture, hydropower generation, fisheries, tourism and development of wetlands in the Kafue Flats (Lochinvar and Blue Lagoon National Parks).

Urban settlements in the region generally display a haphazard pattern of development, which has been characterised by conflicts of land-uses.[30] Agricultural activities have conflicted strongly with conservation activities, especially in the urban areas. Open areas that are conserved for aesthetic purposes have been taken up by illegal agricultural activities particularly, in the low-income residential areas. Cultivation has been carried out very close to watercourses passing through urban areas, increasing siltation in the rivers. For example, in Harare which is built in the Manyame catchment, siltation of reservoirs which supply its water needs, has been reported.

CLIMATE CHANGE

According to the World Wide Fund for Nature (WWF),[31] there are indications of climate change in southern Africa. WWF observes that, "global warming is already affecting much of southern Africa. Records from the countries that make up SADC reveal that temperatures have risen by over 0.5°C over the past 100 years, with the last decade being the warmest and driest ever. Over the past 20 years, there has been noticeably less rainfall, and drought has become an increasingly serious threat. These disturbing changes are happening despite the fact that Africa has contributed little to the build-up of greenhouse gases in the atmosphere."[32]

It has been projected that, "southern Africa will be affected by warming in both summer and winter, that rainfall may diminish somewhat in certain areas, and that soil moisture will probably decrease more generally with doubling of carbon dioxide."[33]

In general, rainfall reduction in the Zambezi Basin has been significant over northern Botswana and Zimbabwe. It is believed that the dry conditions of the 1980s and 1990s are evidence of inter-decadal variability although the effects of abrupt warming in tropical sea surface temperatures in the mid-1970s and an increase in the frequency of *El Niño* events cannot be ruled out.[34]

Box 2.2: Rainfall regimes in southern Africa since 1800

PERIOD	EVENT
1800-30	Southern African rivers, swamps and other water sources dried up. Some well-watered plains turned to semi-arid karroo.
1820-30	This was a decade of severe drought throughout Africa.
1844-49	Southern Africa experienced five consecutive drought years.
1870-90	This period was humid in some areas and former Lake Ngami filled in the Northwest of Botswana.
1875-10	There was a marked decrease in rainfall over southern Africa and 1910 experienced a severe drought.
1921-30	This was a period of severe droughts in the southern African region.
1930-50	Southern Africa experienced dry periods alternating with wet ones, and in some years the rains were very good. The 1946–47 season experienced a severe drought.
1950s	There was abnormally high rainfall in some parts of the region. However, the equatorial region experienced below normal rainfall.
1967-73	This six-year period was dry across the southern African region.
1974-80	This period of six years was relatively moist over much of southern Africa. In 1974 the mean annual rainfall was 100 percent above normal throughout the region.
1981-82	Most of southern Africa experienced drought.
1982	Most of sub-tropical Africa experienced drought.
1983	This was a particularly bad drought year for the entire African continent.
1985	Conditions improved.
1986-87	Drought conditions returned.
1991-92	Southern Africa, excluding Namibia, experienced severe drought.
1992-93	Conditions slightly improved, but the previous year's drought effects continued.
1993-94	Conditions improved.
1994-95	Many SADC countries hit by worst drought in memory, surpassing the effects of the 1991–1992 drought in some parts of the region.
1995-96	Widespread rains in most parts of the SADC region, prompting forecasts of a bumper agricultural yield.
1996-97	Normal rainfall for most of the region.
1997-98	Normal rainfall throughout the region including the northeast although impacts of *El Niño* were significant.
1998-99	Near normal rain in most of the region except in the northeast, namely Tanzania, where below normal rainfall was experienced.
1999-00	Most of the SADC countries received normal to near normal rainfall except Tanzania which experienced dry conditions. Widespread floods devastate several countries in southern Africa with Mozambique the most affected.

Source: Adapted from SADC/IUCN/SARDC, *Water in Southern Africa, Maseru/ Harare, 1996*

Agriculture

Changes in climate will influence high variability in crop outputs from arable agriculture, both rain-fed and irrigated crops. Livestock production will be affected as well through problems associated with poor and variable range productivity and this would lead to desertification processes. Wildlife populations would be decimated through recurrent drought and expansion of population settlements. In this regard, it is suggested that studies be carried out to include the following:[35]

- Preparation of soil and population maps for the different countries of the region.
- Conducting sample surveys of areas planted with crop and the related yields and rainfall data.
- Promoting the use of Geographical Information System (GIS) techniques to obtain optimal results on smallholder vulnerability and so develop effective mitigation responses to climate change.

Livestock is one of the major sectors of agriculture in the basin states especially in Botswana and Zimbabwe. The success of the livestock industry depends much more on the availability of pasture that in turn is dependent on availability of water resources. Climate variability has often led to the incidence of droughts in southern Africa with varying periodicity and intensity. It has been suggested[36] that observations on the impact of climate be made to allow for strategies that would ameliorate any problems associated with livestock in the event of poor climatic conditions.

Irrigation agriculture may suffer due to climate change in the event that less precipitation is received which happens to be the influencing factor to normal runoff from which irrigation water is drawn.

Environment

Since the basin's drainage system is heavily dependent on rainfall, poor rainfall regimes have impacted on rivers and streams as well as on the lakes and reservoirs in the basin. Significant widespread effects of climate variability are likely, with the most severe effects occurring in the tropics. Flooding of coastal areas due to sea level rise will lead to forced displacement of tens of millions of people. Changes in global aggregate food production are uncertain, but agriculture in the tropics and sub-tropics will suffer adverse effects. Changes in the worldwide hydrological cycle will reduce the availability of freshwater in water scarce regions. Financial, institutional, and human resource shortages will make the developing countries most vulnerable to climate change.

Scientists believe that if present trends continue, both Greenhouse Gases (GHG) concentrations and the impact of climate change will increase significantly in the 21st century. If there is no specific human strategy for mitigating climate change, typically the atmospheric concentration of carbon dioxide will increase. This increase will of course depend on the rates of global economic and population growth, energy prices, and the application of new technologies. Studies indicate that global warming would result in changes in precipitation and wind patterns, frequency and intensity of storms, ecosystem stress and species loss, reduced availability of fresh water, and a rising global mean sea level.[37]

In terms of wetlands, climate change may result in excessive inflows in the face of good rains and flooding and this can increase the storage capacities of wetlands, providing a haven for the animal and plant life. In the face of drought conditions, reduced inflows into wetland systems would result in the dearth of water necessary for their maintenance and this would have a ripple effect on those that depend on them.

Changes in climate will also have a serious impact on the availability of groundwater especially when there is less than normal precipitation. This would result in the lowering of water tables and drying up of boreholes as has been evidenced in the basin during droughts.

The impact of climate change on pollution is most disastrous as air, water and land would all be impacted upon. The Zambezi Basin states rely to some appreciable degree on hydroelectric power for their energy needs. However, the majority of the rural communities still use woodfuel. In the event of drier climatic conditions, hydropower generation would be impaired resulting in more and more use of biomass and thermal power generation. Forest fires would also be on the increase, releasing large amounts of carbon dioxide into the atmosphere.

Pollution of land under climate change is a scenario that relates more to conditions of dry climate. Under circumstances of water scarcity, most waste disposal facilities would not function to their capacity. Solid and liquid waste would be disposed off on land, polluting both land and air.

MANAGEMENT ISSUES

Climatic and hydrological data

There is a good network of hydrological monitoring stations which needs to be maintained. In addition, a database, ZACBASE, has now been established and needs to be supported by the riparian states.

The Zambezi Basin countries collect both meteorological and hydrological data which is used at the national level or fed into regional data systems such as those of the Drought Monitoring Centre in Harare. Climatic data includes minimum and maximum temperature, solar radiation, relative humidity, evaporation and wind speed. The hydrological component demands for water level measurements, discharge measurements of rivers and streams, water quality measurements and groundwater level monitoring.

Photo: M Chenje

Satellites are critical to tracking weather changes on a daily basis.

However, both climatic and hydrological data in the Zambezi Basin states is at different levels of processing. Some of it still may be raw or has been processed manually or electronically, analysed and archived for use. In other cases, the data is compiled into hydrological yearbooks and shared with other similar institutions in the SADC region and provided on demand.

The other important element in data collection has to do with instrumentation, human resources and other facilities that allow for quality control. Rainfall monitoring is the most extensive in the basin but lack of systematic observations is the main constraint in the gathering and use of rainfall data. Apart from the networks run by the meteorological departments of the basin states, some rainfall observing stations are also maintained by some government organisations such as agriculture, forestry and others, including the private sector.

Modern observing systems and techniques such as the use of radar and satellite complement the conventional "standard raingauge". Monitoring the movement of the major rain bearing systems such as the ITCZ, Congo Air Boundary (CAB) and tropical cyclones is constrained in the basin, due to declining monitoring station-networks, particularly for the upper air. The upper air temperature data is most affected because radiosonde station networks rarely work and due to lack of technical resources and consumables.

Satellite-based observing systems experience longer periods of idle-stay due to lack of maintenance and consumables. The same is true for radar systems in the basin countries. Therefore, inadequate and inaccurate measurements make forecasting of rainfall even more difficult.

Systematic rainfall observations are important as they allow for timely forecast of food as well as cash crop harvests, which are central to the socio-economic well-being of people in the basin and the reduction of poverty. The establishment of the Drought Monitoring Centre in Harare fulfils this condition, serving the SADC region, including the Zambezi Basin states with timely and relevant information, bulletins and predictions on impending droughts or other climatic variations in the region.

Such information has been useful at both the regional and national levels and the proposed strengthening of the Drought Monitoring Centre[38] is welcome. This provides for improved and additional services to all socio-economic sectors in the Zambezi Basin states. However, the Drought Monitoring Centre needs to strengthen its ties with the SADC Environment and Land Management Sector (ELMS) which is responsible for the Convention to Combat Desertification.

Rainfall data and output volumes from irrigation agriculture are available in the basin states to facilitate the conducting of studies on irrigation requirements. Droughts over the years have shown that output of irrigated crops has been low while in good years, output has been good. These lessons should provide for alternative strategies for irrigation in conditions of water scarcity.

Such strategies are not far-fetched. In some years, the Zambezi Basin states have experienced exceptionally high floods to the extent that only their bad effects have been recognised and yet they could also provide a guarantee for more water during periods of need. Construction of dams in series on catchments prone to flooding would allow for storage of large volumes of water, for irrigation and other purposes. Such structures can also hold back large volumes of water that can otherwise cause extensive flooding and damage. This, however, has to be weighed against the impact on river flows and habitat destruction.

LINKAGES TO OTHER CHAPTERS

Chapter 1: Regional Overview: People and Environment
The physical features and climate of the Zambezi Basin determine human activities, influencing the people's ability to contribute to sustainable development.

Chapter 3: Water and Wetland resources
The physical features and climate of the basin determine the hydrological regime of wetlands. For instance, bedrock is necessary to hold the water of a wetland from percolating further into the ground thereby "draining" the area.

Chapter 4: Biological Resources and Diversity
The different ecosystems and the biological resources are well adapted to the physical features and climate of the basin which determine these ecosystems.

Chapter 5: Agriculture
Agriculture is responsible for some changes in the climate and physical features of the basin. Bare lands are more subjected to excessive evaporation and loss of soil moisture and wind erosion.

Chapter 6: Industry
Resource-based economic growth in the basin states has resulted in significant industrial development, particularly in the mining, manufacturing, tourism and agricultural sectors. The resultant infrastructural development, urbanisation, industrial waste and pollution impact the climate and physical features of the basin.

Chapter 7: Energy
The natural resources of the basin such as coal, vegetation, water and the topography provide the people of the basin with their energy requirements. The geology of the basin is the source for coal, while vegetation provides biomass and the topography provides ideal hydroelectric power generation location. Climatic changes have an adverse effect on natural energy resources such as biomass and water availability.

Chapter 8: Tourism
The rolling hills, majestic mountains, falls, cataracts, rapids, clear waters of the lakes and rivers of the basin are an attraction to the tourist and contribute to the socio-economic development of the eight basin states. A good climate is crucial to tourism growth.

Chapter 9: Pollution
Irresponsible mining promotes visual pollution through open casts and mine dumps. Tailings and wastes end up in natural fluvial systems causing pollution. The removal of natural features such as forests and the development of supporting infrastructure coupled with adverse climatic conditions wipe out the natural beauty of the basin and bring in visual pollution.

Chapter 10: Poverty
Poverty forces people to overexploit natural resources. Concerned only about their immediate survival, the poor often destroy the physical environment, for example, through illegal quarrying and gold panning.

Chapter 11: Gender and Women's Roles
The physical features of the basin contribute to defining the gender responsibilities of the peoples. For instance, men living by the lakeshore devote much of their time to fishing while women take on to collecting water, firewood, valley cultivation and attending to other domestic chores.

Chapter 12: Environmental Management and Regional Cooperation
The physical setting of the basin, which binds eight SADC states, provides for growing cooperation in many different sectors.

Chapter 13: Trends and Scenarios
The location of the Zambezi Basin, vis-à-vis the eight basin states, has defined trends in different areas of endeavour and will continue to do so for many decades to come.

ENDNOTES

1 Denconsult, *ZACPLAN: Sector Study No. 2, Water Consumption and Effluent in the Food and Agriculture Sector including Crops, Livestock and Fisheries,* First Draft Report, March 1998
2 ibid
3 ibid
4 ibid
5 ibid
6 ibid
7 Moyo, S., O'Keefe P., Sill M., *The Southern African Environment: Profiles of the SADC Countries,* Earthscan Publications Ltd, London, 1993
8 Chenje, M., Sola L., Paleczny D. (eds.), *The State of Zimbabwe's Environment 1998,* Ministry of Mines, Environment and Tourism, Harare, 1998
9 ibid
10 op cit 7
11 Munyanyiwa, H., "Proterozoic Stratigraphy of Southern Africa: Special Reference to Zimbabwe", (unpublished paper) Harare, 1998
12 ibid
13 op cit 7
14 op cit 7
15 Torrance, J. D., Malawi, Rhodesia and Zambia: In Griffiths, J. F. (ed.), "Climates of Africa", *World Survey of Climatology,* Vol. 10. Elsevier, New York, 1972
16 Denconsult, *Zambezi River Basin Atlas, ZACPRO 6.1,* SADC/Denconsult, Gaborone/Aalborg, 1998
17 ibid
18 JICA/GRZ, "The Study on the National Water Resources Master Plan in the Republic of Zambia", Interim Report, Lusaka, 1995
19 World Meteorological Organisation; "Global Climate System Review", In J. M. Nicholls (ed.), *World Climate Data and Monitoring Programme,* WMO Publication No. 856 Geneva, December 1993-May 1996
20 Denconsult., *ZACPLAN Sector Studies, Introductory Volume,* Final Report, SADC/Zambezi River Authority, Lusaka, 1998
21 Muchinda, M., "Climate and Drainage in the Zambezi Basin", For SARDC, April 1999
22 Dandaula, H., "Climate and Drainage in the Zambezi Basin", For SARDC, April 1999
23 Mwasile, C. B. and M. Lindunda, "Electricity Generation and Distribution in Drought Conditions"; IUCN Workshop, Pamodzi Hotel, 13 - 15 December, Lusaka, 1995
24 Kaluwa, P. W. R., "Presented notes for the Scientific Advisory Committee Meeting", Harare, Zimbabwe, August 1999
25 Pettersen, S., *Introduction to Meteorology,* Second Edition, Mc Graw-hourill Book Company, INC, New York, 1958
26 UNEP, *Global Environment Outlook 2000,* Earthscan Publications, London, 1999
27 ibid
28 op cit 8
29 op cit 7
30 SARDC-IMERCSA, "The State of the Environment in the Zambezi Basin: Proceedings of a regional workshop held at Mazvikadeyi", Zimbabwe, July 20 - 24, 1998, SARDC/Sida, Harare, 1998
31 www.panda.org
32 www.panda.org
33 Kinuthia, J. H., *Global Warming and Climate Impacts in Southern Africa: How Might Things Change?* Kenya Meteorological Department, Nairobi
34 Mason, S. J., "Regional Manifestations of Climate Variability in Southern Africa", SADC/NOAA/NASA Workshop on Reducing Climate-related Vulnerability in Southern Africa, Victoria Falls, 1996
35 op cit 33
36 Downing, T. E., "Climate Impact Assessment in Central and Eastern Kenya: Notes on Methodology", In: D. Wilhite, W. Easterlings and S. Wood (eds.), *Planning for Drought,* Boulder, Westview Press, Colorado, 1987
37 Ominde, S. H. and C. Juma (eds.), "Introduction", In: *A Change in the Weather, African Perspectives on Climate Change.* ACTS Press, Nairobi, 1991
38 World Meteorological Organisation; "Report on Sustainability of Drought Monitoring Capacity in southern Africa", WMO, Geneva, May 1998

REFERENCES

Arnestrand, M. and G. Hansom, "Management of Scarce Water Resources in Southern Africa", Report No. D3. Centre for International Technical and Educational Cooperation, The Royal Institute of Technology, Stockholm, 1993

Bannink, B.A., "The Zambezi: Reliable Water for All, Forever? A Case Study of Integrated Sustainability Assessment", *Zambezi Freshwater Resources Sustainability Assessment,* Bilthoven, The Netherlands, 1996

Bonkoungou, E.G., "Drought, Desertification and Water Management in Sub-Saharan Africa", In: Benneh, G., W. B. Morgan, and J. I. Uitto (eds), *Sustaining the Future: Economic, Social, and Environmental Change in Sub-Saharan Africa,* New York, 1996

Cane, M.A., Eshel, G. and R. W. Buckland, "Forecasting Zimbabwean Maize Yield Using Eastern Equatorial Pacific Sea Surface Temperature", *Nature,* No. 370 1994

Drought Monitoring Centre., "Ten-day Drought Watch for South Africa", WMO/UNDP Project RAF/88/044, Harare, Zimbabwe 1991-1993

du Toit, R.F., "Hydrological changes in the Middle-Zambezi system", *The Zambezi Science News,* Vol. 17, 1983

Freeman, T. and M. Bradley., "Temperature is Predictive of Severe Malaria Years in Zimbabwe", *Transactions of the Royal Society of Tropical Medicine and Hygiene,* No. 90, 1996

Garanganga, B. J., "Drought Monitoring Centre - Activities, Outputs and Linkages into other Early Warning Systems", WMO, WCASP No. 27, Appendix E, 1993

Gibberd, V., Rook, J. Sear, C. B. and J. B. Williams, *Drought Risk Management in Southern Africa: The Potential of Long Lead Climate Forecasts for Improved Drought Management,* Natural Resources Institute, Chatham Maritime, UK, 1996

Glantz, M., R. Katz and N. Nicholls, (eds.), *Teleconnections: Linking Worldwide Climate Anomalies,* Cambridge University Press, Cambridge, UK, 1991

Hulme, M., Conway, D. Joyce, A. and H. Mulenga, "A 1961-90 Climatology for Africa South of the Equator and a Comparison of Potential Evapotranspiration Estimates", *South African Journal of Science* (in press), 1996

JICA/GRZ, "The Study on the National Water Resources Master Plan in the Republic of Zambia", Interim Report, Lusaka, Zambia, 1995

Kasimona, V. N., "Present Planning in Zambia for the Future Use of Zambezi River Waters", In: Matiza, T. and P. Dale (eds.), *Zambezi Basin Water Projects, Proceedings of a Workshop held at Kasane, Botswana: 28 April - 2 May 1993,* IUCN, Gland, Switzerland, 1995

Lyons, S.W., "Origins of Convective Variability over Equatorial Southern Africa During Austral Summer", *Journal of Climate,* 4(1) 1991

Masika, R. S., "Rainfall patterns in Eastern Africa", *Drought Network News,* International Drought Information Centre, University of Nebraska, Nebraska, 1992

Mason, S. J., J. A. Lindesay, and P. D. Tyson, "Simulating Drought in Southern Africa Using Sea Surface Temperature Variations", *Water South Africa,* No. 20, 1994

Mason, S. J., "Regional Manifestations of Climate Variability in Southern Africa", SADC/NOAA/NASA Workshop on Reducing Climate-Related Vulnerability in Southern Africa, Victoria Falls, Zimbabwe, 1996

Matarira, C. H., and L. S. Unganai., "A rainfall prediction model for southern Africa based on the Southern Oscillation phenomena", FAO/SADC Regional Early Warning Systems Project, Harare, Zimbabwe, 1995

Mohamed, S., "Malawi's Electricity Supply Under Threat", *The Sun,* February 24 - March 2, Lusaka, Zambia, 1997

National Academy of Sciences, *Learning to Predict Climate Variations Associated with El Niño and the Southern Oscillation,* National Academy Press, Washington, D.C. USA, 1996

Ogallo, L. A., "Validity of the ENSO-Related Impacts in Eastern and Southern Africa", In: *Useable Science: Food Security, Early Warning, and El Niño*, UNEP, Nairobi, and NCAR, Boulder, Colorado, 1994

Pallett, J. (Ed.), *Sharing Water in Southern Africa,* Desert Research Foundation of Namibia, Windhoek, Namibia, 1997

Ropelewski, D. F., and M. S. Halpert., "Global and Regional Scale Patterns Associated with El Niño/Southern Oscillation", *Monthly Weather Review,* No. 115 1987

SARDC/IUCN/SARDC, *State of the Environment in Southern Africa,* SADC/IUCN/SARDC, Maseru/Harare, 1994

Torrance, J. D., "Malawi, Rhodesia and Zambia", In Griffiths, J. F. (ed), *Climates of Africa: World Survey of Climatology,* Vol 10, Elsevier, New York, 1972

Tyson, P. R., *Climate Change and Variability in Southern Africa,* Oxford University Press, Oxford, U. K., 1986

Unganai, L.S., "Drought and Southern Africa: A Note from the Harare Regional Drought Monitoring Centre", *Drought Network News,* International Drought Information Centre, University of Nebraska, USA, 1994

WMO, "Global Climate System Review, December 1993 - May 1996", J. M. Nicholls, (ed.), *World Climate Data and Monitoring Programme,* WMO-No. 856, 1998

Zinyowera, M. C., and L.S. Unganai, "Drought in Southern Africa, An Update on the 1991-1992 Drought", *Drought Network News,* International Drought Information Centre, University of Nebraska, USA, 1993

Map 3.1: The wetlands of the Zambezi Basin

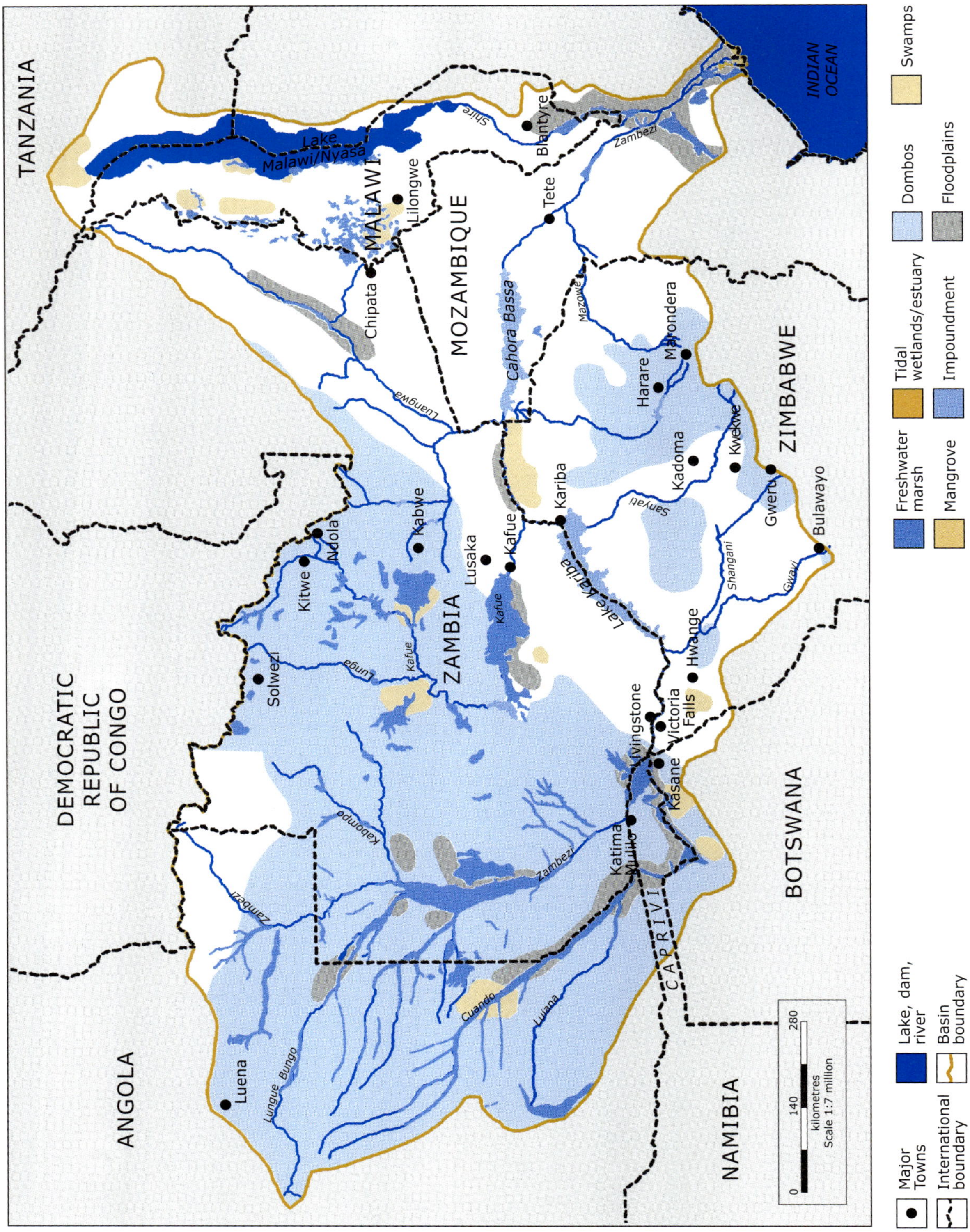

Source: Zambezi River Basin Atlas, SADC/DENCONSULT/ZRA, Lusaka, 1998

3

WATER AND WETLAND RESOURCES

Photo: IUCN

The waters of the Zambezi Basin play an important economic and social role in the riparian states. Water dominates many activities in the basin for industry, domestic use, hydropower generation, irrigation and transportation. The economic development of the riparian countries is to a great extent dependent on the availability of water resources. Water is also the driving force in wetland ecosystems, playing a critical role in the biosphere. The ecological value of water and wetlands is found in the ecological functions that these systems play in the environment.

Photo: M Chenje

Water and wetland resources of the Zambezi Basin are the lifeblood of both human activities and the environment.

Research on the ecological functions of wetlands shows that wetlands play very important roles in nutrient retention and export, groundwater recharge and discharge, flood control and flow regulation, sediment retention, erosion and salinity control, water treatment, microclimate and ecosystem stabilisation.

WATER RESOURCES

Water distribution and losses

The distribution of water resources in the basin is very uneven and is mainly influenced by rainfall and other environmental factors. The development of reservoirs also contributes to the uneven distribution of water.

The basin's water resources are found in natural lakes constituting Lake Malawi/Nyasa, and other smaller lakes, the artificial lakes including Lake Kariba and Cahora Bassa and the many rivers and streams. Groundwater is also an important source of water in the basin.

Lake Kariba, the largest artificial lake in the basin, covers an area of 5,100 sq km and has a total storage capacity of 160 cu km (160,000 million cu m), while Cahora Bassa downstream in Mozambique, has a surface area of 2,665 sq km and stores 52 cu km of water. The two dams on either side of the Kafue Flats in Zambia, Itezhi-tezhi and Kafue Gorge store 5.6 cu km and 740 million cu m, respectively. There are 30 dams which hold more than 10 million cu m each. Ten of them hold more than 250 million cu m, including Mazvikadeyi Dam in Zimbabwe, which can hold 365 million cu m and covers an area of 23 sq km when full.1 Water losses through evaporation are high especially from large reservoirs and lakes. Table 3.1 shows some of the current and planned water storage reservoirs and their estimated evaporative losses.

Photo: M Chenje

The demand for water is increasing to satisfy various needs, including domestic.

Table 3.1: Water storage and evaporation losses from major hydropower schemes in the basin

Reservoir	Average Inflow (m^3/s)	Live Storage (million cu m)	Annual Evaporation (million cu m)
Nkula	420	3	1
Tedzani	420	2	2
Cahora Bassa	2,800	51,750	5,799
Itezhi-tezhi	280	5,000	660
Kafue	307	700	1,472
Mulungushi	13	300	50
Lusemfwa	29		72
Lusiwasi	3	72	5
Kariba	1,730	64,800*	8,923
New/planned schemes			
Kapichira	420	11	4
Lower Fufu	50	1.4	1
Mepanda Uncua	3,370		164
Rumakali	13	255	20
Lower Kafue	307	2.2	1

* Total capacity of Kariba varies between 160-180 cu km
Source: SADC ZACPRO 6 Sector Studies 1998.

Demand for water

The demand for water in southern Africa is increasing as human population grows and the countries become more industrialised and urbanised. Already countries such as South Africa have identified that their internal water supplies (including imports from Lesotho) will not meet demand by the year 2030. The city of Bulawayo in Zimbabwe will need to augment supplies soon. Botswana may import water by 2020.

While Angola and Zambia have adequate water supplies for the future, the rest of the basin states are either facing problems or will be, in less than three decades. Namibia has no perennial inland rivers and has great difficulties in mobilising available water to meet its current demand. Botswana is a dry country surviving on groundwater, which is insufficient to meet the demand of the growing population. While Zimbabwe has reasonable water resources at present, the country will suffer water stress by the year 2025 along with Mozambique and Tanzania. Malawi currently faces water problems and by the year 2025 the country will be beyond the water barrier.[2]

in urban centres and for irrigation schemes is a must and urgently called for in the countries of the Zambezi Basin if the resources are to be sustainably managed. Appropriate water conservation measures need to be put in place by all the basin states.

Table 3.2: Water demand for selected uses in the basin states in 1995, (million cu m/year)

Country	Domestic & Industrial	Stock	Mining & Energy	Irrigation	Nature	Total
Angola	1,720	272	15	750	-	2,757
Botswana	175	44	65	47	6	337
Malawi	730	23	5	1,820	-	2,578
Mozambique	135	65	10	3,000	-	3,210
Namibia	200	70	15	248	5	538
Tanzania	1,690	70	10	10,450	-	12,220
Zambia	532	60	20	1,580	-	2,192
Zimbabwe	697	30	30	4,980	-	5,737

Source: SADC/IUCN/SARDC., Water In Southern Africa. Harare/Maseru, 1996

The numerous demands placed on water resources are, in many cases, conflicting and tend to create tensions and conflicts. Such conflicts over water resources in the Zambezi Basin are increasing[3] as more and more areas are being opened up for economic development. The conflicts include:

- Equitable share of resources between two or more riparian states.
- Water apportionment between large commercial and smallholder commercial farming.
- Access to and use of water between individual commercial farmers within a given river basin or between different types of agricultural use.

The conflicts are also between:

- Agricultural users and urban, industrial and mining users.
- Communities and government.
- Local authorities and water polluters.
- Men and women around the issues of access to water resources.
- Government ministries and between differing policy approaches.

Therefore, the potential impacts of these increasing demands should be recognised and active steps must be taken to ensure more sustainable resource utilisation. The implementation of water demand management practices

Consumptive uses of water

Water withdrawals

For most of the towns that depend directly on the surface water resources of the Zambezi River and its tributaries for water supply, the water is usually pumped or gravitated to treatment plants and used directly. Levels of abstraction vary from place to place. Upstream of the Victoria Falls, water abstraction levels are 0.60 cu m per second in Zambia and 0.17 cu m per second in Zimbabwe. Between the Batoka Gorge and Lake Kariba, major abstractions are from Zimbabwe with levels of up to 1.58 cu m per second. Around Lake Kariba, abstractions from the Zambian side are at 4.16 cu m per second, while Zimbabwe draws 0.03 cu m per second. Downstream of Lake Kariba, more water is abstracted on the Zambian side where rates of 1.26 cu m per second have been recorded.[4] Similarly the towns of Mongu, Senanga, Katima Mulilo, Sesheke, Victoria Falls, Hwange, Kariba, Livingstone, Siavonga, Chirundu and Tete depend on direct abstraction from the Zambezi River for their water supplies. Many more big and small urban settlements depend on the tributaries of the Zambezi for water supply.

Critical to pumping for urban water supply, irrigation and other consumptive uses is the amount of water required by the environment downstream. The water present as compensation flow, needs to satisfy the demand downstream such as for vegetation and the ecosystem, as well as for other needs by consumers.

Irrigation and domestic use

A total of 1,716 sq km are under irrigation in the basin with 47.1 sq km in the Upper Zambezi, 653.6 sq km in the Middle Zambezi and 1,014.8 sq km in the Lower Zambezi.[5] Irrigation consumes about 1,500 million cu m of water annually with more than 50 percent of the water available in the basin being used for irrigation.

Table 3.3: Estimated irrigation water use by country (million cu m/year)

Country	Current	2005	2015
Angola	6.262	9.630	220.000
Botswana	6.656	10.030	10.050
Malawi	390.639	401.000	438.000
Mozambique	44.886	56.222	202.825
Namibia	5.255	10.400	12.900
Tanzania	21.619	26.350	32.100
Zambia	397.490	425.290	476.920
Zimbabwe	604.108	726.200	798.500
TOTAL	1,476.915	1,665.122	2,191.295

Source: ZRA, ZACPRO 6 Sector Study 2 Draft
Report: Water Consumption and Effluent,1998

Industry

Demand for water by industry is concentrated mainly in urban areas related to manufacturing and in rural and semi-urban areas for mining. Major industries in the basin include those that process food and beverages, and the manufacture of textiles and clothing, wood and wood products, paper, chemicals, rubber and plastic, metal products and others.

Water resources use by the manufacturing industry in some countries is difficult to quantify. In the case of Malawi, for instance, "the statistics are not recorded and kept."[6] Water demand by the manufacturing industry in Zambia, is estimated to rise from the 1990 figure of 222,400 cu m/day to 446,100 cu m/day in the year 2015.[7]

Inter-basin water transfers

In the case of Botswana, the country's National Water Master Plan does not spell out any substantial volumes of water that would be required from the Zambezi River prior to 2020.[8] In Zimbabwe, the proposed Matabeleland Zambezi Water project would require 40 million cu m of water per year to supply the domestic water needs of the city of Bulawayo, while annual irrigation needs in Matabeleland would be 100 million cu m.[9] This would require an abstraction rate of 3.17 cu m per second and it would, therefore, not be feasible to take this all from the Zambezi River.[10]

The Department of Water Affairs in South Africa, also has designs on the water resources of the Zambezi River and has proposed an inter-basin water transfer from the Zambezi to supply shortfalls in South Africa.[11] The proposed abstraction rate of 95 cu m per second at Katima Mulilo would require a continuous power supply of 1,300 MW[12] to transfer 2,995 million cu m of water per year from the Zambezi. This would be equivalent to seven percent of the mean annual runoff and is almost a third of the low flows of 300 cu m per second.[13]

Non-consumptive uses of water

Non-consumptive uses of water are also likely to increase in the basin, especially as the economies of the riparian countries improve and there is more recognition for diversification. This would imply expansion in industries that do not consume water such as tourism, hydropower generation and others.

Photo: M Chenje

Although hydropower generation is considered a non-consumptive use of water, large dams such as the Kariba facilitate high evaporation losses.

Table 3.4: Major hydropower facilities in the Zambezi Basin

Country	Power Plant	Generation (megawatts)	Storage (million cu m)
Malawi	Nkula Falls	84	Flow of river
	Tedzani	40	Flow of river
Mozambique	Cahora Bassa	2,075	52,000
Zambia	Mulungushi	20	250
	Victoria Falls	108	Flow of river
	Lunsemfwa	18	700
	Kafue Gorge	900	740
	Kariba North	600	160,000
Zimbabwe	Kariba South	666	160,000
	*Batoka Gorge	1,600	1,700

*Planned Hydropower Schemes
Source: ZRA, ZACPRO 6 Sector Studies 1998

Hydropower

Although hydropower itself is considered non-consumptive, high evaporation losses result from the large storage reservoirs that are created to ensure a steady water flow through the turbines. Evaporation losses from the basin are about 14.3 million cu m, 95 percent of which is from Lake Kariba and Cahora Bassa.

Tourism and recreation

Water-based tourism has in recent years become an important activity in the Zambezi Basin. The waters of the main river systems and wetlands provide excellent habitat for wildlife, the backbone of the tourism industry. All tourist activities rely on the Zambezi Basin, retaining its wilderness aura and continuing to offer suitable habitat for the variety of game. Tourist infrastructure has been developed throughout the basin, especially at Kasane, Victoria Falls, Livingstone, Kariba, and Siavonga.

Fisheries

Fishing methods vary from traditional baskets and traps, seine and drag nets to commercial fishing vessels on Lake Kariba and Lake Malawi/Nyasa, and more recently on Cahora Bassa. Environmental considerations are necessary for the sustenance of the fishing industry. Due to the special requirements necessary for spawning and maintaining their health, fish require permissible velocities, quantity and quality of water. Industrial raw effluent disposal into water systems has had negative consequences on the fisheries of some of the sub-basins of the Zambezi in recent years.

WETLANDS MANAGEMENT

Water availability in the Zambezi Basin is closely linked to its 66,000 sq km of wetlands or surface waters. These floodplains, lakes and reservoirs provide extensive storage holding about 100,000 million cu m of water a year and the total volume of water available excluding net losses is about 110 cu km annually.[14]

Wetlands need seasonal, regular and episodic flooding to function. The timing and duration of floods are as important as the volume of water needed to maintain a healthy wetland. Any alteration in the timing or duration of these natural floods alters the productivity of the system, with adverse consequences particularly to floodplain fisheries. The need to maintain or simulate natural flood regimes should be accounted for when, for instance, constructing hydropower schemes and designing large ecosystems themselves.

Box 3.1: Defining wetland

The Convention on Wetlands of International Importance, otherwise also known as the Ramsar Convention, defines a wetland as an area of marsh, fen, peat land or water, whether natural or artificial, permanent or temporary, with water that is static or flowing, fresh, brackish or salty, including areas of marine water the depth of which at low tide does not exceed six metres. This definition encompasses all surface freshwaters as well as shallow seas. Therefore, the entire length of the Zambezi River and its tributaries, the swamps, portions of the major and artificial lakes associated with it as well as the mangrove swamps and extensive floodplains of the Delta and its distributaries are considered wetlands.

The health of a wetland system can be seen from its productivity. Wetland habitats, productivity and biodiversity as well as ecological processes are dependent on the availability of water. It has been estimated that the minimum water requirement to maintain the wetlands is 3,840 cu m per second,[15] a figure based on average evapotranspiration from the wetlands.

Wetlands play a crucial role in maintaining water quality and regulating river flow. Wetlands can be seen as water managers, providing a series of "plugs" within a river system that maintain, cleanse, and regulate water.[16] An example of these functions is demonstrated by the floodplains of the Upper Zambezi which collect and hold the water in the wet season and slowly release it throughout the year. All the flora and fauna of the wetlands are part of and adapted to the natural seasonal flow regime of the river system. For example, dry season migration of herbivores to wetland ecosystems is an annual feature especially in the Zambezi Floodplain where wildebeests migrate between Zambia and Angola in search of water and pasture.

In the case of the Zambezi River, there is a short period of high flow during the summer rainy season. The river floods its banks and inundates the Zambezi Floodplain. These "sponges" of the Zambezi system gradually release the waters to maintain a constant supply of base flow downstream over longer periods. During summer, and with the constant supply of base flow within the system, a rich grassland is maintained for wildlife and cattle grazing.

Construction of rural as well as semi-urban water supply schemes is also possible and made much easier in areas around wetlands, where groundwater is close to the surface. Being an important natural resource, water has been the main determining factor in rural and urban settlements. Nearly 20 million people, more than 50 percent of the population in the basin, are concentrated around wetlands and many of them are dependent on hand dug wells for the supply of potable water.[17]

Types of wetlands

Under the sector studies for the ZACPLAN, the wetlands of the Zambezi Basin have been classified as impoundments, natural lakes, freshwater marshes and marine wetlands.

Impoundments

This category includes the artificial reservoirs created upon dam construction to supply water for industry, agriculture, energy generation, recreation and other functions. This category groups Lakes Kariba and Cahora Bassa. Also included in this category are the two reservoirs on either side of the Kafue Flats in Zambia, the Itezhi-tezhi and the Kafue Gorge.

Natural lakes

The largest natural freshwater lake in the basin is Lake Malawi/Nyasa that covers 28,000 sq km and is Africa's third largest after Lakes Victoria and Tanganyika. The catchment area draining into the lake is 98,000 sq km and includes most of Malawi and three percent of Tanzania. The Shire River catchment area, which links the lake to the Zambezi River, covers 23,000 sq km.

Freshwater marshes

Freshwater marshes include the Zambezi Floodplains of Zambia which store about 8,600 million cu m at low flow and as much as 27,000 million cu m during peak floods. Other major wetlands of the Zambezi Basin which fall under this category are the Caprivi wetlands, covering 4,000 sq km of open water and floodplains; Lukanga Swamps, 2,000 sq km; the Elephant Marsh/Lower Shire, 1,000 sq km; the Kafue Flats, 6,5000 sq km and the Marromeu wetlands of the Lower Zambezi, 5,000 sq km.

Swamps are also part of this group of wetlands and they support animal and plant life, and are a source of water for small-scale agriculture. There are many seasonal and perennial swamps in the basin, some of which are associated with *dambos*.

Marine wetlands

The coastal areas of Mozambique are relatively protected from the open Indian Ocean wave action by the island of Madagascar. The Mozambican coast is considered as a low energy shoreline characterised by longshore transport and deposition of large quantities of riverine sediments

Photo: Ziana

Reservoirs are some of the wetlands found in the Zambezi Basin.

carried by several large rivers which drain into the Indian Ocean. These factors contribute to the extensive establishment of mangrove vegetation along the entire coast of the country. Table 3.6 presents the status of the mangroves of the marine wetlands in the Zambezia Province of Mozambique.

Table 3.5: Water balance statistics for the Zambezi Basin

Country	Rainfall Range (mm)	Average Rainfall (mm)	Average Rainfall (km^3)	Potential Evaporation Range (mm)	Total Surface Runoff (mm)	Total Surface Runoff (km^3)
Angola	25-1,600	800	997	1,300-2,600	104	130
Botswana	250-650	400	233	2,600-3,700	0.6	0.35
Malawi	700-2,800	1,000	119	1,800-2,000	60	7.06
Mozambique	350-2,000	1,100	879	1,100-2,000	275	220.0
Namibia	10-700	250	206	2,600-3,700	1.5	1.24
Tanzania	300-1,600	750	709	1,100-2,000	78	74.0
Zambia	700-1,200	800	602	2,000-2,500	133	100.0
Zimbabwe	350-1,000	700	273	2,000-2,600	34	13.1

Source: Pallet, J. (Ed) *Sharing water in Southern Africa,* DRFN, Windhoek, 1997.

Benefits and services provided by wetlands

Water and wetland resources of the Zambezi Basin provide numerous benefits to the riparian states for social and economic development. Their management is important to sustain the various sectors of the economy that are dependent on the water resources and the wetlands. Some of the benefits and

Box 3.2: The wetlands of the Zambezi Basin

The Upper Zambezi

The Upper Zambezi catchment covers an area of 507,600 sq km, representing 39 percent of the total Zambezi River catchment area. It extends from the source in the Central African Plateau in north-western Zambia to the Kazungula/Kasane intersection of four basin states, Botswana, Namibia, Zambia and Zimbabwe. It includes large tributaries in Angola such as the Cunado (Kwando), the Luiana, and the Lungue Bungo, some 340 km of river in Zambia, including the Zambezi Floodplains, the Kwando/Linyanti/Chobe system that forms the Namibian-Botswana border and the extensive floodplains of eastern Caprivi in Namibia. The drop in altitude in this section is just over 600 m.

The Middle Zambezi

This section drains much of central and eastern Zambia, and over 60 percent of Zimbabwe. It includes the capital cities of Lusaka and Harare in its catchment and is the most intensively used and industrialised section of the river.

The middle Zambezi wetlands in Zambia, include the Lukanga (2,000 sq km) and Busanga swamps on the Upper Kafue River, the Kafue Flats (6,500-7,000 sq km), the two large impoundments Itezhi-tezhi and Kafue Gorge on either side of the Kafue Flats and the more seasonal Luangwa River. The turbid Luangwa River which is subject to seasonal flush flows, floods extensive grasslands when it rains and dries out to a series of pools and oxbows in the dry season.

The most obvious wetland in the Middle Zambezi is Lake Kariba which covers 5,100 sq km and forms 200 km of the border between Zambia and Zimbabwe. On the Zimbabwean side, the main tributaries to the Zambezi River are the Gwayi River, the Munyati and Sanyati rivers, and the Manyame River that has its source near Harare. Lake Chivero and Manyame Dam, on this river supply water to Harare. Other than Kariba, there are 20 large dams in the Zambezi Basin in Zimbabwe, mostly used to supply irrigation water.

The Lower Zambezi

The lower Zambezi River descends from Cahora Bassa to the Indian Ocean. The main wetlands are the marshes of the Lower Shire in Malawi, the rivers in southern Tanzania that drain into Lake Malawi/Nyasa and the floodplains at the Zambezi Delta in Mozambique. The delta and estuary backup over 100 km inland and extend about 120 km along the seafront[53] covering 15,000 sq km.[54] Before the construction of the Kariba Dam, the Zambezi Delta floodplains covered 18,000 sq km.[55] The Marromeu Complex along the southern bank of the lower Zambezi River is an important wetland in both ecological and socio-economic terms and covers 5,000 sq km. But since the construction of Cahora Bassa Dam in 1975, the Marromeu floodplains have only been flooded significantly on three occasions.[56]

Source: Bethune, S., "Water and Wetlands in the Zambezi Basin", for SARDC, 1999

Table 3.6. Mangrove areas and changes that have occurred between 1972-90

Province	Mangrove area 1972 (sq km)	Mangrove area 1990 (sq km)	Depleted Mangrove areas (sq km)	New area of Mangroves (sq km)	Deforestation rate over 18 years (%)
Zambézia	1,594.2	1,557.6	37.7	1.06	2.4

Source: Adapted from: Saket, M. and R.V. Matusse., *Study for the determination of the rate of deforestation of the mangrove vegetation in Mozambique.* FAO/PNUD Project MOZ/92/013. National Directorate for Forestry and Wildlife, Ministry of Agriculture and Fisheries, Maputo 1994

services provided by the water resources and the wetlands of the basin include:

- Energy
- Aquifer recharge
- Flow regulation
- Soil, sediments and nutrients
- Plant and animal products
- Drought survival
- Navigation and communication
- Water quality
- Conservation
- Tourism and recreation
- Aesthetics.

Energy

The high runoff and 1,000 metre fall in elevation from the source to the sea make the Zambezi River well suited for hydropower generation. This is also true for some of the tributaries. At present, 15 power stations generate 32,800 GWh a year in the Zambezi Basin.

However, reliable electricity generation in the basin is threatened by low water levels caused by frequent droughts. Generation levels can fall by as high as 12.2 percent depending on the severity and persistence of the drought. The high sediment levels carried by some of the rivers, such as the Luangwa in Zambia, can affect the size and life expectancy of reservoirs created for power generation.

Aquifer recharge

Some of the water held back by wetlands percolates into the ground and contribute to the groundwater supplies, recharging adjacent aquifers. The contribution of wetlands to groundwater recharge in the Zambezi Basin has not been fully investigated. However, studies elsewhere show that wetlands are important sources of groundwater. According to these studies, floodplains, marshes and swamps contribute to groundwater recharge depending on their location and the structure of the receiving aquifer. Due to lack of scientific research and information on the role of the Zambezi Basin wetlands to aquifer recharge, it is impossible to determine the amount that wetlands contribute to aquifer recharge. There is, therefore, need for research to determine the interactions between wetlands and groundwater.

Flow regulation

Wetlands regulate flow, attenuating floods and in the case of floodplains, store water during the wet seasons, slowly releasing it throughout the drier periods. This helps to maintain flow in perennial rivers such as the Zambezi and its tributaries. This function of wetlands is dependent on a number of parameters such as the size of the wetland, number of wetlands located in the basin, soil type, and depth of and the vegetation within the wetland, among others. Vegetation and plants in marshes and swamps control erosion as roots hold the soil and trap sediments. Well-vegetated riverbanks recover rapidly after flood damage while denuded areas are badly eroded and often never recover. Wetlands, therefore, play an important role in flood control and erosion prevention. Well-vegetated rivers and floodplains are excellent flood absorbers.

Soil, sediments and nutrients

Wetlands have been used for centuries for flood recession agriculture. Early civilisations and settlements especially in the Middle East, thrived on floodplain agriculture. The variations in inundation levels of wetlands result in fertile soils that attract human settlement.

In the Zambezi Basin, wetlands ecosystems support a large population of people who depend on them for their livelihood. The variations in flooding create fertile soils used to support intensive agriculture. The basin floodplains such as the Zambezi Floodplains, East Caprivi wetlands, the Elephant Marsh and the Kafue Flats have a history of long settlements and this is partly attributed to the fertile soils of these areas. In the Zambezi Floodplains, the rich soils created by the floods of the Zambezi River are reserved for the Royal Lozi people while other tribes

The Zambezi Basin wetlands support wildlife and millions of people.

are restricted to the infertile upland areas of the basin. The Zambezi Floodplains supports people whose agriculture depends on the fertile alluvial deposits from the Zambezi.

About 520,000 sq km of land is annually cultivated within the basin, a proportion of which is subsistence floodplain farming dependent on the seasonal replenishment of nutrients within wetlands. The fertility and the residual moisture found in these areas guarantee a good annual crop of maize and rice.

According to research carried out in the Zambezi Floodplains and elsewhere in Zambia, flooding regimes of the Zambezi and other rivers have created four patterns of agriculture that comprise the seepage gardens, the *sitapa* garden, the *mazulu* and *Lishanjo* gardens.[18] The seepage gardens, locally known as *matongo,* are very reliable in terms of moisture availability. Soil nutrients are also more readily available, as the soils contain some loam and higher organic material. The most common crop grown in these gardens is maize, although millet and tuber crops, such as cassava and sweet potato are also grown.

The *sitapa* gardens are flooded annually since they are located in the floodplain, valley bottoms and to some extent in *dambos*. They contain fertile nutrients in sediments brought from upstream by rivers. The sediments and alluvial soils replenish the floodplains and are important for geomorphological processes such as the creation of sandbars at river mouths. Poor landuse practices can, however lead to the loss of too much sediment. Soil erosion, particularly in the grasslands of western Zambia, croplands of the Luangwa Valley and in central Zimbabwe, is considered one of the most serious environmental challenges of the Zambezi Basin.[19]

The *sitapa* gardens are cropped with early maize and rice. The *mazulu* gardens, mounds found in the floodplains, are believed to originate from a combination of termite activity, deposition of river sediments, and human activity. These areas are cultivated with maize while sorghum is grown on poorer ones. The *Lishanjo* gardens are made of peat derived from the decomposed remains of plant growth, with the Waterberry as the dominant tree species around them. Although these soils are fertile, workability is a major problem since they need to be drained.

Special landuse patterns, dependent on the fertile alluvial deposits, also in the other parts of the basin where floodplains are present, including the Chobe/Caprivi floodplains, Lower Shire areas and Zambezi Delta. For instance, the fertile floodplains of the East Caprivi support a total of 134,708 cattle and 18,470 goats.[20]

Plant and animal products

The wetlands of the Zambezi Basin support a diversity of plant and animal species. Wetlands are principal habitats for fish species, providing cover, suitable breeding and feeding grounds.[21] Wetlands support subsistence, artisanal, ornamental and commercial fisheries. The most notable commercial fisheries are on Lake Malawi/Nyasa, Zambezi Delta/Sofala Bank and Cahora Bassa, and at Itezhi-tezhi, Kafue and Lake Kariba.

The fisheries of these areas depend on the availability of relatively nutrient-rich water provided by the wetlands. On Lake Kariba, the main commercial fishery activity is the offshore sardine fishery. This freshwater sardine or *kapenta*, has also established itself in Cahora Bassa where its exploitation is still limited. However, potential catches are estimated at 15,000 tonnes per year. By 1993, the annual yield from Lake Kariba was about 30,000 tonnes valued at US$55 million.[22]

Fish are a major protein source in the basin states. There are 85 fish species in the Upper Zambezi alone and the potential yield is estimated at 14,000 tonnes per year, while fish catches average 7,500 tonnes annually.[23] The Zambezi Floodplains fisheries support about 300,000 people[24] and act as a food store for a great variety of fish-eating birds. Like elsewhere in the basin, fish are an important protein source in the eastern Caprivi, with some 82 species recorded from the Zambezi.[25]

Earlier fish yields from the eastern Caprivi, between 1970s and early 80s, are estimated to have been as high as 800 tonnes annually for Lake Liambezi[26] and an additional 700-900 tonnes a year from the Zambezi, Chobe and Kwando rivers. This is the period when conditions were substantially wetter. Recently, reduction in floodplains due to inadequate floods, particularly of the shallow seasonal breeding areas and overfishing, have contributed to a decline in the fishery. The annual value of the fish catch in Caprivi is estimated at US$1.8 million[27] and total fish production about 1,500 tonnes.[28] There are other fisheries within the basin, which have not been costed. These include riverine fisheries and those in other reservoirs.

Photo: D Martin

The Zambezi Floodplains fisheries support about 300,000 people and act as a food store for a great variety of fish-eating birds.

At the delta, a thriving shrimp fishery contributes significantly to the Mozambican economy. In Angola, it is estimated that 50,000 people are seasonally employed in the floodplain fisheries near the headwaters of the Zambezi.[29] In Malawi, the major fishing industry is on Lake Malawi/Nyasa where the "Chambo" fish dominates the industry. Lake Malawi/Nyasa, has well over 500 species, mainly endemic cichlids. Elsewhere in the country, the Lower Shire yields 10,000 tonnes of fish annually.

Wetlands generally support a great variety of other trades and resources. Rural communities living alongside the wetlands of the Zambezi Basin are dependent on wetland biota. Wetland resources can, to some extent, buffer the effects of poverty. Poor rural communities living close to wetlands and having access to a variety of "free" wetland resources such

Table 3.7: Average annual fish output from the major wetlands of the basin (tonnes)

Wetland	Area (Sq km)	Catch (Inland Fishery)	Catch (Pelagic Fishery)
Zambezi Floodplains		7,500	
Eastern Caprivi	4,000	1,500	
Lukanga swamp		1,421	
Kafue Flats	6,500	7,063	
Itezhi-tezhi lake	370	641	
Lake Kariba	5,100	2,578	21,000-26,000
Lake Lusiwashi	80	927	
Cahora Bassa	2,665		6,000
Lake Malawi/Nyasa	30,800		50,000-64,000
Lake Malombe	390	10,000	
Lower Shire	500-1,000	4,000-17,000	

Source: Denconsult 1998b, Mapila 1998, Koekemoer 1998

as reeds, thatching grass, timber, fish, crabs and edible plants, are often healthier and have a better quality of life than similar subsistence communities who do not have access to wetland resources. Dry season floodplain grazing is vital to maintaining large national cattle herds. Reeds, grasses and sedges are harvested and used as building materials or for making baskets and fishing traps, for example. Trees provide wood for canoes and poles as well as edible fruit. Dry floodplain grazing is an important wetland occupation in most Zambezi Basin countries where livestock farming takes up a large share of the national economy.

most of the rural communities living in the Zambezi wetlands use dugout canoes to transport people and goods. At present there are several busy river ferry crossings on the Zambezi River, two in Angola, two in Mozambique, two border crossings at Wenela between Namibia and Zambia, four in Zambia and at Kazungula on the border between Zambia and Zimbabwe. Some shipping, mainly recreational, takes place on Lake Kariba and there is a postal service from Mongu to Kalabo on the Zambezi Floodplains. Frequently used waterways are between Kalabo and Mongu, Kalabo and Makoma, Libonda and Sishekamu, Nangula and Ushae.

Drought survival

Prolonged droughts, which are common in southern Africa, negatively impact on the ecology of wetland ecosystems. They lower the water table, reducing the area of wetlands. Such occurrences can have negative impacts on human water supply, fish and other wildlife. Wetlands are important in sustaining life during droughts. Both people and wildlife converge on wetlands for water and food during droughts. For example, during the 1969–70 drought that affected Zimbabwe, 84 percent of the farmers with *dambo* fields were able to support their families.[30]

Photo: D Martin

Some sections of the Zambezi River are navigable, supporting boating and tourism.

Seniors in the Zambezi Floodplains say some of the permanent lagoons in the area support large populations of people during drought years. An example is a lagoon called "Nandombe" (which in Lozi means saviour) whose fish resources support people in and outside the floodplains during times of drought and food shortages. Similarly, river corridors also provide water, food and shelter for people, livestock and wildlife during droughts. In some cultures, wetland grazing is kept in reserve for periods of prolonged water scarcity.

Navigation and communication

As early as even before the days of explorers such as David Livingstone, the Zambezi has offered navigation from the sea to the interior. Some sections of the river are navigable, large boats can be used to transport people and goods locally and commercially in the Upper and Lower Zambezi, on Lake Malawi/Nyasa and Kariba.

Although large boats, cannot navigate many sections of the Zambezi River due to rapids or low water levels,

On Lake Kariba, the Kariba Ferry provides the main transport system linking Binga and Kariba, while on the Zambian side, the Kariba Waterways operates between Siavonga and Chepepo, and Siavonga and Shenga. In Zambia, a number of pontoon ferries operate on the Zambezi River at Chavuma, Ngoma, Sandaula/Malumbu in Mongu, Katima Mulilo/Sesheke ferry, and the Kalongola/Senanga Ferry pontoon. In Lukanga, waterways exist between Waya and Chilwa Island, Chilwa Island and Chitanda. Other crossings are at Caia and Songo in Mozambique.

The Lower Zambezi is navigable for a distance of 570 km upstream from the Indian Ocean to where the Mepanda Uncua dam is planned. In the past, molasses were ferried from the Sena sugar estate to the coast. Potential exists of transferring coal by barge from Moatize in Tete Province to the coast, but this would require an increased water flow from Cahora Bassa dam to maintain the required flow of 2,000 cu m per second and a minimum depth of three metres.[31]

In Malawi, boat transport is used on Lake Malawi/ Nyasa and the Shire River. The Ilala and Mtendere ply the lake from Monkey Bay in the south to Karonga in the north, calling at Chipoka, Nkhotakota, Likoma and Chizumulu Islands, Nkhata Bay, Usisya, Mlowe, Chitimba and Chilumba on the way.[32]

Despite the existence of these waterways, infrastructure is not well developed in most parts of the basin. The deposition of sediments in the rivers, lakes and swamps in the basin make navigation difficult. This is a major problem in the Zambezi Floodplains where communication between and among villages is dependent on water transport. Dredging of the waterways has been a routine activity in this area and yet with the economic hardship often faced by developing countries, many channels are silting and getting clogged up by aquatic vegetation.

Water quality

Water is the solvent and medium, in most of the chemical reactions occurring within the environment. Water quality can decline due to pollution, eutrophication and siltation. Wetlands play a very important role in maintaining water quality because of their ability to function as filters for removing pollutants and sediments from flowing water.[33] They act as natural filters, trapping sediment and nutrients and improving the water quality downstream. Water purification is an essential wetland function where sediments are deposited, minerals such as phosphates and plants take up nitrogen compounds, and organic pollutants decompose.

There has not been any systematic work to determine the water purification functions of the basin's wetlands. However, a research carried out by the Environmental Council of Zambia (ECZ) on the Lukanga swamps on the Kafue River has shown that the quality of water upstream of the swamps is poorer than that observed downstream. Research on constructed wetlands in southern Africa and East Africa shows that wetlands are efficient wastewater treatment "plants" as they play these functions through plant primary productivity, sedimentation and accumulation of sediments, anaerobic conditions and decomposition. Wetlands also remove toxic substances such as heavy metals and various kinds of pesticides from water. Excess nutrients are also removed by wetlands during plant growth.

Conservation

The natural beauty and rich biodiversity of wetlands make them an important focus for conservation. Eight percent of the Zambezi Basin consists of protected areas. Many of the national parks and conservation areas in the basin are wetlands. Most of the wetlands support large numbers of wetland birds, consisting of more than 500 species.

Despite the numerous and important products and values provided by wetlands they have not been accorded the conservation status that they require. Only one Ramsar site exists in the whole basin, and this is the Lochnivar Wetlands in the Kafue Flats. Of the eight riparian states, only five are members of the Ramsar Convention. Tanzania ratified only in 1998 while Angola, Mozambique and Zimbabwe have not. The convention accords international conservation status to important wetlands.

Wetlands are an important focus for conservation, tourism, recreation and education. The ecological value of the Zambezi Basin wetlands has been estimated to be US$15,000 per hectare per year. The mangrove wetlands and estuaries of the delta are worth between US$10,000–100,000 per sq km per year.[34]

Table 3.8: Water quality stations in the Zambezi Basin

Country	Name of Stations
Angola	Zambezi at Lumbala
Malawi	Phwezi, S53 Roadbridge, Monkey Bay, Lake Malawi/Nyasa Outflow at Mangochi, Tengani (Lower Shire).
Mozambique	Cahora Bassa Inflow and Outflow, Lake Cahora Bassa, Mazowe/Zambezi confluence, Mopeia.
Namibia	Kongola.
Tanzania	Kikonge and Kyela/Natural Bridge.
Zambia	Chavuma Falls, Kalabo, Siakasumbi, Watopa Pontoon, Senanga, Lumbungu, Chafumba Pontoon, Kafue Hook Bridge, Kasaka, Mfuwe, Great East Road Bridge.
Zambia/Zimbabwe	Kariba Inflow and Outflow.
Zimbabwe	Kamativi, Sanyati Bay, Copper Queen.

Source: Undated ZACPLAN Document

Tourism and recreation

Tourism and recreation are a major economic activity within the Zambezi Basin largely dependent on water and wetland resources. The direct and indirect functions that water and wetlands have on the biodiversity of the

Table 3.9: Results from water bird surveys of the Marromeu Complex, March 1995

Species	Count
Eastern White Pelican (*Pelecanus onocrotalus*)	9
Pinkbacked Pelican (*Pelecanus rufescens*)	65
Goliath Heron (*Ardea goliath*)	10
Purple Heron (*Ardea purpurea*)	2
Blackheaded Heron (*Ardea melanocephala*)	1
Egrets (*Egretta spp.*)	289
Cattle Egret (*Bubulcus ibis*)	2,975
Black Stork (*Ciconia nigra*)	6
Abdimin's Stork (*Ciconia abdimii*)	30
Yellowbilled Stork (*Mycteria ibis*)	41
Marabou Stork (*Leptoptilos crumeniferus*)	52
Saddlebilled Stork (*Ephippiorhynchus senegalensis*)	36
African Openbilled Stork (*Anastomus lamelligerus*)	1,896
Woolynecked Stork (*Ciconia episcopus*)	6
Hamerkop (*Scopus umbretta*)	2
Glossy Ibis (*Plegadis falcinellus*)	73
Sacred Ibis (*Threskiornis aethiopicus*)	35
Spurwinged Goose (*Plectropterus gambensis*)	101
Knobbilled Duck (*Sarkidiornis melanotos*)	7
Whitefaced Duck (*Dendrocygna viduata*)	57
Fulvous Duck (*Dendrocygna bicolor*)	4
African Fish Eagle (*Haliaeetus vocifer*)	11
African Marsh Harrier (*Circus ranivorus*)	12
Wattled Crane (*Bugeranus carunculatus*)	156
Grey Crowned Crane (*Balearica regulorum*)	25
Longtoed Plover (*Vanellus crassirostris*)	21
Greyheaded Gull (*Larus cirrocephalus*)	3
TOTAL	5,925

Source: Beilfuss, R. D., Warwick T. and N. N. Gichuki (eds.). *Proceedings of the African Crane and Wetland Training Workshop, 8-15 August, 1993*, Maun, International Crane Foundation, 1996

Photo: M Chenje

Photo: P Wade

Photo: M Chenje

Photo: M Chenje

Wetlands are an important focus for conservation, tourism and recreation.

Case Study: The Kariba Dam — A damning lesson

The Kariba Dam, between Zambia and Zimbabwe, is an artificial ecosystem whose destructive history is today's instructive scientific lesson.

Completed in 1959 at a time when Environmental Impact Assessment (EIA) was unheard of, the Kariba Dam is a lasting symbol of how narrow economic growth interests impose more costs than benefits on both people and ecosystems.

Representing an unparalleled engineering feat, which made it the largest dam in the world at the time, the Kariba Dam was also an unparalleled ecological disaster, flooding more than 5,000 sq km.

In the 40 years since it was built, the lessons learned from the destruction of people's lives and thriving ecosystems have spawned new challenges. The lake, which is 250 km long and 40 km across, is a complex system not only in terms of its ecology but also jurisdictional mandates. Its impact extends from the individual to the community, from the local to the national, from the trans-boundary to the global. The institutions involved in its management are many and varied.

The history of the Kariba Dam underpins the challenges it faces today. About 57,000 local people on both sides of the Zambezi River in Zambia and Zimbabwe were displaced. More than 5,000 animals were rescued as the waters of the dam rose. The rescue of the animals was dubbed Operation Noah, invoking biblical comparisons to Old Testament floods during the time of Noah. More was spent on saving animals than on humans and the ecology.

However, wildlife habitats and old growth forests were not given the same attention as the animals, which were rescued. A technical fix to transport the animals to higher ground failed to recognise the interdependence between animals and their habitats.

A research study conducted since the dam was completed quotes one of the displaced people, Philemon Munkuli: "The time of removal was a time of slavery for us because we were forced to clear the roads, which would take us away from our homes."

The BaTonga people in both Zambia and Zimbabwe were forcibly removed from the valley and resettled on unproductive, marginal lands, far from their river culture. Today, the majority are still in abject poverty, depending on government handouts in many cases.

Despite the fact that Kariba was built to provide hydro-electric power, the majority of the rural communities in the Zambezi Valley have no electricity. For example, Zambia derives 62 percent of its net energy needs from woodfuel while Zimbabwe uses 57 percent.

"Less than two percent of the population (of Sinazongwe District in Zambia), about 97,000 people, have access to electricity," complains Patson Chizebuka, secretary of the Sinazongwe District in Zambia. "Electricity is not benefiting local people, it is not even benefiting schools, not even the chief's palace."

The situation is no different for rural communities in Zimbabwe. However, for Kariba town, the dam has brought many benefits. While electrical generation was the sole impetus for development of the dam, the past 40 years have also seen the introduction of other thriving economic activities, including tourism and fisheries. In fact, both have long since surpassed power generation as the most important economic activities.

Kariba Town Clerk Godfrey Magombedze says fishing contributes about 45 percent to the town's economy and tourism 40 percent, while power generation is only 10 percent. Other mainly related activities add the remaining five percent.

"The planner who planned Kariba only saw power generation," says Magombedze. "He did not see fishing. He did not see tourism, that is the problem we have now."

The town is facing growing pains, chief of which is waste and sewage management. While the town had a population of about 7,000 in the late 1970s, the figure had shot up to 30,000 in 1999. The population growth has placed heavy pressure on services such as sewage disposal systems. There are reports of untreated sewage being pumped into the lake with disastrous effects on water quality.

The most affected suburb in Kariba is Mahombekombe, which was built about 40 years ago to accommodate 3,000 dam construction workers. It now has 9,000 residents. The town council is upgrading the sewage system, water facilities and roads under a US$1.18 million-programme funded by the World Bank. The programme is scheduled for completion by December 2000.

In addition to this programme, Kariba plans to control population growth over the next decade or two to ensure that it does not lose its lustre as a tourism destination. The population is projected to be 40,000 by 2015.

Magombedze says the challenge is to ensure that people co-exist with wild animals, one of the largest draw cards for tourists.

"We have to live with game and the challenge is to maintain that. The less game we see, the less tourism we have, and the less tourism, the less jobs, and the less jobs, the less food," he says. "We are trying to get our people to understand these linkages."

As an ecosystem, Lake Kariba can be regarded as alien territory due to two thriving alien species. These are the water hyacinth, a native of South America and *kapenta,* a fish native to Lake Tanganyika.

According to Zimbabwe's 1998 State of the Environment Report (SOE), the water hyacinth was first recorded in Africa in 1879 and in Zimbabwe, in 1937. It has been spreading in the basin's water bodies ever since.

The water hyacinth, an indicator of pollution in a water body, has invaded many parts of the lake, the Zambezi River and its tributaries. Ninety-five percent of the weight of the plant is water, contributing immensely to evapotranspiration. Water hyacinth is not only a factor in terms of evapotranspiration but also a threat to boating, angling and water skiing — major recreational activities on Lake Kariba. The weed has proved to be, so far, an unbeatable challenge to control it by manual, mechanical, chemical or biological methods.

While the water hyacinth is a threat, both commercially and ecologically, the *kapenta* has spawned a thriving fishing industry, which is the major inland fishery of Zambia and Zimbabwe. Annual *kapenta* fish catches in Zimbabwe range between 15,000–20,000 tonnes. In 1996, landed *kapenta* fish in Zimbabwe was valued at about US$9.1 million.

Despite the fact that they are a commercial success, there are fears that the fisheries are in decline due to overexploitation of the resource.

Both the water hyacinth and *kapenta* fisheries are some of the major issues facing people in both Zambia and Zimbabwe. At a regional meeting in 1999 which brought together government officials, local government leaders, traditional chiefs and the private sector, water weeds and Kariba fisheries ranked high on the list of pressing concerns.

The Kariba area also suffers from over-management due to the existence of several institutions, some with overlapping and even conflicting responsibilities. The institutions include urban and rural district councils in both countries, the national parks and wildlife departments on both sides, the Zambezi River Authority (ZRA), the SADC Lake Fisheries Project, and the University of Zimbabwe Lake Kariba Research Station. The latter three are involved in various areas of research, including water pollution and fisheries.

Examples of jurisdictional overlaps involves the Kariba Town Council and the Department of National Parks and Wildlife in Zimbabwe.

The town council has no jurisdiction over activities on the lake, as it is a designated national park. The town council cannot enforce anti-pollution control measures against houseboats on the lake.

The town council cannot also levy charges to boat-owners using its roads. This is the responsibility of the parks department.

Magombedze says this means Kariba residents are subsidising tourism when they should be deriving more revenue from this industry.

Institutional overlaps transcend governmental departments. Traditional communities on both sides of the lake feel marginalised in the management of the resource.

"We have two governments and we have two parliaments which make decisions related to the lake," complains Chizebuka. "We have chiefs, but they are not consulted. For every law we pass (related to the management of resources in the Kariba area), we must make sure that we first of all consult the people. What we have are chiefs that are not consulted."

Chief Sinazongwe complains that lack of coordination in the management of the lake, particularly fishing seasons, undermines efforts to manage the resource sustainably. He says the Lake Kariba fisheries industry is never closed, encouraging fish-poaching throughout the year.

The challenge in tackling these and other problems lies in involving stakeholders at different levels from local and national to regional and global.

By Munyaradzi Chenje

basin promote tourism activities, particularly hunting, photography, bird watching, swimming, bungee jumping, yachting and angling. Tourism in the Chobe National Park in Botswana, the Liwonde, Lake Malawi and Lengwe National Parks in Malawi, Gorongoza and Marromeu in Mozambique, East and West Caprivi Game Reserves in Namibia, the Kafue National Park, Liuwa National Park, West Zambezi Game Management Area, Luangwa Valley, Sioma Ngwezi and Lower Zambezi National Parks in Zambia and the Victoria Falls, Mana Pools and Lake Kariba in Zimbabwe is dependent on wetland ecosystems.

The Victoria Falls is one of the major tourist attractions in southern Africa expected to attract between 500,000-825,000 visitors a year by 2005.[35] Activities include wildlife viewing and photography; boating, including white water rafting; angling; hunting and bungee jumping from the Victoria Falls Bridge.

The Botswana part of the basin receives about 100,000 visitors annually and in 1996, the area realised more than US$9 million from occupancy in hotels and lodges.

Aesthetics and cultural values

Many wetlands in the Zambezi Basin possess unique attributes that cannot be described in terms of use, ecological and hydrological values. The aesthetic appeal and recreational potential of wetlands are not easy to quantify in financial terms, but cannot be ignored. The natural beauty, impressive scenery and large numbers and variety of wildlife, and good recreational fishing potential make the Zambezi Basin wetlands popular tourist destinations and good earners of foreign revenue.

Threats to wetlands

Threats to the Zambezi Basin wetlands include reduced flows caused by droughts and water abstractions, aquatic weed infestation, pesticides especially *dichloro-diphenyl-trichloroethane* (DDT), infrastructure development such as dams, overexploitation of resources due to human pressure, uncontrolled fires, pollution and deforestation, among others. These can lead to a number of ecological problems such as habitat change and species loss, and can affect the socio-economic well-being of the peoples of the basin.

Wetlands such as the Lower Shire River in Malawi, the Zambezi Delta in Mozambique, the Kwando/Linyanti/Chobe system draining down into the Caprivi, and the Zambezi and the Kafue floodplains of Zambia are among the most environmentally sensitive areas in the basin. All are rated as "very high"[36] in terms of environmental sensitivity.

Case studies such as from the Eastern Caprivi, Kafue Flats and the Zambezi Delta clearly illustrate that water levels and flooding patterns are major factors in determining the ecological diversity and productivity of wetlands. Alterations, whether natural, or due to infrastructure development, affect the reproduction, recruitment, growth and mortality of wetlands and wetland-dependent biota.

There are indications of declining productivity in the Zambezi Floodplains due to a combination of reduced flows and increased exploitation of natural resources as a result of increasing population pressure.

Further downstream, the proposed abstraction of water to supply the city of Bulawayo and for irrigation in Zimbabwe, could adversely impact the ecology of downstream wetlands such as the Mana Pools. At Livingstone, Victoria Falls and Tete, raw sewage is discharged into the Zambezi River, causing pollution and posing health risks to downstream users. The growth of problem aquatic plants such as Kariba weed and water hyacinth, is correlated to increased nutrient levels in rivers, sewage effluent, as well as fertiliser-rich runoff from agriculture. New dams such as the hydropower development proposed at Batoka Gorge and downstream of Cahora Bassa need to be carefully assessed in terms of their potential impact on river ecology. The existing large dams have already drastically reduced the vital floodplain areas in the Zambezi Basin.

On Lake Kariba, there is a problem with aquatic weeds, originally Kariba weed and more recently water hyacinth. Aquatic weeds and eutrophication are serious problems in the reservoirs supplying water to the city of Harare. Also of concern is that the fishing regulations for Lake Kariba are different in Zambia and Zimbabwe. For example, in Zimbabwe gill net mesh may not be smaller than 100 mm while in Zambia the size limit is 75 mm.

As the most industrialised of the basin countries, Zimbabwe also faces the worst water pollution problems. DDT, heavy metals and toxins from mining, agro-chemical industries and sewage, contaminate tributaries of the Zambezi flowing from Zimbabwe. Uncoordinated gold panning in Zimbabwe not only causes toxins such as mercury and cyanide to pollute the rivers but also damages river banks with resultant erosion and siltation.

In the Lower Zambezi and the Lower Shire River, wetlands are under pressure due to dense human settlement

Map 3.2: Ecologically sensitive riparian and river-dependent habitats in the Zambezi Basin 1995

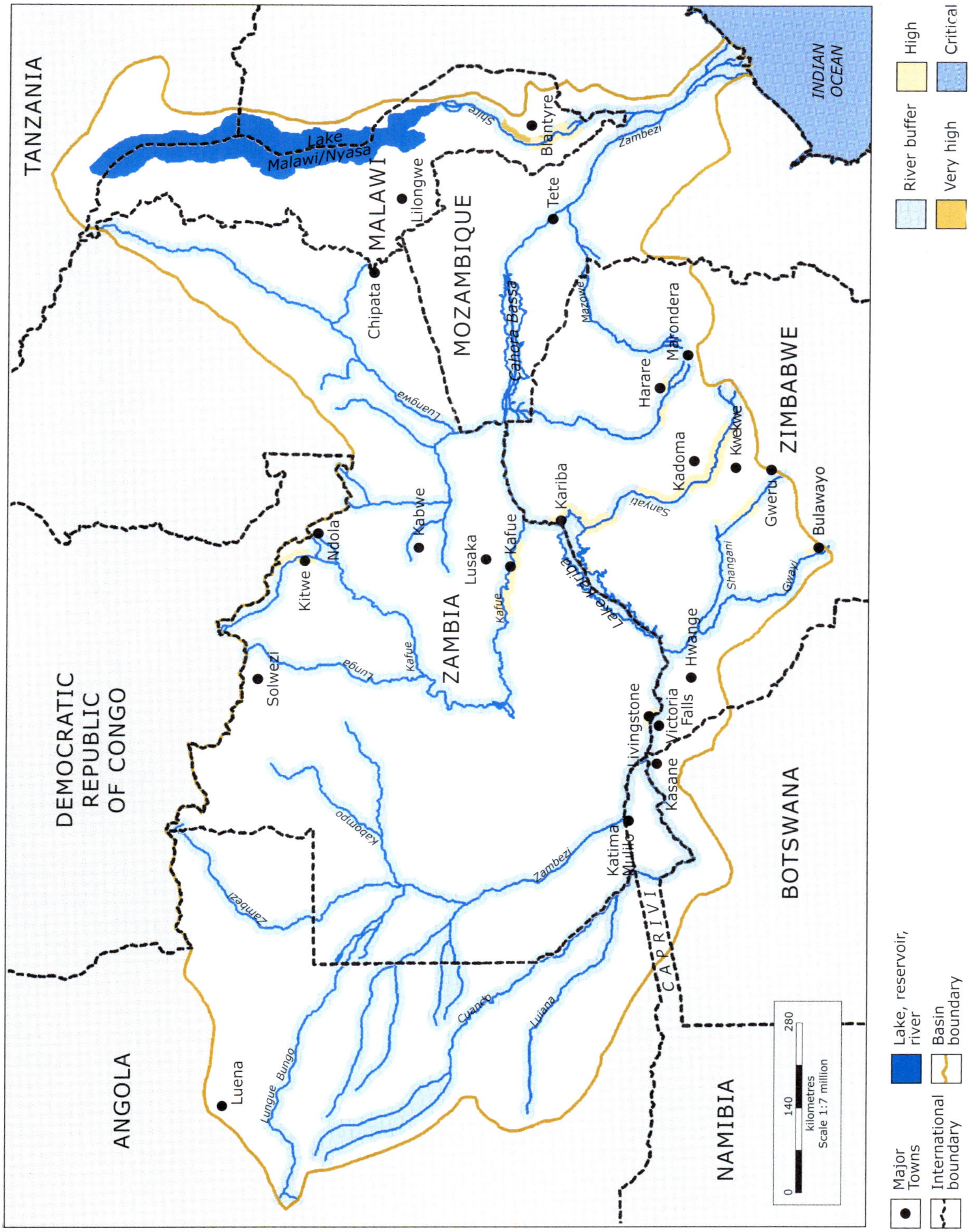

Source: Zambezi River Basin Atlas, SADC/DENCONSULT/ZRA, Lusaka, 1998

Box 3.3: Threats to the Kwando/Linyanti/Chobe and floodplains in Eastern Caprivi

The wetland area in eastern Caprivi was estimated to cover some 11,200 sq km in 1985, about a third of the land area of eastern Caprivi. More recently it has been calculated that the open water and floodplains cover a total area of only 4,000 sq km. The main Zambezi floodplain lies east of the Katima Mulilo/Ngoma road and is inundated from about April to September each year. The Cuando/Linyanti/Chobe system includes permanent wetlands known as the Linyanti swamps, which extend on either side of the Botswana/Namibia border. The main threats to the wetlands of eastern Caprivi are reduced flows in the Zambezi and its tributaries, the Kariba weed, the increasing use of pesticides, road building activities and overexploitation of wetland resources due to increasing human pressure, while future threats include intensive agriculture and pollution.

Reduced flows

In the past 15 years, the wetland area has been reduced by consistently lower flows also noted for both the Upper Zambezi and the Kafue basins. Effects of the low flows have been the drying up of Lake Liambezi between 1985 and 1989, the reduced wetland area since 1985 and the fact that in recent years there has been no flow in sections of the Linyanti River. Sections of the Mamili National Park are drying out and lechwe numbers have declined.

Aquatic weeds

One of the main challenges in the eastern Caprivi wetlands has been to control the invasive Kariba weed. With the introduction of the biological control agent *Cyrtobagous salvinae,* a weevil specific to the Kariba weed, the weed eradication programme has been successfully achieved in Namibia. Another aquatic weed, Nile lettuce or *Pistia stratiotes,* is also known to occur in the Kwando/Linyanti swamp areas but has not caused serious problems.

Pesticides

DDT is still used routinely by the Ministry of Health and Social Services in Namibia to kill malaria mosquitoes. It is known to concentrate in predatory fish and to be harmful to piscivorous (fish-eating) birds and has been found in the eggs of reed cormorants and darters, often near sites where spraying equipment is cleaned. DDT, Dieldrin and Endosulphan, the latter two previously used against tsetse fly and locust sprays have an impact on non-target insects, and insect eating mammals such as bats. Similarly, the molluscides used to combat bilharzia in the Kwando River in recent years probably killed other snail species as well.

Road building activities

Recent roadworks, such as the Trans-Caprivi Highway have taken into account the need to allow natural seasonal flow to pass under roads. The new road from Katima Mulilo to Ngoma follows the edge of the eastern floodplain and will need adequate culverts to allow free flow of floods through the Bukalo channel to the Lake Liambezi depression.

The Caprivi killfish, that occurs in temporary pans in Caprivi, is considered endangered due to its limited distribution. Threats to its habitat include road building. During the dry season gravel from the pans is used for the maintenance of dirt roads. These fish survive desiccation by aestivating as eggs.

Overexploitation of resources

Recent estimates are that 83,000 people live in the eastern Caprivi of which 18,000 are in Katima Mulilo. The population is largely rural and economically dependent to a large extent on natural resources. The birth rate of four percent is one of the highest in the basin. It has been observed that the main causes of environmental degradation in the eastern Caprivi wetlands are: overpopulation that leads to ever-increasing use of natural resources; poverty, which means a reliance on natural resources; and the loss of traditional management practices governing the use of those resources. Overgrazing is further identified as one of the major threats to the eastern Caprivi wetlands. Current stocking rates are three times the carrying capacity of the land. In 1991, approximately 96,000 cattle grazed in an area with a carrying capacity estimated at 30,000–40,000 herd. By 1996, stock numbers were 124,000 cattle and 6,600 goats in the eastern Caprivi and now there are 134,708 cattle and 18,470 goats.[57]

Dense settlement and high stock numbers have also increased contamination of shallow drinking water wells, particularly where these are not fenced off from stock or are sited too near to pit latrines. Piped water supplies, hand pumps and better sanitation practices and community awareness programmes are helping to improve the situation. The Department of Water Affairs is establishing water point committees in most villages in the Caprivi and implementation includes sanitation training. However, harvesting and clearing of riparian trees and reeds for agriculture, for fuel, construction material, boats, and carvings are causing increasing concern.

Some 15 percent of households in the eastern Caprivi rely on fishing as an important income source, 20 percent of the population is actively involved in fishing and 82 percent of them regularly market at least some of their catches.

Ninety fish species have been listed in the Upper Zambezi and a total of 102 species for the eastern Caprivi wetlands. In recent years, the fishery has become increasingly threatened by a combination of natural drying of the wetlands and overfishing caused in part by improved "modern" fishing gear such as gill nets and dragnets. Over 90 percent of fishermen now use nets rather than traditional fishing gear. Fish sizes are perceived to have decreased in recent years and a greater proportion of the fish are now species with a high fecundity and short life span. These changes are typical of the impacts of the change from large mesh (150 mm) nets to those with a mesh size of 35 mm or less (mosquito nets and even shade-cloth). A recent survey of fish in the eastern Caprivi rivers concludes that there are signs of overexploitation in the Zambezi River near Katima Mulilo and in the floodplain back water Lake Lusikili.

Intensive agriculture

The eastern Caprivi is considered to have good potential for irrigation. At present the only irrigation scheme is near Katima Mulilo where 2.4 sq km is used to grow tobacco and maize. The agricultural scheme uses less than 3 million cu m of water a year but there is renewed interest in a large-scale 1,050 sq km-sugar cultivation in an area north west of Lake Liambezi. The main environmental concerns of large irrigation schemes are the resultant increase in pesticides and fertilisers, nutrient-rich run off into the surrounding wetlands, draining of wetlands and floodplains for cultivation, decreasing river flow downstream of abstraction points and an increase in human/wildlife conflicts.

A large sugar irrigation scheme first proposed in 1993 is once more under investigation. Earlier estimates placed the water requirement at 250 million cu m per year. The current estimate is 230 million cu m per year, to be supplied via the Bukalo channel and it is estimated that some 20 percent of this water would be "lost" before reaching the proposed sugar estate. In 1996, the total area cleared for cultivation in the eastern Caprivi was 1,719 sq km of which almost half was in floodplain (673 sq km) and riverine woodland (212 sq km) areas. Yields tend to be low and variable, 70 to 445 kg per hectare for millet and sorghum and 30 to 700 kg per hectare for maize. Fertiliser is usually not applied and cleared land is often abandoned after a few years and new areas cleared.

Source: Bethune, S., "Water and Wetlands in the Zambezi Basin". For SARDC, 1999

which in turn is threatened by flooding and wild animals, particularly hippos and crocodiles. Evidence exists that water regulation from Cahora Bassa has substantially reduced the Zambezi Delta wetland area, affecting wetland productivity in Mozambique and even offshore shrimp catches. Salt-water intrusion is a serious problem in the delta due to reduced freshwater flows and the lack of substantial seasonal floods. New infrastructure development proposed downstream of Cahora Bassa will worsen the already declining ecological situation in the delta.

Population pressure

Population pressure on wetland ecosystems often leads to habitat change and loss of species. In the Caprivi, recent estimates indicate that 83,000 people live in the wetland

Photo: M Chenje

Aquatic weeds are easily transported on the water from one place to another, exposing more areas to invasion.

Box 3.4: Threats to the Kafue Flats

The most important wetlands in the Middle Zambezi are those associated with the Kafue River. The Kafue Flats are the subject of a detailed ongoing community based WWF — Zambia wetlands project, initiated in 1986. In many ways, they present a microcosm of the various resource conflicts and environmental threats faced by the basin.

Pollution

The Kafue River and its floodplains are highly utilised and heavily industrialised and therefore subject to pollution from mining, industry, agriculture and municipalities. Preliminary findings of a joint study by the University of Zambia and the Swedish Lulea University of Technology show high levels of heavy metals particularly copper in the Kafue River near the Copperbelt. The introduction of the 1990 Environmental Protection and Pollution Control Act has gone some way towards controlling pollution, but monitoring and enforcement remain a problem. Aquatic weeds are also a menace.

Water supply

The Kafue River system has a catchment area of 154,000 sq km and a mean annual flow of 350 cu m per second. The Kafue Flats have a floodplain area of 6,500 sq km. The floodplains are sandwiched between two large lakes, the Itezhi-tezhi with a storage capacity of 5,600 million cu m and the smaller, deeper, Kafue Gorge with a capacity of 740 million cu m. Water from the system is used to supply the city of Lusaka and Kafue, the Nampunde copper mine, industries such as the fertiliser factory, as well as irrigation schemes. Kafue water is needed to irrigate 160 sq km of sugarcane at the Nakambala Sugar Estate and a further 60 sq km of other crops elsewhere. The resultant conflict for water resources in the Kafue Flats is perhaps best illustrated by the results of an investigation which revealed that the 1982 allocations for hydropower, irrigation and water supply exceeded the low flow volumes of the river.

Hydroelectric power generation

The Itezhi-tezhi Dam maintains a constant flow of 183 cu m per second in the Kafue River for power generation at Kafue Gorge. This regulation has smoothed flow, reducing peak floods and increasing minimum flows. The ecological impact has been to increase the permanently inundated area, while dramatically reducing the seasonally inundated or floodplain area. This has reduced the available dry season floodplain grazing and fish production in the area. A quarter million cattle graze on the Kafue Flats. A combination of hydrological changes associated with river regulation and overfishing has caused a serious decline in the fish resources of the Kafue. In the 1970s, fishers from elsewhere in Zambia moved into the area and annual yields of 7,700 tonnes were caught. Indiscriminate fishing activities together with the reduced floodplain area and hydrological changes adversely affected the spawning, growth and survival of fish stocks especially bream. As a result of declining fisheries, the fishers have since moved away from the area. In 1977, there were 2,634 fishers in the Kafue Flats and by 1984 only 1,157 remained, the rest moving on to Lake Kariba.

Overexploitation of wildlife resources

In addition to high grazing and fishing pressure, other wetland resources such as game and building materials are subject to increasing human pressure as settlement increases. Near the National Park and Game Management Areas there are conflicts with wildlife as well as conflicts with park authorities regarding illegal hunting and utilisation of wetland resources.

Source: Bethune, S., "Water and Wetlands in the Zambezi Basin". For SARDC, 1999

areas, exerting immense pressure on the wetland resources. In the Zambezi Floodplains, the substantial rapid increase in human population has negatively impacted on the habitat for waterfowl among other impacts. People are settled on river levées, along oxbow lakes and lagoons, and on termite mounds.[37]

Pollution and the drying of floodplains have also caused habitat damage and species loss, especially in the Caprivi. Since the beginning of the 1990s, the Kwando River has been experiencing reduced water flows, resulting in wetland loss and reduction in wetland-dependent species. Overfishing in the same area has also led to the depletion of some fish species. Overexploitation of riverine vegetation such as reeds and grasses is threatening wetland ecosystems, ultimately leading to habitat loss.

Box 3.5: Threats to the Zambezi Delta

There is evidence that water regulation from Cahora Bassa has substantially reduced the Zambezi Delta wetland area and wetland productivity in Mozambique and even offshore shrimp catches. Salt-water intrusion is a serious problem in the delta on account of reduced freshwater flows and the lack of substantial seasonal floods. New infrastructure development proposed downstream of Cahora Bassa will worsen the already declining ecological situation in the delta.

The cumulative impact of Kariba and Cahora Bassa dams on the floodplains of the Zambezi Delta are a cause for concern and researchers are currently investigating the feasibility of ecological flood releases to rehabilitate the Zambezi Delta. At a workshop held on the sustainable use of the Cohora Bassa Dam and the Zambezi Valley in Songo, Mozambique in 1997, the participants agreed that the controlled releases from Cahora Bassa should simulate the natural seasonal and annual flow regime of the Zambezi River. Overhunting in the delta, the development of sugar irrigation schemes and harvesting of the mangroves are some of the threats to the Zambezi Delta.

Photo: M Chenje

Pollution of wetlands is a threat to species and has led to massive fish deaths in some water bodies such as Lake Chivero in Zimbabwe.

A comparison of annual flows in the Zambezi River measured at Victoria Falls since 1907 shows four distinct phases, low flows from 1907–23, with an average flow of 756 cu m per second, higher flows from 1924–45 with an average of 941cu m per second and very high flows from 1946-80 with an average rate of 1,392 cu m per second and much reduced flows since then, with an average flow rate of 750 cu m per second.[38]

As human population densities in the Zambezi Basin increase, the available per capita supply decreases, forcing more and more people to move into floodplains to settle. The most striking example of this is in the Lower Shire River where even the smallest island is now cultivated and conflicts with wild animals such as hippos and crocodiles are a serious problem. Due to high population and livestock concentrations around wetlands, there is widespread human encroachment of wetlands, leading to habitat loss.

Most of the available information on wetland productivity within the Zambezi Basin is based on fish production. Throughout the basin, fisheries are under threat. The main threats to fisheries are hydrological changes either natural or due to large dam construction and overfishing. This has been most serious in Lake Malombe where catches have declined steadily since 1990 and in the Lower Shire where catches have shown a dramatic decrease since 1992.[39]

Other threats include flow fluctuations and regulation caused by hydropower schemes and the resultant impact on floodplains downstream such as in the Kafue, below Kariba, in the Lower Shire and below Cahora Bassa. Hydrological changes in the Zambezi Basin have affected the ecological character of many wetlands. The environmental impacts of these dams on the ecology and water needs of the downstream communities especially in Mozambique have not been fully assessed.

Water quality degradation

The three main threats to water quality are siltation due to deforestation and resultant erosion; intensive irrigation; and pollution from cities and industries. Large areas of the Zambezi Basin are subjected to erosion as more land is cleared and existing arable land is poorly managed. The silt finds its way into the rivers and wetlands. In Angola, slash-and-burn agriculture, poor soils and fuelwood harvesting contribute to erosion, especially near towns such as Luena, Cazombo and Lumbala.[40]

The areas such as Sucoma and Dwangwa in Malawi and Nakambala in Zambia, which have been identified as threatened by high environmental impacts from agricultural activities in the basin are associated with sugar estates.[41] The main threats arise from pollutants such as fertilisers, herbicides and insecticides, which wash into water systems. Similarly, industrial pollution impacts originate from the main mining areas such as the Copperbelt in Zambia and Hwange in Zimbabwe, around industrial towns and large cities such as Harare, Lusaka and Tete. Hotspots for water quality degradation are the Manyame and Kafue sub-catchments where effluent from urbanisation and mining originate.

Pollution from industrial and domestic sewage, has in recent years been increasing due to growing urban centres along the Zambezi River, such as Mongu, Senanga, Katima Mulilo, Kasane-Kazungula, Sesheke, Livingstone, Victoria Falls, Kariba, Siavonga, Chirundu and Tete. Most of these urban centres including cities such as Blantyre, Bulawayo, Harare and Lusaka and the mushrooming tourist facilities along the river, discharge effluents into the river and its tributaries.

Poor sanitation in densely settled areas such as the Zambezi Floodplains, East Caprivi, Chobe Enclave, Lower Shire and Kafue Flats has created problems of human waste disposal, a major health risk for the people.

Invasion of alien species

The invasion of aquatic weeds is a major threat to the water resources and wetlands of the basin. Alien aquatic plants such as water hyacinth and Kariba weed are a serious menace to the Zambezi Basin wetlands. Water hyacinth is identified as a serious environmental problem in Kariba,[42] Kafue, Lake Chivero and other water bodies. Problems caused by floating aquatic weeds include access problems for fishers as the plants get entangled in fishing nets. Where plant growth is particularly dense, the weed interferes with boating and may even prevent sunlight reaching submerged macrophytes. There is an ongoing controversy regarding whether or not evaporative losses are increased by transpiration from these weeds as on the one hand they shade the water surface reducing evaporation from open water and on the other, they transpire.

Photos: M Chenje

Water hyacinth has become one of the major threats to the Zambezi Basin wetlands.

Opportunity cost of ecosystem change

Although work has started in the Zambezi Basin under the IUCN Zambezi Basin Wetlands Conservation and Resource Utilisation Project (ZBWCRUP), there appears to be serious gaps on the opportunity cost of ecosystem change. However, the opportunity cost of ecosystem change can be illustrated where the construction of large dams for hydropower has caused ecosystem changes. The first change is the alteration of a flowing river section into a reservoir or lake. This can have both positive and negative impacts, on the positive side is the assured availability of water for a constant supply of hydropower and the development of a lacustrine fishery as happened on Lake Kariba.

On the negative side, is the irreversible loss of riparian land inundated by the impoundment and the changes in biota from those adapted to seasonal flooding to those that can survive quiet, deep waters. Downstream of the dam, the impacts tend to be largely negative, the more steady outflows from hydropower plants tend to eliminate seasonal flooding so essential to wetland functioning and reduce the overall size of seasonal wetland areas. This in turn reduces overall productivity with economic impacts on the fisheries that these wetlands support. Other less quantifiable "costs" include impacts on subsistence recession floodplain gardening and the availability of reeds, sedges and wood for use as building materials.

Large dams negatively impact the nearshore marine environment where declining shrimp catches in Mozambique are attributed to reduced flows and less nutrients reaching the sea. It is necessary to weigh the economic benefits of development schemes against the opportunity costs lost to the environment and people.

WATER RESOURCES MANAGEMENT

Water resources management in the basin can be related to the evolution of central administrative structures within the countries. Prior to the entry of colonial administration, people exerted less pressure on the water resources due to the small population. In times of water scarcity following droughts, the people migrated to areas where water was available.

The birth of institutional structures responsible for various sectors arose with the introduction of colonial administration that was characterised by the development of urban centres. These centres offered various services to people including the development of industrial activity apart from being the seat of administration. Institutions were put in place responsible for planning, development, management and maintenance of water supply infrastructure. The post-colonial period has continued with the adopted systems from colonial administrations with emphasis on enforcement of these policies and strategies for water resources management.

Human resources

The development and management of water resources and policy enforcement require adequate resource backup both in terms of financial and human resources. In a situation where adequate infrastructure is available to supply water to consumers, an equally efficient and effective labour force with relative expertise, experience and skills should also be developed. The labour force should keep pace with the growth of demand for the service both for the urban and the rural areas.

Training should involve planning, decision-making and appropriate pricing levels to improve the performance of the water utility. Innovations, technology choice, research and development are all relevant and important areas in which human resources need to be developed.

Institutions

Water resources management at the national level is the responsibility of government ministries and departments. These institutions are responsible for the planning, development, supply and maintenance of water and water infrastructure, including the development of appropriate policies and strategies.

Planning for water services encompasses management at the basin and sub-basin level, the development of water supply infrastructure, water resources monitoring and provision of water to both urban and rural consumers.

While a number of these functions remain the task of the public sector in most of the basin states, some of the responsibilities have been delegated to the private sector in countries like Malawi. Policy development and enforcement in Malawi remains the responsibility of government while the functions of water resources monitoring, management and supply are under the regional and urban water boards. Regionally and nationally, water conservation needs to be part of water resources management policies and until a stronger legislative backbone supports the concept, water conservation successes will be small and mostly in drought prone areas. However, even with the establishment of effective policies, success will be elusive unless there is a commitment to enforcement. Policies or regulations not only have to be established, but supported from the highest levels and with realistic budgets allocated for their enforcement, including the development of human resources.

Once the need for water conservation is accepted and policy commitments have been made, the financial resources must be made available and people trained or educated for significant progress to occur. Water conservation measures require initial capital investments, as well as operation and maintenance (O&M) budgets depending on the technology used.

Photos: M Chenje

The need for water conservation is still to be widely accepted at all levels of society. Oblivious of the need to conserve water a gardener (top photo) uses a hose to "sweep" debris.

Legislation

Legislation provides the enabling legal environment to assure that future (and hopefully present) consumers maximise their reuse potential. Appropriate legislation exists in all of the basin countries with regard to water resources development and management. Throughout the basin, urban centres, particularly the major cities, are experiencing phenomenal growth in population which, in a majority of cases, is not matched with available infrastructure. Such a disparity is often blamed for the limited access to safe water and the illegal water connections, which make it difficult to account for all the water supplied.[43]

Policies in the water sector within the basin cater for water supply, sanitation, ownership, as well as for the environment. For instance, Article 95 of the Namibian Constitution, ratified in 1990, states that, "the state shall aim at the maintenance of ecosystems, essential ecological processes and the biological diversity of Namibia, as well as the utilisation of living natural resources on a sustainable basis for the benefit of all Namibians, both present and future."

Article 100 of the same Constitution states that, "land, water and natural resources shall belong to the State if they are not otherwise lawfully owned."[44] For legislation to be effective in the management of water resources, it is important that there be clear objectives, a sustainable long-term sector policy that defines rights, rules and regulations and adequate enforcement of the rules and regulations.

Legislation can mandate the type of low-water use fixtures that can be placed in all new construction. For existing fixtures, repair and replacement programmes can be successful tools of water conservation and management. With a vigorous campaign, overall in-house water demand could be reduced by as much as 50 percent through use of suitable water saving fixtures. Water resources management requires a measure of education, awareness and full participation by all to help reduce the amount of water used, as opposed to water required. Some important areas that legislation needs to address are:

- Banning the use of baths in new hotels;
- Updating municipal water supply and drainage regulations;
- Modifying flush toilets and showerheads;
- Water wastage;
- Controlling water pollution;
- Controlling groundwater abstraction by private owners;
- Monitoring and compliance.

Water resources monitoring

Water resources monitoring is an important element in water resources management. A large network of hydrometric and meteorological stations exists within the basin countries for monitoring rainfall, river flows, evaporation, reservoir and lake levels and other parameters.

Apart from monitoring the resource, legislation needs to consider the monitoring of usage in homes, industry, agriculture and other sectors. It is also important that water use by the environment is taken into consideration.

Policy enforcement

The existence of policies both at the national and regional levels is inadequate in itself. Enforcement of policy allows for the users of water to comply with the existing rules and regulations that promote better resource management and use. The drilling of private boreholes and construction of dams by private enterprises and individuals without obtaining a right to do so from the policy implementers, for instance, hampers proper water accounting. Wastage through leaks, faulty equipment and other means can be addressed through constant enforcement of the existing policy where the user pays for the water. Policy initiatives could also consider incentives for industry and domestic consumers by reducing tax for categories that comply with better water use practices.

Water pricing

Water pricing in the basin states varies with consumer categories. In Zimbabwe for instance, water is currently under priced. The main pricing mechanism is the national "blend" price whereby the historical cost of the several government-constructed dams and associated works was averaged to calculate this price. The problem with the current system is the use of national blend prices. This measure barely adheres to the "user pays principle". The system also involves extensive cross-subsidisation, with some consumers paying more than the costs, while others pay much less. As part of the current reforms in the water sector in Zimbabwe, the government has opted for the "user pays principle" as a pricing mechanism to be implemented in the future.[45]

While tariffs for water deter reckless use of the resource, their administration sometimes fails to encourage consumers to conserve water. For example, bulk meter readings (although easier to manage) do not instil a sense of responsibility in consumers. Legislation governing use of water is also available, although poorly enforced. On a positive note, water recycling, as a management strategy, is common to many cities in the basin states. Much of the recycled water is used for industrial cooling; irrigation of parks, sports fields and construction.

Before any attempt can be made to "manage" a river or wetland system, it is important to recognise that natural phenomena and development activities alter the state of the environment. Agreeing on the "desired state" of particular wetlands, is an essential first step in setting goals that enable proactive basin management. These "desired states", for example for wetlands, should be identified in terms of water quality, productivity, biodiversity, and floodplain area.[46]

Several issues and problems affect the effective management of the basin's water resources. These include pollution, eutrophication, siltation induced by poor land uses, changed flow regimes, aquatic weeds, increased competition for water, potential degradation and loss of wetland ecosystems, poor planning and sectoral approaches.

A number of riparian states have embarked on national water reform programmes whose main objectives are to

Photo: SARDC

Enforcement of policy allows for the users of water to comply with regulations that promote better resource use.

improve water resources management through institutional and legislative reviews. Malawi, Namibia, Zambia and Zimbabwe are all involved in water resources reform programmes whose main objectives are to improve the management of water and ensure equitable distribution. Zimbabwe has already enacted the 1998 Water Act which, among other things, provides for the establishment of catchment committees. In addition to national programmes, there are a number of basin-wide programmes such as the ZACPLAN, whose main objective is to effectively manage the water resources of the Zambezi Basin.

Public participation, awareness and empowerment

Throughout the Zambezi Basin, there is a move towards greater involvement of communities in the management and conservation of water and wetland resources. In Malawi and Namibia, successful water management arrangements have been put in place through the establishment of water point committees while in Mozambique, there are irrigation cooperatives.

Community Based Natural Resource Management (CBNRM) is gaining momentum in most basin countries. Community controlled hunting areas managed by Village Trust Committees in Botswana and community conservancies such as Salambala in Namibia, Administrative Management Design Programme for Game Management (ADMADE) in Zambia and the Communal Areas Management Programme For Indigenous Resources (CAMPFIRE) project in Zimbabwe have gained political acceptance and have placed resource ownership back in the hands of the communities most dependent on natural resources.

Public participation, awareness and empowerment is quite evident with the wetland programmes and projects taking place in the basin. Under the Zambezi Basin Wetlands and Resource Utilisation project facilitated by the World Conservation Union (IUCN ROSA), the local communities are involved in the design and implementation of the programmes. Communities manage projects such as the Chobe Enclave Conservation Trust in Botswana and the Tchuma Tchato in Mozambique, the Integrated Rural Development and Nature Conservation programme in East Caprivi in Namibia, and other programmes in the Kafue Flats of Zambia.

There is an increasing trend in public participation in the water resource issues of the riparian states. As prescribed in the Dublin Principles, stakeholder participation is an important prerequisite to the success of many water programmes. Many countries in the basin have heeded this call and are seriously involving stakeholders in water resources planning and development. The policy to decentralise water management to the catchment level means that communities have to be made aware of the importance of water and empowered to carry the work at that level.

Photo: M Chenje

Public participation and empowerment are important in water and wetland resources management.

Map 3.3: Ecologically sensitive riparian and river-dependent habitats in the Zambezi Basin 2005

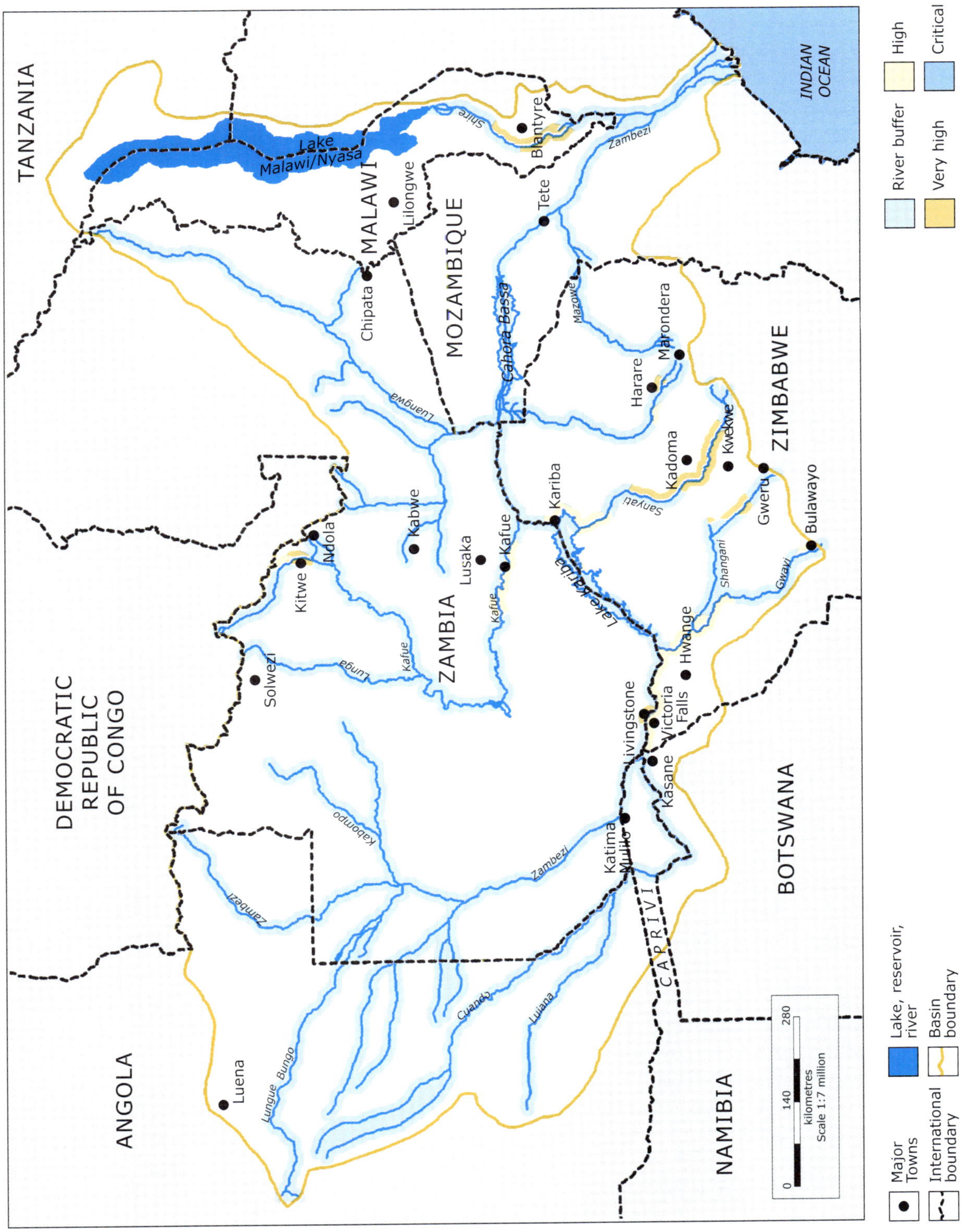

Source: Zambezi River Basin Atlas, SADC/DENCONSULT/ZRA, Lusaka, 1998

Regional cooperation in water resources management

Within the Zambezi Basin, the water and wetland resources are shared in several ways. Within a country, water is shared by upstream and downstream users. For example, the City of Blantyre draws its water from the Shire River and almost from the same point of abstraction, the Electricity Supply Commission of Malawi (ESCOM) uses the remaining flow to generate power. The water then flows downstream to the Lower Shire where it is required for the irrigation of sugar and as compensation flow for the wetlands and smallholder wetland cultivation. All these users share the water resource along the length of the river and the activities of an upstream user affect those downstream. For example, excessive withdrawals at Walkers Ferry, especially during years of low flows, would impact on the irrigation requirements at the Sugar Corporation of Malawi estate at Nchalo and the fisheries industry of the Elephant Marsh downstream.[47]

The Shire River example also illustrates sharing by different users, urban and rural, different types of uses, consumptive (irrigation) and non-consumptive (hydropower generation), as well as sharing between people and environment.

The Zambezi system offers an excellent example in southern Africa of sharing the water resources of regional importance. To foster a spirit of good neighbourliness and to avoid international disputes, it has become necessary to consider the joint management of the resources of the Zambezi Basin. This is why the Zambezi System Action Plan (ZACPLAN) was endorsed by the basin states in 1987.

A number of mechanisms for managing shared transboundary resources have been developed and these include international conventions and agreements, bilateral agreements, regional groupings such as the SADC itself, river basin agreements, authorities and commissions. Chapter 12 on Regional Cooperation, deals with this issue in greater detail.

Bilateral and multilateral instruments and programmes

Bilateral or multilateral instruments and/or programmes are also used by a number of countries to effectively manage water resources.

The IUCN/SADC IFFW Regional Wetlands Conservation Programme

The Inland Fisheries, Foresty and Wildlife sector (IFFW) Regional Wetlands Conservation Programme was initiated in 1990 to promote effective conservation and wise use of southern African wetlands. Among its objectives are to:

- promote public awareness;
- facilitate wetland inventories;
- establish training needs and help improve technical expertise;
- improve community participation in wetland management; and
- give attention to the economic values of wetlands and promote sustainable development in wetland areas.

International Crane Foundation

The International Crane Foundation (ICF) is responsible for the welfare of cranes throughout the world and is involved both in captive breeding of cranes and surveys of the rarer crane species and their habitats. It is also active in wetland conservation in southern Africa. In recent years, important studies have focused on the Zambezi Delta in Mozambique. Some of the results of this work were presented at a workshop on the sustainable use of Cahora Bassa Dam and the Zambezi Valley held in October 1997 at Songo.[48]

Wetlands International Waterfowl Research (IWRB)

The African Waterfowl Census has been conducted in Africa since 1990 and covers some 30 African countries, including all the Zambezi Basin countries except Angola. Bi-annual surveys are conducted to determine both the numbers and species of birds in each of the main wetlands in each country. The census is coordinated by Wetlands International based in Wageningen, in The Netherlands.[49]

World Wide Fund for Nature (WWF)

The ongoing WWF Zambia wetlands project initiated in 1986 focuses on community participation in wildlife management in the Kafue and Bangweulu wetlands and has improved standards of living in the communities dependent on these wetlands.[50]

Through the LIFE project, WWF is active in the eastern Caprivi, in Namibia, where it supports the Integrated Rural Development and Nature Conservation (IRDNC) project to promote the establishment of community wildlife conservancies.[51] The first community conservancy at Salambala was established in August 1998.

LINKAGES TO OTHER CHAPTERS

Chapter 1: Regional Overview: People and Environment
Due to rising human population and growth in industry within the basin, the demand for water is expected to rise. Increased pressures on water resources and wetlands are anticipated as people encroach on aquatic ecosystems for resources to support them.

Chapter 2: Physical Features and Climate
Physical features of the Zambezi Basin are important characteristics in that they influence the climate and drainage of the basin. The water resources and wetlands of the Zambezi River Basin are wholly dependent on the region's climate. Changes in climate, therefore would have direct impact on the water resources availability and wetland area.

Chapter 4: Biological Resources and Diversity
The biodiversity of the wetlands is an important regional asset as it promotes the tourism industry apart from offering other benefits such as food, building material and income. Their sustainability is at the mercy of the way the wetland resources are used and managed.

Chapter 5: Agriculture
Agriculture, especially irrigation, takes up a large share of water consumed in the basin states. While it forms the main activity of the majority of the people of the basin, the water resources and the wetlands of the basin could be negatively impacted upon through population pressure.

Chapter 6: Industry
Industry's impact on the water resources and wetlands of the basin is through pollution as waste is discharged into watercourses, causing the resource to be of poor quality.

Chapter 7: Energy
Economic performance of the basin states depends on having adequate and reliable energy, and hydropower offers the greatest opportunity to realising this goal. While there are negative impacts on the environment as a result of hydropower schemes, countries should strive to minimise these through environmental impact studies prior to the development of any scheme and offer remedies where any deficiencies occur.

Chapter 8: Tourism
The tourism industry depends in part on the water and wetland resources of the basin. Mismanagement of the water and wetlands will lead to drastic negative impacts on the industry.

Chapter 9: Pollution
Pollution from mining, agriculture, industry and other sources is threatening the quality of water and the wetlands of the basin. Enforcement of regulations on pollution control must be strengthened.

Chapter 10: Poverty
Poverty may be seen as the cause and the result of environmental degradation. To secure food, habitat and income in a basin of high population growth rates and diminishing per capita land area, people continue to invade the wetland resources for support.

Chapter 11: Gender and Women's Roles
Gender concerns and perspectives should be integrated in water resources and wetlands management within the basin and women must be part of the planning and development process of the resources.

Chapter 12: Environmental Management and Regional Cooperation
The availability and abundance of water and wetland resources provides assurance for fewer conflicts within countries and the basin. Planning for water is a must for the basin states as population increases.

Chapter 13: Trends and Scenarios
As water and wetland resources are directly dependent on the basin's climate, there are equally two scenarios for the basin: either the same amount of water resources and large expanses of wetlands or less water and the drying-up of wetlands. Current projections to the future though, reveal that there is more likely to be water stress in some countries of the basin in the next 20-25 years. Planning for the future water resources should, therefore, be more urgent now than ever before.

END NOTES

1 Pallet, J. (ed.) *Sharing Water in Southern Africa*, DRFN, Windhoek, Namibia, 1997
2 Ohlsson, L., *Water and Security in Southern Africa*, Sida Publications on Water Resources No.1, Department of Natural Resources and the Environment, S-105 25 Stockholm, Sweden, 1995
3 Matiza-Chiuta, T., "Water and Wetland Resources", For SARDC, April 1999
4 ibid
5 Denconsult., "Water Consumption and Effluent in the Food and Agriculture Sector including Crops, Livestock and Fisheries", ZACPLAN Sector Studies, First Draft Report, SADC/Zambezi River Authority, Lusaka, Zambia 1998
6 Government of Malawi, "Country Situation Report-Water Resources", Lilongwe, December 1997
7 Government of Zambia, "Country Situation Report-Water Resources", Lusaka, December 1997
8 Sekwale, M., "Botswana National Water Master Plan", in: Matiza, T., Crafter, S. and P. Dale (eds.) *Proceedings of a Workshop held at Kasane Botswana 28 April to 2 May 1993*, IUCN, Gland, Switzerland, 1995
9 Mpande, R. L., "The Proposed Matabeleland Zambezi Water Project", in: Matiza, T., Crafter, S. and P. Dale (eds.) *Proceedings of a Workshop held at Kasane Botswana 28 April to 2 May 1993*, IUCN, Gland, Switzerland, 1995
10 ibid
11 Basson, M.S., "South African Water Transfer Schemes and their Impact on the Southern African Region", in: Matiza, T., Crafter, S. and P. Dale (eds.) *Proceedings of a Workshop held at Kasane Botswana 28 April to 2 May 1993*, IUCN, Gland, Switzerland, 1995
12 van der Riet, cited by Basson, M.S., "South African Water Transfer Schemes and their Impact on the Southern African Region", in: Matiza, T., Crafter, S. and P. Dale (eds.) *Proceedings of a Workshop held at Kasane Botswana 28 April to 2 May 1993*, IUCN, Gland, Switzerland, 1995
13 Heyns, P., "Irrigation Needs in the Eastern Caprivi Region of Namibia", in: Matiza, T., Crafter, S. and P. Dale (eds.) *Proceedings of a Workshop held at Kasane Botswana 28 April to 2 May 1993*, IUCN, Gland, Switzerland, 1995
14 Denconsult., *ZACPLAN Sector Studies, Introductory Volume, Final Report*, SADC/Zambezi River Authority, Lusaka, Zambia, 1998.
15 ibid
16 Lindley, D., "Wetlands and our Survival", unpublished paper presented at the Globe Regional Conference on the Environment "Partnership for Sustainability", held at the South African Parliament, Cape Town 4-5 September 1998
17 op cit 3
18 op cit 3
19 op cit 14
20 Bethune, S., "Water Resources". in: Heyns et al (eds.)., *Namibia's Water. A Decision Maker's Guide*, Department of Water Affairs and DRFN, Windhoek, Namibia 1998
21 Chabwela, H. (ed.) "Status of Wetlands of Zambia, Management and Conservation Issues", unpublished report produced for the Environmental Council of Zambia, Lusaka, Zambia 1994
22 Masundire, H. M. and Matiza, T., "Some Environmental Aspects of Developments in the Zambezi Basin Ecosystems", in: Matiza, T., Crafter, S. and P. Dale (eds.) *Proceedings of a Workshop held at Kasane Botswana 28 April to 2 May 1993*, IUCN, Gland, Switzerland, 1995
23 Denconsult., "Water Consumption and Effluent in the food and Agriculture Sector including Crops, Livestock and Fisheries", *ZACPLAN Sector Studies, First Draft Report*, SADC/Zambezi River Authority, Lusaka, Zambia 1998
24 Bethune, S., "Water and Wetland Resources", For SARDC, April 1999
25 Curtis, B., Roberts, K. S., Griffin, M., Bethune, S., Hay, C. and H. H. Kolbelg., "Species Richness and Conservation of Namibian Freshwater Macro-invertebrates, Fish and Amphibians", *Biodiversity and Conservation No. 7*, 1998, pp447-466
26 Van der Waal., "Aspects of the Fishery of the Eastern Caprivi, Namibia", *Madoqua* 17, 1990 pp 1-16
27 Day, J., "The Status of Freshwater Resources in Namibia", *Research Discussion Paper No. 22*, Directorate of Environmental Affairs, Ministry of Environment and Tourism, Windhoek, Namibia, 1997
28 Koekemoer, J. H., "Internal Report on the Eastern Caprivi Surveys, 1993, 1994, 1995 and 1997, Chapter 9: Socio-economics", Ministry of Fisheries and Marine Resources, Inland Fisheries, Windhoek,Namibia, 1998 (Unpublished)
29 op cit 3
30 op cit 3
31 op cit 14
32 Laisi, E., Personal observation, 1976-1995
33 Chabwela, H. (ed.) "Status of Wetlands of Zambia, Management and Conservation Issues", unpublished report produced for the Environmental Council of Zambia, Lusaka, Zambia 1994
34 op cit 14
35 Nyakunu, E. P., "Tourism and its Impacts on the Environment: Lessons from the Zambezi River Basin", Paper presented at the Workshop on the State of the Environment Reporting Programme in the Zambezi Basin, held from 20-24 July, Mazvikadeyi, Chinhoyi, Zimbabwe 1998
36 op cit 24
37 Mendlesohn, J. and C. Roberts., *An Environmental Profile and Atlas of Caprivi*, Directorate of Environmental Affairs, Namibia, Gamsberg Macmillan Publishers, Windhoek, Namibia, 1997
38 Mendlesohn, J. and C. Roberts., *An Environmental Profile and Atlas of Caprivi*, Directorate of Environmental Affairs, Namibia, Gamsberg Macmillan Publishers, Windhoek, Namibia, 1997
39 Mapila, S. A., "The Fisheries Industry: Lessons from the Zambezi River Basin", Paper presented at the Workshop on the State of the Environment Reporting Programme in the Zambezi Basin, held from 20-24 July, Mazvikadeyi, Chinhoyi, Zimbabwe 1998
40 op cit 24
41 op cit 14
42 Chenje, J., (ed.) "State of the Environment Reporting Programme in the Zambezi Basin: Inception Mission Report", IMERCSA/IUCN/Sida, Harare, November 1998
43 Macy, P., *Urban Water Demand Management in Southern Africa: The Conservation Potential*, Sida, Harare, 1999
44 Government of Namibia, "Country Situation Report-Water Resources", Windhoek, May 1998
45 Macy, P., *Urban Water Demand Management in Southern Africa: The Conservation Potential*, Sida, Harare, 1999
46 Rogers, K. H. and R. Bestbier., *Development of a Protocol for the Definition of the Desired State of Riverine Systems in southern Africa*, Department of Environmental Affairs and Tourism, Pretoria, South Africa, 1997
47 Laisi, E., Personal experience

48 Beilfuss, R. D., and C. M. Bento., "Impacts of Hydrological Changes on the Marromeu Complex of the Zambezi Delta, with Special Attention to the Avifauna", Paper presented at the Workshop on the Sustainable Use of the Cahora Bassa Dam and the Zambezi Valley, 29 September to 2 October, 1997,Songo, Mozambique, 1997

49 Dodman, T., de Vaan, C., Hubert, E. and C. Nivet., "African Waterfowl Census 1997", *Wetlands International*, Wageneingen, The Netherlands, 1997

50 Matiza-Chiuta, T., "Raising Awareness for the Sustainable Utilisation of Wetlands in Southern Africa", in: Van der Walt, M. M. and G. I. Cowan (eds.) *Proceedings of the Southern African Sub-regional RAMSAR Meeting held in Pretoria, South Africa from 2- 6 February 1998*, South African Wetlands Conservation Programme, Department of Environmental Affairs and Tourism, Pretoria, South Africa, 1998

51 Jon Barnes., Personal Communication

REFERENCES

Ashley C., Muller, H. and M. Harris., "Population dynamics, the environment, and demand for water and energy in Namibia", Research Discussion Paper 7, Directorate of Environment Affairs, Ministry of Environment and Tourism, Windhoek, 1995

Barnard P, Bethune, S and H. Kolberg., "Biodiversity of Terrestrial and Freshwater Habitats" in: Barnard, P. (ed.)., *Biological Diversity in Namibia: A Country Study*, National Namibian Biodiversity Task Force, Windhoek, Namibia, 1998

Basson, M. S., "South African Water Transfer Schemes and their Impact on the Southern African Region" in: Matiza, T., Crafter, S. and Dale, P. (eds.). *Proceedings of a workshop held at Kasane, Botswana 28 April to 2 May 1993*, IUCN, Gland, Switzerland, 1995

Chenje, M. and P. Johnson (eds.)., *State of the Environment in Southern Africa*, SADC, IUCN, SARDC., Maseru/Harare 1994

Chenje, M. and P. Johnson. (eds)., *Water in Southern Africa*, SADC,IUCN,SARDC. Maseru/Harare 1996

Department of Water Affairs and Forestry., "White Paper on a National Water Policy for South Africa", DWAF, Pretoria, South Africa, 1997

Durham, D.S. "The Use of Zambezi Water for Urban Supplies in Zimbabwe: Reflection on Current Demands", in: Matiza, T., Crafter, S. and Dale, P. (eds). *Proceedings of a workshop held at Kasane, Botswana 28 April to 2 May 1993.* IUCN, Gland, Switzerland, 1995

Economic Commission for Africa (ECA)., "Inter-country Cooperation for Integrated Development of Water Resources of the Zambezi River Basin", Unpublished Report ECA/RCID/48/97 Addis Ababa, Ethiopia, 1997

Gammelsrod, T., "Variation in Shrimp Abundance on the Sofala Bank, Mozambique and its relation to the Zambezi river runoff", *Estuarine, Coastal and Shelf Science,* 1992a

Gammelsrod, T., "Improving Shrimp Production by Zambezi River Regulation", *Ambio,* 21:145-147 1992b

Gibson, J., "Proposal for a Southern African Wildlife Sanctuary in the Wetlands associated with the Source of the Zambezi", in: Matiza, T., Crafter, S. and P. Dale (eds.). *Proceedings of a workshop held at Kasane, Botswana 28 April to 2 May 1993*, IUCN, Gland, Switzerland, 1995

Grobler, M. and Ferreira, J. "The Dying of Lake Liambezi", *Custos,* 1990, 19:39-47

Hay, C.J., van Zyl, B.J., van der Bank, F.H., Ferreira, J.T. and G.J. Steyn. "The distribution of freshwater fish in Namibia", *Cimbebasia.* (In press)

Howard G.W., "Floodplains: Utilisation and the Need for Management", in: Matiza, T. and Chabwela, H.N. (eds.) *Wetlands Conservation Conference for Southern Africa, Proceedings of the SADCC Wetlands Conference for Southern Africa, held in Gaborone, Botswana 3-5 June 1991*, IUCN, Gland, Switzerland, 1992

Kasimona, V.N. and J.J. Makwaya., "Present Planning in Zambia for the future use of Zambezi River waters", in: Matiza, T., Crafter, S. and Dale, P. (eds.) *Proceedings of a Workshop held at Kasane, Botswana, 28 April to 2 May 1993*, IUCN, Gland, Switzerland, 1995

Kimberley, M., "The Next Step-A Zambezi Basin Convention?", in: Matiza, T., Crafter, S. and Dale, P. (eds.) *Proceedings of a workshop held at Kasane, Botswana 28 April to 2 May 1993*, IUCN, Gland, Switzerland, 1995

Massinga, A.V., "Dam Developments and their Environmental Effects", in: Matiza, T. and Chabwela, H.N. (eds.) *Wetlands Conservation Conference for Southern Africa, Proceedings of the SADCC Wetlands Conference for Southern Africa, held in Gaborone, Botswana, 3 to 5 June 1991*, IUCN, Gland, Switzerland, 1992

Matiza-Chiuta, T., "Raising Awareness for the Sustainable Utilisation of Wetlands in Southern Africa", in: Van der Walt, M.M. and Cowan, G.I. (eds.) *Proceedings of the Southern African Sub-Regional RAMSAR Meeting held in Pretoria, South Africa from 2 to 6 February 1998,* South African Wetlands Conservation Programme, Department of Environmental Affairs and Tourism, Pretoria, 1998

Matiza, T., Crafter, S. and Dale, P. (eds.) *Water Resource Use in the Zambezi Basin, Proceedings of a Workshop held at Kasane, Botswana, 28 April to 2 May 1993*, IUCN, Gland, Switzerland, 1995

Matiza, T. and H.N. Chabwela (eds.) *Wetlands Conservation Conference for Southern Africa, Proceedings of the Southern African Development Coordination Conference held in Gaborone, Botswana, 3 to 5 June 1991*. IUCN, Gland, Switzerland, 1992

Mukosa, C., Pitchen, C. and C. Cadou, "Recent Hydrological Trends in the Upper Zambezi and Kafue Basins", in: Matiza, T., Crafter, S. and P. Dale, (eds.) *Proceedings of a Workshop held at Kasane, Botswana 28 April to 2 May 1993*, IUCN, Gland, Switzerland, 1995

Mumeka, A., "The Water Resources of the Kafue Flats and Bangweulu Basin", in: Jeffery, R.C.V., Chabwela, H. N., Howard, G. and Dugan, P.J. (1992), *Managing the Wetlands of Kafue Flats and Bangweulu Basin, Proceedings of the WWF-Zambia Wetlands Project Workshop held at Kafue National Park, Zambia 5 to 7 November, 1986*, IUCN, Gland, Switzerland, 1992

Pinay, G., "Hydrobiological Assessment of the Zambezi River System: A Review", Working Paper WP-88-089, International Institute for Applied Systems Analysis, Laxenburg, Austria, 1988

Rogers, K. H. and R. Bestbier., "Development of a Protocol for the Definition of the Desired State of Riverine Systems in South Africa", Department of Environment Affairs and Tourism, Pretoria, 1997

Rukira, L., "Country Position Paper-Namibia", Paper presented at the ZACPRO 6 Workshop on the Integrated Zambezi River Basin Action Plan in Livingstone, Zambia, Hydrology Division, Department of Water Affairs, Namibia, Unpublished Report, 1994

Schlettwein, C. and S. Bethune., "Aquatic Weeds and their Management in Southern Africa: Biological Control of *Salvinia molesta* in the eastern Caprivi", in: Matiza, T. and H. N. Chabwela (eds.) *Wetlands Conservation Conference for Southern Africa, Proceedings of the SADCC Wetlands Conference for Southern Africa, held in Gaborone, Botswana 3 to 5 June 1991*, IUCN, Gland, Switzerland, 1992

Schlettwein, C.H.G., Simmons, R.E., Macdonald, A. and H. J. W.Grobler., "Flora, Fauna and Conservation of East Caprivi Wetlands", in Simmons R.E., Brown, C.J. and M. Griffin (eds.) *The Status and Conservation of Wetlands in Namibia, Special Wetlands Edition, Madoqua,* 1991

Skelton, P.H., *A Complete Guide to the Freshwater Fishes of Southern Africa*, Southern Book Publishers, Halfway House, Pretoria, South Africa, 1993

Subramanian, S.P., "A Brief Review of the Status of the Fisheries of the Bangweulu Basin and the Kafue Flats", in: Jeffery, R.C.V., Chabwela, H. N., Howard, G. and P. J. Dugan. *Managing the Wetlands of Kafue Flats and Bangweulu Basin, Proceedings of the WWF-Zambia Wetlands Project Workshop held at Kafue National Park, Zambia, 5 to 7 November 1986*, IUCN, Gland, Switzerland, 1992

Tapfuma, V., "The Batoka Gorge Hydroelectric Project: Design and Potential Environmental Impacts", in: Matiza, T., Crafter, S. and P. Dale (eds.) *Proceedings of a workshop held at Kasane, Botswana 28 April to 2 May 1993*, IUCN, Gland, Switzerland, 1995

Tarr, P., "Environmental Assessment in Southern Africa: Is There a Need for a Regional Approach?", Unpublished paper presented at the Globe Regional Conference on the Environment "Partnerships for Sustainability" held at the Parliament of South Africa, Cape Town, from 4-5 September 1998

Taylor, E.D. and N.V. Clarke., "Does *Salvinia molesta* Still Pose a Threat to Biodiversity in Namibia's Eastern Caprivi Wetlands?", *Madoqua,* (in press)

Timberlake, J., "Biodiversity of the Zambezi River Basin", Paper presented at the SARDC Regional Workshop on the State of the Environment in the Zambezi Basin, Mazvikadeyi, Chinhoyi, Zimbabwe, 20 to 24 July 1998

Timberlake J., "Biodiversity of the Zambezi Wetlands: Review and Preliminary Assessment of Available Information", Phase I, Final Report, February 1998, IUCN Zambezi Basin Wetlands Conservation and Resource Utilisation Project, IUCN, Harare, Zimbabwe, 1998

Tvedten, I., Girvan, L.A., Maasdorp, M., Pomuti, A. and G van Rooy., "Freshwater Fisheries and Fish Management in Namibia: A Socio-economic Background Study", Social Sciences Research Division Report No. 12, Multi-disciplinary Research Centre, University of Namibia, Windhoek, 1994

Welcomme, R., *The Fisheries Ecology of African Floodplains*, CIFA Technical Paper 3, Rome, Italy, 1975

White, L., *Bridging the Zambezi: A Colonial Folly*, MacMillan Press, Hampshire, United Kingdom, 1993

Zulu, E.H., "Environmental Impact of Mining in the Zambezi Basin", Paper presented at a Regional Workshop on the State of the Environment in the Zambezi Basin, Mazvikadeyi, Chinhoyi, Zimbabwe, 20 to 24 July 1998

4

BIOLOGICAL RESOURCES AND DIVERSITY

Since time immemorial biological diversity (biodiversity) has been central to human survival. Food, many industrial materials and medicines are provided by plants, animals and microorganisms. Scientists believe that at least 80,000 of the worlds' plant species can be eaten.[1] In addition, plants provide people with useful material for construction and for processing into finished goods while animals supply meat and leather. Antibiotics like penicillin come from microorganisms. Natural vegetation helps ecosystems to protect water resources by regulating and stabilising run off, and acting as a buffer against floods and droughts. It also plays an important role in the breakdown and adsorption of pollution, recycles nutrients, and prevents landslides. Social benefits include both recreational and cultural values, and activities such as tourism depend heavily on healthy ecosystems rich in biodiversity.

Photo: IUCN

The birds of the basin are comparatively well-known. There are about 700 species of birds recorded of which only 15-20 are endemic to the basin.

Box 4.1: Defining biodiversity

Biodiversity describes the richness and variety of the natural world and is an abbreviated form of the words "biological diversity". It has been described by the 1992 Convention on Biological Diversity (CBD) as the variation between ecosystems and habitats; between different species; and the genetic variation existing within species.

Biodiversity describes the various forms of life on earth such as people, plants, animals, birds, fish, amphibians and reptiles. Such forms of life are supported by a complex interaction of prevailing elements, including the availability of water, soils, temperature regimes, topography and geology.

Species diversity refers to the number and variety of species that occur in a geographical region or ecosystem. The more species there are, the wider the range of natural resources available.

Genetic diversity refers to the variation of genes within species. Genetic diversity allows different animals and plants to flourish in different conditions. For example, elephants in Namibia are most adapted to desert conditions in that country than those in the Zambezi Valley. Different varieties resist different pests and diseases, thrive in different climates and soils and mature at different times.

Ecosystem diversity refers to the variations of ecosystems. Different ecosystems fulfil different important functions. Forests, for example, help prevent soil being washed away by rain and rivers, safeguard against flooding, absorb carbon dioxide and produce oxygen; while wetlands and the organisms they contain help purify water by trapping sediment, nutrients, and harmful bacteria.

Biodiversity is a measurable attribute that can be surveyed and quantitatively described. The units of measurement, however, are not tonnes per hectare but units of diversity such as numbers of species or processes; generally it is measured using species number. It is a biological measurement, not a measure of utilisation or of utility.

Chenje,M., Sola, L., and Paleczny, D. (eds.), *The State of Zimbabwe's Environment 1998*, Government of the Republic of Zimbabwe, Ministry of Mines, Environment and Tourism, Harare, Zimbabwe, 1998, p.273.

Industrial-scale agriculture and forestry, overexploitation, pollution, introduced species, dam construction, and wasteful consumption are some of the threats to biodiversity in the Zambezi Basin.

BIODIVERSITY IN THE BASIN

Many of the patterns seen today in the distribution of species, genes and ecosystems in the Zambezi Basin are a result of historical processes that have occurred over millions of years. An understanding of present-day biodiversity, therefore, is based on knowledge of changes in the climate, hydrology and geomorphology, which have occurred in the region over thousands of years, as well as the modifications to the environment caused by human activity. Many areas that today appear to be "natural" bear the marks of human habitation, crop cultivation, and resource harvesting.[2]

Biomes of the Zambezi Basin

In broad terms, there are four main biomes, or large ecological regions characterised by similar climate, vegetation and species, across the basin: Congolian, Zambezian, Montane and Coastal.

Congolian biome

The Congolian biome is the area associated with the headwaters of the Zambezi in northwestern Zambia and northeastern Angola with a moister and warmer climate than that of the rest of the plateau portion of the basin. Here the vegetation and species are a mix of those found in the forested Congo Basin and in the less tropical, more wooded Zambezi Basin.

Zambezian biome

The Zambezian biome covers most of the basin, about 95 percent, and comprises woodland, grassland, swamp and lakes. The climate is strongly seasonal with a marked dry season. This biome is sometimes subdivided into moister areas with miombo broad-leaved woodland, and drier areas with mopane or acacia woodland.

Montane biome

The Montane biome lies between 1,800-2,000 metres above mean sea level and is cooler, wetter, often shrouded in mist, and with a much more temperate climate. Species found in this biome include grassland herbs and introduced species such as pine and wattle.

Coastal biome

The Coastal biome is essentially that small part of the basin where climate is modified by proximity to the coast – the delta area and its immediate surroundings. The dry season is not so marked, temperatures do not fluctuate greatly, and habitats include dry forest and grasslands. Most species found here are widespread along the East

Map 4.1: Zambezi Basin Biomes

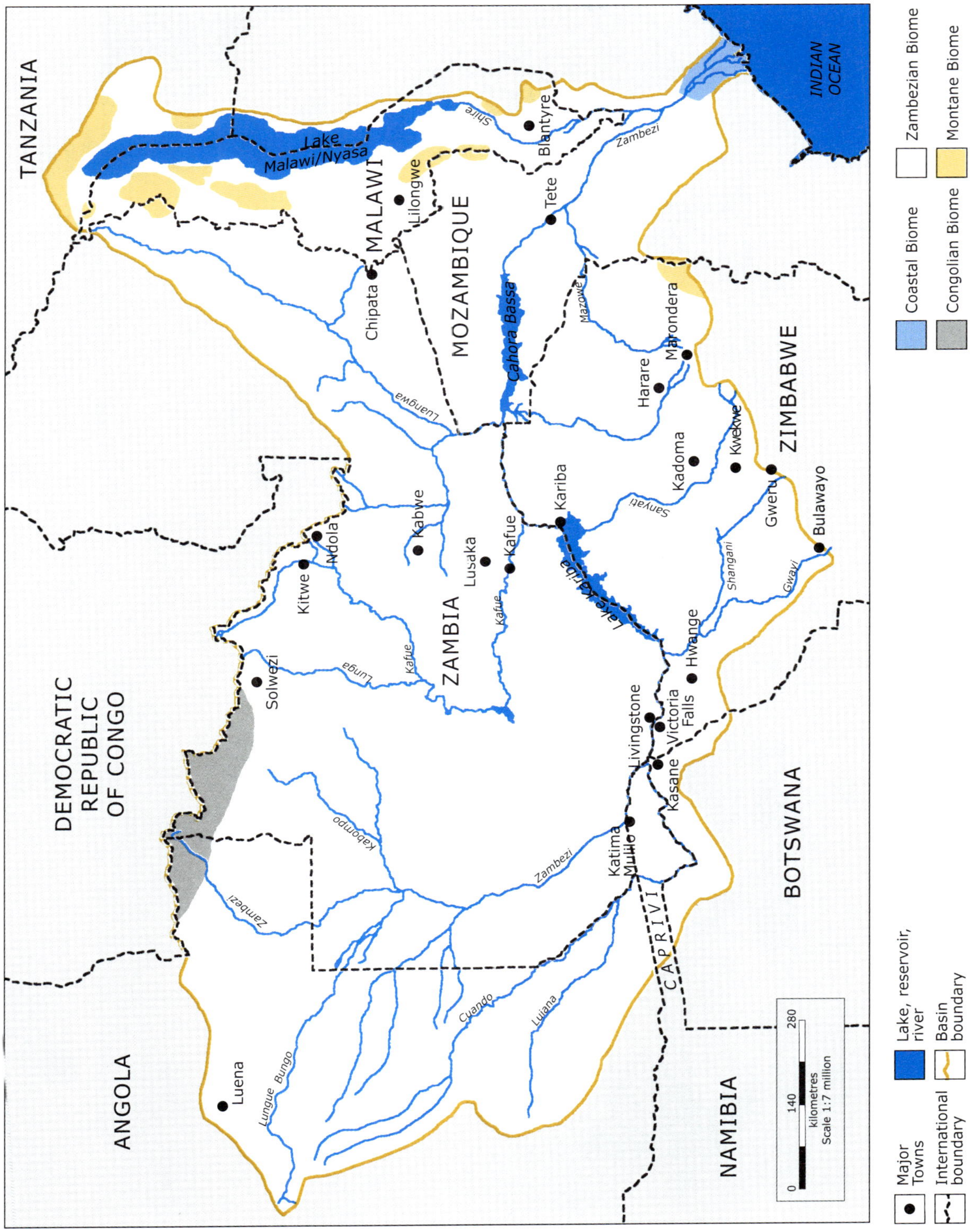

Source: Zambezi River Basin Atlas, SADC/DENCONSULT/ZRA, Lusaka, 1998

Map 4.2: Zambezi Basin – Broad vegetation groups

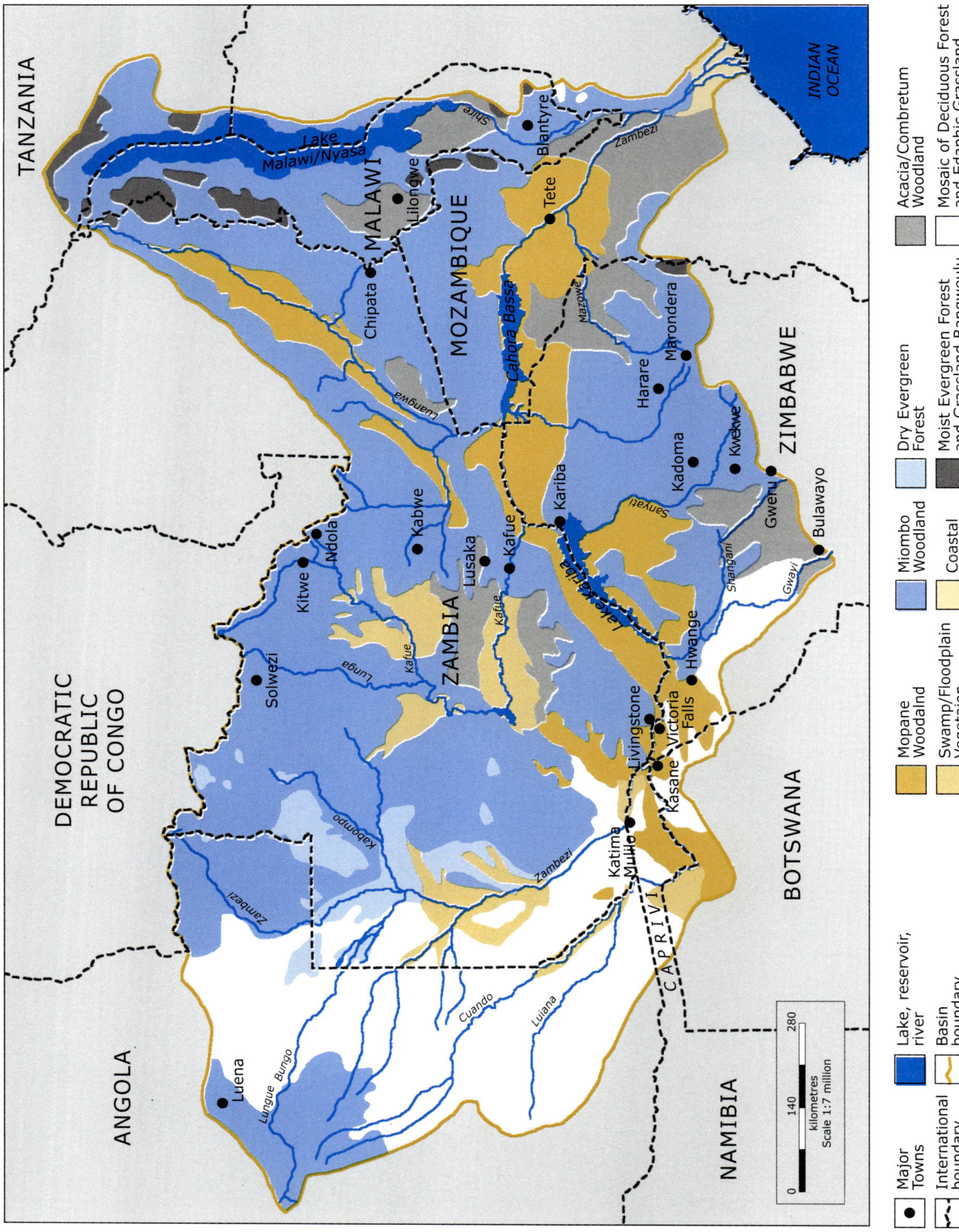

Source: Zambezi River Basin Atlas, SADC/DENCONSULT/ZRA, Lusaka, 1998

African coast from Somalia to northern KwaZulu-Natal.

To these four biomes, Lake Malawi/Nyasa can be added as a fifth, but only as regards solely aquatic organisms such as fish associated with this deep-water lake.

Vegetation

The vegetation of the Zambezi Basin is varied, as would be expected of such a large area spanning mountaintops to the coast.

In terms of broad vegetation types, almost half of the basin is classified as wetter or drier miombo woodland, part of the Zambezian biome.[3] Miombo is a type of woodland dominated by trees of the genera *Brachystegia, Julbernardia* or *Isoberlinia* with a well-developed grass layer. Other widespread vegetation types are mopane woodland dominated by mopane, mosaics of various types of woodland, dry forest, including that dominated by Zambezi Teak, with grassland, and open woodland dominated by various species of *Acacia*.

Photo: IUCN

The vegetation of the basin is varied with miombo woodland common throughout.

Both miombo and mopane woodlands are not particularly species-rich, nor do they contain many species of restricted distribution. Therefore, conservationists sometimes overlook them. But this hides their major economic value to those dependant on natural resources for survival.

The vegetation of the Montane biome is both species-rich and shows a marked difference from surrounding areas. It is restricted, in the basin, to the Misuku Hills, Nyika Plateau, Shire Highlands and Mount Mulanje in Malawi, and the Nyanga Mountains in Zimbabwe. Mount Gorongoza in Mozambique, and the Bvumba and Chimanimani Mountains in Zimbabwe are just outside the basin. Vegetation types found in these upland areas are rainforest, montane grassland and a fynbos-like shrub land on the highest plateaux. This biome, with its specific range of habitat characteristics and island-like nature, supports some endemic plant and animal species.

Other vegetation types of restricted distribution and high conservation interest are the dry forests of Upper Zambezi Valley, dominated by the tree *Cryptosepalum pseudotaxus*, and the hot, dry parts of the Lower Zambezi and Shire valleys, characterised by *Combretum* shrubs and trees such as *Xylia torreana*, *Pterocarpus antunesii* and *Pteleopsis* species.

Swamp and floodplain vegetation types are, from a biodiversity viewpoint, not particularly rich and many of the species are widespread in similar environments across the continent. The floodplains of the Upper Zambezi though, have been a focal point of much evolution and species radiation.

Knowledge on the distribution of plant species across the basin is not comprehensive, primarily as large areas are still very poorly recorded, especially in the upper reaches and in the coastal plain. However, between 6,000-7,000 plant species are in the Zambezi Basin.[4]

Wetland plants

Plant species found in swamps and permanently wet places tend to be highly adapted, and readily dispersed by water and birds, therefore, often widespread. In seasonally flooded areas, such as *dambos* and floodplains, species are less highly adapted and are also less likely to be distributed widely by birds or water currents. The ancient floodplains of the Upper Zambezi have been the site of speciation in groups such as bulbous plants and the so-called "underground trees" or suffrutices.[5] The latter are species of woody plant which have evolved from various tree species, independently of each other, to overcome the rooting impediments of a poorly drained subsoil and frequent fires and frosts. Instead of growing a large woody, above-ground trunk, the trunk has been effectively forced underground and form branches extensively just beneath the soil surface. The great majority of the 98 recognised species in southern Africa, including *Parinari capensis*, *Annona stenophylla*, *Cryptosepalum exfoliatum*,

Trichilia quadrivalvis and *Syzygium guineense* subspecies *huillense*, are found in the floodplains of the old Upper Zambezi or in associated areas of Kalahari sands.[6]

Photo: IUCN

In permanently wet places such as swamps, wetland plants tend to be highly adapted, readily dispersed by water and birds, and therefore, often widespread.

Woodland plants

Woodland covers most of the Zambezi Basin and it is in woodland, or previously wooded areas, that the majority of the basin's people live. The major woodland tree species are *Brachystegia*, *Julbernardia*, Mopane and Zambezi Teak. Locally, other genera such as *Acacia*, *Combretum* and *Terminalia* are important.

Species composition depends primarily on climate and soil type. In the lower, drier areas of the valley floors on clay-rich soils, mopane is dominant. Miombo woodland comprises most of the remaining woodland area with *Brachystegia spiciformis* being particularly widespread, and *Julbernardia globiflora* in drier areas south of the Zambezi or on stony soils. *Julbernardia paniculata* and other species of *Brachystegia* are common under wetter, warmer conditions.

Montane and forest plants

The area of the basin covered by montane vegetation, rainforest and dry forest or thicket is comparatively small, but these areas contain almost half of the plant biodiversity, and most of the species of restricted distribution.

Montane vegetation has been likened to islands in a sea of savanna woodland. Perhaps because of this, there has been a certain amount of speciation and there are a number of plant species restricted to, or endemic, to certain mountains – particularly the Nyika Plateau and Mount Mulanje in Malawi. Many of these species are grassland herbs from widespread genera, and are not normally woody plants. The fynbos-like shrublands on the highest plateaux in the Nyanga area have a lot of similarities to the fynbos of the Cape, although they have nowhere near that level of species diversity. Species found in the Montane biome are generally restricted to these habitats. Owing in part to the inhospitable environment, there are no species of economic interest and few threats, other than from invading trees such as pine and wattle. But still the conservation interest and importance is very high.

Rainforest is only found in the Zambezi Basin where local rainfall or climatic conditions allow it to develop. Such conditions are found in sheltered gullies or on the gentle windward slopes of mountains. Rainforests are very rich in plant diversity, both herbaceous species and trees, and because so much of the original forest extent has been cleared for agriculture, they are of major conservation interest.

Another category of forest is that fringing rivers and drainage lines throughout the woodland areas. Most of the broad seasonal rivers have a fringe of large trees such as Winter Thorn, Natal Mahogany, *Acacia* and Waterberry, but in many places these fringes are being destroyed by clearance for agriculture or woodfuel. Riparian fringes are an important wildlife habitat and corridor and also protect riverbanks from erosion. Most of the plant species found there are widespread in similar habitats elsewhere although some are curiously restricted in distribution.

At the headwaters of the Zambezi in Mwinilunga District in Zambia and in northeastern Angola, the gallery or riverine forests have a different species composition to those further downstream with species more typical of the Congo Basin such as *Marquesia*, *Berlinia giorgii* and *Lannea antiscorbutica*. These species are of great conservation interest from a basinwide perspective as they are so limited in occurrence here. The different species composition of this area is also mirrored in the biodiversity of other groups such as birds, small mammals, dragonflies and butterflies.

At the mouth of the Zambezi and along tidal creeks are mangrove swamps comprising specialised trees such as Red and White Mangroves (*Rhizophora mucronata* and *Avicennia marina*) and *Lumnitzera racemosa*. These species are widespread along the East African coast, but are of significant biodiversity interest from a basin-wide viewpoint. Mangrove areas are not species-rich owing to the high level of adaptation required to a saline environment.

Non-flowering plants

Non-flowering plants refer to ferns, mosses and algae. Recent studies commissioned by the World Conservation Union (IUCN) in Zambia, show that algae diversity has been extensively studied in lakes Kariba, Bangweulu and Mweru, although the latter two are outside the basin. The total diversity reported in lakes Bangweulu and Kariba is 163 species. All the species found in Lake Bangweulu are also found in Lake Kariba, which contains additional 87 species of algae.[7]

The majority of the approximately 270 species of fern and fern-allies[8] are confined to moist, generally montane or forest habitats; there are probably only about 200 species in the Zambezi Basin itself. Very few species, such as *Thelypteris confluens*, which is found in wetland areas, and bracken found in montane areas and at forest margins, are widespread.

Like ferns, mosses are mostly confined to forests and other moist shaded environments. Apart from one checklist (Zimbabwe), and some studies in Zambia, very little seems to have been recorded on moss biodiversity.

The only studies of algal biodiversity, and then very preliminary, have been done on Lake Kariba, Lake Chivero near Harare, Lake Malawi/Nyasa, and a couple of studies in the Caprivi Strip in Namibia.

Vertebrates

The Zambezi Basin shows similarities in its fauna to many parts of the savanna or woodland zone of southern and eastern Africa. There are no major centres of endemism or particular "hot spots" for most groups, apart from Lake Malawi/Nyasa for aquatic organisms, although the montane and moist forest habitats support some species not found elsewhere in the basin.

Mammals

The Zambezi Basin is renowned for its assemblage of large mammals such as elephant, buffalo, giraffe, lion and, until recently, white rhino. Compared to East Africa, there is a preponderance here of browsers rather than grazers as the basin is mostly woodland rather than the highly productive grasslands associated with the East African Rift Valley. The basin has a total number of 200 mammals.[9]

Some of these species, such as the colobus monkey, samango monkey, blue duiker and various rodents, are restricted to moist forests in the eastern part of the basin, while the tree pangolin is only found in the gallery forests of the Zambezi headwaters. A group of particular interest is the lechwe antelope. Although often considered as subspecies, it is likely the three living groups should be considered full species.[10] The black lechwe is endemic to the Lake Bangweulu area in northern Zambia, as was the now extinct Roberts' lechwe, which was part of the palaeo-Upper Zambezi. The Kafue lechwe is restricted to the Kafue Flats in southern Zambia. The Red lechwe is more widespread, but is principally confined to the Zambezi Basin. It would appear that an early form of the lechwe, which inhabited the wetlands of the old basin, split up a few hundreds of thousands of years ago into different populations, which later became isolated and started to speciate.

Table 4.1: Number of species in the Zambezi Basin

Group	Number of species (c.)	Areas of high diversity
Flowering plants	6,000-7,000	Barotseland, Mt Mulanje, headwaters,
Non-flowering plants	200 ferns	Mt Mulanje, Nyanga
Mammals	200	Middle Zambezi, Luangwa, headwaters
Birds	700	Kafue
Reptiles	200	Barotseland, Lake Malawi/Nyasa, headwaters,
Amphibians	90	Barotseland, Lake Malawi/Nyasa, headwaters,
Fish	165 + more than 500 in Lake Malawi/Nyasa	Lake Malawi/Nyasa
Dragonflies	210	headwaters
Butterflies	1,100	Eastern mountains, headwaters
Freshwater molluscs	100	Lake Malawi/Nyasa

Source: Timberlake, J., "Biodiversity of the Zambezi Basin", for SARDC, 1999.

Photo: M Chenje

Photo: M Kullberg

A herd of elephants crossing the Zambezi River in search for food and a young lion in Matusadona National Park, Zimbabwe, eating an impala.

Photo: M Kullberg

A Lilacbreasted Roller, strategically placed, waiting for the right moment to catch an insect.

Birds

The birds of the basin are comparatively well-known. There are about 700 species[11] of birds recorded of which only 15-20 are endemic to the basin, including the Black-cheeked Lovebird and the Slaty Egret. Of these, a total number of 167 are considered to be wetland-related, while birds confined to montane forests include Swynnerton's Robin, Chirinda Apalis and Roberts' Prinia. In the Zambezian biome, the division between wetter miombo woodland and the drier mopane and *Acacia* woodlands is clearly reflected in bird species composition.

The wetlands and other habitats of the Zambezi Basin are important sites for Afrotropical migrants consisting of about 70 species and Palaeoartic migrants, about 90 species. These are species that come to southern Africa to breed, or those that breed in the northern summer in Europe and Asia, and then spend the northern winter in Africa as non-breeding migrants, respectively. Examples of Afrotropical migrants of particular interest and concern are the Rock Pratincole, African Skimmer and Carmine Bee-eater along the Zambezi River, and the Blue Swallow on montane grasslands. Palaeoarctic migrants of interest and concern include the White Stork, Harriers, Lesser Kestral, Corncrake and Black-winged Pratincole. Perhaps the bird species that best epitomises the conservation concerns in the basin is the Wattled Crane – about 95 percent of the world population of 13,000-15,000 birds are dependent on the floodplains and dambos of the basin and nearby areas.[12]

Fish

The fish species of the basin have been fairly well documented, principally owing to their economic interest, although details of the fish diversity of the Zambezi Delta are still scanty. There are about 165 species of freshwater fish recorded from the basin in addition to more than 500 endemic species (mostly cichlids) in Lake Malawi/Nyasa. Of this total, 42 percent are found only in the Upper Zambezi system (including the Kafue), 36 percent are found only in the Middle and Lower Zambezi, while the remainder are common to both.[13] The lungfish and eels are all found only in the Lower and Middle Zambezi.

With the exception of the Lake Malawi/Nyasa cichlids, which are often endemic to just one stretch of shoreline or island, most fish species are relatively widespread. Only 15 species have a limited distribution.[14] From a conservation point of view, Lake Malawi/Nyasa is not just of basin-wide importance but significant worldwide importance. How such explosive speciation came about, much of it

Photo: M Chenje

Some of the many fish species in the Zambezi Basin, such as the kapenta form the basis for the commercial fish industry.

perhaps only in the last few hundred years,[15] is not clear, but the large rise and fall in lake level over recent millennia, isolating populations previously freely interbreeding and the high reproductive rate are probably key features.

Reptiles and amphibians

Knowledge on the biodiversity of reptiles and amphibians of the basin is comparatively good, although there are still large gaps in the knowledge of distributions. There are 200 species of reptile and about 90 species of amphibian recorded. Only very few are endemic to the basin or areas in it. As with the fish, the Upper, Middle and Lower Zambezi have clear differences in species composition, with 44 percent of the 160 wetland-related species confined to the Upper Zambezi, 22 percent confined to the Middle/Lower Zambezi, and the remainder common to both.

Photo: M Kullberg

The Zambezi River is a suitable habitat for crocodiles since its sufficiently warm and provides food and cover.

Invertebrates

Invertebrate biodiversity is not well known in the Zambezi Basin except for groups such as dragonflies, butterflies, freshwater molluscs, dung beetles and grasshoppers/crickets. In addition, groups of particular economic interest – mosquitoes and tsetse fly, locusts, ticks and agricultural pests – have been the focus of detailed research. The number of species of insects and other invertebrates present in the basin is unknown, but it is likely to be in tens if not hundreds of thousands.

Dragonflies are a group of insects that are principally confined to water and wet places where they breed. Of the 210 species recorded from the basin, 136 are widespread in the Afrotropics while 74 species appear to have limited distributions.[16] Collecting has been localised with many large gaps remaining, particularly in the Upper Zambezi Valley, but the richest area in terms of species number and endemics appears to be the Zambezi headwaters. Eleven species of Congolian affinity, apparently endemic, are found there. Dragonflies have little economic significance as such, but are known to be useful indicators of wetland health.

There are around 1,100 species of butterflies recorded from the Zambezi Basin.[17] The richest area is again the headwaters of the Zambezi around Mwinilunga, where 340 species have been recorded. The forest and montane areas, and those where the Congolian elements come into the basin, are the most different and most diverse from the predominantly woodland remainder. The loss of certain species generally indicates that their food plant has been lost from that habitat, indicating environmental dearth.

Freshwater molluscs have been comparatively well-studied across the basin as some species are intermediate hosts of the parasite which causes *schistosomiasis* or bilharzia. Of the 98 indigenous species recorded from the basin, 47 are found in Lake Malawi/Nyasa.[18] Of the 98 species, 28 are endemic to the basin, including 23 endemic to Lake Malawi/Nyasa, showing the biodiversity impor-

tance of this unique lake. The Afrotropics as a whole contain 382 species of mollusc. The patterns of species distribution show similarities to those for fish with a very distinct fauna for Lake Malawi/Nyasa, a distinct East African coastal element, and an Upper and Middle/Lower Zambezi element. The Congo Basin supports a much higher number of freshwater molluscs than the Zambezi, probably because of its greater stability over the last few million years. There are also a significant number of species in common.

Photos: IUCN

The number of species of insects and other invertebrates present in the basin is unknown, but it is likely to be in tens if not hundreds of thousands.

THREATS TO BIODIVERSITY

Biodiversity is under threat in the Zambezi Basin for the same underlying reasons as elsewhere in the world: the conflict between supply and demand. Rapid population growth and the growing aspirations of that population increase the pressures on limited resources. In many places, a greater demand is placed upon species and ecosystems than they are able to meet by themselves at natural rates of increase.

Historically, in the Zambezi Basin, the major changes in biodiversity have resulted from land clearance and from overexploitation of certain species of economic value, and for aquatic and wetland species, the construction of large and medium-sized dams along the Zambezi River and its catchment. Pollution and fire are other threats to biodiversity in the Zambezi Basin.

Overexploitation

Overexploitation is when humans use a particular species or group of species beyond the point where that species can replenish itself. It is usually localised when viewed

Photo: M Kullberg

Dams, such as Kariba, change rivers' flooding regimes, impacting on habitat and species composition.

from a basinwide-scale, but has been severe and more widespread in some cases. Examples include the destruction of large mammals such as elephant in some countries of the basin, rhino and antelope over extensive areas. Particularly during the latter part of the last century and the first few decades of the 20th century; extraction of valued timber such as mukwa, African Ebony and Zambezi Teak; fish from Lake Malombe and Lake Malawi/Nyasa and the Lower Shire, and grass from some grazing lands in Zimbabwe have impacted species and habitats.

Land clearance

Land clearance is when settled human populations expand and require more land for urban development or for arable agriculture, or new technologies are introduced allowing people to farm larger areas. In many cases, especially if fields are stumped and ploughed and the area is on readily-erodible soils, the process is irreversible over human time scales. This means that even when fields revert to bush fallow, the vegetation does not revert to the former structure and species composition, but to a new state. In many parts of the basin, the vegetation is in a stable state, but generally not what it originally was. This can, for example, be seen in the acacia woodlands of Matabeleland in Zimbabwe, in some of the woodlands of the Zambezi Valley, in the sugar plantations of the Zambezi Delta, and possibly in some of the grasslands of the Upper Zambezi. Land is still being cleared for agriculture, particularly in what were marginal agricultural areas in the Middle Zambezi Valley, and also for increasing urban populations in Zambia and Zimbabwe. In Zambia, for example, the World Bank estimates that the conversion from forest to cultivation is expected to increase by 1.5 percent per year.[19]

Dams and modification of hydrology

The construction of dams has had probably the greatest effect on the biodiversity of wetland and aquatic species and on wetland ecological processes. Many dams including large ones such as Kariba, Cahora Bassa, Itezhi-tezhi and Kafue Gorge, and a myriad of smaller ones on almost all tributaries of the Zambezi in Zimbabwe have been built over the years. Hydrology is changed, flooding regimes greatly modified, impacting habitat and species composition. Two new major habitats have been created – a pelagic (open water) environment and a littoral (shoreline) environment. Creation of these new environments has led to previously rare species such as freshwater molluscs, becoming abundant,[20] and to the invasion or introduction of species new to the basin.

As slow-water environments are now available in the Middle Zambezi, whereas before they were not, fish species from the Upper Zambezi are now appearing, resulting in mixing of two previously separate faunas.[21]

This will also obviously have evolutionary consequences. Fears have been expressed on the effects of flood regulation on wetlands. For example, there were many studies in the 1960s and 1970s[22] on the Kafue Flats to determine changes to the Kafue lechwe, waterbirds, fisheries and vegetation. Despite marked changes in relative abundance, it appears that the worst fears expressed have, however, not materialised. The gradual drying up of some of the wetlands of the Zambezi Delta, has been ascribed to regulation of the annual flood by Cahora Bassa and Kariba.

Introduction of alien species

The introduction of alien organisms is a major concern to conservationists. Some alien species can out-compete native species or modify the ecology of an area. Examples include the invasion of pines and wattle trees into the montane grasslands of Mount Mulanje in Malawi and Nyanga in Zimbabwe and, the introduction of the Nile Tilapia fish into the waters of the Middle Zambezi, and Kariba weed into the Chobe system and Lake Kariba. The introduction of the Lake Tanganyika sardine into Lake Kariba in the 1960s has not resulted in loss of biodiversity as a new habitat was created and there were no native species to occupy it. However, fears have been expressed over the possible introduction of *kapenta* into Lake Malawi/Nyasa where it would feed on lake flies which are the main food for some endemic cichlids, endangering these species.[23]

Fire

Fire has modified vegetation structure and species composition in the moister woodlands with a good grass layer, and in particular in the extensive grasslands associated with floodplains. Large areas of Barotseland are burnt regularly, as are the grasslands of the Zambezi Delta. Soil erosion is increased, and plant and animal species, which cannot avoid fire, by flying away, going underground, or regenerating rapidly, are reduced in distribution and abundance.

Photo: M Chenje

Fish deaths have occurred in different parts of the basin and the underlying cause is pollution.

Photo: M Chenje

Burning fields increases soil erosion, reducing plant and animal species that cannot avoid fire.

Pollution

Pollution often has an insidious chronic effect on biodiversity rather than a sudden acute effect. However, industry and extensive agriculture are not as prevalent in the Zambezi Basin as they are in South Africa or Europe, and so the extreme effects often noted elsewhere by environmentalists have not really occurred in the basin.

Increasing use of fertilisers in agriculture, and large urban populations resulting in large volumes of sewage, have, however, given rise to acute problems of water pollution, such as in Lake Chivero outside Harare. There have been some effects noted on birds owing to the use of DDT against tsetse fly in the Zambezi Valley in Zimbabwe.[24] Of perhaps greater concern for conservation in this case, is the increased land clearance and overutilisation of natural resources that result from the introduction of cattle into such areas.

Civil conflict

Civil conflict and war have led to significant ecological damage in and outside protected areas, as well as to the marginalisation of environmental management institutions and conservation programmes. By 1991, the wildlife population of national parks and reserves in Angola had been reduced by civil war to only 10 percent of their 1975 levels.[25]

Threatened species

As a result of the threats discussed above, some species have become extinct, such as the blue wildebeest in Malawi, the tssessebe in Mozambique and the kob in Tanzania,[26] and others face high risk of extinction in the future. Species facing high risk of extinction include, for example, birds such as the Wattled Crane and mammals such as the African Wild Dog or Painted Hunting Dog, the Kafue Lechwe and the Black rhino.

The examples given in Figures 4.1 and 4.2 on numbers of threatened animal and tree species may be higher than indicated because species diversity in the Zambezi Basin is not yet fully documented.[27]

Photo: M Kullberg

A pack of the endangered Painted Hunting Dog spotted in Hwange National Park, Zimbabwe.

Figure 4.1: Number of threatened animal species in southern Africa

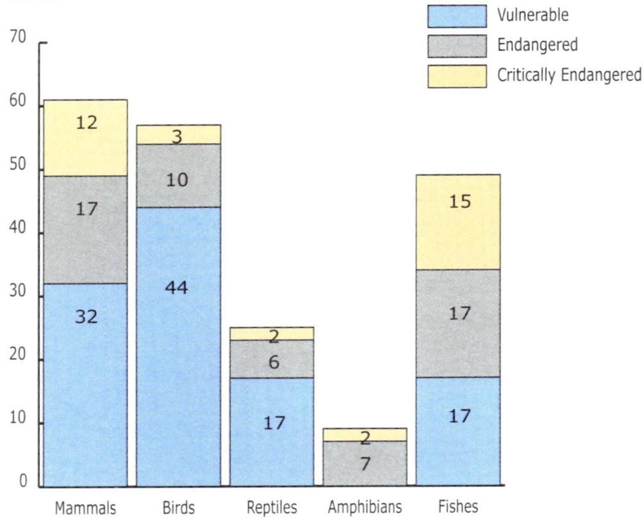

Figure 4.2: Number of threatened tree species in the Zambezi Basin Countries

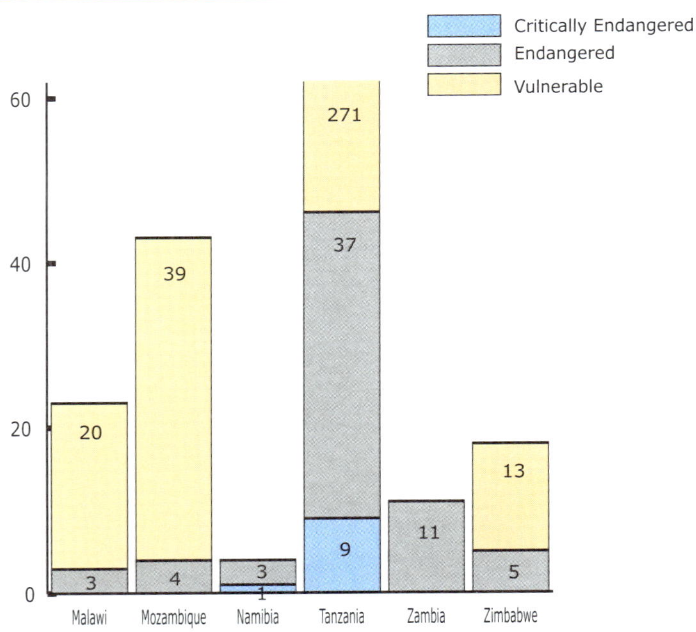

Note:
For Angola information is not available, and Botswana doesn't have any tree species in the categories listed below.

- Critically Endangered (CR) – a taxon facing an extremely high risk of extinction in the wild in the immediate future.
- Endangered (EN) – a taxon facing a high risk of extinction in the wild in the near future.
- Vulnerable (VU) – a taxon facing a high risk of extinction in the wild in the medium-term future.

Source: World Conservation Monitoring Centre, www.wcmc.org.uk

Table 4.2: Some bird species of conservation concern

Species	Where found	Conservation category
White-winged Flufftail	Zambia, Zimbabwe	Endangered
Slaty Egret	Botswana, Namibia, Zambia	Vulnerable
Wattled Crane	Most wetlands; not L. Shire	Vulnerable
Corncrake	Malawi, Zambia, Zimbabwe	Vulnerable
Madagascar Squacco Heron	Malawi, Mozambique, Zimbabwe	Near-threatened
Shoebill	Only Bangweulu in Zambezi Basin	Near-threatened
Lesser Flamingo	All areas in Zambezi Basin	Near-threatened
Great Snipe	Anywhere	Near-threatened
Black-winged Pratincole	Western parts of Zambezi Basin	Near-threatened

Source: Collar, N.J, and Stuart, S.N., *Threatened Birds of Africa and Related Islands, The ICBP/IUCN Red Data Book, Part 1,* Cambridge, U.K., 1985 and Mundy, P.J., personal communication with J. Timberlake.

BIODIVERSITY AND SELECTED SOCIOECONOMIC ASPECTS

Generally, the benefits arising from the conservation of components of biodiversity such as vegetation, vertebrates and invertebrates can be considered in three groups: biological resources, ecosystem services and social benefits.

Vegetation

Important products of plants include, for example, wood both for construction and firewood. To illustrate forests' importance, it is important to note that the net firewood energy requirements of the Zambezi Basin countries is on average more than 70 percent. This shows the heavy dependence of the majority of the people on forests for their energy needs.[28]

Mopane is greatly desired as a strong termite-resistant construction wood and for firewood, while species such as Zambezi Teak, False Mopane or Mchibi and Bloodwood, Mukwa or Kiaat have been heavily exploited for commercial timber (furniture, parquet blocks, railway sleepers, etc.). Other important timber species include Muwanga, Pod Mahogany, African Ebony, Panga Panga and Leadwood.

Woodlands also provide wild foods such as mushrooms and edible insects and the habitat for activities such as beekeeping.

Except from providing valuable trees such as Red Mahogany for timber, forests also provide soil nutrient inputs such as leaf litter. Trees in farming systems may also provide environmental services such as protection from soil erosion. Forests provide materials for traditional medicine and craft-making. Throughout countries such as Malawi, Mozambique, Zambia and Zimbabwe, herbal material such as roots and bark, which is dried or has a long shelf life dominates herbal medicine markets.

Wetland plants of particular economic importance include papyrus, cattail and reeds, all of which are used in construction or craft-making. They are of particular importance in wetland areas owing to the general paucity of trees and wood. The reeds and cattail are heavily harvested for mats and basket making. For example in Malawi, western Zambia and Tanzania, handcrafts based on wetland plants are a thriving business. Species of water lily are also used as a source of food in times of shortage.

There are some wetland plants of negative economic importance because they can block water channels, cause fish mortality and provide suitable breeding grounds for bilharzia-carrying snails and mosquitoes. These include the water hyacinth, water lettuce, Kariba weed and the water fern.

Swamp and floodplain vegetation types cover only a small part of the basin, but can be of significant economic importance for fishing, materials for crafts and for livestock grazing. They are often the focus of human settlements owing to the availability of water year-round, as well as being a regular source of protein from fishing.

Plants are also important for grazing and in the provision of habitat for wildlife. Many of the soils are also suitable for agriculture, and the ability of cut trees to coppice reduces soil erosion and restores fertility through bush fallow.

Mangroves are significant economically as the trees provide good quality construction wood, help reduce shoreline erosion and provide a suitable habitat for fish breeding. Algae are also of major economic significance as they provide food for fish and other aquatic organisms.

Box 4.2: Wild foods and fruits from the Zambezi Basin

Many wild foods are derived from the woodlands and forests of the basin. The miombo woodland produces more than 50 edible plants, including fruits, drupes, flowers and seeds, tubers, bulbs. Wild vegetables, mushrooms and insects are most abundant in the wet season. In the case of caterpillars, branches (and even whole trees in exceptional circumstances) will be cut to gain access to the caterpillars.

The woodlands and forests are rich in variety and quantity of fruit trees. Wild fruits are mostly consumed by children, but also eaten by adults when walking through the bush, herding animals or collecting other products. Many fruits are a major source of iron and some, particularly *Parinari curatellifolia*, have a high crude protein and calcium content. Wild fruit is consumed mainly in the hot dry season and early rainy season before the cultivated crops are ripe. Donkey berry *(Grewia flavescens)* fruits, ground and made into porridge, comprised nearly 25 percent of the food items during the dry season in a remote communal area in northern Zimbabwe after the 1981-82 drought. In Malawi, in the Mangochi and Lower Shire areas, water berry *(Syzygium abysinica)* trees are managed for fruit production.

More than 50 species of trees indigenous to the miombo ecosystem of southern Africa bear edible fruits that provide people and wildlife in the area with vital nutrients. Many of the fruits are reserves of food during seasonal shortages and famines. These species are under threat - rates of deforestation in the miombo woodlands are among the highest in the world.

A variety of products can be obtained from the use of the fruits which include: oil, vinegar, jam, jelly, and wine. Marula sweets and jam can be made from marula fruit. An alcoholic drink (as well as a non-alcoholic one) can be made from the flesh. Marula fruit is also used to make a popular liqueur, *Amarula*.

Moist miombo woodlands have abundant and diverse mushroom populations. In Malawi, 60 species of edible fungi have been documented. In the Boni area, Kasungu district, 26 species of edible mushrooms were identified.

Source: IMERCSA, Draft Manuscript, *Biodiversity of Indigenous Forests and Woodlands of Southern Africa*, 2000.

Table 4.3: Some medicinal plants from indigenous forests

Plant Name	Medicinal Uses	Part Used
Sclerocarya birrea	Bark infusion used with natron for dysentery, tonic, exudate for skin diseases	Fruit, leave, exudate
Maytenus buchananii	Wound healing, ulcers, boils, mouth infections, toothache, dysentery	Bark, leaves, roots
Olive tree	Intermittent fever, laxative, diarrhoea	Leaves, bark, fruits
Pterocarpus angolensis	Treatment of skin problems such as sores and ring worms	Exudate
Acacia mellifera	Anti-venom — provides relief from snake bite	Leaves, bark
Ziziphus mucronata	Treatment of malaria and diarrhoea	Leaves, bark
Afzelia quanzensis	Relieves toothache	Bark
Albizia antunesiana	Prophylactic against colds and coughs	Root
Velvet bushwillow	Treatment of wounds and sores	Leaf paste
Diospyrus mespiliformis	Treatment of ringworm, wounds and sores	Root
Tamarind tree	Leprosy treatment, fevers, laxative, cardiac diseases, constipation, cicatrizer	Fruits, leaves, roots, bark
Natal Mahogany	Parasitic skin infections and inflammation, antiepileptic, bronchial inflammation	Bark, roots

Sources: IWU, M.M., *Handbook of African Medicinal plants*, Boca Raton, CRC Press, 1993. Ngozi, M.A.. "Report on consumption of non-wood forest products from Namibia", Unpublished report, 1996. Chidumayo, E.N., *Handbook of miombo ecology and management*, Stockholm Environment Institute, Stockholm, 1994.

Vertebrates

The economic importance of mammals is obvious – they provide meat in many areas and can be a major draw for tourism in others. Regarding cattle, for example, Malawi, Zambia and Zimbabwe dominate with 90 percent of the animals in the basin.[29]

On the negative side, mammals such as elephant and baboon can do a lot of damage to trees and crops. Elephant, lion and leopard can be a threat to human life and livestock. Some rodents are major agricultural pests, particularly of stored grain or foodstuffs, but healthy populations of small predators such as mongoose and serval cats can keep small mammal and reptile populations under control.

Birds of economic significance include the ducks and geese, many of which are hunted for meat. Some wetland areas of the basin, such as the Kafue Flats, have great concentrations. Francolins and guinea fowl are also hunted for food in many areas. Species of negative importance include the Red-billed Quelea, which can devastate small grain crops in a short period of time. For example, the winter wheat crop in Zimbabwe is always under threat from this species such that massive aerial chemical spraying is undertaken. The impact of the chemicals on the environment is not well-known.

Economically, fish are of major importance in areas close to reservoirs and along major rivers. The Lake Tanganyika sardine, or *kapenta*, now forms the basis for an enormous commercial fishing industry. Some 17,034 tonnes of *kapenta* were, for example, harvested in 1997

Table 4.4: Inland fish production (including capture fisheries and aquaculture)

Country	Current landings (tonnes/year)	Potential production (tonnes)
Angola	40,000	115,000
Botswana	1,900	15,000
Malawi	70,000	120,000
Mozambique	4,000	34,500
Namibia	2,800	6,000
Tanzania	294,780	370,000
Zambia	70,500	80,000
Zimbabwe	28,000	35,000
Total	511,980	775,500

Note that this table provides the total current and potential fish landings in the basin states and that these figures do not necessarily apply to the Zambezi Basin alone.
Source: SADCFISH, SADC Inland Fisheries Sector Policy and Strategy Statement, 1997.

Photo: M Kullberg

Guinea fowl rearing is one of many activities that prevent people from overdepending on timber forest products.

Photo: M Chenje

Raising crocodiles is now a commercial activity in the basin, promoting conservation and generating revenue.

on the Zimbabwean side of Lake Kariba alone compared to 1,172 tonnes caught in 1977.[30] The species was so successful because there was a niche created for it.

Reptiles of economic importance include turtles, some snakes and crocodiles, but there are no economically significant amphibians even though the giant bullfrog is sometimes eaten in Mozambique. Most of the turtle species, particularly the Zambezi Flap-shelled Turtle in Malawi are caught and eaten by people. Some snakes, especially the puff adder, mambas and cobras, are poisonous and cause death and distress to humans, but these and other species also help keep rodent pests under control.

Raising of the Nile Crocodile is now a commercial enterprise in the Zambezi Valley, with eggs often being collected from the wild and a number of one or two-year-old infants being returned to maintain populations in the wild. In some places, crocodiles are considered a pest as they attack small stock and people. Current perceptions by communities in the Lower Shire Valley in Malawi, for example, are quite negative because people see no value in crocodiles. While significant numbers of crocodiles are culled from the area annually, communities presently enjoy no benefits from such activities. Therefore, crocodiles are seen as a menace to community well-being both in terms of risk to human life and dwindling profits from fishing.[31]

Invertebrates

Tsetse flies are a major economic concern as some species carry human sleeping sickness, and most transmit the near-fatal trypanosomiasis to cattle and other livestock. Therefore, in an area where tsetse is present, even at low levels, it is not possible to keep cattle, hence the use of drought power is excluded, except under a drug regime. Knowledge of tsetse fly ecology and distribution is fairly good for much of the basin, but less so in Angola and Mozambique.

The major significance of molluscs is that a few act as hosts for bilharzia, hence their presence in water bodies near where people live is a health hazard. Some of the bivalve molluscs are also important food for certain fish species and turtles.

INITIATIVES AND CONSTRAINTS FOR CONSERVATION OF BIODIVERSITY

The conservation of biodiversity includes sectoral concerns of wildlife, forest and fisheries management. It also includes conservation of agricultural crops and livestock genetic resources and the issue of sustainable and equitable use of these resources, and the active involvement of all stakeholders.

Global multilateral environmental agreements (MEAs)

Although some African countries signed environmental policy instruments or multilateral environmental agreements (MEAs), it was not until the 1992 Earth Summit that most countries signed a number of global MEAs relevant to the African situation.

However, enforcement remains a problem due to constraints such as:[32]

- Inadequate policy framework for implementation.
- Constraints of financial resources.
- Lack of qualified personnel in the different disciplines related to MEAs at the national level.
- Inadequate participation by national stakeholders as well as the general public, in the negotiations of MEAs.
- Failure of some MEAs to reflect national environmental priorities.
- General absence of in-depth understanding of the contents of MEAs.

Box 4.3: Some international conventions on conservation of biological resources

Convention on Biological Diversity (CBD)
The Convention on Biological Resources (CBD) is an international convention with the objective to conserve biological diversity, promote the sustainable use of its components, and encourage equitable sharing of the benefits arising out of the utilisation of genetic resources.

The CBD, which was adopted in 1992, places a duty on parties to conserve biological diversity in their jurisdiction, as well as outside their jurisdiction in certain cases. All eight Zambezi Basin states have ratified the convention.

Convention on the Conservation of Migratory Species of Wild Animal (CMS)
The objective of CMS is to protect those species of wild animals that migrate across or outside national boundaries. Among other things the contracting parties to this convention acknowledge the need to take action to avoid any migratory species becoming endangered.

Convention on the International Trade in Endangered Species of Wild Fauna and Flora (CITES)
The 1973 CITES Convention tackles the problem of endangered species protection by controlling the international market. It is an important component of conservation, especially of heavily poached animals such as the rhino and the elephant.

Convention on Wetlands (Ramsar)
The Convention on Wetlands, signed in Ramsar, Iran, in 1971, provides the framework for national action and international cooperation for the conservation and wise use of wetlands and their resources. Of the eight basin states Botswana, Malawi, Namibia and Zambia have ratified the convention. Lake Chilwa and the Kafue Flats are the only two designated Ramsar sites in the basin.

World Heritage Convention (WH)
The objectives of the World Heritage Convention are to establish an effective system of collective protection of the cultural and natural heritage of outstanding universal value, organised on a permanent basis and in accordance with modern scientific methods. It provides for parties to integrate the protection of their heritage into comprehensive planning programmes.

Under the World Heritage Convention certain sites can be designated as having outstanding universal value. Sites such as the Lake Malawi National Park in Malawi, Victoria Falls in Zambia and Zimbabwe, and the Mana Pools complex in Zimbabwe are designated under the convention.

Botswana and Namibia have not yet ratified the convention.

Source: Chenje, M., *Reporting the Southern African Environment – A Media Handbook*, SADC/IUCN/SARDC, 1998, p.130.

The most favoured means of implementing MEAs and environmental policy in general, at the national level, has been the enactment of legislation and passing of decrees.

Wildlife conservation

The 1999 Protocol on Wildlife Conservation and Wildlife Law Enforcement was adopted by SADC States (see also Chapter 12: Environmental Management and Regional Cooperation) with the objectives to:

- Harmonise relevant sections of "Wildlife Acts" to create an enabling environment for more effective law enforcement procedures.
- Standardise penalties for poaching of, and trafficking in, wildlife and wildlife products.
- Establish extradition procedures for poachers and traffickers.
- Establish wildlife agencies for improved cooperation in law enforcement activities.
- Find ways of sharing information on the movements and whereabouts of poachers and identification of illegal trafficking routes.
- Establish a database on illegal traffickers, their arrests and convictions.
- Undertake joint operations in the field to apprehend poachers, recover and dispose of illegally obtained wildlife products.

To date, much of the effort to conserve the larger forms of animal and plant life can be seen from the large tracts of land designated as protected areas, national parks, safari areas and game reserves. Protected areas cover about 27 percent of the basin area[33] and more than 10 percent of this is protected as national parks.[34]

Map 4.3: Wildlife estates

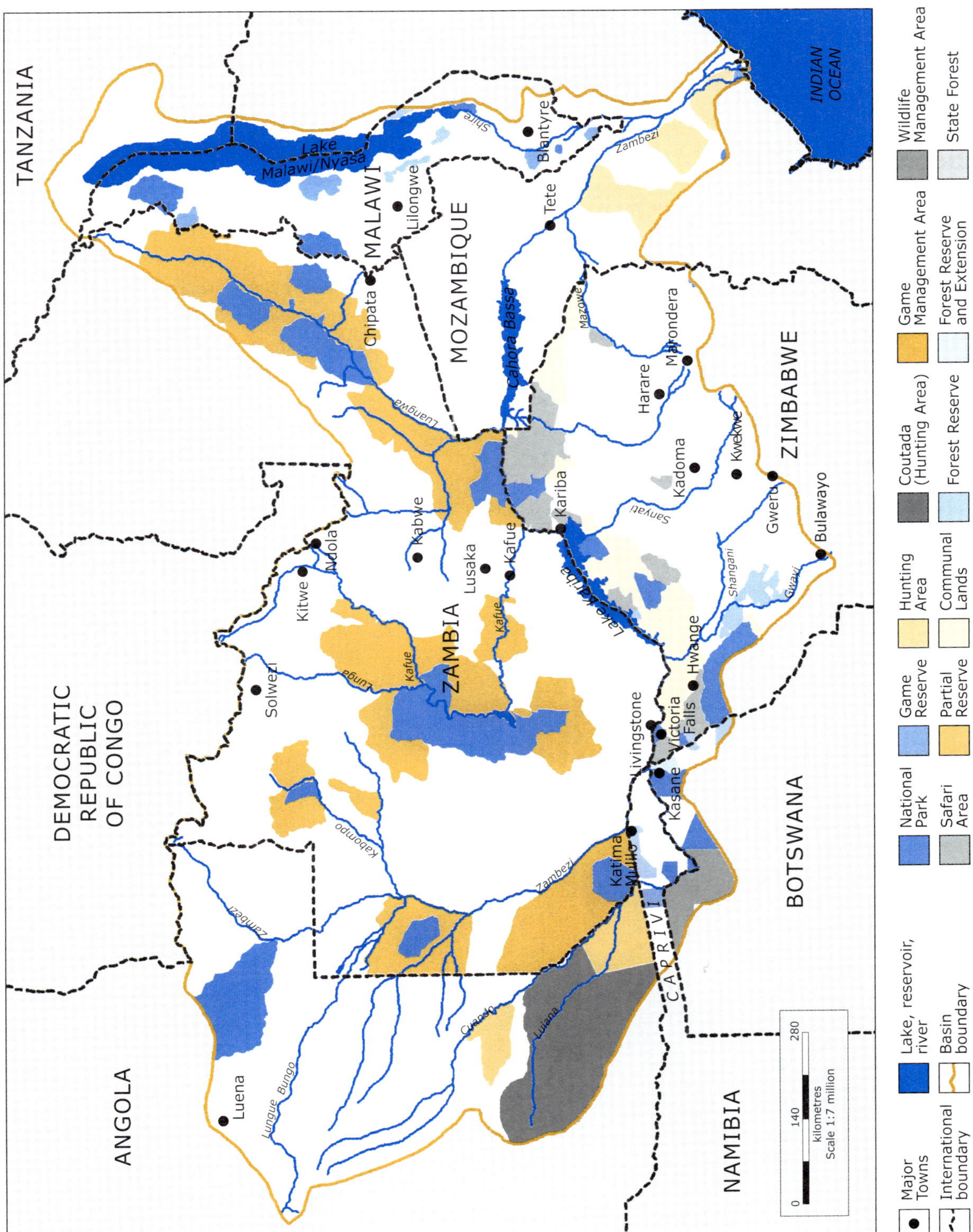

Source: Zambezi River Basin Atlas, SADC/DENCONSULT/ZRA, Lusaka, 1998

However, despite the abundance of wildlife resources in the region there is pressure:

- Grazing livestock are displacing wild animals.
- Expanding agriculture is converting woodlands to farms.
- Dams are altering river characteristics.
- Pollution is increasing.
- Habitats are being harvested for various uses.

Poaching has severely reduced elephant populations in many countries and the black rhino has been decimated in some countries and is virtually extinct now.[35]

Wildlife management has traditionally been top-down with the state owning the wildlife in protected areas and enforcing wildlife legislation through state agencies. Typically, a government department or a semi-government agency that is responsible for enforcing wildlife legislation manages wildlife.

Unfortunately, it has proved ineffective in many places, especially in controlling illegal harvesting of wildlife and facilitating local community development. In the face of a diminishing resource base, conflicts between local people and conservation authorities have escalated, and law enforcement has become less practical and more costly.[36]

Table 4.5: Area in Zambezi Basin conserved as national parks

Country	Proportion of basin (%)	Area under NP (km^2)
Angola	18.3	14,450
Botswana	2.8	10,570
Malawi	7.7	1,425
Mozambique	11.4	n/a
Namibia	1.2	1,334
Tanzania	2.0	0
Zambia	40.7	88,092
Zimbabwe	15.9	20,727
TOTAL	100	136,598
		= 10.3 percent of basin

Source: Timberlake, J.R., *Biodiversity of the Zambezi Basin wetlands: review and preliminary assessment of available information - Phase 1*. Consultancy report for IUCN ROSA, Harare. Zambezi Society, Harare/Biodiversity Foundation for Africa, Bulawayo, 1998.

However, a number of innovative responses to the weaknesses in wildlife management have been introduced. These include game ranching, which allows ownership of wildlife to be transferred from the state to the individual ranchers. An added incentive for game ranching, other than the fact that game ranchers can be used to restock depleted state protected areas, is that its economic potential through tourism, sales of meat and trophies and hunting is high compared with that of domestic livestock.

Four Zambezi Basin states, Botswana, Namibia, Zambia and Zimbabwe are taking part in a project called Regional Development of Community-Based Management and Utilisation of Wildlife which has, among its objectives to devolve authority to manage wildlife to the local communities.

Community Based Natural Resources Management (CBNRM)

Another innovative response to the dwindling wildlife resource has been the introduction in the 1980s, of community-based natural resource management (CBNRM) programmes. CBNRM systems seek to enhance conservation of biodiversity outside protected areas and private lands while at the same time affording rural people benefits from the wildlife resources in their areas.[37]

In the basin, various CBNRM projects exists, except in Angola and Tanzania. The Malawi Natural Resources Management Project is newly born. The Botswana and Namibia Living in a Finite Environment (LIFE) projects are exploring and forging ahead latent partnerships and alliances. Administrative Management Design Programme for Game Areas (ADMADE) initiative of Zambia has also made some significant strides. Communal Areas Management Programme for Indigenous Resources (CAMPFIRE), in Zimbabwe, is the oldest having started in the mid-1980s. The Tchuma-Tchato project in the Tete province of Mozambique was initiated in 1995.

Transboundary conservation areas

Transboundary conservation areas or peace parks are a new initiative to soften political barriers, not only to enhance the management of shared natural resources, but to also keep communities and their cultures,[38] separated by artificial imaginary lines on a map, intact. The major justification for transboundary parks is to harmonise natural resource management policy and practice which tend to be different across national political boundaries.[39,40] Whenever elephants cross a national boundary such as between Zambia and Zimbabwe, they face a whole range of new threats and opportunities. For example, while Zimbabwe allows sport-hunting of elephants, Zambia does not. Transboundary Conservation Areas (TBCAs) are defined as "contractually legitimised co-management entities, which involve recognised land and resource authorities, formed for the use and conservation of natural resources on land in and beyond two or more countries."[41]

Currently, there are about 26 potential TBCAs in the Southern African region.[42] An organisation called Peace Parks Foundation is raising funds for the development of the first seven TBCAs. Six of the identified TBCAs are located in the Zambezi Basin. It is, however, important to point out that TBCAs, which do not address the needs and expectations of people, particularly those adjacent to the parks are bound to fail. The alleviation of poverty in these areas is critical. In addition, uneven development among the countries involved may also limit the potential of TBCAs.

Photos: IUCN

Grazing livestock is a threat to wildlife conservation since they displace wild animals.

Forest conservation

The total forest cover of the Zambezi Basin is 76,310 sq km and the annual rate of deforestation in the SADC region is between 0.75-2.2 percent with Angola and Malawi having the lowest and highest rates respectively. Deforestation, forest degradation and desertification pose major threats to the basin's forest resources.[43]

Appropriate conservation and management of the SADC natural forests can produce high environmental benefits. The most important functions of natural forest resources for the SADC countries' economies are:

- Forest environmental protection.
- Provision of specialised local forest products.
- Support to wildlife activity and ecotourism.[44]

The SADC Forestry Sector aims to promote sustained regional self-sufficiency in forest/wood products, enhance the productive and environmental value of trees, and protect, manage and control forest resources. The sector has a revised Forestry Policy and Development Strategy, approved by the SADC Council of Ministers in September 1997. SADC is also considering ways to make regional contributions to international forestry platforms and agendas, such as Agenda 21 of the Earth Summit (1992) and the Conven-tion to Combat Desertification (1994).

The Zambezi Basin states share concerns regarding irregular or uncertain supply of raw materials, overexploitation, and insufficient reforestation. All national forestry programmes are intended to ensure the conservation and sustainable development of forest resources. Agenda 21's programme objectives include effectively

Table 4.6: Some key institutions dealing with wildlife conservation by country

ANGOLA
- The Direcca Nacional da Agriculture Florestas (DNAF)
- The policy and executive body of the DNAF, the IDF (Forestry Development Institute)
- AAA (Angolan Association for the Environment)
- University of Angola
- IUCN-ROSA

BOTSWANA
- Department of Wildlife and National Parks
- National Conservation Strategy (Coordinating) Agency
- Tawana and Chobe Land Boards

MALAWI
- Department of National Parks and Wildlife
- SADC Wildlife Sector Technical Coordination Unit
- Malawi Wildlife Society

MOZAMBIQUE
- GPZ (Zambezi River Development Authority)
- DNFFB (National Directorate of Forestry and Wildlife)
- IUCN-ROSA

NAMIBIA
- Ministry of Environment and Tourism
- Integrated Rural Development and Nature Conservation
- WWF and Namibia Nature Foundation
- IUCN Wetlands Programme, Eastern Caprivi and Botswana wetlands sub-programme.

ZAMBIA
- Department of National Parks and Wildlife Service (DNPWS)
- EDF (EU funded Environmental Development Fund) & JICA (Japanese Aid) Projects
- IUCN Country Office
- Environmental Council of Zambia (ECZ)

ZIMBABWE
- Department of National Parks and Wild Life Management (DNPWLM)
- Forestry Commission
- WWF Regional Office
- IUCN-Regional Office for Southern Africa (ROSA)
- Biodiversity Foundation for Africa

Source: Denconsult, Sector Studies under ZACPLAN, *Sector Study No. 4, Impacts of Development within Tourism, Recreation and Wildlife*, 1998.

Photo: D Minty

Deforestation poses a major threat to the forest resources in the basin.

ensuring the sustainable utilisation and production of forests' goods and services. Associated objectives shared by most of the countries are sustainable supply of goods and services from their forests, maintenance or enhancement of forest productivity, and rationalisation of forest resource utilisation.[45]

One of the major constraints in the development of sustainable forestry management is that information regarding appropriate forest management systems for the various indigenous forest types is lacking in most of the basin states. This is partly attributed to past national forestry policies which

Table 4.7: SADC forestry policy and legislation

POLICY	ANGOLA	BOTSWANA	MALAWI	MOZAMBIQUE	NAMIBIA	TANZANIA	ZAMBIA	ZIMBABWE
Forest Policy	No	No	Yes	Yes	Yes	Yes	Yes	Yes
Forest Policy reviewed			Yes	Yes	Yes	Yes	Yes	-
Review Forest Policy finished awaiting approval	-	-	Finished and appr. 1996	Finished and appr. 1997	Finished and appr. 1997	No	Finished and appr. 1997	-
Attempt to prepare a Foresty Policy year	1989	1990	-	-	-	Draft Forest Policy 1994	-	-
Similar strategy paper prepared year	1996	-	-	-	-	-	-	-
Actual Forest Legislation/year	1962	1968	1997	1965	1996	1982	1973	1984 /1987
Forest Legislation under review/ last attempt	Yes	Yes	Yes, finished	Yes	Yes	Yes	Yes, finished	Yes
Review Forest Legislation finished, pending approval	Yes, review finished but no further info.	No	Appr, Forest Bill 1997,	No	Draft Forest Act	No	Draft Forest Act 1997	No

Source: IMERCSA, Draft Manuscript, Biodiversity of Indigenous Forests and Woodlands of Southern Africa, 2000.

gave higher priority to the establishment, management and protection of exotic softwood and hardwood forest plantations than to the indigenous forests. The problem can be addressed through forestry research and participatory management systems based on people's needs.[46]

Fisheries conservation

The fisheries of the Zambezi Basin, as elsewhere, are based on natural production processes and are, therefore, highly dependent on environmental conditions affecting the productivity of the water bodies.[47]

The use and management of aquatic resources in the SADC region are guided by policies and strategies that control exploitation and manage the habitats in which the resources are found. A general objective of management is the attainment of the optimum rate of exploitation. How this optimum is defined depends on the objectives of policymakers. In some cases, the objectives might have social considerations such as improvement of socio-economic conditions of small rural communities through the generation of jobs or improvement of income distribution. The design of any fishery management programme is aided by a clear understanding of the goals and targets of that programme.[48]

Inland fisheries is coordinated by the Inland Fisheries Sector Coordination Unit based in Malawi, which addresses the key issues highlighted in the SADC Inland Fisheries Sector Policy and Strategy Statement which are to:[49]

- Establish regional training and educational programmes based on assessed needs, and maximise the use of the institutional resources available in the region.
- Manage the exploitation and use of fish resources from the region so as to optimise sustainable yields and promote regional self-sufficiency in fish supply.
- Protect the inland fisheries resources and regional fisheries from possible negative effects of all actions impacting on the inland ecosystems.
- Enhance the knowledge of the region's fisheries resources available for development and exploitation.
- Reduce post-harvest losses, improve marketing of the fish, and increase the use of fisheries products in diets
- Develop a regional strategic research programme that is responsive to specific needs of management, processing, marketing and utilisation of the fishery resources and to enhance the region's scientific image at the global level.
- Promote the expansion of viable aquaculture schemes among the smallholder communities in the member states.

The SADC Marine Fisheries and Resources Sector, which applies to Mozambique and the Zambezi Delta, tries to increase fish as food in the diet of people and to earn foreign revenue through exports of high-value fish species.[50]

However, several management problems are hindering the sustainable utilisation through fisheries such as, for example:

- Lack of reliable data regarding capture and reliable aquaculture development statistics, which among other things, means that investment opportunities have lagged behind other industries.[51]
- Increasing fish thefts. Over the last few years, there has been an increase in fish thefts where fishing crews sell part of the daily catch and this is never declared at the company base.[52]

Areas of future conservation interest

Areas of particular interest for biodiversity can be selected from a basinwide perspective or from a continental perspective. From a continental viewpoint there are perhaps four areas of particular conservation significance: Lake Malawi/Nyasa, the swamps, floodplains and woodlands of the palaeo-Upper Zambezi, the Middle Zambezi and Luangwa valleys, the Lower Shire Valley and the Gorongoza/Cheringoma/Zambezi Delta area.

Lake Malawi/Nyasa

Lake Malawi/Nyasa is the southernmost of the deep-water lakes associated with the formation of the East African Rift Valley. The Rift Valley lakes are of moderate antiquity and, despite major fluctuations in water level over the centuries[53] have been the sites of much evolutionary radiation, especially in some fish groups and freshwater molluscs. Lake Malawi/Nyasa is the only place in the basin with a natural pelagic (open water) habitat, while Lake Kariba and Lake Cahora Bassa are very recent and artificial. The number of freshwater mollusc species is around 47 with 50 percent endemism. This high species number and level of endemics is not reflected in other groups.

The swamps, floodplains and woodlands of the palaeo-Upper Zambezi

The old, stable plateau of the palaeo-Upper Zambezi, covering the floodplains and swamps of the Upper

Map 4.4: Threats to fisheries in the Zambezi Basin in 1995

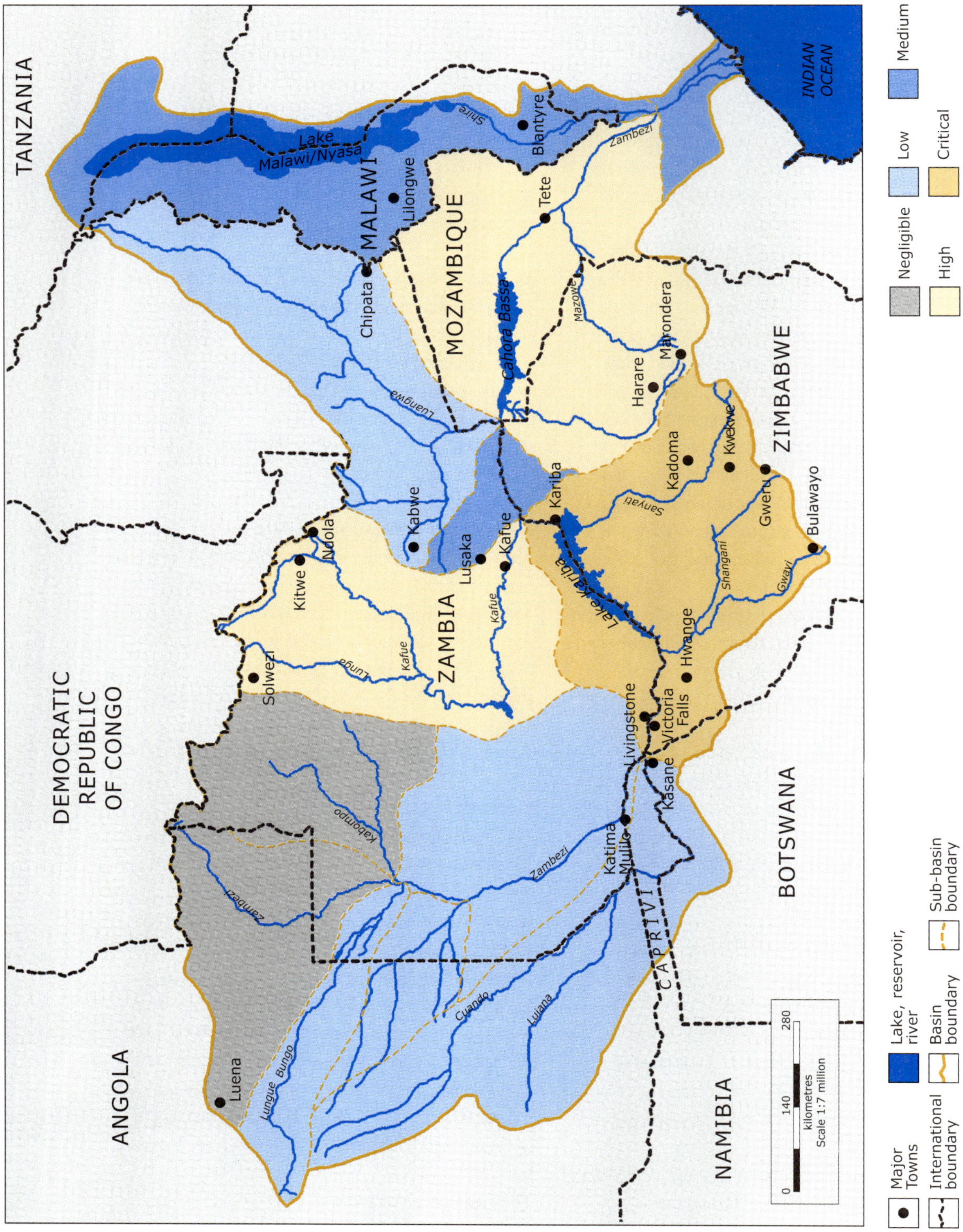

Source: Zambezi River Basin Atlas, SADC/DENCONSULT/ZRA, Lusaka, 1998

Photos: IUCN

The basin's floodplains are rich in animal and plant life.

Zambezi, Busangu, Kafue and Okavango as well as the associated woodlands on higher ground, have also been the sites of evolutionary radiation in a range of groups. The Zambezi floodplains are particularly rich in reptiles and amphibians. These floodplains, dambos and grasslands are some of the most extensive and least disturbed left on the African continent. The miombo, teak and *Cryptosepalum* woodlands on Kalahari sands in the west of the area are also of significant biodiversity interest.

The Middle Zambezi and Luangwa valleys

The section of the Middle Zambezi Valley from Kariba downstream to Cahora Bassa, and the Luangwa Valley in Zambia are two of the last remaining extensive protected areas for large mammals. They contain vegetation typical of hot dry valleys, such as mopane woodland, *Combretum* thickets and riparian fringing woodland. The interest is not in their biodiversity as such, or in unusual species, but in its potential to support reasonable populations of large mammals such as elephant, rhino and buffalo that are under threat elsewhere.

The Gorongoza/Cheringoma/Zambezi delta area

The Mount Gorongoza/Urema Trough/Cheringoma Plateau/Zambezi Delta area is extensive but covers an enormous diversity of habitats from mountain to mangrove,[54] not found in such proximity elsewhere on the continent. Mount Gorongoza, although just outside the Zambezi Basin, supports a wide range of montane and forest species, and the valley floor of the Gorongoza National Park has, in the recent past, supported large numbers of wildlife.

The Cheringoma Plateau, clothed in miombo woodland and dry forest containing many unusual species, rises up on the other side and then gently descends to the extensive grasslands of Marromeu, coastal dunes and mangrove swamps. Over this large area not only can viable populations of a multitude of species survive, but the ecological processes that sustain such a landscape can continue to operate.

At a more detailed level, there are a number of smaller sites of biodiversity interest. The headwaters around Mwinilunga are very rich biologically, containing many species not found elsewhere in the basin. The Kafue Flats are a well studied wetland system that not only supports an endemic lechwe antelope but also hosts seasonally vast numbers of waterbirds, including Palaeoarctic migrants. Mount Mulanje in southern Malawi supports extensive and rich rainforests on its slopes and a number of endemic plants up on the plateau. Although the Okavango Swamps, the largest and richest swamps in the region, are outside the present-day basin, they have been joined hydrologically and are still joined biologically.

FUTURE CHALLENGES

Although studies of the classification of living organisms (taxonomy) has named and classified many of the organisms found in the basin, this knowledge is not evenly distributed across the different biological groups. For example, over 95 percent of flowering plant species present have probably been named, but less than 50 percent of the beetles. Consequently, reasonable statements can be made on the biodiversity of groups such as flowering plants, mammals, birds and fish, but it is very unclear on total diversity and patterns in groups such as bacteria, beetles and other invertebrates.[55]

The biodiversity of the basin as a whole has not been comprehensively documented. Studies have been concerned principally with individual groups or with individual countries rather than the whole basin.

There is, therefore, a need for commitment by the governments of the region to implement and sustain integrated, multiple land-use management policies with rural people in order to realise the full benefits of flora- and fauna-resources.

LINKAGES TO OTHER CHAPTERS

Chapter 1: Regional Overview: People and Environment
The Zambezi Basin plays host to an abundant variety of ecosystems, species and genetic resources, which are critical to the development of the region, raising the standard of living for the people and alleviating poverty.

Chapter 2: Physical Features and Climate
The wide variety of landforms and vegetation zones has resulted in a remarkably high diversity of ecosystems and species in the Zambezi Basin. However, biodiversity can be seriously impacted upon in the event of serious climatic changes.

Chapter 3: Water and Wetland Resources
Water and wetlands are critical to the basin biodiversity sustaining both individual species and ecosystems.

Chapter 5: Agriculture
Threats to biodiversity in the basin include land-use practices such as land clearance for agriculture. The expansion of the human population and the increasing withdrawal on the natural capital is a threat to biological resources.

Chapter 6: Industry
Uncontrolled industrial development, particularly in biodiversity rich areas, can contribute to the destruction of ecosystems and reduction in species distribution.

Chapter 7: Energy
Activities in the energy sector have implications for biological diversity in the basin. Energy developments, such as construction of dams for hydropower have, for example, major effect on biodiversity through modification of habitat and reduction of flooding.

Chapter 8: Tourism
Tourism in the Zambezi Basin is closely related to the rich fauna and flora of the basin. Nature-based tourism, one of the fastest growing industries in the world, depends on healthy ecosystems rich in biodiversity.

Chapter 9: Pollution
Air, water and land pollution is a threat to biodiversity in the Zambezi Basin, and it often has an insidious chronic effect on biodiversity rather than a sudden acute effect.

Chapter 10: Poverty
The sustainable use of biological resources in the Zambezi Basin is crucial for poverty alleviation. The flora and fauna in the basin provide communities with food, medicines and other basic requirements.

Chapter 11: Gender and Women's Roles
Women play a major role in managing natural resources such as soil, water, forests and energy, but this recognition is far from generally accepted. Gender inequalities need to be reduced and women's participation in issues concerning natural resource management should be enhanced.

Chapter 12: Environmental Management and Regional Cooperation
Ecosystems transcend both national and political boundaries and often do occur in two or more countries and are subject to different and sometimes opposing management practices. The conservation and wise use of biodiversity should be placed firmly on the regional cooperation agenda.

Chapter 13: Trends and Scenarios
If development and land use continue to fly in the face of ecological processes then both biodiversity and, in the future, human development will suffer. Without biodiversity there would be no future.

ENDNOTES

1 World Wide Fund For Nature, Biodiversity and people, http://www.panda.org, 1996
2 World Resources Institute, WRI Article, The Diversity of Life, http://www.wri.org, 1992
3 White, F. *The Vegetation of Africa: a descriptive memoir to accompany the UNESCO/AETFAT/UNSO vegetation map of Africa*, UNESCO, Paris, 1983
4 Pope, G., personal communication with Timberlake, J.R.
5 White, F., "The underground forests of Africa: a preliminary note", Gardens' Bulletin 29, Singapore, 1976, p.57-71
6 ibid
7 IUCN, *Zambia Biodiversity Strategy and Action plan, the Country Study Report*, Lusaka, 1998
8 Schelpe, E.A.C.L.E., "Pteridophyta", In: *Flora Zambesiaca*, Crown Agents, London, 1970
9 Brock-Doyle, A., personal communication with Timberlake, J.R., plus additions
10 Cotterill, F., personal communication with Timberlake, J.R.
11 Mundy, P.J., personal communication with Timberlake, J.R.
12 Timberlake, J.R., *Biodiversity of the Zambezi Basin wetlands: review and preliminary assessment of available information — Phase 1*, Consultancy report for IUCN ROSA, Harare Zambezi Society, Harare/Biodiversity Foundation for Africa, Bulawayo, 1998
13 Marshall, B., personal communication with Timberlake J.R., and Skelton, P., "Diversity and distribution of freshwater fishes in east and southern Africa", *Annales de Musée Royal de l'Afrique Centrale, Zoologie,* 275, 1994, p.95-131
14 ibid
15 Owen, R.B., Crossley, R., Johnson, T.C., Tweddle, D., Kornfield, I., Davison, S., Eccles, D.H. & Engstrom, D.E., *Major low levels of Lake Malawi ad their implications for speciation rates in cichlid fishes*, Proceedings of the Royal Society of London, B 240, 1990, p.519-553
16 Fitzpatrick, M., personal communication with Timberlake, J.R.
17 ibid
18 Dudley, C., personal communication with Timberlake, J.R.
19 Chidumayo, E.N., *Miombo Ecology and Management*, Intermediate Technology Publications, London, 1997
20 Kenmuir, D.H.S., "The mussel resources of Lake Kariba", *Transactions of the Zimbabwe Scientific Association 60,* 1980, p.7-10
21 Balon, E.K., "Fishes from the edge of Victoria Falls, Africa: demise of a physical barrier for downstream invasions", Copeia 3, 1971, p.643-660
22 Williams G.J., and Howard G.W., (ed.), *Development and Ecology in the Lower Kafue Basin in the Nineteen Seventies*, Proceedings of a National Seminar on Environment and Change, Kafue Basin Research Committee, Lusaka, 1977
23 Eccles, D.H., "Lake flies, water fleas and sardines — a cautionary note", Biological Conservation 33, 1985, p.309-333
24 Douthwaite, R.J., and Tingle, C.C.D., (ed.), *DDT in the Tropics: the impact on wildlife in Zimbabwe of ground-spraying for tsetse fly control*, Natural Resources Institute, Chatham, 1994
25 UNEP, *Global Environmental Outlook 2000*, Earthscan, London, 1999, p.60
26 ibid
27 ibid
28 SARDC-IMERCSA, "CEP Factsheet, Zambezi River Basin Series No 2, Energy", 1998
29 Denconsult, Sector Studies under ZACPLAN, *Introductory Volume*, 1998, p.19
30 SARDC-IMERCSA, *The Zambezi*, "The Zambezi — a rich source of fish", Vol 1, No 2, 1998
31 SARDC-IMERCSA, *The Zambezi*, "Programme aims to improve crocodile management", Vol 2, No 1, 1999
32 op. cit 25
33 Meynell, P.J., Sola, L., and Nalumino N., *Strategic Environmental Assessment of Developments Around Victoria Falls*, 1996, p.3
34 op. cit 12
35 Dalal-Clayton, B., *Southern Africa Beyond the Millenium: Environmental Trends and Scenarios to 2015*, Environmental Planning Issues No 13, International Institute for Environment and Development: Environmental Planning Group, London, 1997, p.17
36 ibid p.48-49
37 SADC/IUCN/SARDC, *State of the Environment in Southern Africa*, Maseru/Harare, 1994, p.172
38 SADC Natural Resource Management Programme, Resources Africa, SADC-NRMP/WWF/USAID/IUCN, Vol 1, No. 4, Harare, 1997, p.2
39 IUCN, *Regional Networking and Capacity Building Initiative for Southern Africa (NETCAB), Phase 2* Proposal Oct 1998-Sept 2001, IUCN, Harare, 1998, p.32
40 Cumming, D., *Southern Africa: Natural Resources, Landuse and Conservation in relation to Trans Boundary Conservation Areas* (Draft), 1998, p.5
41 Trans Boundary Conservation Area — "Constraints and Possibilities: Community Perspectives", One page loose handout
42 op. cit 40 p.17-19
43 SADC, www.sadcreview.com, 1999
44 Directorate of Forestry, *Namibia Forestry Strategic Plan,* Ministry of Environment and Tourism, Windhoek, 1996, p.4
45 SADC, *Forestry Sector Policy and Development Strategy for the Southern African Development Community*, Lilongwe, Malawi, May 1997, p.6
46 op. cit 43
47 Denconsult, Sector Studies under ZACPLAN, Sector Study 2, 1998
48 SADC/IUCN/SARDC, *Water in Southern Africa*, Maseru/Harare, p.111
49 op. cit 43
50 ibid
51 IMERCSA, "The State of the Environment in the Zambezi Basin", Proceedings of a regional workshop held at Mazvikadeyi, Zimbabwe, 1999, p.119-121
52 Chenje,M., Sola, L., and Paleczny, D. (eds.), *The State of Zimbabwe's Environment 1998*, Government of the Republic of Zimbabwe, Ministry of Mines, Environment and Tourism, Harare, Zimbabwe, 1998, p.304
53 Owen, R.B., Crossley, R., Johnson, T.C., Tweddle, D., Kornfield, I., Davison, S., Eccles, D.H. & Engstrom, D.E., "Major low levels of Lake Malawi ad their implications for speciation rates in cichlid fishes", Proceedings of the Royal Society of London, B 240, 1990, p.519-553
54 Tinley, K.L, *Framework of the Gorongosa Ecosystem*, PhD thesis, University of Pretoria, Pretoria, 1977
55 op. cit 12

REFERENCES

Balon, E.K., "Replacement of Alestes imbiri Peters, 1852, by A. lateralis Boulenger, 1900 in Lake Kariba, with ecological notes", Fisheries Research Bulletin, Zambia 5: 119-162, 1971

Barbosa, L.A.G., *Carta Fitogeográfica de Angola*, Instituto de Investigaçâo Científica de Angola, Luanda, 1970

Douthwaite, R.J. & Tingle, C.C.D. [editors], *DDT in the Tropics: the impact on wildlife in Zimbabwe of ground-spraying for tsetse fly control*, Natural Resources Institute, Chatham, 1994

Eccles, D.H., "Lake flies, water fleas and sardines - a cautionary note", Biological Conservation 33: 309-333, 1985

Heywood, V.H., (ed.), *Global Biodiversity Assessment*, UNEP/Cambridge University Press, Cambridge, 1995

Hughes, R.H., and Hughes, J.S., *A Directory of African Wetlands*. IUCN/UNEP/WCMC, Gland, 1992

Maclean, G.L., *Robert's Birds of Southern Africa*, John Voelcker Bird Book Fund, Cape Town, 1993

Shaw, P.A. & Thomas, D.S.G., "Lake Caprivi: a late Quaternary link between the Zambezi and middle Kalahari drainage systems", Zeitschrift für Geomorphologie N.F. 32: 329-337, 1988

Smithers, R.H.N., *Mammals of the Southern African Subregion*, University of Pretoria, Pretoria, 1983

Thomas, D.S.G. & Shaw, P.A., "Late Cainozoic drainage evolution in the Zambezi Basin: geomorphological evidence from the Kalahari rim", *Journal of African Earth Sciences 7:* 611-618, 1988

Thomas, D.S.G., and Shaw, P.A., *The Kalahari Environment*, Cambridge University Press, Cambridge, 1988

Tinley, K.L., *Framework of the Gorongosa Ecosystem*, PhD thesis, University of Pretoria, Pretoria, 1977

Werger, M.J.A. [editor], *Biogeography and Ecology of Southern Africa*, W. Junk, The Hague, 1978

Wild, H., and Barbosa, L.A.G., *Vegetation map of the Flora Zambesiaca region*, M.O. Collins, Harare, 1967

5

AGRICULTURE

Photo: M Chenje

Agriculture is the mainstay of the Zambezi Basin states, supporting millions of people both as producers and consumers, and contributing immensely to the economies of the states.

Agriculture is the mainstay of the Zambezi Basin states, supporting millions of people both as producers and consumers, and contributing immensely to the economies of the states. On average, agriculture contributes about 34 percent of southern Africa's gross domestic product (GDP), employs 80 percent[1] of the total labour force and accounts for about 26 percent of foreign exchange earnings.[2] In the basin countries, the dominance of agriculture and its contribution to national development cannot be overemphasised. It contributes more than 50 percent of raw materials to industry.[3]

Both subsistence and commercial agriculture in the basin depends on rainfall, which is variable and unreliable across the region. Rainfall ranges from about 10millimetres (mm) in some parts of Namibia to about 2,800mm in some parts of Malawi. In addition to variability between countries, rainfall is also variable within a country and across seasons, making rain-fed agriculture risky. The countries in the basin experience many droughts and floods (see Chapter 2), natural disasters, which often cause wide-scale crop failures.

Drought is also an overriding factor in the degradation of cultivated lands and rangelands in many parts of the basin, impacting on plant cover. Drought increases soil degradation problems, while soil degradation magnifies the effects of drought. As a result of drought during the 1994-95 season, cereal harvests in parts of the basin declined by 35 percent compared to that of 1993-94, with maize harvests alone falling by 42 percent (SADC, 1996).

With the majority of the people in the basin dependent on agriculture and the region's food security worsening, the countries are always facing a serious challenge to achieve food self-sufficiency.

LAND TENURE AND ACCESS

Critical to the success of agriculture is land tenure and access to resources. Land tenure and property rights in the basin are a blend of the traditional systems, the colonial system and the post-independence government policies. Colonialism in all the basin states stripped the people of not only their traditional rights to resources but also to land, forcing them into unproductive areas with poor soils and rainfall. The result has been overcrowding in some areas and low crop yields due to overexploitation of soils. In Malawi, for example, 56 percent of the smallholders cultivated in 1987-88 less than a hectare of land, 31 percent had one-to-two hectares and only 13 percent had more than two hectares.[4] Land pressure is so high in Malawi that it has forced 28 percent of marginal or unsuitable land into cultivation. Smallholder farmers do not practice soil and water conservation technologies, exacerbating the environmental problems.

Photo: M Chenje

Under communal tenure, the fundamental problem is that of land allocation.

Tenure regimes in the basin countries include state, private, communal and open access dating back to the historical acquisition of land. Under state property regime, the main landuse categories are forests, parks and conservation areas where the state either directly manages and controls the use of resources through government agencies or leases the resources to groups or individuals with usufruct rights for a specified period.[5] In open access there is no defined user group or owners hence the benefits are available to anyone.

Communal or common property is private property for a group of co-owners, each of whom has structured rights and duties within the regime. Like private property, strict common property excludes non-owners. Common property regimes are characterised by common interests, common cultural norms, indigenous authority systems and some interaction among community members.[6]

Under communal tenure, a fundamental problem arises with respect to land allocation, access to resources and conflicts between different social groups. The presence of numerous interest groups such as the state, private individuals, landless peasants and the international community all competing for land puts pressure on local institutions thereby compounding the situation.

In Botswana are three categories of land ownership, namely tribal comprising 71 percent of national land, state comprising 24.8 percent and freehold about 4.2 percent. The trends in Botswana indicate a deliberate policy of increasing land ownership by the local people at the expense of state and private freehold.[7] Communal tenure is the most complicated as land is entrusted to the state and traditional leaders.

In Mozambique, all land became state property soon after independence in 1975. However, since 1979 legislation on land was passed and different models that recognise property rights of communities, companies and private individuals have been developed. The new Land Bill of 1997 recognises security of land ownership for the family, which can acquire title deeds. Under this arrangement two methods guaranteeing access to and possession of land were established, that is recognition occupation based on oral testimony and demarcation of community land.[8]

Namibia's land tenure is probably the most complicated in the region. The country does not have an encompassing and comprehensive legislation on land although the Land Policy vaguely makes an attempt. Some overlapping and sometimes contradictory land laws from colonial regimes and government, and traditional laws determine Namibia's land tenure, resulting in inconsistent application.[9]

Map 5.1: Land Cover/Use in the basin

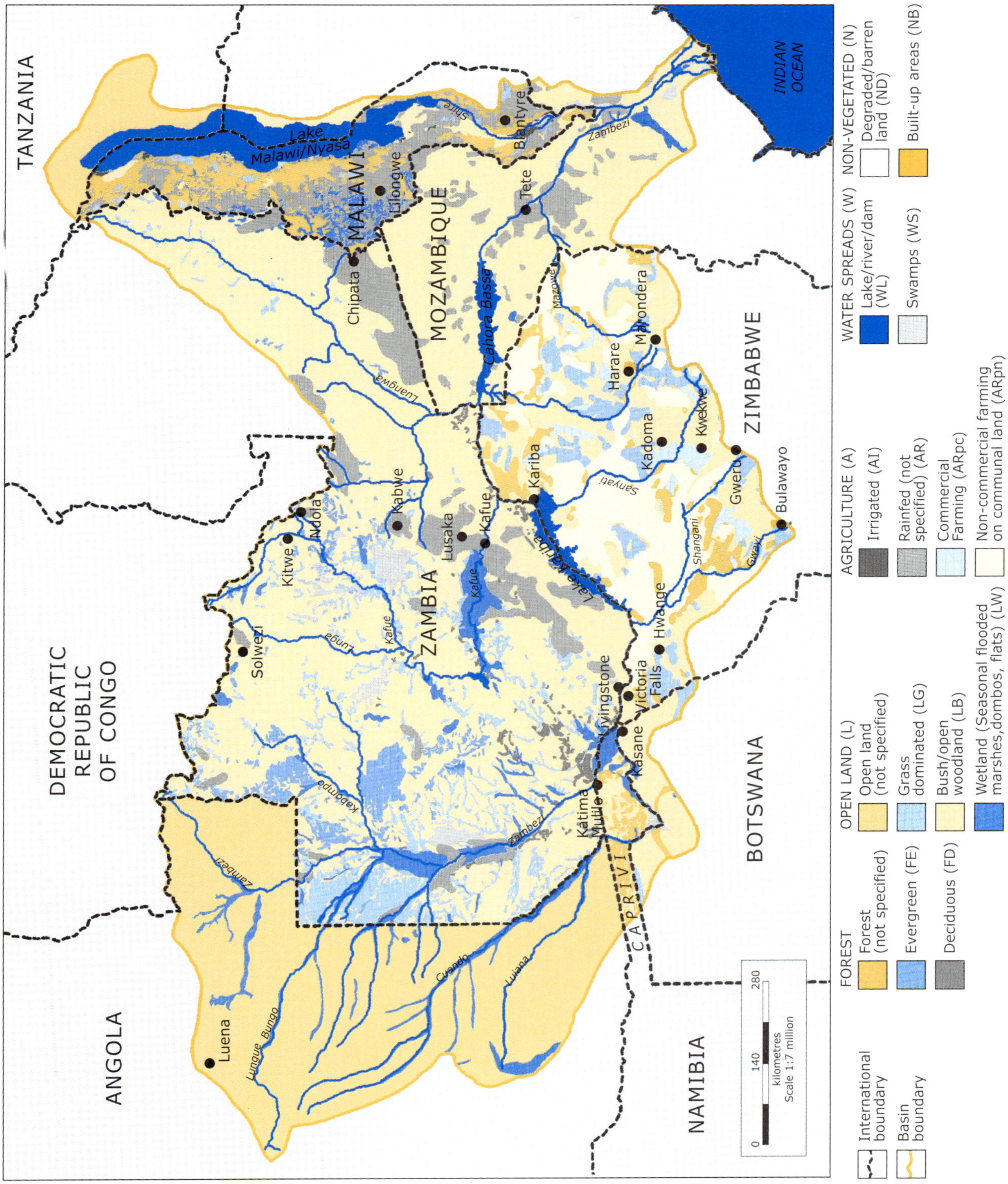

Source: Zambezi River Basin Atlas, SADC/DENCONSULT/ZRA, Lusaka, 1998

Namibia's rural population is about 75 percent and about 90 percent of this population lives in communal areas. Communal and commercial farming lands cover about 44 percent of the country with commercial farming located in the drier south and central part of the country while communal farming land is in the wetter north and east. The current land reform process is proceeding slowly with only about 50 farms (less than one percent of commercial farmland) having been bought for resettlement and only 2,000 people resettled.[10]

Photo: IUCN

Under customary tenure, communal farmers in Zambia receive land for different uses, including cropping.

Zambia has undergone three land reform phases since independence in 1964. The first phase converted all land into state property with no selling rights. From 1985–95 came the second phase, which protected customary land tenure and allowed local authorities to recommend land to be given under title deeds. However, most land is held under customary land tenure due to the slow implementation of the reforms. Under customary tenure, community members receive land for different uses such as residential, grazing and cropping. Grazing land is shared while croplands remain within the family and usufruct rights passed on for generations.[11]

In Zimbabwe, land reform was in two phases. After 1980 reform took the form of resettlement of communal farmers on identified commercial farms on a willing-buyer willing-seller basis. The second phase is being implemented and no payment is received for the designated land. The tenure categories are state, private, communal and open access whose characteristics are as explained earlier.

The increasing land-based conflicts are due to, among other factors, weak natural resource administration institutions (policies and laws), lack of clarity of roles of the responsible institutions and poor enforcement of existing laws and regulations. The social, institutional, administrative and legal issues of tenure determine utilisation, ownership rights and interaction over land. A range of property regimes with different legislation and administrative structures, different resource endowments, population densities, infrastructural development and rules and regulations govern the allocation and management of land in the basin countries.

AGRICULTURAL ACTIVITIES

The basin has 15 agro-climatological zones, which vary in rainfall and growing season. Maize, tea, tobacco, coffee, sugar, wheat and cotton are some of the major crops produced under commercial agriculture in the basin states. Tea, tobacco, coffee, sugar and wheat are generally grown through irrigation but in some cases, these crops are also grown under rainfed agriculture. The commercial agriculture sector is highly mechanised, however much labour is still required and the labour force usually consists of poor rural communities.

Rainfed and irrigated crop production in the basin accounts for 86 percent of about 5.2 million hectares of the yearly-cultivated area in Malawi, Zambia and Zimbabwe. Malawi has 90 percent of its cultivated area in the country in the basin, Zambia 76 percent and Zimbabwe has 56 percent.[12] More than 60 percent of total cultivation in the basin is concentrated in the Tete and Shire River-Lake Malawi/Nyasa sub-basins while 90 percent of total crop production is in five of the 13 sub-basins. Malawi, Zambia and Zimbabwe also dominate in cattle rearing with 90 percent of the animals in the basin.[13]

The basin has various farming systems that include controlled water use from small and traditional irrigation systems for smallholders aimed at increasing food security and dietary diversity, and large commercial/corporate irrigation farms used for perennial mono-cropping, mixed cropping and horticulture. Cash cropping in the basin is largely limited to tobacco, cotton, citrus, coffee, tea, sugar, wheat and barley and horticulture with varying commercial production in the basin countries. Cattle rearng is also important and the most common cash producing enterprise of any farming system throughout the basin, particularly in and around floodplains.[14] The farming systems and agricultural production in the Zambezi Basin are discussed in more detail using a sub-basin approach in the next sections.

Farming systems and agricultural production

The Upper Zambezi

The dominant crop, maize is grown throughout the basin. The highest yields of maize occur in the Upper Zambezi sub-basin and in the highlands of the Lungue Bungo River sub-basin while the lowest hectarage by district is found in the Botswana area of the basin. In the higher rainfall areas, maize is partially replaced by cassava and in the south it is partially replaced by sorghum. The highest yields of cassava in the basin have been observed in the Northwestern province of Zambia while rice is grown extensively in the Upper Zambezi, Luanginga River and in the Barotse sub-basins in areas that are prone to annual flooding.

Angola and Zambia dominate in the Upper Zambezi Basin but the majority of the agriculturally potential lands lie in Angola. These lands, however, are not fully utilised due to the low population in these areas as a result of civil war. The basin lies in two of Angola's agro-ecological regions, the central region, which is suitable for crops such as maize, sisal, sunflower and wheat; and the southern region which is mainly used for grazing and the cultivation of sorghum and wheat. The main landuse in the eastern region of the basin, which is inhabited by the Lunda and Luena tribes, is agriculture but they also do some fishing. Crops grown include cassava, groundnuts, beans, soya, maize and rice.

Agriculture on the floodplains and dambos, which are interspersed with well-wooded interfluves of Kalahari sands, is practised along the narrow, fertile margins between the floodplains and sand. Farms in these areas are fairly small with sizes ranging from two to four hectares. The main crops grown include staple crops, fruits and vegetables that are generally consumed and some sold locally. In the central region of the Angola area of the basin, the terrain is made up of floodplains with only a few areas of dryland, which are suitable for agriculture. The main activities of the Quioca and Luena tribes in this part of the basin are fishing and hunting. The northern area's main landuse practices include agriculture (cassava and peanuts), fishing, bee keeping and the cutting and selling of exotic wood while in the southwest region cultivation of cassava, peanuts, maize and groundnuts is practised by the Luchaze and Bunda tribes.

Cash cropping has been limited to tobacco and cotton production mainly in the Kaomo area. However, attempts have been made to produce cashew nuts along the margins of the main Zambezi floodplain but yields have been very low. Sunflowers and millet are grown generally throughout the Upper Zambezi Basin. The highest hectarage of millet by district has been recorded in the Cuando/Chobe sub-basin in Angola, in the Caprivi Strip, in the Barotse and Luanginga sub-basins. The most productive areas in the Upper Zambezi Basin include the Angolan highlands and parts of Solwezi, Kabompo, Mufumbwe, Kasempa and Kaoma districts where the land is not covered by Kalahari sands. However, rainfall is high in these areas and the soils are generally of low fertility. Rivers and streams in these areas tend to be perennial and the farms are generally of a large size compared to the average for the basin, often exceeding ten hectares.

Photo: IUCN

Maize is the dominant crop and is grown throughout the basin.

Throughout the area in the Upper Zambezi, but particularly in and around the floodplains, cattle rearing predominates and this is the most important cash producing activity in the area. The cattle migrate annually from the floodplains to the higher surrounding land in summer. In high rainfall areas where soil fertility is poor, shifting slash-and-burn subsistence agriculture is giving way to a more settled state as population pressure increases.[15] In the Barotse floodplain, for example, livestock rearing is the major component of the farming system.

Table 5.1: Upper Zambezi Basin areas under cultivation

Sub-basin	Zambia	Angola	Namibia	Botswana	Total
Kabompo	44,458	726	-	-	45,184
Upper Zambezi	8,754	33,766	-	-	42,520
Lungue Bungo	2,423	10,875	-	-	13,298
Luanginga	11,010	6,331	-	-	17,341
Barotse	113,871	462	3,100	-	117,433
Cuando/Chobe	2,793	40,032	11,638	868	55,331
Total	183,309	92,192	14,738	868	291,107

Source: Denconsult, ZACPLAN: Sector Study No.2, Final Report, SADC/ZRA, Gaborone/Lusaka, August 1998

In Botswana, subsistence cultivation of cereals is the major agricultural activity in the area and the rearing of livestock is also carried out. Due to the erratic nature of flooding in the area, commercial farming is not viable; and only one commercial farming operation is situated in Kazungula, and a number of smaller operations such as prisons and brigades irrigation farms.

Within the Caprivi area of Namibia, the only irrigated crops are a few hectares of tobacco and maize. The rest of the crops are rainfed and these include the dryland crops, for example maize, sorghum, millet and the river-fields on the flood-plains where vegetables are grown.[16]

The Middle Zambezi

The Middle Zambezi is dominated by Zambia and Zimbabwe and agriculture in this area includes both subsistence and commercial farming unlike the lower Zambezi. In Zambia, the majority of the land in the Zambezi Basin is for subsistence farming. Both commercial farming and peasant subsistence farming are practised in these areas with no designated separation of the two types of farming. In Zimbabwe, however, there is definite separation of the commercial and communal farming areas.

The most productive area in Zimbabwe is the plateau in the south along the watershed and the west between Harare and Karoi. This entire area is dominated by commercial farms with only a few intervening communal farming areas, and is characterised by high rainfall. The area to the south of the basin, where the majority of the communal areas have been designated, lies in the lower rainfall lands. Farming systems in the communal areas range from small-scale commercial through resettlement to communal and they account for about 40 percent of the area of the Zambezi Basin in Zimbabwe. Animal drought power is extensively used and cattle are an important factor in all farming systems. Maize and sorghum are the dominant food crops, while cotton and sunflower the most important cash crops.

Commercial farms in Zimbabwe account for about 35 percent of the area of the Zambezi Basin in the country and they consist of large farming units, most of which exceed 200 hectares in size. Generally, the population densities are low due to the large farms. Arable farming areas have been opened up in most cases and the remaining land is left in its natural state and used for grazing.

Table 5.2: Middle Zambezi Basin areas under cultivation

Sub-basin	Zambia	Zimbabwe	Botswana	Mozambique	Total
Kafue	372,363	-	-	-	372,363
Kariba	83,714	626,145	200	-	710,059
Luangwa	438,965	-	-	2,883	441,848
Mupata	13,330	5,376	-	-	18,706
Total	908,372	631,521	200	2,883	1,542,976

Source: Denconsult, ZACPLAN: Sector Study No.2, Final Report, SADC/ZRA, Gaborone/Lusaka, August 1998

Photo: IUCN

Peasant subsistence farming is practiced throughout the basin.

Traditionally, shifting slash-and-burn agricultural systems predominate in the extreme northeast and northwest of the plateau in the northern reaches of the Middle Zambezi although population pressure is now imposing more sedentary systems.[17] Highly productive commercial and corporate farming enterprises exist in the central sector of this northern, high-rainfall region.

In the Kabwe, Mukonchi and Chisamba areas in Zambia, more commercial farms are located and cotton is becoming an increasingly important crop among emergent commercial farmers that have relocated from the drier south. West of Mumbwa and adjacent to the Kafue River is the Great Concession commercial farming area which is underutilised due to poor communications and lack of electricity.[18] Producers of horticulture and floriculture crops dominate the Lusaka/Kafue area. The limestone aquifers in the area provide water for irrigation.

Tobacco is the major crop grown on the few commercial farms in the Southern Province of Zambia. Cattle rearing is the main activity carried out by the peasant farmers. Sugar and irrigated wheat and coffee are the main crops grown in the Mazabuka area and the Nakambala sugar estate is located in this area. In the Eastern Province plateau adjacent to the Malawi border, maize is grown widely while cotton, groundnuts and tobacco are the main cash crops. The use of animal draught power is extensive.

The hoe method of cultivation is practised in the Luangwa and Zambezi Valley areas. This method involves cutting down trees and sometimes debarking and then piling the biomass into scattered heaps.[19] Generally, the low-lying rift valley systems are unproductive because of the low and erratic rainfall, high temperatures and very mixed soils that are prevalent. The crops grown include sorghum, bulrush millet and early maturing maize varieties, including vegetables at the end of the rainy season. Two irrigation schemes are found at Sinanzongwe on Lake Kariba and at Chiawa at the Kafue/Zambezi confluence.

Photo: M Chenje

Tobacco is the major crop grown on the few commercial farms in the Southern Province of Zambia.

The Lower Zambezi/L. Malawi/ Nyasa-Shire River Systems

In Tanzania, agriculture is mainly rainfed. However, two irrigation schemes are found in the Mbeya region. The major crops grown include cassava, maize, bananas and rice in the lowland areas of the Kyela plains and cash crops grown include tea, tobacco, coffee, cocoa and maize. One of the major problems contributing to low agricultural production is flooding. This tends to limit the area planted to rice. When floods occur, the planted area is reduced to about 7,000 ha.[20]

The main form of landuse in the Lake Malawi/Nyasa basin is rural subsistence farming. The dominant factor with regard to landuse is the large population of the country that is virtually packed on a relatively small area of land. The commercial estates are designated farming areas where other forms of settlement and farming are

Table 5.3: Lower Zambezi Basin areas under cultivation

Sub-basin	Zimbabwe	Mozambique	Malawi	Tanzania	Total
Shire R./ Lake Malawi/ Nyasa	-	41,069	1,759,506	50,000	1,850,575
Tete	743,242	207,445	-	-	950,687
Zambezi Delta	-	164,832	-	-	164,832
Total	743,242	413,346	1,759,506	50,000	2,966,094

Source: Denconsult, ZACPLAN: Sector Study No.2, Final Report, SADC/ZRA, Gaborone/Lusaka, August 1998

restricted. Commercial estates are few and usually very crop specific. They are mainly used for sugarcane production in the Lower Shire Valley at Nchalo and Dwangwa (on the lakeshore), tea production in the Mulanje and Thyolo districts, rubber in Nkhata Bay and smaller farming units for production of tobacco throughout the southern and central parts of the country.[21]

Cash crops are many and varied. Tobacco is economically the most important and mainly grown by tenants on large and small estates. Other cash crops such as cotton and groundnuts are not tied to the estate farming system to the same extent as tobacco. Most of the land is for rural subsistence farming with individual family units allocated generally less than five hectares for both cultivation and animal rearing. The hectares of all crops grown in the Lower Zambezi Basin by country and sub-basin are given in the table below.

Table 5.4: Land potential and utilisation by country, ('000 square kilometres)

Potential	Gross Area	Arable Estimate	Current Cropping	Balance Uncropped	% Utilisation
Angola	246	32	1	31	3
Botswana	21	-	-	-	18
Malawi	108	23	19	4	83
Mozambique	165	31	4	27	13
Namibia	16	1	-	-	9
Tanzania	23	6	3	3	50
Zambia	575	67	12	55	18
Zimbabwe	217	44	14	30	32
Total	1,371	204	53	150	26

Source: Adapted from Denconsult, ZACPLAN: Sector Study No.2, Final Report, SADC/ZRA, Gaborone/Lusaka, August 1998

The dominant staple food throughout the area is maize, but cassava predominates on the lakeshore districts like Nkhota Kota, Nkhata Bay and in Tanzania. The Southern region of Malawi has the richest soils for agriculture, and plantations take up a major part of that land. Along the Shire Valley, sorghum partly replaces maize and in wet lowland areas around the north of the lake and at Nkhota Kota on the western lakeshore plain, rice predominates. In the south, are large commercial tea and tobacco plantations around Mulanje and the biggest sugar estate in Malawi is also in the south.[22]

In Zimbabwe, agricultural activity occurs in the commercial and communal farming areas as described in the Middle Zambezi. The climatic conditions experienced suit a wide range of agricultural systems and crops. The cooler Eastern Highlands where winter rainfall prevails are suited to the commercial production of fruit, vegetable and tree crops as well as coffee and tea. Maize is the staple food but is augmented in drier areas by sorghum and millet. Cotton is an important cash crop in this region.

The Zambezi Basin in Mozambique is almost entirely used for rural subsistence agriculture. In the delta area, farming is dependent on the available moisture. In areas that are seasonally flooded, relay cropping is practiced with rice, maize and beans. Two sugar estates are located in the delta area and coconuts are grown in areas near the coast. In the hinterland where elevations are higher, cassava takes over from maize and rice as the staple crop and cotton is grown on the alluvial soils at the extreme west end of the delta.[23] The Lower Zambezi Basin is more utilised than the middle and upper basins.

FARMING SYSTEMS

Large corporate perennial mono-cropping

These are large corporately owned estates comprising many thousands of hectares devoted to a single crop. Sugar is the most important crop grown under this system with one scheme in Zambia, two in Malawi and two in Mozambique although currently out of production. In Botswana and Zimbabwe citrus is gaining prominence under this category.

Photo: IUCN

Large estates are involved in perennial mono-cropping, and citrus farming is gaining prominence in Botswana and Zimbabwe.

Map 5.2: Major irrigation schemes in the basin

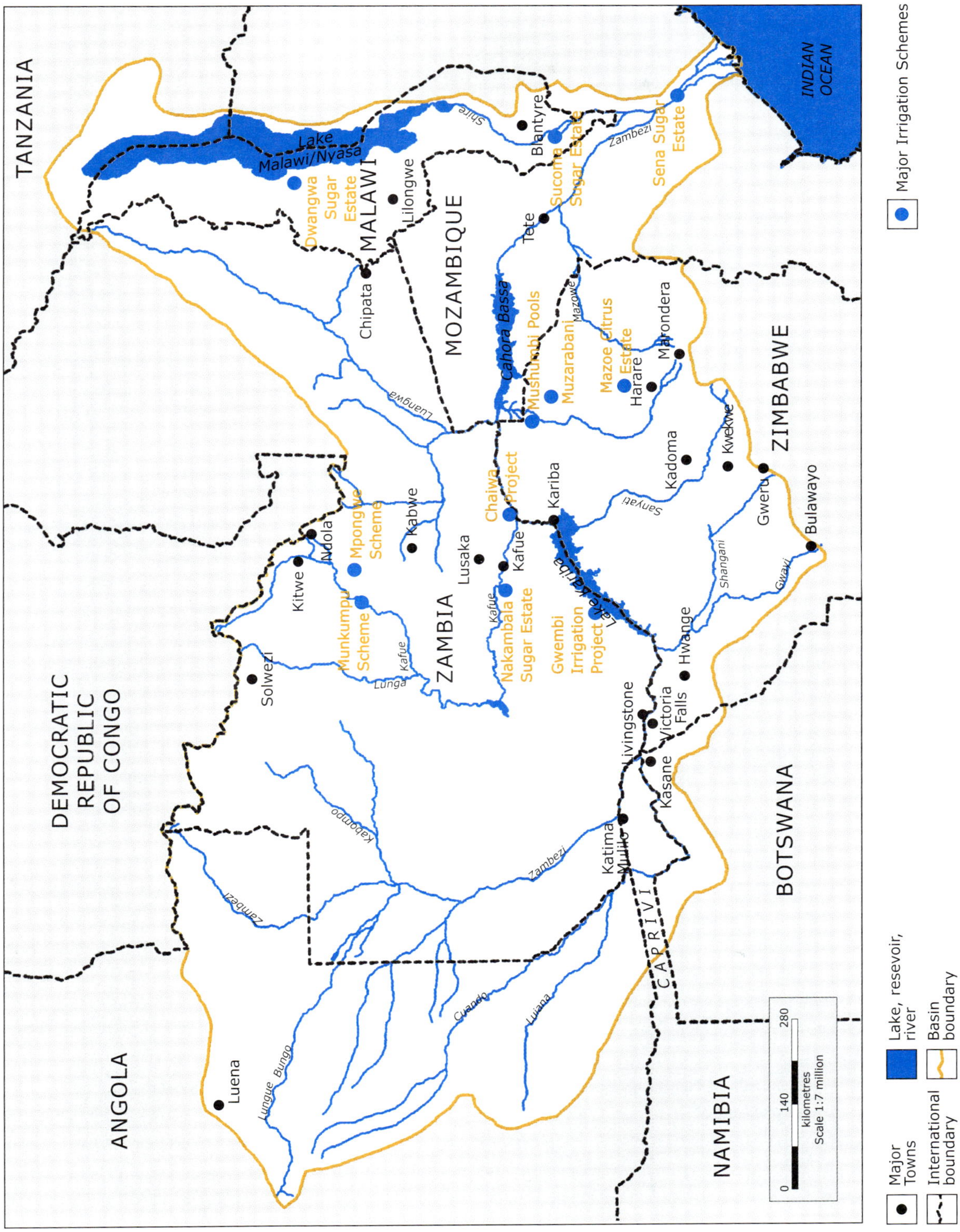

Source: Zambezi River Basin Atlas, SADC/DENCONSULT/ZRA, Lusaka, 1998

This farming system requires large quantities of water that are usually provided from private dams or run-of-river. Flood irrigation, overhead sprinklers and pivots are the most common irrigation methods used.

Table 5.6: Zambezi Basin cattle numbers by country

Country	Upper-basin	Middle-basin	Lower-basin	Total
Angola	151,140	0	0	151,140
Botswana	2,000	0	0	2,000
Malawi	0	0	1,252,566	1,252,566
Mozambique	0	0	115,170	115,170
Namibia	83,425	0	0	83,425
Tanzania	0	0	350,655	350,655
Zambia	592,190	1,515,747	30,626	2,138,563
Zimbabwe	0	1,557,593	1,245,153	2,802,746
Total	828,755	3,073,340	2,994,170	6,896,265

Source: Adapted from Denconsult, ZACPLAN: Sector Study No.2, Final Report, SADC/ZRA, Gaborone/Lusaka, August 1998

Large commercial/corporate summer mixed cropping

Most common in Zambia, Zimbabwe, and to a less extent Malawi, this type of farming has a pre-designed rotation in which about 50 percent of the land is irrigated. The popular irrigated rotation is wheat/soya (two crops per year) and the un-irrigated land dedicated to a maize/soya rotation (one crop per year).[24] However, part of the irrigated land may be devoted to perennial crops such as coffee or citrus.

Table 5.5: Irrigation water use in basin countries (Ml/annum)

Country	Current	2005	2015
Angola	6,282	9,630	220,300
Botswana	6,656	10,030	10,050
Malawi	390,639	401,000	438,000
Mozambique	44,886	56,222	202,825
Namibia	5,255	10,400	12,900
Tanzania	21,619	26,350	32,100
Zambia	397,490	425,290	476,920
Zimbabwe	604,108	726,200	798,500
Total	1,476,935	1,665,122	2,191,595

Source: Adapted from Denconsult, ZACPLAN: Sector Study No.2, Final Report, SADC/ZRA, Gaborone/Lusaka, August 1998

Due to the rising cost of water and its scarcity, farmers are converting to pivot and drip irrigation systems for annual and perennial crops respectively. The overhead irrigation system has proved less efficient, forcing the conversion, especially in Zimbabwe.

Commercial mixed cropping

In the Zimbabwean part of the basin, the major water users are commercial farmers, whose average farm size is about 1,200 hectares. The average cropping pattern is wheat, maize, tobacco, soya, cotton and horticulture. Most farms have water and storage rights. Irrigation water from private dams, boreholes and weirs accounts for about 70 percent while flow rights account for 10 percent with the balance coming from government dams.[25] Beef cattle, dairy, sheep and game form the livestock component of farms under this category.

Commercial horticulture

This type of farming is most common in Zambia and Zimbabwe. It is a specialised commercial mixed cropping system where horticulture and floriculture crops for export are grown under controlled greenhouse conditions. Specialised management, irrigation, pack-sheds and cool stores are critical factors requiring heavy investment.

Flowers (roses), peas, runner beans and others are the main crops grown and field crops are relay planted throughout the winter months.

Government irrigation schemes

Formal irrigation schemes for smallholders are more common in Zimbabwe than other parts of the basin. Access to gravity fed water from perennial sources was the main reason why these schemes were constructed. However, the perennial sources have become scarcer in the basin, making flood and sprinkler irrigation more common. A mixed farming system of maize, beans and vegetables as core crops and fruit trees and livestock are practised. In Malawi, such schemes are dedicated to growing rice twice a year.

Traditional irrigation system

These vary from small plots that are bucket-irrigated from a well to perched water tables exploitation. Traditional irrigation systems are widespread and make up a considerable irrigation capacity when added together. However, this is not well documented. River bank maize, tobacco and vegetable production constitute perched water table irrigation used in dry seasons.

Photo: IUCN

Small plots are usually bucket-irrigated and used to grow vegetables to either sell or family consumption.

The consumption linkages that arise from agricultural income trigger household consumer demands for non-agriculture goods.

The basin's rural population spends more income on local services, durables, livestock and horticultural products. Ensuring equitable distribution of agricultural income gains among rural households can lead to higher employment and income multipliers as well as reduce poverty. Agriculture has been a major source of capital and revenue for non-agriculture sectors, and surpluses are consistently transferred out of agriculture through fiscal, crop pricing and trade policies.[28] Private investors use savings from agriculture as investment fund for non-agriculture pursuits.

The indirect growth stimulated by agricultural growth leads to increases in the demand for agricultural products, in turn stimulating further expansion of food crops and livestock production. The demand effects are greater in poorer countries of the basin because they depend on high marginal budget shares for food. Some of the basin countries, with the exception of Angola, Botswana, Namibia and Zimbabwe, are rated as low-income economies that tend to have a high share of agriculture in gross national product.[29]

Despite playing a dominant role in the economies, addressing poverty, food security and environmental problems, agriculture in the basin states is facing lack of development and stagnation due to poor macroeconomic policies. The 'food production-population imbalance' in the basin countries requires that food supplies grow at four-five percent per year, a rate most countries have been unable to achieve. Civil conflicts and drought have exacerbated the levels of food insecurity in the basin.

The poor macroeconomic environment, debt crisis and deteriorating terms of trade, characterising most basin countries are taxing the sector and making it more difficult for the sector to move faster and commercialise smallholder agriculture.

This type of production includes dambo vegetable growing, tobacco nurseries in Malawi and Zambia. Dambos are subject to seasonal excessive wetness to an extent they influence land-use. Among the uses for dambos developed by traditional communities are: water source for domestic use and livestock, grazing, cultivation of cereals, root crops and vegetable.[26]

AGRICULTURE AND ECONOMIC GROWTH

Agriculture dominates the economies of the basin states with cash crops accounting for at least 60 percent of export earnings.[27] Agriculture is likely to maintain its strategic role in the economy of basin countries into the foreseeable future given its forward and backward linkages to the non-agriculture sectors.

Agricultural growth contributes to overall economic growth in a number of ways. Among others, the sector:

- increases the exports of agricultural products that raise income and foreign exchange earnings;
- increases food supply for domestic consumption and raw materials for domestic industries that in turn reduce the price of wage goods; and
- promote growth in manufacturing.

AGRICULTURE AND THE ENVIRONMENT

Awareness of the limited capacity of regional ecosystems to support and sustain ever-growing human economic activity has increased over the last 10 years. In most basin countries, the capacity to sustain economic activity is approaching its limits as evidenced by widespread environmental degradation.[30]

Agricultural intensification can lead to mismanagement of inputs such as fertilisers, pesticides such as methyl bromide and irrigation water with serious environmental problems such as water logging and salinisation of irrigated land, chemical contamination of water, pesticide poisoning and destruction of species. Although isolated cases have occurred, the basin has largely avoided most of the above problems by the mere fact that none of the countries widely participated in a modern green revolution.

Environmental degradation in the basin is a visual testimony of the self-destructiveness of an impoverished agricultural sector. The 'food production-population imbalance' of the basin is leading to production increases through opening up of new and sometimes marginal land as well as intensification of agricultural production. Without adequate agricultural yields to secure their livelihoods, farmers are expanding into environmentally fragile areas. Fallow periods have dropped drastically, usually unaccompanied by better fertility management, resulting in poor yields. This in turn increases weeds, soil acidity and erosion.[31]

In Botswana, most of the land is devoted to cattle and wildlife grazing with over half of the rural population owning cattle. Land degradation due to overgrazing and prolonged droughts is a serious problem and attempts to contain it have been largely unsuccessful to date.[32]

One of the more serious environmental problems in the basin's agriculture is the mismanagement of resources in extensive farming systems. Rainfed farming has resulted in the:

- conversion of primary forest to agriculture;
- loss of biodiversity;
- climate change;
- exposure of fragile soils;
- soil erosion and lowland flooding;
- degradation of watershed protection areas; and
- shorter fallows with loss of soil nutrients that increase pressure on common property resources such as woodlands and grazing areas.

Indigenous institutions that regulate and manage resources have broken down, leading to open access regimes, declining ecosystems resilience and reduced ability to recover from droughts.

Photo: M Chenje

Rising population and severe land shortage have seen cultivation expanding to marginal lands and steep slopes.

In Zimbabwe, where more than 69 percent of the people live in the rural areas, the inequitable distribution of land has led to problems such as high population densities, high rates of soil erosion, deforestation, land fragmentation into uneconomic units, low productivity, overstocking and overgrazing. More and more people are being forced to settle in riverbeds, mountaintops, grazing areas and fragile lands, exacerbating environmental problems in the country.

Resource degradation as a result of poverty and population pressure has reached serious levels. Malawi is one such country with a rapidly rising population and severe land shortage where cultivation has expanded to marginal lands and steep slopes. Soil erosion rates have increased in the central and southern regions of the country while shorter fallow periods and mono-cropping are reducing soil fertility and agricultural productivity in densely populated areas.[33]

Competition for land and water resources between sectors and production systems is projected to intensify in the basin. Food supplies in the basin are dependent on irrigated cereals production that is expected to grow despite the high cost of irrigation. The growing water demand for irrigation, industrial and domestic use is due to increased competition for water, which is pushing up the price beyond a level profitable for staple food production.[34] Future water supply is threatened by qualitative problems arising from agriculture, posing a health threat to crop yields, people, livestock and wildlife.

Rising salt concentrations in irrigated areas, fertiliser and pesticide contamination of surface and groundwater as well as discharge of effluents are some of the environmental threats from agriculture. In the basin, any impact due to fertilisers, pesticides and herbicides mostly results from excessive use or poor application. Fish deaths and other aquatic life impacts are indicators of chemical pollution of rivers.

Zimbabwe still imports most chemicals that are either banned or severely restricted for health or environmental reasons, cited in the Prior Informed Consent (PIC) Code of Conduct adopted by the Food and Agriculture Organisation (FAO) in 1985. The government still uses DDT as a control mechanism, having bought 10 tonnes in 1997 to control bubonic plague in Matabeleland North Province.[35] However, the commonly used modern pesticides sold by commercial companies break down into harmless or less toxic products.

Biomass burning from deforestation and savanna fires account for about 30 percent of the total amount of carbon dioxide emitted by human activities.[36] The basin is already vulnerable to climate variations that are manifested in arid areas with low rainfall. Further expansion of the agricultural sector can lead to forest clearance and burning of trees and shrub layer which will release greenhouse gases into the atmosphere. Methane from livestock and wet rice fields as well as nitrogen oxides from crop fertilisers are other agricultural sources of greenhouse gases. However, in terms of global warming, basin countries contribute negligible amounts.

In Zambia, deforestation is a key environmental problem. Seven million cubic metres of fuelwood are consumed annually and indigenous forests are the major source of sawn timber, poles and mining timber.[37] Shifting cultivation has resulted in land clearance for agriculture and reduction of indigenous species cover.

Livestock has two major impacts on water quality, namely nutrient enrichment and sedimentation of rivers resulting from land degradation caused by overgrazing from overstocking. In communal areas of the basin countries, an increase in sediment concentration in surface waters and siltation of reservoirs, weirs and rivers is evident as has been observed in the Mazowe Valley. Cattle and goats are the dominant livestock whose population pressure ensures that they are grazed on hills or wet areas unsuitable for crop production. Goats are more efficient

Box 5.1: Impacts of overstocking

Overstocking represents a major factor contributing to the deterioration of land quality in the communal lands and the people to poverty. The communal land producers have always used the maintenance of large and different types of livestock as a survival strategy. The level of stocking is determined by economic objectives rather than the ability of the land to support large herds. For example, between 1980–91, communal cattle numbers in Zimbabwe increased from about 2.9 million to 4.0 million, recording an average annual growth rate of three percent, despite the severe effects of the 1982–84 drought. The irony is that large herds of livestock in a small area, among other things, compact the soil, contributing to land degradation. The result is that the degraded land fails to support them, limiting the survival options.

In the Mazowe valley, for example, the major impact of livestock on water quality is not due to nutrient enrichment of the water but to sedimentation and soil erosion caused by overstocking. Attempts to limit livestock populations to safeguard the environment have not been very successful, resulting in environmental degradation in many areas, including Omay, Hurungwe, Mukumbura, in the Zambezi Valley.

Source: Various

Table 5.7: Agriculture's impacts on water quality

Sub-basin	Fertiliser	Pesticide Herbicide	Pesticide Disease Control	Erosion/ Siltation	Livestock
Cuando/Chobe/Barotse/Luanginga/ Lunguo Bungo/Upper Zambezi/ Kabompo	0	0	0	0	0
Kafue	1(c)	0	0	1 (s)	0
Kariba	0	0	1 (s)	2 (s)	0
Luangwa	1(c)	0	0	1 (s)	0
Mupata	1(c)	0	1 (s)	2 (s)	0
Shire River/Lake Malawi/Nyasa	1(c)	0	0	2 (s)	0
Tete	1(c)	0	1 (s)	1 (s)	0
Zambezi Delta	0	0	0	0	0

Notes: (c) stands for commercial agriculture and (s) for subsistence agriculture.
Ranking of Impact: 0=No impact; 1=Slight impact; 2=Moderate impact; 3=Significant impact.
Source: Adapted from Denconsult, ZACPLAN: Sector Study No.2, Final Report, SADCZRA, Gaborone Lusaka, August 1998

Photo: J Chenje

The eradication of the tsetse fly in some parts of the basin has led to agricultural expansion in marginal lands.

grazers and browsers than cattle and can have a serious impact on vegetation if overstocking occurs.[38]

Rapid population growth, agricultural income stagnation and environmental degradation of rural areas are part of a nexus that is emerging as a paradigm. The three go together, and solutions to any one of them will involve solutions to the other two as a consequence or cause.[39] Rural women's access to education, health and agricultural production services, improved property rights, including land tenure and pricing of common pool resources and urban development as an outlet for rural labour, are some of the policy instruments that need further research and consideration to the above nexus.

MANAGING AGRICULTURE AND THE ENVIRONMENT

In response to productivity declines due to increased population pressure, soil exhaustion, unreliable rainfall and other factors, some farmers have developed special management techniques to combat specific environmental constraints. These techniques include the choice of crops and development of new varieties such as drought resistant species, reduced tillage; soil conservation and maintenance of fertility such as terracing, composting and crop rotations.[40]

Countries of the basin should ensure that management of natural resources contributes to improved productivity and income while making sure that agriculture does not undermine the diversity and richness of those natural resources. It is important to:

- maintain and increase the productivity of all forms of land and water utilisation for the benefit of the basin people;
- ensure conservation of natural resources primarily for sustainable production; and
- maintain and enhance the quality of the basin's environment.

A coordinated basin agricultural policy is needed to fight hunger, poverty and environmental degradation.

Under the framework of Southern African Development Community (SADC), basin countries have had reasonable experience and success in dealing with food security issues by depoliticising and opening up public debate. The 1992 drought was the worst in SADC's history. It was through regional cooperation in food security and transportation that the region averted famine by importing record levels of food aid, a demonstration of successful regional strategy.[41]

SADC has pioneered regional cooperation in food security, research, environment, land management, livestock production and disease control, fisheries, forestry and wildlife that are coordinated by various units such as the SADC Environment and Land Management Services, Food, Agriculture and Natural Resources and programmes like the Sorghum and Millet Improvement Programme (SMIP) and institutions such as the International Crops Research Institute for the Semi-Arid Tropics (ICRISAT).

These agricultural research and natural resource centres and programmes are working with international institutes such as the Consultative Group on International Agricultural Research (CGIAR) and the International Centre for Research in Agroforestry (ICRAF).

The 1996 SADC Policy and Strategy for Environment and Sustainable Development lists water quality, sustainable agriculture and food security, among others, as priority issues around which policies and programmes are built.

FUTURE CHALLENGES

The time has come for member states to enter into agreements that enhance basin agricultural trade and free mobility of resources. While states experience surpluses and deficits in commodities such as maize, beef, vegetable and processed foods, intra-basin trade is small. Protocols on common agricultural policies, environment, chemicals, seed certification and other areas should be adopted. Lack of comprehensive agricultural policies coupled with lack of capacity and over reliance on foreign experts provides a rationale for building and strengthening regional institutions that will move agriculture as an essential step in industrialisation.[42]

Policymakers should be aware of the dangers of agricultural intensification by avoiding pesticide subsidies, inefficient irrigation management systems and under pricing of water that encourages inputs mismanagement. Although intensification of agriculture is essential to relieve pressure on natural resources, it must be based on appropriate technology and farming systems that reform property rights to assure secure rights and strengthen community rights over farmland and common property resources. More effective community-based organisations for managing common property resources, undertaking soil erosion control and moisture conservation programmes should be established and entrenched where they exist.[43]

Regional food security in the basin countries can be enhanced through transport integration and food aid management, information exchange and early warning on climate, drought management, mitigation and preparedness.

LINKAGES TO OTHER CHAPTERS

Chapter 1: Regional Overview: People and Environment
Agriculture is one of the most important sectors on which the majority of the basin population depends. Poor land management can, however exacerbate the impoverishment of the region.

Chapter 2: Physical Features and Climate
Conversion of marginal land, overgrazing and deforestation are some of the agricultural activities that are degrading the physical features of the basin's environment and need addressing.

Chapter 3: Water and Wetland Resources
The intensification of agriculture through irrigation and fertilisers in the basin will raise the demand for water and result in salinisation of soils in hot dry areas. The pressure of land shortage is resulting in the conversion of wetlands, particularly *dambos,* into drinking holes for livestock and crop production areas.

Chapter 4: Biological Resources and Diversity
Forests, biodiversity and climate change are interlinked. The opening up of forests for both settlement and agriculture results in biodiversity loss, soil exposure, high evaporation and impact on carbon sinks.

Chapter 6: Industry
Agriculture contributes more than 30 percent to GDP, remains the most important foreign exchange earner, contributes more than 50 percent of raw materials to other industries and will maintain a strategic role in the economies of the basin states.

Chapter 7: Energy
For the majority of smallholder rural farmers in the basin, biomass energy remains a cheap and only viable source of energy for tobacco curing and tea drying. If not properly managed, both agriculture and non-renewable forms of energy can cause environmental degradation through deforestation and pollution.

Chapter 8: Tourism
Tourism is providing a viable land-use option in poor marginal areas that cannot sustain crop production. Conservancies that have been established alongside livestock production in some parts of the basin are a good example. However, the conversion of cropland into game ranches has negative implications on food security and self-sufficiency although, in some cases, the economics of game-farming have proved better land-use options.

Chapter 9: Pollution
Fertilisers are most commonly used in the basin's agriculture. Excessive use contributes to groundwater pollution. Aerial spraying for pest control is a major source of air pollution that has affected wildlife, fish and people in the basin.

Chapter 10: Poverty
Poverty is the major cause of hunger and malnutrition in the basin. Declining real world prices for agricultural exports, drought and declining per capita real incomes are some of the underlying causes of poverty among the basin population. Household and national strategies that address food availability and access to ensure food security and poverty alleviation are necessary.

Chapter 11: Gender and Women's Roles
Smallholder agriculture needs selective and appropriate mechanisation in order to remove drudgery and excessive competition for women and children's labour.

Chapter 12: Environmental Management and Regional Cooperation
The lack of comprehensive agricultural policies and over reliance on foreign experts form a good basis for building and strengthening regional institutions aimed at research and development, agricultural policy analysis and training.

Chapter 13: Trends and Scenarios
The number of people living in poverty in the basin has been increasing over the past years. It is estimated that this trend will grow. Rapid population growth and worsening environmental degradation are bound to increase, placing an enormous burden on social services and the environment.

ENDNOTES

1 Southern African Development Community (SADC), *Food, Agriculture and Natural Resources*, SADC Secretariat, Gaborone, 1-4 February, 1996
2 Mkandawire R. and K. Mtlosa., *Food Policy and Agriculture in Southern Africa*, SAPES Trust, Harare, 1993
3 Rukuni M., "Getting Agriculture Moving in Eastern and Southern Africa and a Framework for Action" in: IFPRI, *A 2020 Vision for Food, Agriculture and the Environment in Sub-Saharan Africa*, International Food Policy Research Institute, Washington D.C., June, 1995
4 Government of Malawi, *State of the Environment Report for Malawi 1998,* Environmental Affairs Department, Lilongwe, 1998
5 Katerere, Y. and E. Guveya, "Seeking New Perspectives in Land and Natural Resource Management", workshop paper, ZERO - Regional Environmental Organisation, Maseru, July 1-3, 1998
6 Bromley, D.W., 1995b, in: Katerere, Y. and E. Guveya, "Seeking New Perspectives in Land and Natural Resource Management", workshop paper, ZERO - Regional Environmental Organisation, Maseru, July 1-3, 1998
7 White, R., "Land Issues and Land Reform in Botswana", in ZERO, *Enhancing Land Reforms in Southern Africa; Reviews on Land Reform Strategies and Community Based Natural Resources Management*, ZERO, Harare, 1998
8 Negrao, J., "Land and Rural Development in Mozambique", in IUCN-ROSA, *Environmental Strategies for Land Tenure and Community-based Natural Resources Management in Southern Africa*, Harare, 1998
9 Blackie, R., "An Overview of Land Tenure in Namibia" in IUCN-ROSA, *Environmental Strategies for Land Tenure and Community-based Natural Resources Management in Southern Africa*, Harare, 1998
10 ibid
11 ZERO - Regional Environmental Organisation, *Enhancing Land Reforms in Southern Africa; Case Studies on Land Reform Strategies and Community Based Natural Resources Management*, ZERO, Harare, 1998
12 Denconsult, ZACPLAN: *Sector Studies No.2, Final Report*, SADC/ZRA, Gaborone/Lusaka, August, 1998
13 ibid
14 ibid
15 Denconsult, ZACPLAN: *Sector Studies, Introductory Volume, Final Report*, SADC/ZRA, Gaborone/Lusaka, August, 1998
16 Chenje J., *State of the Environment Reporting Programme for the Zambezi Basin, Inception Mission Report*, SARDC-IMERCSA/IUCN/Sida, Harare, November 1998
17 Op. cit 15
18 ibid
19 Op. cit 16
20 ibid
21 Op. cit 15
22 Personal communication with Malawi Trade Attaché, 1999
23 Op. cit 15
24 ibid
25 ibid
26 Owen R., Verbeek C., Jackson J. and T. Steenhuis (eds.), *Dambo Farming in Zimbabwe: Water Management, Cropping and Soil Potentials for Smallholder Farming in the Wetlands*, University of Zimbabwe, Harare, 1995
27 Abdulai A. and C.L. Delgado (eds.), "Re-establishing Agriculture as a Priority for Development Policy", in IFPRI: *Sub-Sahara Africa; Food, Agriculture and the Environment,* Discussion Paper 3, International Food Policy Research Institute/ Swiss Development Cooperation, Washington,/ Zurich, October, 1995
28 ibid
29 Badiane O. and C. L. Delgado (eds.), "A 2020 Vision for Food, Agriculture, and the Environment", in IFPRI: *Sub-Saharan Africa, Food, Agriculture and the Environment,* Discussion Paper 4, International Food Policy Research Institute, Washington, D.C., June, 1995
30 Southern Africa Development Community (SADC); *Productivity–Key to Sustainable Development in SADC*, SADC Consultative Conference Theme Document, Windhoek, February, 1997
31 Op. cit 29
32 SADC ELMS, *The Fundamentals of Natural Resources Policy Analysis for the SADC Region*, SADC ELMS, Maseru, 1997
33 ibid
34 Alexandratos N., (ed.); *World Agriculture: Towards 2010, An FAO Study*, Food and Agriculture Organisation, John Wiley & Sons, New York/Chichester, 1995
35 Op. cit 12
36 Op. cit 34
37 Op. cit 32
38 Op. cit 12
39 Delgado C. L., "Africa's Changing Agricultural Development Strategies: Past and Present Paradigms as a Guide to the Future; Food, Agriculture and the Environment" Discussion Paper 3, International Food Policy Research Institute, 2020 Vision Initiative Workshop on Issues Facing African Countries, Saly, Portudal, 14-17 December, 1994
40 Hailu Z. and A. Runge-Metzger; *Sustainability of Land Use Systems: The Potential of indigenous Measures for the Maintenance of Soil Productivity in Sub-Saharan African Agriculture*, GTZ, Eschborn, 1993
41 Op. cit 3
42 ibid
43 Op. cit 27

REFERENCES

Alexandratos N., (ed.), World Agriculture: Towards 2010, An FAO Study, Food and Agriculture Organisation/John Wiley & Sons, New York/Chichester, 1995

Chenje J., State of the Environment Reporting Programme for the Zambezi Basin, Inception Mission Report, SARDC/IUCN/Sida, Harare, November 1998

Denconsult, ZACPLAN, Sector Studies, Introductory Volume, Final Report, SADC/ZRA, Gaborone/Lusaka, August 1998

Denconsult, ZACPLAN, Sector Studies No. 2, Final Report, SADC/ZRA, Gaborone/Lusaka, August 1998

Hailu Z. and A. Runge-Metzger, Sustainability of Land Use Systems: The Potential of Indigenous Measures for the Maintenance of Soil Productivity in Sub-Saharan African Agriculture, GTZ, Eschborn, 1993

International Food Policy Research Institute, "Sub-Saharan Africa, Food, Agriculture and the Environment: A 2020 Vision for Food, Agriculture, and the Environment", Discussion Paper 4, Washington D.C., June, 1995

International Food Policy Research Institute, "Sub-Sahara Africa; Food, Agriculture and the Environment", Discussion Paper 3, Swiss Development Cooperation, Washington, D.C./Zurich, October, 1995

IUCN ROSA, Environmental Strategies for Land Tenure and Community-Based Natural Resources Management in Southern Africa, IUCN ROSA, Harare, 1998

Katerere, Y., and E. Guveya, "Seeking New Perspectives in Land and Natural Resource Management", workshop paper, ZERO – Regional Environment Organisation, Maseru, July 1-3, 1998

Mkandawire R. and K. Mtlosa, Food Policy and Agriculture in Southern Africa, SAPES Trust, Harare, 1993

Moyo, S., The Land Question in Zimbabwe, SAPES Trust, Harare, 1995

Owen R., Verbeek C., Jackson J. and T. Steenhuis (eds.), Dambo Farming in Zimbabwe: Water Management, Cropping and Soil Potentials for Smallholder Farming in the Wetlands, University of Zimbabwe, Harare, 1995

SADC ELMS, The Fundamentals of Natural Resources Policy Analysis for the SADC Region, SADC ELMS, Maseru, 1997

Southern African Development Community (SADC), "Food, Agriculture and Natural Resources", SADC Secretariat Consultative Workshop, Johannesburg, 1-4 February 1996

ZERO - Regional Environment Organisation, Enhancing Land Reforms in Southern Africa; Case Studies on Land Reform Strategies and Community-Based Natural Resources Management, ZERO – Regional Environment Organisation, Harare, 1998

6 INDUSTRY

The SADC region is endowed with abundant natural resources that form the basis for resource-based industries, particularly manufacturing. Other important economic activities include mining, agriculture, tourism and fisheries.

Industrial development in the Zambezi Basin started at the turn of the 20th century with the drifting of white settlers from South Africa towards the north, settling in the present-day Zimbabwe and Zambia. The development of the railway from South Africa into Botswana and further north facilitated the movement of people and goods, opening up new areas and the establishment of administrative and commercial centres. The settlers found diamonds, gold, copper, nickel and coal, among others, in

Photo: M Kullberg

The opening up of new areas led to the establishment of administrative and commercial centres, such as Kariba (above) which has grown as a result of the damming of the Zambezi River.

Botswana, Namibia, Zambia and Zimbabwe, influencing mining settlements such as in the Copperbelt in Zambia and some parts of Zimbabwe.

Today, the major industrial settlements are:

- Blantyre and Lilongwe in Malawi.
- Tete in Mozambique.
- Chililabombwe, Chingola, Chipata, Kabwe, Kafue, Kitwe, Livingstone, Lusaka, Luanshya, Mazabuka, Mufulira, Ndola and Solwezi in Zambia.
- Bulawayo, Chegutu, Gweru, Harare, Hwange, Kadoma, Kwekwe, Marondera and Norton in Zimbabwe.

There are no established industrial settlements of major significance in the Zambezi basin in Angola, Botswana, Namibia and Tanzania.[1]

While diversity in economic activity is critical to economic growth and reduces overdependence on one sector, such as agriculture, industrialisation and urbanisation also negatively impact on the environment. This is particularly the case where environmental management measures and institutions are weak. Lack of investment in cleaner production and incentives to minimise pollution exacerbate the problems.

MAJOR INDUSTRIES IN THE BASIN

Mining

The mining sector plays a significant role to the economies of most of the basin countries, contributing about 10 percent of GDP and accounting for 60 percent of foreign exchange earnings. Among others minerals, SADC countries are leading suppliers of asbestos, chromite, coal, copper, diamond, gold, nickel and the platinum metals and zinc to the world minerals markets.[2] Anglo-American and De Beers of South Africa are some of the major mining corporations in the SADC region, producing around 32 percent of the total value of SADC mineral output.[3]

However, the mining sector is going through a lean spell owing to falling world prices of minerals, particularly gold, nickel and copper, whose price was projected to be an average of 12 percent lower in 1998 than in 1997. Increasing production and labour costs, the rationalisation and privatisation of the industry have worsened the situation. For example, in Zimbabwe seven mines have closed down, the largest being the Hartley Platinum Mine near Chegutu.[4]

While mining activities often have a negative impact on the natural environment, the impacts are regarded as minimal, except where more environmentally harmful techniques are practised, for example, the use of mercury in gold extraction.

The principal mining countries in the basin are Angola, Namibia, Zambia and Zimbabwe. Botswana, Malawi, Mozambique and Tanzania have little or non-existent mining activity within the basin. The major minerals produced in the basin are copper, cobalt, diamonds, gold, coal, asbestos, nickel, chrome, lead, zinc, pyrite and silver.

Mining, apart from contributing revenues to the economies of the basin states, has played a major role in providing and extending infrastructure, boosting other sectors of the economy. Railways, roads, power generation facilities and water supply plants have been built primarily to service the mining operations in different parts of the basin.

Copper and cobalt are mainly found in the Copperbelt Province of Zambia. Here, mining is the second largest employer and accounts for over 60 percent of the country's foreign exchange requirements.[5] The minerals produced come from various mines such as Konkola in Chililabombwe, Nchanga in Chingola, Chambishi in Chambishi, Mufulira in Mufulira, Nkana in Kitwe, and Luanshya and Baluba mines in Luanshya. In addition to these mines, there are three smelters in Mufulira, Kitwe and Luanshya.

New policies in Zambia to liberalise the economy, have led to significant developments in exploration and development of new mines, and the ownership structure of the mines. As a result of these policies and

Photo: M Chenje

Mining is the backbone of most basin countries, contributing about 10 percent of GDP and accounting for 60 percent of foreign exchange earnings.

Table 6.1: Overview of mining in the basin

COUNTRY	COMMENT
Angola	No significant development in the basin area. Sporadic occurrence of artisan workings.
Botswana	No significant development in the basin. Sporadic occurrence of artisan workings.
Malawi	Some mining activity in the basin in Northern and Southern Malawi.
Mozambique	Some mining activity in the basin around Tete area, mainly coal. Operations severely curtailed as a legacy of the war.
Namibia	No significant development in the basin. Sporadic occurrence of artisan workings.
Tanzania	No significant development in the basin. Sporadic occurrence of artisan workings.
Zambia	Extensive mining activity in the basin. The scale of operations is generally small except on the Copperbelt where mining is a major industry with associated infrastructure and secondary growth.
Zimbabwe	Extensive mining activity in the basin. The scale of operations is generally small, however, and environmental impacts comparatively localised.

Source: Denconsult, ZACPLAN Studies: Sector Study No. 7, ZRA, Final Report, August 1998

other regulatory conditions, there has been an upsurge in both exploration and mining activities. The commissioning of the Dunrobin gold mine near Mumbwa and the re-opening of the Bwana Mkubwa copper mine in Ndola are good examples.

Apart from copper and cobalt, the basin is endowed with many other minerals that contribute significantly to the economies of the basin countries. For example, in Zimbabwe, about 90 percent of the minerals is exported. Mining contributes about 45 percent to the country's total export earnings and eight percent to GDP.

Most of the mineral resources in the country are found in the Great Dyke. Over 40 minerals are mined here and the major ones, in terms of both value and volume are gold, asbestos, nickel, chromite, iron ore, coal,

Photo: Thompson Publications

The Great Dyke contains some of the world's largest platinum resources, second only to the Bushveld Complex in South Africa.

lithium, graphite, copper, tin and recently platinum group metals, diamond and black granite. Other major industrial minerals include feldspar, fireclay, kyanite, magnesite, quartz, tungsten and tin.

In Zimbabwe alone, there are over 1,400 operating mines with 86 percent being gold mines made up mainly of medium to small-scale operators scattered all over the country. The remaining 14 percent of the mines comprise chrome-mining operations on the Great Dyke, nickel, asbestos, copper, coal, phosphate, black granite and limestone mines, and quarries.[6]

The Great Dyke also contains some of the world's largest platinum resources, second only to the Bushveld Complex in South Africa. The platinum resources, currently estimated at about 1,300 megatonnes of mineralisation and containing 2,800-8,500 tonnes of platinum and gold, provide an excellent potential.

There is a high exploration potential for both precious and industrial minerals within the basin especially on the Great Dyke. The greenstone belts with their vast gold reserves offer an attractive exploration target for potential investors. The country's gold content per square kilometre is one of the highest in the world and provides potential for further exploration and mine development.

In addition to tens of thousands of jobs and contribution to economic growth, mining has also contributed to the:

- Development of roads.
- Provision of housing.

- Provision of energy such as electricity, releasing the pressure on forest products.
- Provision of other services such as water, schools, sanitation and clinics.

The provision of such services is illustrated by the mining industry in Zimbabwe where Rio Tinto Zimbabwe is developing a six million tonne-a-year opencast mine in the northwest of the country at Gokwe, that will be used to supply a proposed 1,400 MW power station for the country's power utility, Zimbabwe Electricity Supply Authority (ZESA). The station is due to come into operation before 2006.[7] This means that more rural centres such as growth points can benefit from the scheme as more power is made available and distribution lines are brought closer to remote areas.

Photos: M Chenje

Mining has contributed to the development of roads and other infrastructure such as sanitation.

Gold is Zimbabwe's major foreign currency earner, bringing the country more than US$750 million annually. In 1998, production reached 25,146 kg, making Zimbabwe one of the 10 largest producers of gold in the world. About 25 percent of Zimbabwe's gold is derived from panning operations. The Small-Scale Miners' Association has more than 5,000 members and employs up to 100,000 people.[8]

Even though Malawi does not have a huge mining industry, prospecting has indicated that there are some minerals that could be exploited commercially. At the moment, mining is restricted to coal, mainly from the Mchenga mine. Malawi's privatisation commission estimates that the mine has 2.3 million metric tonnes of reserves, with a further potential of about 20 million metric tonnes of coal deposits. The mine currently produces about 90,000 metric tonnes annually.[9] The development of the mining industry at the moment focuses on the exploitation of monazite. The mine, at Kangankhunde, contains large reserves of monazite and strongtiantite.[10]

In Mozambique, coal is exploited at Moatize near the town of Tete where two billion tonnes have been delineated. Other areas north of Cahora Bassa have been investigated and good reserves of coal have been found. Between 1981-84, exploration of the area confirmed other coal basins known as the Luia Dome and Metangua Rift, which are recommended for further exploration.

The Metangua Rift is characterised by cretaceous deposits extending for about 100 km southeast from Estima. This area has an advantage over other areas northwest of the Zambezi River because of good road connections, electricity and proximity to the Moatize deposits.

In Mozambique coal seams, which occur in the lower part of the sequence, are among the largest and most important mineral resources. Also occurring in the Karoo are tillites, sandstones and schists. Within the Karoo terrain, immediately north of Tete and Moatize, is a large window of Precambrian gabbroic rocks with low-grade iron-titanium deposits.[11]

Prospects for gold exist in the Tete Province, including strategic minerals such as of the platinum group metals, chromium, manganese, cobalt and fluorite. Until recently, mining in Mozambique played a relatively minor role in the development of the country's economy due, among other reasons, to poor security, incomplete geological data, an undeveloped legal framework and overall business environment, and lack of internal and external financial resources. With peace in the country, Mozambique offers great prospects for minerals development.

Manufacturing

The actual number and distribution of industries in the Zambezi Basin is unclear due to lack of data. However, thousands of smaller factories exist but statistics are difficult to obtain. The current activities concentrate on manufacturing goods using raw materials from agriculture, livestock and forestry to satisfy both domestic and export markets.

The major activities consist of manufacturing of textiles and garments, leather, sugar, foods, dairy products, beers and soft drinks. These activities are concentrated in larger urban centres, particularly in Bulawayo, Chipata, Gweru, Harare, Kabwe, Kadoma, Kafue, Kitwe, Kwekwe, Livingstone, Lusaka, Luanshya, and Ndola where also some larger chemical, steel and specialised manufacturing industries are located.

In the long-term, the contribution of the manufacturing sector to the GDP in the basin countries is likely to increase from current levels due to increased investment in the productive sectors. Several manufacturing enterprises have been earmarked for technical modernisation and privatisation as part of the economic reforms being implemented throughout the region.[12]

The industrial sector in Angola has been devastated by the civil war and manufacturing in the Zambezi Basin part of the country is minimal.

Mozambique has several industries, which manufacture chemicals, automobile tyres and tubes, textiles, garments, builders' hardware, wall clocks, instruments, glassware, furniture, and others, indicating a diverse manufacturing base. However, many of the activities are not based in the Zambezi Basin part of Mozambique but further south

Namibia's manufacturing sector is relatively small compared to the other sectors of the economy and it contributes little to the country's GDP. The government emphasises the development of the small and medium-sized enterprises (SMEs) and has devised several supportive measures, including the provision of seed capital, micro-loans and credit guarantee schemes. The Export Processing Zones incentives programme introduced in 1995 is attracting some foreign investment in a range of non-traditional manufacturing activities. However, the share of manufacturing in the basin is small.

Zambia is highly industrialised, accounting for 60 percent of the work force. Industry accounted for more than 25 percent of GDP in 1996. Recent economic liberalisations has added to the pressure on the sector, as imports of manufactured goods are now easier and subject to

Table 6.2: Distribution of major industries within sub-basins

Sub-basin	No of industries	Location
Tete	10	Harare
Luangwa River	8	Chipata, Kabwe
Kariba	35	Bulawayo, Gweru, Kadoma, Kwekwe, and Livingstone
Kafue River	40	Kafue, Kitwe, Luanshya, Lusaka, and Ndola

Source: Denconsult/ZRA, ZACPLAN; *Sector Study Number 7, Final Report,* ZRA, Lusaka, August 1998

Table 6.3: Growth of manufacturing value-added (MVA) (% annual growth rate)

	1990	1991	1992	1993	1994	1995	1996
Angola	-	-	-	-	7.6	-	-
Botswana	11.7	6.6	6.4	1.0	1.3	4.3	-
Malawi	11.3	3.0	3.5	-1.0	-4.0	-0.7	-1.8
Mozambique	-	-	-	-6.7	-6.0	13.3	13.6
Namibia	5.0	5.4	4.9	6.0	6.9	7.5	-3.7
Tanzania	4.1	4.3	1.9	2.1	-3.8	1.6	4.8
Zambia	7.8	-7.6	-4.2	5.6	-9.0	-4.5	2.5
Zimbabwe	4.4	2.3	-9.5	-8.3	10.0	-14.0	-9.6

Note: Minus sign stands for values between 0 and 0.5
Source: http://www.sadcreview.com

Table 6.4: Share of manufacturing sector to GDP (%)

	1990	1991	1992	1993	1994	1995	1996
Angola	n/a	n/a	n/a	16.0	3.4	4.5	n/a
Botswana	6.0	4.0	4.0	6.2	6.0	6.0	n/a
Malawi	14.0	13.0	15.0	12.0	13.7	13.9	14.0
Mozambique	n/a	n/a	11.3	8.1	7.2	13.9	14.2
Namibia	n/a	n/a	7.4	9.2	11.2	11.2	10.1
Tanzania	7.8	8.2	8.1	8.0	6.8	8.2	8.3
Zambia	43.0	36.0	36.0	23.0	24.7	26.5	25.5
Zimbabwe	n/a	n/a	22.8	20.2	20.7	18.6	17.6

Note: n/a - figures not available
Source: http://www.sadcreview.com

lower tariffs than before. Export subsidies in some of Zambia's neighbours impose still more pressures. Nevertheless, Zambia remains one of the few countries in sub-Saharan Africa with an indigenous manufacturing base, and with a pool of qualified engineers and technicians at competitive wage levels. Existing businesses in the sector include:

- Chilanga Cement;
- Lever Brothers;
- Metal Fabricators of Zambia;
- Nitrogen Chemicals of Zambia;
- Zambia Breweries; and
- Zambia Sugar

Photo: M Chenje

However, most of Zambia's exports have been in primary form, and there is significant scope for increasing value-added within the country. It is anticipated that once the privatisation of Zambia Consolidated Copper Mines (ZCCM) is complete, the manufacturing sector will see an increased demand for locally manufactured products, as the copper conglomerate's successors seek local ancillaries and vendors.[13]

There are opportunities in most sectors to fill gaps in the production chain. For example, Zambian tobacco is presently being exported for processing and re-imported for cigarette manufacture, losing the opportunity to add value to the domestic crop. Similarly, although cotton is a mainstay cash crop for many of Zambia's 300,000 smallholders, facilities for converting the cotton to lint are inadequate and the bulk of the national crop is sold at minimal market prices and processed elsewhere.

A wide range of other products are also being produced successfully. These include biscuits, jam, cheese, salami and tinned foodstuffs and there is considerable scope to expand the range. The entire market for other consumer goods is wide open. Apart from furniture and plastic items, all Zambia's non-grocery goods are imported. Although Zambia has good deposits of phosphates and other agro-materials, around 85 percent of fertilisers used in agriculture and other applications, are imported.

The manufacturing sector in Zimbabwe is well diversified in terms of products. Manufacturing is the leading sector in the country, contributing more than 20 percent of GDP and employing 17 percent of the labour force. Manufactured goods include metal and metal products, food and drink, textiles, clothing, footwear and chemicals. The manufacturing sector has become an important player in the country's export activities, with markets in Africa, America, Europe, and Asia. The sector's exports are a reflection of the agricultural and mining base of the economy. The principal manufactured exports include; ferro-alloys, clothing, metal products, chemicals, and plastics and cotton-lint.

Photo: IUCN

The basin is home to various industries with manufacturing products ranging from chemicals to foodstuffs.

With the advent of the Export Processing Zones (EPZ) in Zimbabwe, the share of manufacturing in merchandise exports is set to grow. Table 6.5 provides a list of goods manufactured in Zimbabwe and their production trend by value betweeen 1992–94.

Agriculture

Agriculture is the single largest employer of all the industries of the basin states, contributing substantially to their economies in foreign currency earnings. About 70 percent

Table 6.5: Manufacturing production in Zimbabwe by value (Z$ m)

Sector	1992	1993	1994
Foodstuffs and stock feeds	5,123.2	5,329.6	6,746.3
Drink and tobacco	2,073.1	2,523.4	3,208.4
Textiles (including ginning)	1,772.3	2,584.1	2,607.7
Clothing and footwear	1,175.8	1,394.6	1,732.2
Wood and furniture	570.1	691.5	1,136.5
Paper, printing and publishing	1,020.2	1,132.9	1,553.7
Chemical and petroleum products	2,822.8	3,153.6	3,314.8
Non-metallic mineral products	640.1	740.2	1,076.6
Metals and metal products	4,674.7	4,107.1	5,662.7
Transport equipment	1,315.4	1,387.5	1,454.4
Other manufacturing groups	230.2	261.3	323.3
Total	21,417.9	23,305.8	28,816.6

Source: www.sadcreview.com

of the population and labour force in the SADC member states are dependent on agriculture for subsistence and also for employment and income.[14] The agricultural industry also provides raw material for manufacturing and for processing into finished products.

Agro-industry (intensive commercial processing of agricultural produce) is not a widespread phenomenon in the basin although it exists to a greater or lesser extent in Malawi, Zambia and Zimbabwe. Most irrigation agriculture in the Zambezi Basin is relatively small-scale and is utilised by smallholders and subsistence farmers. However, there is larger scale irrigation applied to commercial agricultural schemes in Malawi, Mozambique, Zambia and Zimbabwe. Overall, Zimbabwe exhibits the highest incidence of organised irrigation agriculture in the basin.[15] The agricultural sector is discussed in greater detail in Chapter 5.

Tourism

Tourism is now one of the basin's most important industries. Nature-based tourism, eco-tourism, is undergoing the fastest growth. Ecotourism emerged as a feasible option for both conserving the natural and cultural heritage of basin states and contributing to sustainable development.

The tourism industry experienced steady growth, recording a nine percent growth in arrivals in 1997. In 1998, tourist arrivals were expected to grow by between 7-8 percent over the 1997 arrivals.[16]

In 1997 the region, excluding Mozambique, received 10.6 million tourists and the tourism receipts increased from US$2.6 million in 1994 to US$4.1 million in 1997. The SADC Tourism sector currently has 12 projects with a total cost of US$5.188 million of which US$2 million worth of on-going projects will be implemented.[17]

A study of the tourism sector operating in the Zambezi Basin is vital since tourism is fast becoming a major source of foreign exchange in many basin countries. The explosive growth of the eco-tourism market during the 1980–90s has created the potential for unprecedented development of the economic, social, and infrastructural components of the basin environment.

The attractions to the major markets in Europe, North America and Japan include the abundant wildlife in game reserves and national parks, the special natural habitats and features such as Victoria Falls and Lake Malawi/Nyasa, and the region's rich historical and cultural heritage. The tourism sector is covered in more detail in Chapter 8.

Photo: N Greaves

The explosive growth of the tourism market during the 1980-90's has created the potential for unprecedented development of the economic, social, and infrastructural components of the basin environment.

Map 6.1: Major industrial towns

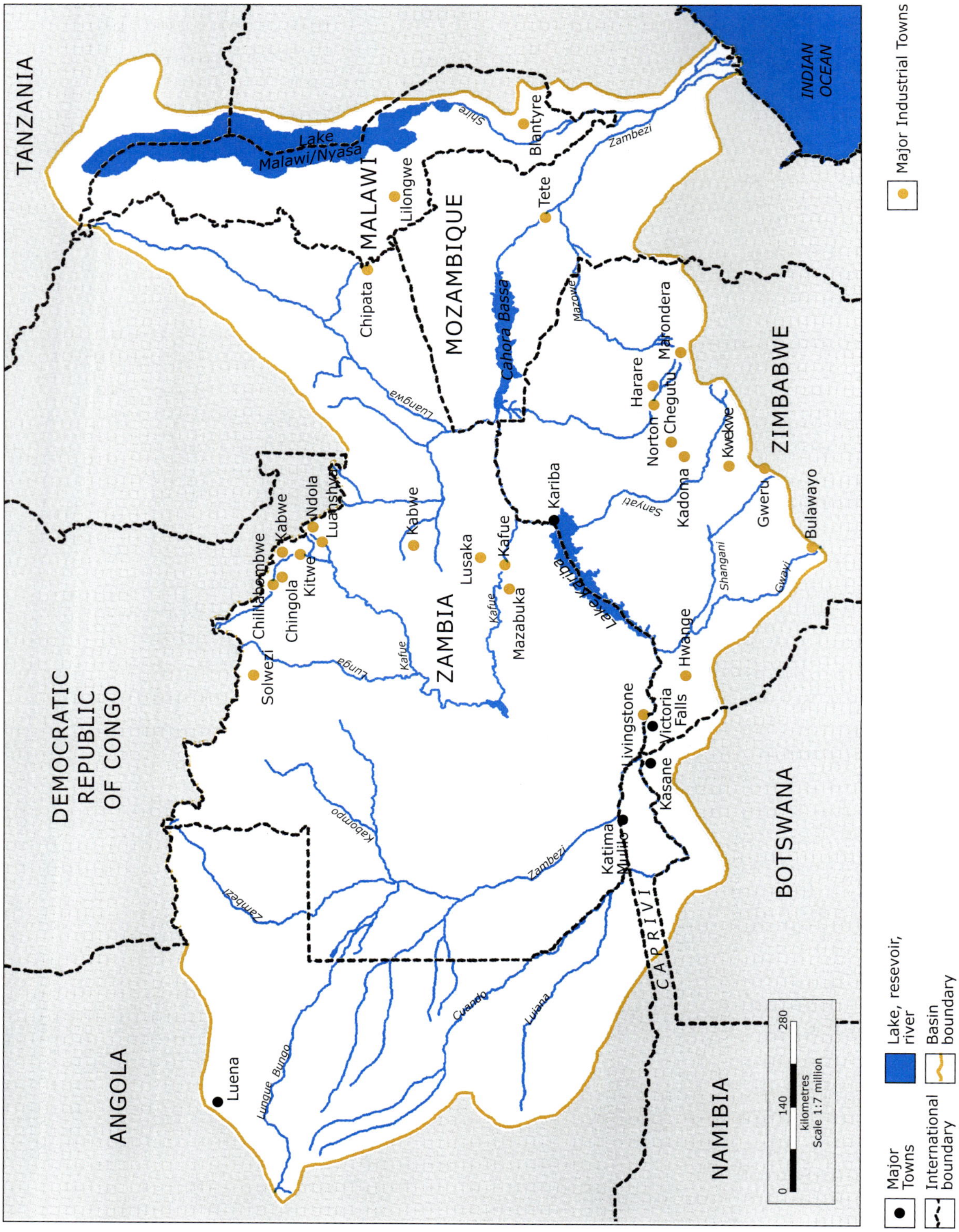

Source: Zambezi River Basin Atlas, SADC/DENCONSULT/ZRA, Lusaka, 1998

Table 6.6: Tourism receipts in the basin countries (US$ m)

Country	Year 1994	1995	1996	1997
Angola	13	10	9	9
Botswana	124	162	178	181
Malawi	5	6	7	7
Mozambique	n/a	n/a	n/a	n/a
Namibia	220	224	208	210
Tanzania	192	258	322	360
Zambia	43	47	60	65
Zimbabwe	125	154	219	250
Total	722	861	1,003	1,082

Note: n/a - not available
Source: WTO - Africa Trends, 1998.

Fisheries

The Zambezi Basin is one of the richest sources of freshwater fish in the southern African region due to its dense network of rivers and significant share of world-renowned lakes. The fisheries sector of Malawi, Mozambique, Zambia and Zimbabwe, in particular, depend on the Zambezi River and its tributaries for a sizeable contribution to their national fish outputs.

Commercial fishing activities in the basin are mainly restricted to larger water bodies such as Lake Kariba and Lakes Malawi/Nyasa and Malombe, with generally low environmental impacts. Traditional artisan fisheries do not generate significantly adverse environmental impacts, except where human populations have swelled to such an extent that excessive pressure is exerted on the natural resource base.[18]

Photo: M Chenje

Kapenta fishing on Lake Kariba is one of the major inland fisheries in the basin, employing thousands of workers and earning millions of dollars annually.

Beneficiaries of the basin's fish include subsistence fishers of the basin, small-scale cooperative fishers on the large lakes, and large-scale commercial fishers also on the large lakes. The scale of fishing at the grassroots level is usually under reported, and thus the extent of its contribution to the national annual output is underestimated. However, the inland fisheries sector provides close to one million jobs in the SADC region.

The demand for fish by the basin's growing population is high. Fish supply per capita is estimated at nine kilogrammes (live weight equivalent) per person

Table 6.7: Overview of fisheries in the basin

COUNTRY	COMMENT
Angola	No commercial fisheries and limited artisan fishing in the basin area.
Botswana	No commercial fisheries and limited artisan fishing in the basin area.
Malawi	Well developed commercial and artisan fisheries on Lake Malawi uyand Lake Malombe. Some artisan fishing on the Shire River as well as other perennial rivers in the basin.
Mozambique	Commercial and artisan fisheries on Lake Cahora Bassa. Some artisan fishing on the Zambezi River as well as other perennial rivers in the basin.
Namibia	No commercial fisheries and limited artisan fishing in the basin area.
Tanzania	Limited commercial and artisan fisheries on Lake Malawi. Some artisan fishing on perennial rivers in the basin.
Zambia	Well developed commercial and artisan fisheries on Lake Kariba. Some artisan fishing on Lake Itezhi-tezhi, Zambezi River, Kafue River, Zambezi Floodplains and other perennial rivers in the basin.
Zimbabwe	Well developed commercial and artisan fisheries on Lake Kariba. Some artisan fishing on the Zambezi River as well as other perennial rivers in the Zambezi Basin.

Source: Denconsult, ZACPLAN, Sector Study No. 7, final Report, ZRA, August 1998

annually, far below the World Health Organisation (WHO) recommended minimum fish supply per capita value of 13–15 kg/person/year.[19] Table 6.7 provides a synoptic overview and status on fisheries in the basin. The fisheries industry is also discussed in Chapter 3.

Small-scale industrial activities

Small-scale industries are growing fast in the basin states. Due to poverty people are resorting to establishing small ventures to support themselves. These include among others, brick making, selling firewood and charcoal, carpentry and smithies. Brick making has taken its toll on forests near urban areas as brick makers cut down trees to burn the bricks. IUCN-The World Conservation Union estimates that constructing 350 one-roomed brick houses means clearing 20 ha of woodland.

PRESSURES ON THE ENVIRONMENT

A thriving and vibrant industry is a pre-requisite in every country for provision of goods and services, employment and sustained economy. The industrial demand for raw materials and uncontrolled waste production are increasingly unsustainable.[20] Industrial processing activities, products and wastes alter natural raw materials and environmental quality, posing serious implications for human and environmental health.

Under the SADC framework, basin countries continue to face serious dumping problems owing to their limited technical capacity to deal with industrial waste. There is also the issue of porous borders or pollution problems of a transboundary nature that are difficult to handle.

Photo: IUCN

The industrial demand for raw materials and uncontrolled waste production are increasingly unsustainable.

Improved industrial investment and trade policies as well as regulations are some of the strategies that have been employed at national level to address porous borders. However, much remains to be done at the regional level to monitor and control environmental pollution problems.[21]

Environmental problems associated with mining

Coal dumps are major or potential sources of air pollution in the region, particularly in Malawi, Mozambique, Tanzania, Zambia and Zimbabwe. These are located under highly reactive sandstones and mudstones, which can ignite spontaneously.

In Kabwe (Zambia), copper smelting releases 200,000 tonnes of sulphur dioxide and dust containing traces of zinc, lead and copper, affecting people and vegetation.[22] Half of the children in cities with copper smelting facilities suffer from respiratory diseases. Vegetation is also affected by the release of sulphur dioxide. Air quality is affected by emissions from smelters, and to a lesser extent refineries of compounds of carbon, sulphur and nitrogen and toxic metal particles. There are also indirect emission effects from the use of fossil fuels as energy, and from potentially hazardous dusts and gases released in the workplace. Air pollution levels in the region are a problem, affecting mainly urban areas, which have heavy concentrations of industrial development.

Gaseous pollution from mining is found on the open cast coalmines due to the spontaneous combustion of coalfaces. It has, for example, been experienced from the Maamba Collieries in Zambia that spontaneous combustion is always initiated from the high sulphur content of mudstones above the main seam (five to seven percent sulphur), where it starts quickly, particularly during the rains. However, washed coal (0.6 – 0.7 percent sulphur) can take six to12 months before combustion starts.[23]

The mining sector is also responsible for water pollution through the discharge of a variety of highly toxic metals. Mercury, for example, is used to separate gold from sand, and can lead to serious illnesses and death if washed into watercourses and later consumed.

Also associated with coal is the accumulation of low-grade fines. Most of these fines are stored in slurry dams, leading to acidic effluents. Previous attempts at some mines, in particular, Maamba coal mine, to deal with this material by, for instance, briquetting, have failed because of the lack of markets. There is need to reassess this problem on a regional basis.

Photo: M Chenje

While water quality monitoring is a continuous exercise, effluent control is difficult.

Coal mining operations can have severe acid mine drainage problems due to the sulphur content. Mining and mineral processing in the Zambezi Basin produces waste products and ecological disruption, which may generate potential environmental hazards at each stage of the metal production process. Tributaries passing through the site of extensive mineral winning activity feed major rivers draining into the Zambezi River. The major ones being the Kafue River which drains the entire copper mining region of Zambia and the Mazowe River which drains most of the mining operations in the Shamva region of Zimbabwe.

While the Kafue River drains the whole of the Copperbelt Province and receives the total industrial and domestic liquid waste of the major industrial area, it is also the main source of potable water to approximately three-quarters of the population of this region.[24] Although water quality monitoring is a continuous exercise, effluent control is regarded as difficult.

The main problem facing the mining industry is its inability to comply with the standards for suspended solids particularly in the effluent released from the Nchanga Division of the Zambia Consolidated Copper Mines (ZCCM).[25]

Exploration, mine development and dumping of barren overburden or waste, can degrade the habitats of local flora and fauna, and prohibit alternative landuses such as forestry or agriculture. Examples include toxic serpentinite dumps from chromite mining in Zimbabwe, and the spontaneous combustion of coal dumps, at Maamba Colliery in Zambia and Wankie Colliery in Zimbabwe. Water quality may also be affected by naturally occurring acid mine drainage from mines and waste piles. This is a particular problem for coalmines in Botswana, Zimbabwe and in the Zambian Copperbelt, which is rich in acid generating sulphide ores.

Environmental problems associated with mining operations in the basin states include the effects of heavy metals that are a major source of contamination, causing pollution of soils and water. Heavy metals are usually found as trace elements in the soil. Many of the heavy metals, which enter rivers and streams are adsorbed to particles or precipitate. The sediments and water column, control the distribution and transportation of metals in aquatic environments, respectively. Some of the heavy metals are useful in small quantities to both plants and animals, including humans. They can, however, be toxic and lethal to humans even at low concentrations because of their tendency to accumulate in the body.

A large proportion of waste materials produced by mining operations lies between two extremes of fineness and coarseness. In the extraction of ores, the mined material may be ground finely with particle sizes below 0.2 mm. The tailings that are dumped after removal of metals have fine particle size and have reasonable physical characteristics that could permit plant growth and retain adequate amounts of water. Yet, such material will have been formed from hard, unweathered rocks and so may not contain much of the finest clay fraction. There is also neither organic matter nor microbial activity. As a result, the dumps may be very unstable, easily blown by wind when they are dry and eroded by heavy rain when they are wet. Much of this instability is a result of the inability of natural vegetation to regenerate and not because of their structure. Part of this instability is also due to topography, since erosion will be greatest on the steep slopes.[26]

In Zimbabwe, coal dust levels were set at internationally acceptable standards, of below 0.5 mg/cu m. Health problems associated with excessive exposure to dust, include silicosis, asbestosis, and bronchial carcinoma.[27]

Mozambique's large coal reserves at Moatize will see the development of new mines including two open cast ones, and it is estimated that this would lead to the removal of 15 million cu m of earth to extract six million cu m of the mineral.[28] This is likely to cause air pollution problems in Moatize and possibly in Tete, depending on wind direction.[29]

Some waste materials contain sulphides such as pyrite and also carbonates, particularly of calcium or magnesium. When pyrite oxidises to give sulphuric acid, it is neutralised by the carbonates to give calcium and magnesium sulphates. These are not toxic at normal concentrations but they are soluble and in arid climates, they may accumulate causing salinity in the surface layers of the soil as moisture evaporates. They may also form an accumulation of crystalline salts on the surface of the soil or waste. In humid climates, the salts are leached away as they are formed.[30]

The most serious environmental problem is caused by acid mine drainage in mineral exploitation. Acid mine drainage is very often a multi-factor pollutant and the importance of each factor will vary both within and between affected systems. There is likely to be acid mine drainage where coal, sulphur, copper, zinc, silver, gold, lead and uranium, among others, have been mined. In an abandoned mine site, acid mine drainage can affect groundwater, waste rock stockpiles and tailings dumps.

For example, extensive mining and metallurgical operations of copper mines produce huge amounts of waste rocks from underground and open pit mines, tailings and slag heaps from concentrators and smelters. Waste rock dumps and tailings dumps occupy more than 10,000 ha on the Copperbelt in Zambia. Acid and heavy metal leachate from the tailings dumps contaminate local soils and can infiltrate into the groundwater.[31] The mining operations cause land stability problems due to subsidence and open pits. Tailings dumps are by far the biggest source of soil contamination.

In steel manufacturing, raw materials used include lime, iron ore and coal. The crushing and cleaning of the limestone and ore produces sludge and its disposal poses problems. In the processing of the ore, pollutants such as coke, sulphuric acid and tar are produced.[32] Chapter 9 discusses further other environmental problems due to mining.

Ecologically sensitive areas

Before the Kafue River enters the mining area, it is in its natural state. In this region, the water quality is characterised by low suspended solids, turbidity and alkalinity. The water hardness varies from soft to moderately hard. The river is also low in nutrients like nitrates and phosphates. When the river passes through the mining locations, changes occur in total suspended solids (TSS), sulphates and heavy metals. This trend is, however, reversed by the time the river passes the Machiya ferry and Itezhi-tezhi.

The river goes through some natural purification after leaving the mining area. The total dissolved solids (TDS) composition of the river reflects the nature of the activities used in controlling releases from the purification process and as such comprises sulphate, calcium, magnesium and other hardness constituents formed by the addition of lime to control pH of such releases to precipitate the heavy metals. The addition of lime has increased sulphates in the river systems, which have proved difficult to handle.[33]

It has been observed that some countries are often more concerned with attracting foreign investment at the expense of the environment. The following example from Zambia seems to support this observation:

> "It must be recognised that environmental control techniques impose, in general, an additional cost burden on the operator and that the preservation of the environment at the expense of development or, in the extreme, at the risk of shutting down an industry could have extreme negative effects on the welfare of a large proportion of the community."[34]

The quote above clearly shows that environmental management is complex, and various issues have to be taken into account before the best option can be identified.

Photo: Thompson Publications

It has been observed that some countries are often more concerned with attracting foreign investment at the expense of the environment.

Table 6.8: Water quality of the Kafue River upstream and downstream of ZCCM

Parameter	Above ZCCM 1990		Below ZCCM 1987	
	Mean	Maxima	Mean	Maxima
Ph	7.3	7.9	7.7	8.3
TSS	2.4	2.74	14	44
Sulphate	20		29	134
Cobalt	0.2		<0.01	<0.01
Copper	0.02	0.031	<0.01	0.01

All values except pH are expressed in mg/l

Source: Mbinji, W., *"Environmental Issues in the Copper Mining Industry of Zambia,* Zambia Consolidated Copper Mines Limited", Kalulushi, 1991

Environmental problems associated with manufacturing

Sources of industrial effluent range from pulp-and-paper mills to fertiliser factories and granulation plants, abattoirs, textile manufacturing using dyes and noxious cleaning chemicals, cloth manufacturers, various chemical industries, water treatment plants, steel industries and others. These industrial processes produce large quantities of different kinds of pollutants.[35]

The United Nations Environment Programme (UNEP) has recognised that the "industries in Africa have only limited access to environmentally-sound technology, and historically have aimed at maximising production, with little or no concern for the environment. Many industries are still unaware of the potential benefits of cleaner production and relevant legislation is usually non-existent. Countries are often more concerned with attracting foreign investment which could be deterred by making production standards too restrictive."[36]

However, effort is being taken to address the issue, with some multinational and local corporations are beginning to adopt precautionary environmental standards.[37] Industry is the primary source of pollution in the Zambezi Basin for producing:

- huge amounts of liquid effluents that often find their way into water bodies;
- solid waste, which is often dumped illegally on the land; and
- emissions that pollute the air.

Agro-industry related environmental impacts accrue from:

- Location of plant or facility on or near sensitive habitat.
- Location of agro-industry along watercourses, leading to their eventual degradation.
- Water pollution from discharge of liquid effluents.
- Gaseous, particulate and odour emissions from processing operations.
- Accidental releases of potentially hazardous solvents, acidic and alkaline materials.
- Environmental deterioration (erosion, contamination of water and soil, loss of soil fertility, disruption of wildlife habitat) also results from intensification of agricultural landuse to supply the processing industry.

Photo: Thompson Publications

Many industries are still unaware of the potential benefits of cleaner production and relevant legislation is usually non-existent.

Pollution from industrial wastes and effluents

The main water pollution problems from industries are generally confined to local areas. At present these are concentrated in 13 major towns: Harare, Chipata, Kabwe, Bulawayo, Gweru, Kadoma, Kwekwe, Livingstone, Kafue, Kitwe, Luanshya, Lusaka, and Ndola, but single industries located in the countryside are likely to significantly affect streams or rivers receiving the wastes.

The main industrial pollution problems are organic wastes from sources such as breweries, abattoirs, other meat industry and sugar industry. Moreover, the industries add to the other pollution sources of nitrogen and phosphorus such as domestic wastewater and fertiliser runoff, contributing to the eutrophication of surface waters.

Runoff from factory dumps has created serious water pollution problems in the basin. This ends up flowing into rivers supplying the urban areas with water. People who have no access to clean water sources resort to directly utilising water from some of these rivers. A case in point is Harare where the Mukuvisi River, which drains into the Manyame, is generally considered as the most heavily polluted river system in Zimbabwe. This is because the river flows through both industrial and residential areas of Harare, collecting waste discharges along the way. In its upper reaches, the river receives effluents from a phosphate fertiliser manufacturing plant. Most of the pollutants that find their way into the river are soluble, including pyrites cinder, aluminium sludge and fluoride-rich scrubbing alcohol.[38]

A study carried out in 1997 on the environmental status of Lake Chivero, indicates that the massive fish deaths, which occurred in the lake in March-April 1996 were a result of massive pollution from sewage and industry.[39] The lake was also, at one time, choking from the water hyacinth. This was again attributed to serious pollution from industry and agricultural activities upstream of the lake. Harare sits on the Manyame catchment of which Mukuvisi River is part and all pollutants are washed into Lake Chivero and ultimately transported to the Zambezi by the Manyame River. The same is true for some of the urban centres in Malawi, Mozambique, northern Namibia and Zambia.

Box 6.1: Effects on water quality by industry

Breweries and soft drink industries:
Among the larger industries, 23 breweries and soft drink factories have been identified in the major industrial towns. The water quality impacts from the breweries are mainly related to the discharge of wastewater containing organic solids and solubles such as yeast, alcohol and barley. These solids reduce oxygen content and increase available nutrients, contributing to eutrophication in the receiving water. For the soft drink industries, the wastes mainly contain sugar and fruit residues. The use of caustic soda for cleaning bottles raises the pH in the effluents, affecting the water quality locally around the discharge point.

These industries discharge a considerable load of organics, which is likely to affect the water quality due to oxygen depletion near the discharge point. Fish kills have also been reported near breweries. The impacts from such effluents can be severe oxygen deficit if the receiving river is small compared to the discharge.

Sugar industries:
Currently, five sugar refining factories are located in the Zambezi Basin – two in Malawi (99,000 and 110,000 tonnes/year), two in Mozambique (84,000 and 81,000 t/year), and one in Zambia (150,000 t/year).

The main impact from the sugar industry derives from the loading of the receiving water with organic wastes, resulting in excessive oxygen demand and contributing to eutrophication. The organic wastes contain cane wash, cellulose matter, cane juice and molasses. Very high amounts of BOD (3,000-6,000 mg/l) are discharged if no effluent treatment exists.

Plant oil and soap industries:
These industries process oil-containing seeds (cotton seeds, sunflower seeds, groundnuts, sesame) into edible or industrial oil products. Some of these products are further processed into soap. The oil and soap industries are spread throughout the basin, typically located in the areas of raw material production. Two larger oil industries are located in Lusaka and one in Kabwe.

The processes include mechanical crushing of seeds to press out oil, neutralising the acids in extracted oil using sodium hydroxide, scrubbing the oil using water to remove impurities, bleaching/decolourising the oil using activated carbon. The manufacturing of soap involves further adding of sodium hydroxide.

The wastes from this type of production contain high amounts of dissolved and suspended organic material, and consequently affect the receiving water body by increased oxygen demand and eutrophication. The wastes often contain sodium hydroxide, increasing the alkalinity of the water. Although the actual amounts of these wastes have not bee quantified, the wastewater is likely to affect the recipient water for some distance from the outlet.

Meat:
Except for the abattoirs, which are located in most urban centres throughout the basin, the major meat processing industries are concentrated in the industrial towns. These industries are slaughterhouses and meat cleaning and packing factories. The wastes contain high amounts of organic material deriving from blood, meat, fat, grease and casings. Some of the factories also discharge salt and chlorine.

The wastes are generally discharged through the public sewers without any treatment, but no quantity estimates of effluents are available. It is, however, well known that untreated wastewater from this type of industry may create severe local effects on the receiving water bodies.

Dairy industries:
Seven major milk-processing factories are located in Kitwe, Bulawayo, Gweru, Chipata, Kadoma, Lusaka, and Harare. The wastes contain mostly milk from spillage and washouts. Additionally, a number of milk collecting centres are located around the basin, but no quantification of effluents has been obtained.

Leather tanning:
Tanning enterprises are often small and located in small towns in cattle ranching areas. However, two large-scale tanneries are found in Kabwe and Bulawayo. The leather tanning process involves a variety of chemicals which characterise the waste effluents such as: arsenic, DDT, zinc chloride and various dichlorobenzenes, chlorine and sodium fluoride, caustic soda, sodium sulphide, sulphuric acid, hydrochloric acid and chromium. Such a mixture of toxic chemicals will, if discharged in significant amounts, cause various lethal or chronic effects on the flora and fauna in the receiving water. These chemicals can also render the water locally unsuitable for use.

Textile industries:
Eight large textile manufacturing industries are found in the basin. Two are located in Kadoma, two in Livingstone and one each in Kabwe, Ndola, Kafue and Kitwe. The wastes contain various chemicals used mainly for the dying and printing processes: caustic soda, hydrogen peroxide, sodium silicate, and dyes. Other chemicals used include solvents for the dyes such as aromatic hydrocarbons (benzol, toluol, xylol, naphthas and aromatic amines) and solvents used for cleaning machinery. Many of these chemicals are toxic or carcinogenic and can, when discharged in significant quantities, create severe health risks for people using the polluted water.

Other major industries:
One large paper industry is located in Lusaka. Wastewater from such industries is normally rich in fibres and various chemicals used for purposes such as bleaching.

Some major steel industries are also found, particularly in Kwekwe. The environmental problems related to this type of industry are mainly due to the slag produced during the iron refining processes. The slag deposits contain a variety of metals, including cadmium and chromium, which are washed out by rainwater.

Semi-industrial activities:
In addition to the major industrial activities as mentioned above, there are a number of semi-industrial activities which may affect the water quality in the vicinity of their location or in the urban sewerage systems. Among these are:

- Battery manufacturing/repair (producing acidic waste containing lead)
- Garages (discharging oil products)
- Gas stations (leaching oil products)
- Petroleum storage facilities (spillage and leakage of oil products)

Presently, the significance of these pollution sources may be considered low due to the generally low economic activity in the basin countries. However, with continued economic growth, water quality impacts from these activities will be an issue at the local level.

Source: Denconsult/ZRA, *ZACPLAN: Sector Study Number 7, Final Report*, ZRA, Lusaka, August 1998

In the case of Mozambique, the level of environmental problems from the industrial sector is low due to the low level of industrialisation and production.[40] However, it has been noted that, "the engineering and petroleum refining industries are situated in the major cities (of which Tete is one), close to bays, into which they discharge effluent without any treatment whatsoever."[41]

The main impact of the contamination of water resources has been on the health and damage of aquatic habitats resulting from the pollution. People who drink contaminated water, particularly with heavy metals, risk suffering serious long-term health problems.

Waste management is a big problem in the basin. Approximately 95 percent of all the waste in the Zambezi Basin countries is disposed off on land. The level of urbanisation in the basin worsens waste disposal problems. Harare, for example, generates about 124,000 tonnes of solid waste annually. Besides poor methods of disposing of the waste, another problem associated with such huge amounts of waste is that some of it comprises of plastics and metal cans, which cannot decay.[42]

Photo: P Wade

Huge amounts of waste, some of which comprises of plastics and metal, are being generated and are difficult to manage.

Environmental impacts related to fisheries

The main threats to fisheries are hydrological changes, which may either be natural or due to the construction of large dams and from overfishing. Overfishing has, for example, been most serious in Lake Malombe where catches have declined steadily since 1990 and in the Lower Shire where catches have shown a dramatic decrease since 1992.[43]

Other threats include;

- Overexploitation of fisheries stock and long-term degradation of the resource base.
- Capture of non-target species and habitat damage through use of certain equipment and fishing practices.
- Pollution from oil and fuel spills and leakage, and from bilge flushing of commercial fishing vessels.
- Propeller turbulence and anchor damage to benthic flora and fauna.
- Navigational hazards from abandoned or lost fishing nets and traps.
- Possible use of explosives and poison with consequent negative impact on non-target aquatic biota.
- Introduction of exotics, leading to degradation of indigenous stocks.
- Growth of communities reliant on one primary product with consequent risks for that community if the resource base collapses such as Lake Malombe, in Malawi.
- Communities vulnerable to an influx of outsiders with capital able to finance more intensive fisheries, displacing traditional artisan fishermen.[44]

INSTITUTIONS AND LEGISLATION

Laws exist in all the basin states for regulating industrial activities but these are administered by different ministries and departments. In Zambia, for example, 28[45] laws, including the constitution, cover environmental management in that country. Effluent discharge regulations under the 1991 Local Government Act, which is administered by the minister of local government, overlap with the regulations provided for under the 1993 Environmental Protection and Pollution Control Act, which is administered by the Environment Council of Zambia. Also in Zambia, the 1993 Investment Act, which provides for investment in manufacturing, mining, agriculture and transport communications, does not explicitly provide for natural resource management or pollution control. The act is administered by the minister of commerce, trade and industry.

In addition to obvious problems related to different mandates and competing sectoral interests, some of the laws undermine each other. Good examples include Zambia's 1985 Petroleum (Exploration and Production) Act and Zimbabwe's Mines and Minerals Act. The Zambian law does not provide for environmental impact of petroleum exploration or production activities.

The Mines and Minerals Act in Zimbabwe overrides all laws dealing with natural resources management.[46] Once a mining permit has been granted under the Act, there are few restrictions on exploration of the mining rights. As a result, extensive timber felling without any obligation to re-forestation takes place, poaching by mine workers,

Table 6.9: Major types of hazardous wastes and their impact on the environment

Industry	Hazardous Wastes	Impacts on the environment
Tanneries	Chrome salts, arsenic, hides, dyes	1 Long-term human health 2 Impact on aquatic environment
Agro-industries	Disinfectants, herbicides, pesticides.	1 Result in respiratory diseases 2 May kill aquatic life
Brick-making	Chromium, fluoride, sulphur dioxide	1 Respiratory diseases 2 Kills off surrounding vegetation
Clothing and textile	Cyanide dyes, oils	1 Dyes lethal to human and aquatic life 2 Oils may form film on water killing aquatic life
Glass and ceramic making	Arsenic, barium, manganese, selenium	1 Kills both humans and aquatic life 2 Cardiovascular diseases 3 Skin cancer
Dry cleaning	Solvents, bleach, hydro-chlorides,	1 Disruption of pH buffer systems and disordering ecological balances
Metal finishing	Hydro-chlorides, chromium, cadmium, zinc, solvents	1 Can result in cardiovascular diseases 2 Killing of aquatic and human life 3 Neurological disorders 4 Disruption of pH buffer systems
Transport and workshops	Oils, solvents, batteries (lead, cadmium, chromium, copper, hydrogen)	1 Form anaerobic film on water, killing aquatic life 2 Cardiovascular diseases
Paper and printing industry	Methanol, mercury, zinc, wax	1 May poison aquatic and human life
Chemical industries	Various types of chemical	1 Can affect aquatic and human life

Source: Adapted from Hamza, 1991

dumps and non-compliance with quittance requirements abound.[47]

In Malawi, there are six major environmental laws but issues such as waste management and sanitation, air and noise pollution, and toxic substances and chemicals are not covered by any law.[48] The government has committed itself to the review of policy and development of appropriate laws.

Generally, the failure to good environmental reporting by industry in the basin is due to a number of reasons that include:[49]

- Poorly equipped training institutions.
- Limited number of trained trainers.
- Fewer developed databases.
- Attitudes within enterprises.
- Lack of resources.

National initiatives in the Zambezi Basin need to set environmental targets, and environmental issues must be considered in the context of economic performance and pave the way for harmonising financial targets, budgets and benefits with environmental health.

REGIONAL INITIATIVES

Mining

The basin's contribution to the world supply of minerals continues to decline mainly due to lack of investment in the mining sector in most basin countries. The rationalisation and restructuring of the industry has compounded the situation, resulting in massive job losses. The resulting problems have serious socio-economic, environmental and political implications. Basin countries need to develop post-mining economic programmes if post-mining sustainable development is to become a reality. In line with the general objectives outlined in the SADC Treaty, countries in the region have approved the SADC Protocol on Mining, which provides the legal basis for basin countries and the SADC states in general, for cooperation in the mining sector.

The SADC Mining Sector Coordinating Unit (MCU) has recognised the importance of the Zambezi Basin and potential impacts that could arise from the various mining operations, which are either closed or still operating. The unit has taken steps to create an activity that will identify impacts of mining in the Zambezi Basin through a SADC

Box 6.2: Objectives of project on mining in the basin

The objectives of the project are to:

1 Establish the effects of mining and its effluent on groundwater and water quality of the rivers of the Zambezi Basin (past and present), to include the wetlands.
2 Identify and quantify effluent to the Zambezi Basin arising from mining operations in the region and effect on the ecological systems.
3 Identify and quantify groundwater pollution and depletion arising from mining activity.
4 Create a database of information on groundwater pollution and mine effluent and their effects on the surrounding environment.
5 Identify technical and financial constraints that have hindered or constrained groundwater quality maintenance and effective effluent control.
6 Formulate guidelines on SADC mine effluent, groundwater maintenance quality standards and treatment techniques.

Source: SADC, *Mining, Sector Profile*, Lusaka, 10-12 February, 1999

project on the environmental impacts of mining and related industries in the Zambezi River Basin.

Private sector participation

To strengthen links between the private sector and governments, the private sector will, under the auspices of the Mining Industry Association of Southern Africa (MIASA), participate in SADC mining sector working group, sub-committee and technical committee meetings. They will also be accorded a forum at which they will interact with the Committee of SADC Mining Ministers.

Photo: Thompson Publications

There is considerable work to be undertaken to quantify the nature and extent of environmental degradation caused by metals production within the SADC region.

Environmental sustainability

There is considerable work to be undertaken to quantify the nature and extent of environmental degradation caused by metals production within the SADC region. Currently there exist only isolated case studies and little systematic analysis of the problem.

In line with the broad aim of the SADC Treaty to achieve sustainable utilisation of natural resources and to effectively protect the environment, the mining sector has set out to achieve the following objectives:

- Fostering regional integration within the mining sector.
- Promoting investment in the sector.
- Integrating mining into the economies of the region.
- Maximising benefits from mining within the region.
- Promoting sustainable and environmentally friendly mining industries.

The mining sector has identified areas for regional integration, which include information exchange, enhancement of technological capacity, establishment of common standards, promotion of private sector participation, development of small-scale mining, environmental protection, and occupational health and safety. There is a shortage of available capital, skills and regulatory requirements for mines or governments to undertake the type of systematic base-line studies required to ascertain the true extent of environmental hazard and human health effects.

Without that evidence, it is increasingly difficult to estimate the costs of natural resource degradation associated with mineral exploitation, and to determine how to share those costs between the polluter, state and community. As systematic evidence emerges, it is becoming clear that these costs are high. These costs have largely been measured in terms of the expense involved in remedial treatment of degraded water quality, for example, or compensation for damage by toxic dust to local farmland.

Manufacturing

The SADC Industry and Trade Committee of Ministers has directed that efforts on industrial development focus on regional industrial policy and strategies to deal with issues of:

- industrial investment potentials in the region;
- effectiveness of current policies and strategies;
- balanced development; and
- the effects of trade liberalisation on local industries.[50]

The signing of the SADC Trade Protocol has provided a framework for improving and expanding intra-SADC trade. However, while the protocol addresses trade obstacles relating mainly to market access, its effectiveness will depend on the elimination of other obstacles, including: an inadequate production base; inadequate trade policies; inadequate market access; insufficient commercial skills; under-developed financial infrastructure; inadequate trade statistics; and inadequate information and communication facilities.[51]

The SADC Trade Protocol, signed in August 1996, has since been ratified by three Zambezi Basin states, namely Botswana, Tanzania and Zimbabwe. The protocol was not yet effective in February 2000.

Environmental Management Systems (EMS)

Quality environmental management is fast becoming a fact of business life and the International Organisation for Standardisation (ISO) 9000 and 14000 series of standards are now passports to international trade, even though they are voluntary. These series are generic system standards intended for implementation by industry, be they in manufacturing or services, public or private.

Environmental Management Systems (EMS) based on the ISO 14000 standards are management tools enabling industry or organisation of any size and type to control the impact of its activities, products or services on the environment. EMS have a structured approach to; setting environmental objectives and targets, achieving them, and demonstrating that they have been achieved.[52]

The standards don't set specific environmental performance policies and targets, but enable organisations to manage the impact of their activities on the environment. External regulations or organisations that serve on the various sub-committees set the environmental policies, objectives and targets relating to their situation and context.

The main cause of many industrial related environmental problems in the Zambezi Basin is the persistence of policies, which largely neglect and fail to avoid harmful impacts on the environment and natural resource base.

Photo: Thompson Publications

There is need to review all relevant laws governing environmental and natural resource use and management to ensure that they conform with international developments over the past few years.

Some industries in the region have realised the symbiotic relationship between the economy and ecology as well as benefits that arise from responsible environmental management and performance. Through their national standards associations, basin countries have developed environmental management systems. Zimbabwe was the first African and basin country to adopt ISO 14001 and 14004.[53] The main objectives of the ISO 14000 series of standards are to:

- Enable organisations to manage their operations in such a way that the actual or potential detriment to the environment is minimised.
- Achieve and maintain compliance with the laws, regulations and organisation's policy.
- Recognise the need for sustainable development over a long period of time.

The national standards associations in the basin countries are working with industry in, among other areas;

- preparing and publishing standards specifications,
- providing laboratory tests for raw materials and manufactured goods,
- providing a World Trade Organisation Technical Barriers to Trade (WTOTB) enquiry point,
- promoting use of standards by industry, registration of organisations, complying with international standards, and
- ISO series certification.

In Zimbabwe, for example, the Standards Association of Zimbabwe (SAZ) works with four categories of standards — waste and effluent water and test methods, water quality, air quality and radioactivity. SAZ has five committees dealing with waste and effluent water, air quality, radiation protection, mirror committee (to the international committee on environmental management ISO 14000 standards) and lastly, fuels (CH20).[54] By end of January 2000, Zimbabwe had put in place a total of 977 industrial-related standards of which 104 of them relate to the environment.[55]

All the Zambezi Basin countries are participating in a regional initiative, SADC Standards (SADCSTAN) to establish cooperation on standardisation through their respective national standards associations.[56] Among other issues, SADCSTAN seeks to harmonise member countries' standards on commonly traded products in the region. So far three areas needing cooperation and harmonisation have been identified — waste management, water quality and air quality. Zimbabwe's SAZ volunteered to coordinate the regional air quality programme under SADCSTAN.

Under SADCSTAN, 11 projects dealing with common and masonry cements, pneumatic tyres for passenger and commercial vehicles, fish and fish products, plugs and sockets and electrical installations and buildings are being implemented. Zambia currently chairs SADCSTAN while South Africa provides the Secretariat for the 11 projects.

Potential benefits that are associated with effective environmental management systems include;

- conservation of input materials and energy,
- meeting customers' environmental expectations,
- improved environmental performance,
- maintaining good public/community relations,
- technology development and transfer,
- obtaining insurance at reasonable cost,
- enhanced corporate image among legislators, regulators, customers and the general public,
- improved market share,
- ability to dispose of waste, and
- cost control.

Through UNEP and the United Nations Industrial Development Organisation (UNIDO), industry is also embarking on a Cleaner Production (CP) concept to enhance environmental management and performance by reducing waste and using environmental-friendly methods in processing. UNEP and UNIDO established Cleaner Production Centres (CPC) around the world of which three are in Africa.

CPC-Zimbabwe serves the southern African region by conducting cleaner production audits and disseminating information on the concept.[57] Plans are currently underway to implement more regional projects and establish more centres in the region.

Among other regional projects being implemented by CP-Zimbabwe are Blantyre Cold Storage Commission meat processing in Malawi and sugar refinery in Zambia.[58] Both projects are aimed at waste and pollution reduction through use of appropriate technology.

Industries of the basin face the challenge of not just implementing environmental management systems but also ensure their supplies are from compliant sources, both locally and abroad.

FUTURE CHALLENGES AND OPPORTUNITIES

Industrialisation offers countries in the region the opportunity to diversify their economies so that they are not dependent on any one sector. There is need to attract more investment by ensuring an appropriate environment climate that takes into account political stability, economic incentives, which include tax breaks for companies, which invest in cleaner production. Close ties between governments and the private sector can help ensure that industrialisation does not impose environmental costs on the people of the basin. Already some countries in the basin are showing the benefits of collaboration between government and private sector in their efforts to reduce the use of ozone depleting substances. This is provided for under the 1987 Montreal Protocol on Substances that Deplete the Ozone Layer.

There is need to review all relevant laws governing environmental and natural resource use and management to ensure that they conform with international developments over the past few years. There is also need to incorporate relevant international agreements into regional protocols and national laws so that they can be enforceable. The basin states should look at:

- Which areas of legislation need changing to suit modern sectoral operations.
- Problems identified and related to landuse and natural resources management.
- Guidelines in connection with pollution control.
- Economic incentives aimed at encouraging investment in cleaner production.
- Raising awareness among consumers so that their choice can encourage industry to adopt environmentally friendly production processes.

Map 6.2: Environmental impacts from industrial activities in 1995

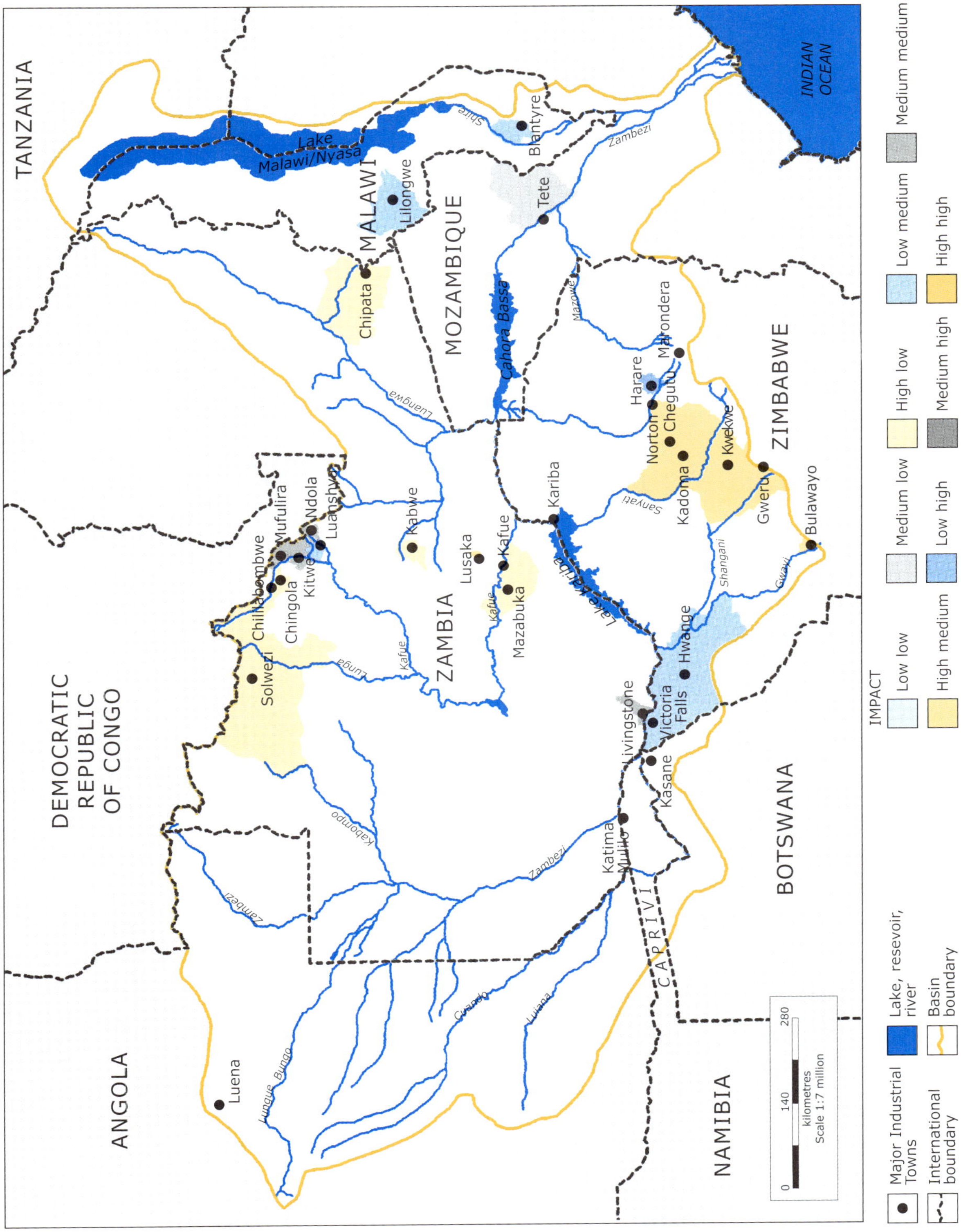

Source: Zambezi River Basin Atlas, SADC/DENCONSULT/ZRA, Lusaka, 1998

Map 6.3: Environmental impacts from industrial activities in 2005

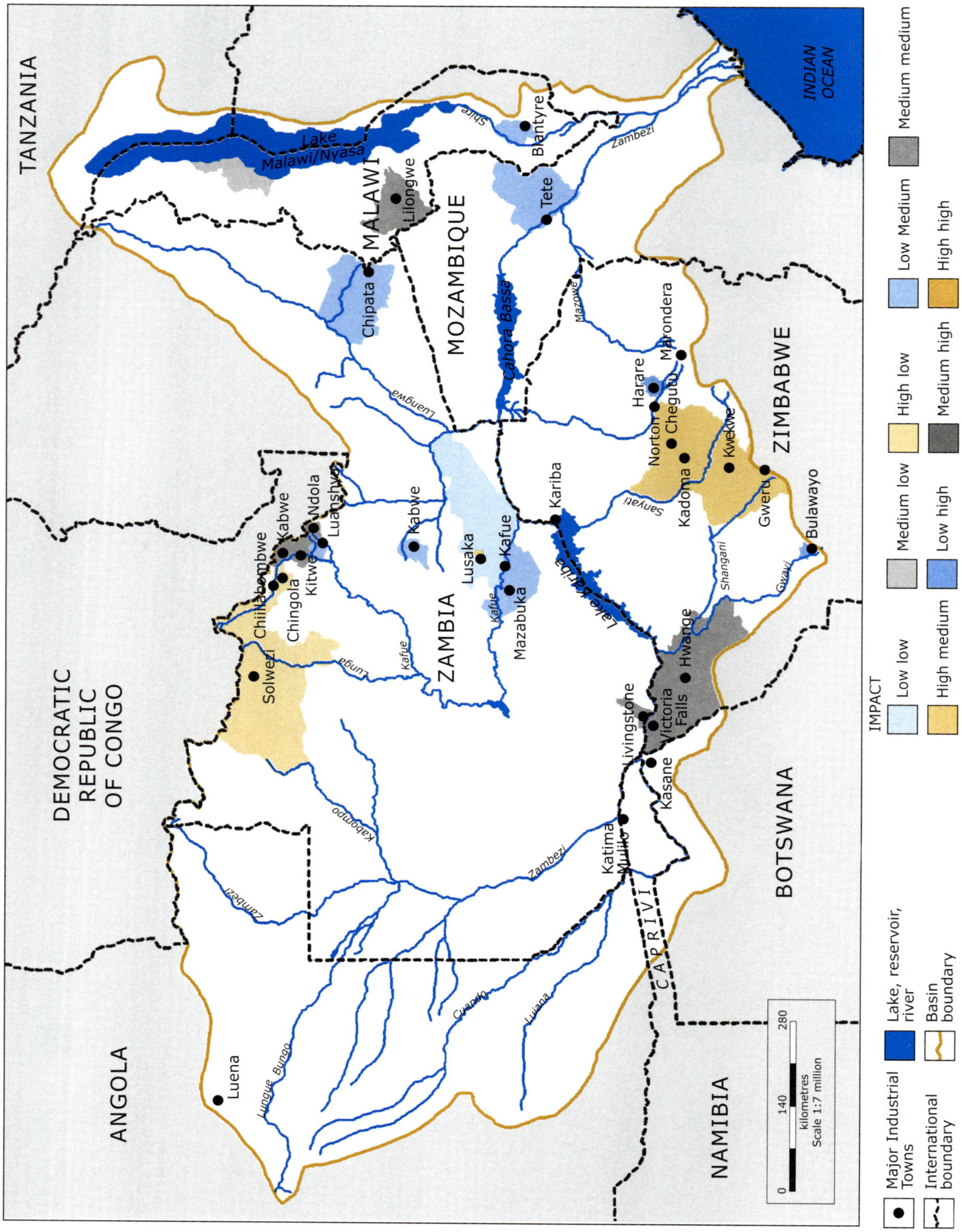

Source: Zambezi River Basin Atlas, SADC/DENCONSULT/ZRA, Lusaka, 1998

LINKAGES TO OTHER CHAPTERS

Chapter 1: Regional Overview: People and Environment
A vibrant industrial sector supports the economies of the basin states, offering employment and promoting economic growth. The major industrial activities in the basin are agriculture, manufacturing, tourism, and mining.

Chapter 2: Physical Features and Climate
Industry in the basin states depends, to a large extent, on the physical features and climate. Land features influence mining and settlements while climate and soils are major factors in agriculture.

Chapter 3: Water and Wetland Resources
The water and wetland resources of the basin states support a diversity of industrial activity. Agriculture, fisheries, manufacturing, power generation, tourism and other such industries, heavily depend on water resources.

Chapter 4: Biological Resources and Diversity
The performance of the industrial sector in the basin states also depends on the biological resources. These resources are used as raw material for manufactured goods, and include animal and plant life.

Chapter 5: Agriculture
Agriculture is the single largest employer of all the industries in the basin states, contributing substantially to their economies in foreign currency earnings. The agricultural industry also provides raw materials for processing into finished products.

Chapter 7: Energy
Industrial activity in the basin states is dependent upon good supply of energy resources. Mining of coal and the presence of suitable physical features and abundant water supplies, offer inputs for the generation of thermal and hydropower for the industrial sector.

Chapter 8: Tourism
The tourism industry is one of the fastest growing industries in the basin, bringing in foreign currency earnings. While employing a large number of people, there is room for improvement in dealing with poverty especially in rural areas so as to assist other industries in absorbing skilled and unskilled labour.

Chapter 9: Pollution
The industrial sector is one of the largest polluters of the environment. Gaseous emissions from industry pollute the air while liquid and solid waste also pollutes the soil and water resources.

Chapter 10: Poverty
While the industrial sector, especially agriculture, mining and manufacturing has contributed to the welfare of workers, millions of people in the basin still live in poverty. The diversification of the economy could help poverty alleviation efforts through the creation of more jobs.

Chapter 11: Gender and Women's Roles
Industries such as agriculture and manufacturing have made strides in addressing gender imbalances. However, more needs to be done to give equal opportunities to men and women to enhance social equality, equity and poverty alleviation and social and economic development.

Chapter 12: Environmental Management and Regional Cooperation
With diversified economies that include agriculture, mining, manufacturing and others, regional cooperation is necessary to facilitate the exchange of goods and services, skills and expertise for industry and related markets. The negative impacts of industry on the environment also require coordinated efforts as pollution traverses international borders.

Chapter 13: Trends and Scenarios
Pollution and dumping of waste continue to be serious problems due to limitations in expertise and technical capacity, and with the growth in industry, these problems may become insurmountable. Efforts need to be made by each basin state to strengthen its technical capacity in dealing with pollution and waste management.

ENDNOTES

1 SADC-WSCU/DANIDA, *Zambezi River Basin Atlas*, Zambezi River Authority, Lusaka, 1998
2 SADC, "Mining: Towards a Common Future", Consultative Meeting Report, SADC, Mbabane, 17-19 February 2000
3 Warhurst, A. (1992). "The Political Economy of Mining and Environment: The Policy Challenge of Environmental Management in the Southern African Development Community", Mining and Environment workshop, SADC MCU, Lusaka, 1-3 December 1992
4 op. cit 2
5 SADC MCU, *Review of the Mining Industry in SADC*, SADC MCU, Lusaka, 1998
6 Statistical data from member states of SADC, project profiles: 1994 SADC/EU Mining Investment Forum, SADC MCU, Lusaka
7 http://mbendi.co.za/indy/ming/mingzic1.htm
8 ibid
9 ibid
10 ibid
11 Zulu, E. H., "Mining in the Zambezi Basin of Mozambique", SADC Mining Sector Coordinating Unit, Lusaka, undated
12 http://www.sadcreview.com
13 ibid
14 ibid
15 Denconsult, ZACPLAN: *Sector Study Number 7, Final Report,* ZRA, Lusaka, August 1998
16 SADC, "Tourism: Towards a Common Future", Consultative Meeting Report, SADC, Mbabane, 17-19 February 2000
17 ibid
18 Denconsult, ZACPLAN: *Sector Study Number 4, Final Report,* ZRA, Lusaka, August 1998
19 SARDC-IMERCSA, "The Zambezi - a rich source of fish", *The Zambezi Newsletter*, Vol. 1, No. 3, Harare, 1998
20 UNEP/UNDP, "Industries and Enforcement on Environmental Law in Africa, Industry Experts Review Environmental Practice", Nairobi, August 1997
21 SADC, "Industry and Trade: Towards a Common Future", Consultative Meeting Report, SADC, Mbabane, 17-19 February 2000
22 Zulu E.H., "Environmental Impacts of Mining in the Zambezi Basin", Paper presented at the Regional Workshop on the State of the Environment in Zambezi Basin, Chinhoyi, July, 1998
23 ibid
24 Mbinji, W., "Environmental Issues in the Copper Mining Industry of Zambia, Zambia Consolidated Copper Mines Limited", Kalulushi, 1991
25 ibid
26 op. cit 22
27 ARA-TECHTOP, *Environmental Pollution: Key Issues for Mining and Industry in Zimbabwe*, Symposium organised by the ARA-TECHTOP Environmental Economics and Resource Management Division, Harare, 1991
28 NORAD/Environment Working Group, *Mozambique: The Present Environmental Situation 1990*," EWG, Maputo, 1990
29 ibid
30 op. cit 22
31 op. cit 27
32 SARDC-IMERCSA, Pollution, *Zambezi River Basin Factsheet* No. 11, Harare, 1999
33 op. cit 22
34 op. cit 24
35 op. cit 32
36 United Nations Environment Programme, *Global Environment Outlook-2000*, Earthscan Publications Limited, London, 1999
37 ibid
38 Mathuthu, A.S; Zaranyika, and F.M. Jonnalagadda, "Monitoring of Water Quality in Upper Mukuvisi River in Harare, Zimbabwe", *Environmental International*, Vol.19, p.51-61, 1993
39 Moyo, N.A.G., *Lake Chivero: A Polluted Lake*, University of Zimbabwe Publications, Harare 1997
40 op. cit 28
41 ibid
42 op. cit 32
43 Bethune, S., "Wetlands and Water Resources Management in the Zambezi Basin, An overview", for SARDC, 1998
44 Denconsult, *ZACPLAN: Sector Study Number 7, Final Report,* ZRA, Lusaka, August 1998
45 Chipungu, P.M. and D.M. Kunda, (eds.), *State of the Environment in Zambia*, ECZ/CIDA/NORAD, Lusaka, 1994, p 73.
46 Moyo, M., "Environmental Management in Zimbabwe: The Way Forward", Paper presented at the symposium organised by the ARA-TECHTOP Environmental Economics and Resource Management Division, Harare, 1991
47 ibid
48 *State of the Environment Report for Malawi 1998*, Environmental Affairs Department, Government of Malawi, Lilongwe, 1998, p 10.
49 Mombemuriwo, I., "The Perception of the Private Sector Views Regarding State of the Environment Reporting", Paper presented at the Consultative Meeting to develop a Plan for the State of the Environment Assessment and Reporting for Southern Africa, Harare, 1999
50 SADC, "Industry and Trade: Towards a Common Future", Consultative Meeting Report, SADC, Mbabane, 17-19 February 2000
51 op. cit 12
52 http://www.iso.ch
53 M.P. Mutasa, personal communication., Standards Association of Zimbabwe, Harare, February 2000
54 ibid
55 ibid
56 ibid
57 M. Mubayiwa, personal communication., Cleaner Production Auditor, Cleaner Production-Zimbabwe, Harare, February 2000
58 ibid

REFERENCES

African Development Bank, *Country Environmental Profile: Mozambique*, Environment and Social Policy Working Paper Series No. 21, Abidjan, 1995

Bachs, A., Makuku, S. J. and J. Maviya, "Industry and Development in Zimbabwe", Department of Natural Resources, Harare, 1992

Chanda, D., "Technology Transfer within SADCC: Prospects and Problems", University of Zambia, Lusaka (undated)

Christiansson, C. and J. Ashuvud, "Heavy Industry in a Rural Tropical Ecosystem", University of Stockholm, Stockholm, 1985

Denconsult, "ZACPLAN, Sector Study Reports Numbers 4 and 7", ZRA, Lusaka, August 1998

Government of Botswana, *The Contribution of Wildlife to the Economy of Botswana*, Ministry of Commerce and Industry, Gaborone, 1988

Government of Namibia, "The Reconstruction and Development of Namibia", Volume 1, Windhoek, 1990

Government of the Republic of Zambia/IUCN, *The National Conservation Strategy for Zambia*, GRZ/IUCN, Lusaka, 1985

Kowero, G.S., *An Economic Evaluation of the Mobile Sawmilling industry in Tanzania*, African Rural Social Science Series, Research Report No. 8, Winrock International Institute for Agricultural Development, Morrilton, Arkansas, 1990

Lopes, H., Jergman, K. and U. Aswathanarayana (eds.), "Environment, Technology and Sustainable Development", Proceedings of the African International Conference, Maputo, November 25-29, 1991, Maputo, 1992

Mbinji, W., and J. A. Charman., *Environmental Issues in the Copper Mining Industry of Zambia*, ZCCM, Environment Control Department, Kalulushi, 1991

Mwaura, P. and F. M. Kamau., "An Overview of the Forest Industry in Eastern and Southern Africa", *Unasyiva*, Vol. 42, 1991

Scandinavian Institute of African Studies (SIAS)., *Manufacturing Industry and Economic Development in the SADCC Countries*, SIAS, Uppsala, 1984

UNEP/UNDP, "Industries and Enforcement on Environmental Law in Africa, Industry Experts Review Environmental Practice", Nairobi, August 1997

Werner Biermann., *The Problems of Industrialisation in Tanzania*, Journal of African Studies, No. 3, 1987

World Commission on Environment and Development., *Our Common Future*, Oxford University Press, Oxford/New York, 1987

WRI/UNEP/UNDP/The World Bank., *World Resources 1998-99*, Oxford/New York, 1998

Yeager, R., "Democratic Pluralism and Ecological Crisis in Botswana", *Journal of Developing Areas*, Vol. 23, No. 3, April 1989

Photo: P Wade

Of late, the region has expanded use of new renewable energy technologies, predominantly the use of solar energy and wind power.

7 ENERGY

Energy resources found within the Zambezi Basin include both renewable ones such as biomass and hydropower, non-renewable ones such as coal and petroleum as well as new energy technologies like solar and wind power.

About 74 percent of the total energy requirements of the eight Zambezi River Basin countries come from biomass fuels such as wood, charcoal, and plant residues, and is mostly used for domestic purposes and in agriculture for curing tobacco. Biomass fuels are also used for brick manufacturing; lime production; fish smoking; and coffee- and tea-drying.

Electricity is either hydro or thermal based, making it renewable or non-renewable, respectively. The estimated hydropower potential of the Zambezi River is 20,000 megawatts (MW) of which 4,684 MW has been developed.[1] Potential hydropower plants with a total installation capacity of more than 13,000 MW have also been identified in the basin.[2] Coal and coal-bed methane are used to generate thermal power in the region, which boasts abundant coal reserves.

Of late, the region has expanded use of new renewable energy technologies, predominantly the use of solar energy and wind power, but also energy sources such as small hydro generation plants and biogas. Rural communities in and outside the basin have welcomed these technologies. Zimbabwe, for example, has benefited from the Global Environment Facility's Solar Photo Voltaic Pilot programme, which has seen some 10,000 units installed in rural areas. Wind driven pumps are utilised by rural and commercial farmers to pump water for livestock and to a limited extent, to supplement irrigation.

Photo: M Chenje

Thermal power generation is a major source of energy in the basin, particularly in Zimbabwe.

However, production and use of all forms of energy have associated environmental consequences. Fuelwood burning, for example, contributes to deforestation, one of the factors that lead to soil erosion, siltation and ultimately land degradation. Further, destroying forests reduces the region's capacity to curb climate change since forests act as sinks of carbon dioxide.

Just as much as there is a need for electricity from hydropower plants, constructing dams may have widespread negative environmental impacts. These include the displacement of people and animals; burial of heritage sites, plants and animals. Other impacts include the depletion of wetland habitats and aquatic life; increase in waterborne diseases; sedimentation, siltation and siltation-induced flooding, especially if Environmental Impact Assessments (EIAs) are not properly carried out. For example, Lake Kariba alone displaced over 57,000 people while the Cahora Bassa in Mozambique forced the relocation of about 25,000 people. Animals and plants were not spared either.

At the global level, the burning of fossil fuels is believed to be the cause of climate change and general global warming. Regionally, problems associated with mining and use of coal include fine dust, gases and smoke which pollute the air, while the excavated spoils piled up in dumps are unsightly and a hazard.

Nevertheless, energy is needed for economic development and is required in many industrial processes and for domestic use.

In terms of energy consumption patterns, two factors are strongly related to the consumer behaviour — demographic and economic. The majority of the people in the basin are rural based and only a small number of them have access to commercial energy sources. The relatively low industrialisation level of the Zambezi Basin states explains the low energy consumption levels. In most countries, shortages of foreign currency also limits access to energy-consuming equipment and prevent electricity supply utilities expanding networks sufficiently to reach all potential customers.[3]

NON-RENEWABLE ENERGY RESOURCES

Non-renewable energy sources used in the Zambezi Basin include fossil fuels such as coal, petroleum and gas. Most of the thermal-based generation plants use coal as an input fuel. There are no nuclear power plants in the Zambezi Basin.

Photo: M Chenje

About 74 percent of the basin population depends on forests and plant residues for their fuelwood needs.

Table 7.1: Total energy consumption in the Zambezi Basin states 1985-95

Country	Commercial Energy (peta-joules) 1995	Commercial Energy Percent change since 1985	Traditional Fuels (peta-joules) 1995	Traditional Fuels Percent change since 1985
Angola	26	(15)	59	50
Botswana	-	-	-	-
Malawi	12	38	103	(22)
Mozambique	16	8	170	21
Namibia	-	-	-	-
Tanzania	34	24	341	29
Zambia	50	(10)	135	29
Zimbabwe	128	(7)	74	7
South Africa*	3,659	(16)	151	13
USA**	92,275	35	3,853	265

Notes: * For comparison within the SADC region
** For comparison with developed country

Total energy consumption includes commercial energy and traditional fuels. Commercial energy refers to apparent consumption and is defined as domestic production plus net imports, minus net stock increases, and minus aircraft and marine bunkers. Commercial energy consumption includes energy from solid, liquid, and gaseous fuels, plus primary electricity. Traditional fuels includes estimates of the consumption of fuelwood, charcoal, bagasse, and animal and vegetal wastes.

Source: WRI, 1998-99 World Resources, A Guide to the Global Environment, World Resources Institute, New York, 1998.

Several explorations have been carried out in the basin to locate economically mineable oil and natural gas reserves, but these have not yielded anything. Coal-bed methane, which is found in most coal deposits, remains a potential gaseous fuel. Zimbabwe, for example, is actively investigating its potential.[4]

Box 7.1: Power rationing may be introduced

Power rationing could be introduced nationwide soon if the Zimbabwe Electricity Supply Authority (Zesa) fails to get enough foreign currency to pay for imports and the diesel shortage persists.

Zesa yesterday acknowledged that although its cash-flow situation was improving, its financial position had not improved sufficiently for it to wipe out a huge debt accrued over the years because of under-pricing of tariffs.

Unavailability of foreign currency and diesel in the country had worsened the situation and the utility might be forced to cut supplies to residential areas during peak periods. The shortage of diesel would constrain the internal general capability at the country's thermal power stations, especially Hwange. Zesa normally maximises on internal generation, should there be a problem with power imports.

But this would be done, causing as little disruption to the country's economy as possible. In this line, industrial areas have been left out of the load-shedding schedule announced by Zesa.

"This is a preparatory measure, I would like to make it clear that Zesa will only resort to load-shedding as a last resort, and all efforts are being made to avert such a situation," said the Zesa consumer services director, Engineer Joshua Chifamba.

Zesa imports about 55 percent of the country's electricity requirements from Electricidade de Mozambique, Eskom of South Africa, Snel of the Democratic Republic of Congo, and Zambia, at a cost of over $220 million (US$ 5.8 million) a month. The schedule would see residential areas alternately being disconnected from the national electricity grid twice during the five-day working week. This will occur once during the morning peak period, beginning at 6 - 10 am, and the afternoon peak period from 6 - 10 pm. The country's peak demand is around 2,000 megawatts, more than half of which is met through imports. It will be necessary to disconnect some consumers to save the system from collapsing once the available energy becomes less than the demand.

"The schedules have been designed to minimise disruptions in the economy. Maximum demand customers, unless the situation becomes very serious, would not be disconnected as much as possible," said Engineer Chifamba.

Zesa has accumulated a huge debt over the last few years, during which it has been selling power at below cost. "We have been negotiating with the utilities from whom we buy power to convert our debt into a loan, which we will service over a period of time," said Engineer Chifamba.

Some of the utilities from which Zesa imports power have been threatening to disconnect Zimbabwe from the inter-connected grid owing to non-payment of tariffs. Zesa has suspended most capital projects, deferred payment to creditors and slashed its overtime bill, among other cost-cutting measures aimed at improving the authority's cash flow position.

Since the government agreed to the quarterly tariff adjustments, Zesa's cash-flow situation has been improving. The consumer services department has been encouraging Zimbabweans to conserve energy as much as possible, both in the home and in industry.

Source: The Herald, Wednesday 26 January, 2000, Harare

Table 7.2: Non-renewable energy resources

Country	Coal reserves (mn tn)	Oil (mn tn)	Natural gas (mn cu m)
Angola	*	156	372,000[1] 402,000[2]
Botswana	3,300	0	0
Malawi	16	0	0
Mozambique	240	0	41,000
Namibia	0	0	134,000
Tanzania	200	0	35,000
Zambia	30	0	0
Zimbabwe	1,700	0	50,000,000[3]
Total	5,486	156	50,984,000

Notes: *No assessments have been made but other sources suggest abundance of coal in Angola.

1 Associated, 2 Non associated and 3 Coal bed methane

Source: Maya, R.S., (Eds), Options for Greenhouse Gas Mitigation in the power sector in SADC, Phase I Report, Inventory of GHG emissions from the power sector, Mond Books, Harare, 1999.

Coal

Virtually all known, exploitable coal reserves for Malawi, Mozambique, Zambia, and Zimbabwe are found within the Zambezi Basin. Angola and Namibia do not have known coal deposits. Coal reserves within the basin are utilised mostly for thermal power generation, domestic uses such as cooking and heating, industrial uses such as smelting and making chemicals, and commercial uses such as tobacco curing and railway traction. Coke oven gas is also produced from coal burning and is used in industry and, of late, has replaced diesel as a start up in thermal power generation.

The total coal production in the SADC region increased from 212.24 million tonnes in 1996 to 223.98 million tonnes in 1997.[5]

Box 7.2: Coal reserves in the Zambezi Basin

Coal deposits occur in several locations in Malawi with proven reserves of 16 million tonnes and unassessed reserves in excess of 800 million tonnes. Despite the great potential of coal, current demand is less than one percent of the total energy demand and is mostly by industrial steam boilers. Malawi obtains 40 percent of its coal from in-country production in the Northern region. The Kaziwiziwi mine has since been exhausted and closed in 1990. Current production is from the Mchenga mine and future plans are to exploit the coalfields of the Lower Shire valley.

In Mozambique, coal deposits abound, although proven reserves are only about five million tonnes and probable reserves are estimated at 82 million tonnes. Estimates of possible reserves exceed three billion tonnes. Production has been centred on Moatize in Tete Province whose capacity has doubled in recent years. Exports of coal amounted to only 17,000 tonnes, however, as a result of disruptions in the transport system.

Tanzania is endowed with coal reserves estimated to be as high as 1,200 million tonnes of which 300 million tonnes are proven. Presently, there is only one operating mine in Tanzania, at Kiwira, which is northwest of Lake Malawi/Nyasa near Mbeya. The Mchuchuma/Katewaka coal field is located 750 km inland in the southwestern part of Tanzania. Geological exploration is ongoing to establish coal reserves, characteristics, and to establish a mining plan and pricing.

Zambia's single coal mine at Maamba exploits the near out-crop areas of thick coal development in part of the Siankondobo coalfield, north of Lake Kariba. Major faults divide the coalfield into two main structural units known as the Kazinze and Izuma basins. In the Izuma Basin, single thick seam is developed, while in Kazinze, the seam is split into two components by a laterally persistent sandstone band.

The Maamba Colliery was originally designed to produce over one million tonnes of saleable coal annually. Its actual capacity has never approached this design target and has actually dropped from about 750,000 to 350,000 tonnes. Zambia Consolidated Copper Mines (ZCCM) Nkana and NCZ are Maamba Colliery's largest customers. Most of the coal from Maamba is pyritic inorganic sulphur and only about 40 percent has inherent sulphur that cannot be completely washed.

Zimbabwe has a total of 10.6 billion tonnes of coal in 21 deposits. Coal deposits occur in the younger rocks at the northern and southern edges of the Basement shield. Proven resources can last 107 years and the total reserves over 2,000 years at present production rates, of about five million tonnes per year.

Zimbabwe has two coalmines, one at Wankie Colliery with production capacity of six million tonnes per year and another at Sengwa with a production capacity of 200,000 tonnes per year. About 60 percent of Hwange coal is used as run-of-mine steam coal at Hwange Thermal power plant. The rest is used for industrial and domestic purposes. The mine at Sengwa has since been closed due to viability problems, but is set to resume production as the proposed Sengwa Thermal power station becomes operational. Sengwa coal has less sulphur (0.4 percent) than Hwange coal (2.66 percent).

Source: Mhone, G. and Nguira, A., "Malawi", in: Sam Moyo, Phil O'Keefe and Michael Sill (eds), *The Southern African Environment, Profiles of the SADC Countries*, , Earthscan Publications, London, 1993.

Lopes, H., "New and Renewable Sources of Energy (NRSE) in Mozambique", SADC TAU, Luanda, Angola, September 1989.

Black & Veach International, National Development Corporation, "Mchuchuma/Katewaka Coal Power Plant Feasibility Study", US Trade and Development Agency, Washington D.C., 1996.

World Bank and Department of Energy/Zambia, "Joint UNDP/World Bank Energy Sector Management Assistance Program, Energy Sector Strategy", World Bank, Washington D.C., September 1988.

Yamba, F.D., Chidumayo, E.N., Jain, P.C., Mbewe, D.N., Mbewe, D.J. and Mwale, T., Zambia "Country Study on Climate Change, Volume 1", Inventories, Ministry of Energy and Water Development/Zambia, GTZ, Lusaka, November 1995.

ZESA Corporate Planning Department, Unpublished file material, 1999.

Petroleum and natural gas

Although there are no petroleum plants within the basin, indications are that the thick extensive sedimentary rocks have a potential for hosting petroleum deposits and natural gas. This has led oil companies to undertake extensive exploration in Malawi, Mozambique and Zimbabwe.

Petroleum products are used in all sectors, including the domestic sector. In the household sector, kerosene is the most frequent commercial fuel used for lighting and cooking. In most countries, about 80 percent of rural people's lighting needs are met by kerosene.

Regarding natural gas, a Zimbabwean company, Shangani Energy Exploration, has been exploring for coal-bed methane reserves since 1990. Much of the explorations have been in the Gwayi area where a resource of 50 billion cu m of methane has been established within a relatively small area. This gas could be used for generating electricity, fertiliser production or as a transport fuel.

Thermal power

Thermal power is electricity produced by burning fossil fuels, including coal, coalbed methane, natural gas and diesel to get steam to turn turbines. However, it is coal that is mostly utilised by the basin countries in thermal power production. The major thermal plants within the basin are located in Zimbabwe.

Zimbabwe's major thermal power plant is Hwange Power Station with a capacity of 920 MW. Three other thermal power stations are in Harare (135 MW), Bulawayo (120 MW) and Munyati (120 MW).[6] A power station at Sengwa coalmine, with a capacity of 1,200 MW, has reached advanced stage of development, involving mainly independent power producers. Several diesel power plants, with a total capacity of about four MW, are scattered around the country. These diesel plants are mainly used in isolated systems at rural institutions or as backup for some establishments.[7]

Thermal plants in Malawi include Chichiri gas turbine (15 MW) in Blantyre, diesel plants in Lilongwe (4.3 MW) and Mzuzu (1.8 MW) which all feed into the national grid. The diesel and gas turbine plants serve as standby power supply. Chitipa is supplied by an off-grid diesel plant of 300 kW capacity.

Zambia has few thermal power plants as it has developed a lot of its hydro resources. Diesel thermal

Box 7.3: Prospects for oil in Lake Malawi

Exploration and research under the waters of Lake Malawi/Nyasa has shown that the lake, the third largest in Africa and one of the world's largest freshwater lakes, has shown potential for its rocks containing hydrocarbons which could be drilled for oil.

Charles Kaphwiyo, director of the Malawi Geological Survey, however, said although geographical exploration so far has confirmed existence of the oil-bearing sedimentary rocks, real proof can only be ascertained by drilling a three-kilometre stratum to collect samples for laboratory study. "Such drilling would be located off-shore," he noted.

Kaphwiyo said at least three explorations have been carried out on Lake Malawi/Nyasa so far and preliminary studies have shown that hydrocarbons, the chemical combination of hydrogen and carbon potential for petrol, exist in the lake. The first exploration was done in 1981 by a team from Duke University in North Carolina, United States.

Kaphwiyo said the exploration, which covered the whole lake, dubbed "Calendar Lake" since its surface area is about 584 by 83 km, revealed thick sequences of the sedimentary rock on its bed. "The discovery raised the status of the lake as a target for hydrocarbons exploration," he added.

Shell Exploration BV, also known as Royal Dutch Shell, was the first international oil company to be granted an oil exploration licence. It carried out an air-borne magnetic survey over the lake in 1982 and confirmed the existence of the hydrocarbons in the deep waters in the northern part of the lake. However, Kaphwiyo said the latest study between 1985-87 discovered the potential for hydrocarbons in the shallow waters in the central parts of the lake, especially around the central lakeshore districts of Nkhota Kota and Salima.

Ten international oil companies sponsored the latest study. The venture was funded by the World Bank. If the exploration results come to fruition, it will considerably help Malawi's agro-based economy, which has had quite a beating of late with the failing prices of its chief cash crop, tobacco.

By Raphael Tenthani, PANA Correspondent, September 14, 1999
Source: Pan-African News Agency (PANA)

Figure 7.1: Use of petroleum products in the transport sector (TJ)

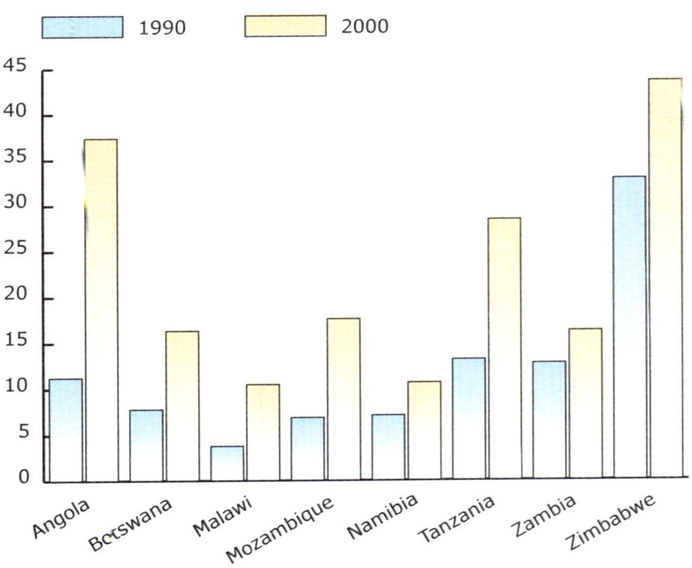

Source: United National Environment Programme (UNEP), Climate Change Mitigation in Africa, Proceedings of International Conference, Victoria Falls, Zimbabwe, 18-20 May, 1998. Southern Centre/Riso, Denmark, 1998.

Photo: M Chenje

Many of the thermal power plants in the basin are in Zimbabwe, including in its capital, Harare.

plants from isolated systems and standby plants give a total of 10 MW. The Zambia Consolidated Copper Mines (ZCCM) owns four gas turbines with a total installed capacity of 80 MW, located in the Copperbelt. These gas turbines are kept on cold standby.[8]

RENEWABLE ENERGY RESOURCES

The Zambezi Basin is endowed with renewable energy sources, comprising mainly biomass and hydropower. The exploitation of new renewable energy technologies using solar and wind power or geothermal, small hydro or biogas as a source of power is however, only just beginning although it has great potential.

About 74 percent of the net energy supply in the Zambezi Basin states is derived from biomass in the form of fuelwood, charcoal, crop and wood residues, and dung. Biomass energy is important for cooking, heating and lighting, especially for the poor and rural folk.

There are substantial biomass resources in the Zambezi Basin and proper use of these resources could mean significant savings of foreign currency for the implementing country in terms of imports of other commercial fuels. For example, ethanol blending with gasoline could mean a reduction in the import of petroleum fuels. Alcohol can also be utilised directly in cars or be used in power stations to generate electricity.

Fuelwood

The amounts of fuelwood energy derived from the basin by each country are not readily available. However, information on fuelwood consumption compiled for selected basin countries is indicative of the consumption patterns.

The reasons for the excessive use of fuelwood are several:

- Since forests are a renewable resource, they can be exploited in perpetuity, if properly maintained. Furthermore, for now woodfuel is the most reliable and affordable source of energy for poor households.
- Woodfuel is an indigenous source of energy. Its use does not require foreign exchange to acquire it.
- In most areas, fuelwood is not purchased but just collected. This "free resource" enables poor households to channel most of their incomes to other needs.
- In some countries, particularly those in which charcoal use is prevalent, trading in charcoal is a major source of income for some households. In Zambia, for example, the charcoal industry generates about US$30 million annually. About 60,000 people directly depend on it for the bulk of their income.[9]

No forestry inventory has been done in Angola since colonialism ended in 1975. However, it is estimated that Angola has 50,000–70,000 sq km of forest, representing about 40–60 percent of land area. The Ministry of Agriculture estimates that there is a minimum reserve of 17.45 million cu m of trees, which would allow 20 years rotative annual cut of 850,000 cu m. The subproducts could be used as woodfuel.

Botswana has abundant energy resources in the form of wood. Although in some areas wood is harvested at unsustainable rates for the country as a whole, it is in surplus. Only 1.8 million tonnes of the annual biomass increment of 50 million tonnes is harvested.[10] In Malawi, 90 percent of total energy being supplied is biomass mostly in the form of wood. Biomass and biogas together make up about 92 percent of all energy consumed.[11]

Mozambique's vegetation is predominantly forest and savanna. Recent assessments identified 50,000 sq km (six percent of national territory) of medium to high productive forest, and 150,000 sq km of low productivity forest. Open forest occurs throughout the country and

Figure 7.2: Percentage of net energy supplies in basin countries

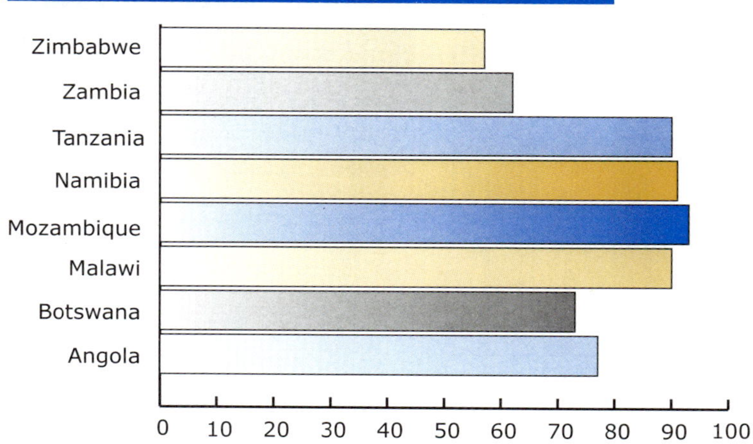

Source: SARDC-IMERCSA, CEP Factsheet, Zambezi River Basin Series: No 2, Energy, Harare, 1998.
The Figures from Angola and Mozambique: Kaale, D., Woodfuel use in Southern Africa, Paper presented at the SADC Community Building Workshop in Windhoek, Namibia, 10-14 May, 1994, 1994.

is widespread north of the Zambezi River.[12] About 80 percent of Mozambique's population lives in rural areas and depends on wood for cooking as well as heating water for domestic use, room heating and drying of food stuffs. More than 90 percent of the families exclusively use fuelwood for cooking, and less than 10 percent use charcoal.

Photo: D Minty

Photo: CURE Resource Centre

Charcoal and firewood provide most of the energy needs of the majority of the people in the basin countries.

In Namibia, there is indiscriminate clearing of trees in forest reserves for fuelwood and other purposes without "any allowance for regeneration" and this is particularly so in the northern region.[13] The predominant desert conditions also aggravate this problem. Rural households use up to 90 percent of firewood for their energy needs, resulting in massive deforestation.

About 25 million cu m of woodfuel is consumed in Tanzania with 97 percent of this being used as firewood in rural areas. Like in other basin countries, consumption levels exceed regeneration and plantation rates. Charcoal is also widely used, especially in urban areas where 70 percent of the urban population use charcoal for energy. In Zambia, fuelwood, which is used predominantly as a household cooking fuel, is the country's single largest source of energy. Woodlands and forests are estimated to cover about 100,000 sq km, which is about two-thirds or 66 percent of Zambia's total land area.

Wood is also the single largest source of energy in Zimbabwe, supplying about 57 percent of total energy consumed in the country. More than six million tonnes of wood are consumed annually, supplying mainly rural communities and the urban poor.[14]

Crop and other wastes

Available crop residues in the Zambezi Basin countries include those from major cash crops such as maize, sisal, sugar cane, tobacco, rice, coffee, and coconut. Timber wastes are in the form of sawdust, thinnings and off cuts. These wastes can either be burnt directly for heat, or gasified to produce both power and heat. If compacted, it is also possible to produce briquettes out of these wastes.

The Zambezi Basin is also capable of producing biogas from the dung produced by the large livestock population, organic waste and agricultural residues. However, only a fraction of the total potential can be realised since only cattle or pigs in an enclosure can provide waste for biogas production.

Ethanol

Ethanol is usually produced as a byproduct of sugar production and is, among other things, used for blending petrol and as a spirit in hospitals. Malawi is the only country that produces ethanol at a largescale within the basin, totalling about 6.33 million tonnes annually. However, there is great potential for ethanol production from crops such as sugar cane, cashew nuts, manioc, maize, citrus trees and other related crops, especially in Zambia and Mozambique.[15]

Table 7.3: Existing hydropower developments

Year	Facility	River	Sub-basin	Country	Capacity (MW)
1924	Mulungushi	Mulungushi	Luangwa River	Zambia	20
1944	Lunsemfwa	Lunsemfwa	Luangwa River	Zambia	18
1958	Kariba South	Zambezi	Kariba	Zimbabwe	666
1966	Nkula A	Shire	Shire River-Lake Malawi	Malawi	24
1970	Lusiwasi	Lusiwasi	Luangwa River	Zambia	12
1971	Kafue Gorge	Kafue	Kafue River	Zambia	900
1972	Victoria Falls A-C	Zambezi	Kariba	Zambia	108
1975	Cahora Bassa	Zambezi	Tete	Mozambique	2,075
1976	Kariba North	Zambezi	Kariba	Zambia	600
1977	Tedzani I&II	Shire	Shire River-Lake Malawi	Malawi	40
1992	Nkula B	Shire	Shire River-Lake Malawi	Malawi	100
1995	Wovwe	Wovwe	Shire River-Lake Malawi	Malawi	5
1996	Tedzani III	Shire	Shire River-Lake Malawi	Malawi	52

Source: Denconsult, Sector Studies under ZACPLAN, Sector Study 5, Final Report, Lusaka 1998, p.12.

Hydropower generation

The Zambezi Basin is a source of hydroelectricity in the southern African region. In southern Africa, it is only surpassed by the Congo River, which potentially holds Africa's largest hydro resource. The Zambezi River and its tributaries have been exploited for their hydro potential by most of the basin countries such as Zambia, Zimbabwe, Malawi and Mozambique.[16] A total of 42 potential hydro sites in the Zambezi Basin (12,892 MW) occur in Angola, Malawi, Mozambique, Zambia and Zimbabwe.[17] Other countries of the SADC region outside the basin benefit from direct import of electricity from the river's power stations and through the SADC Power Pool (SAPP). An example is South Africa, which imports electricity from the Cahora Bassa Dam in Mozambique.

As shown in Map 7.1 and Table 7.3 there are no existing hydropower plants found in the Angolan, Botswana or Namibian part of the Zambezi Basin. Of the 4,620 MW available, about five percent is in Malawi, 45 percent in Mozambique, 36 percent in Zambia and 14 percent in Zimbabwe.[18]

The Cahora Bassa hydropower plant, with a capacity of 2,075 MW, is the largest in SADC, followed by the Inga in the Democratic Republic of Congo, with 1,771 MW. Under current arrangements, Zimbabwe can import from Cahora Bassa up to 400 MW monthly, and both South Africa and Mozambique use the remainder.

About 40 possible new hydropower plants with a total installation capacity of more than 13,000 MW have been identified in the basin. About 85 percent of this capacity is on the Zambezi River itself, while the remainder is on its tributaries. Close to half of the mapped potential is in Mozambique, about 25 percent in Zambia and about 20 percent in Zimbabwe. A little more than five percent of the potential is divided between Angola, Malawi and Tanzania.[19]

Due to the war situation in Angola, the need for rehabilitation of the power generation, transmission and distribution system is great. The priorities within the electricity sector will, therefore, probably not include development of hydropower schemes in the Angolan part of the Zambezi Basin.[20]

In Malawi, a new hydropower plant, Kapichira I, with an installed capacity of 64 MW is under construction and was due to be commissioned in 1999. A second phase, Kapichira II, is expected to be commissioned in 2002. The Kholombidso project was initially proposed in the 1950s, with control of Lake Malawi/Nyasa and navigation as main purposes, and irrigation and hydropower generation as spin-offs. A potential has been identified in South Rukuru/North Rumphi (Lower Fufu) and Songwe rivers, tributaries to Lake Malawi/Nyasa in the Northern Region.[21]

Photo: IUCN

The Zambezi Basin is a major source of hydropower, supplying countries even outside the basin.

Map 7.1: Existing hydropower developments

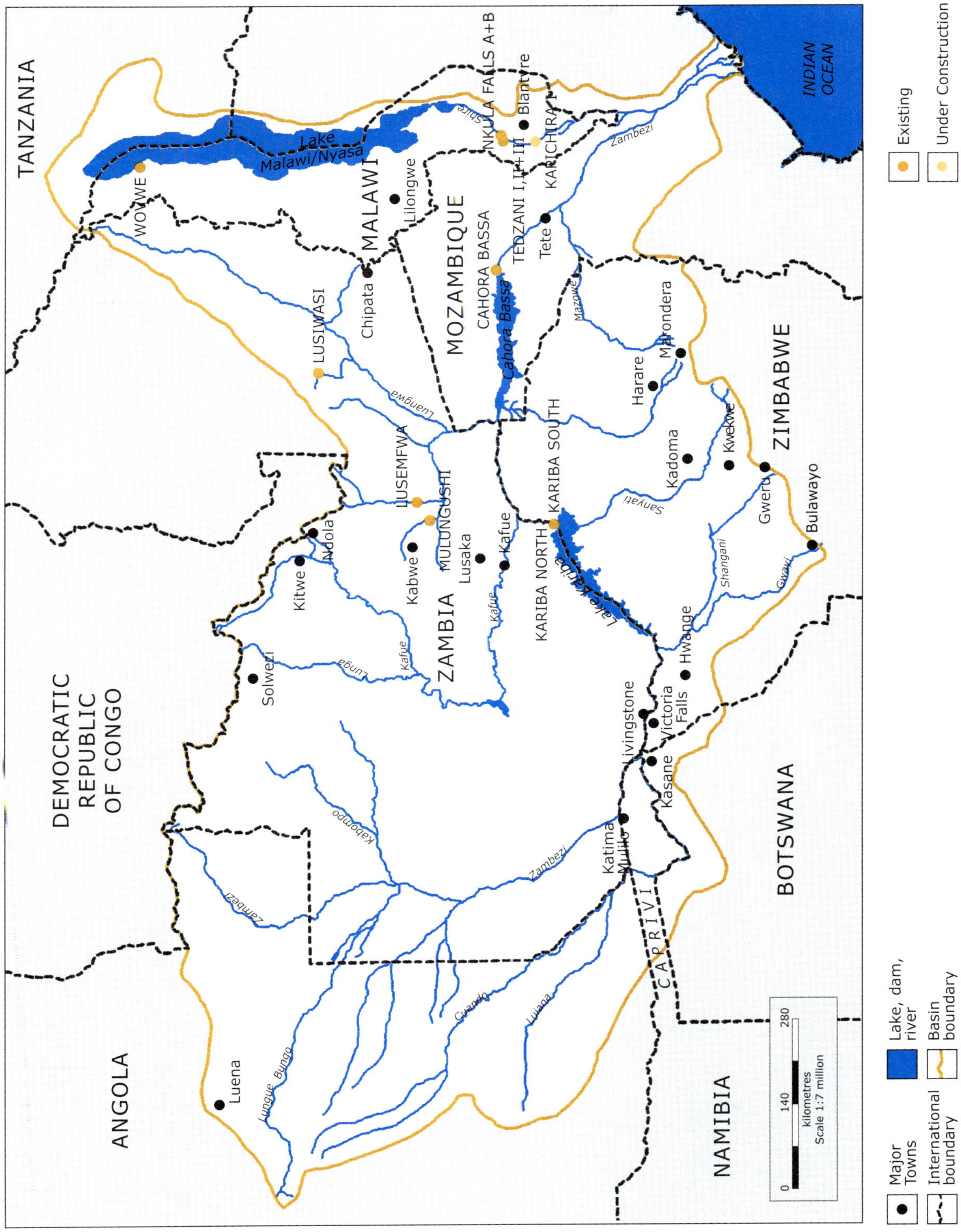

Source: Denconsult/NNR, Zambezi River Basin Atlas, SADC-WSCU, Danida, Lusaka, 1998, p.83.

Map 7.2: Potential hydropower developments

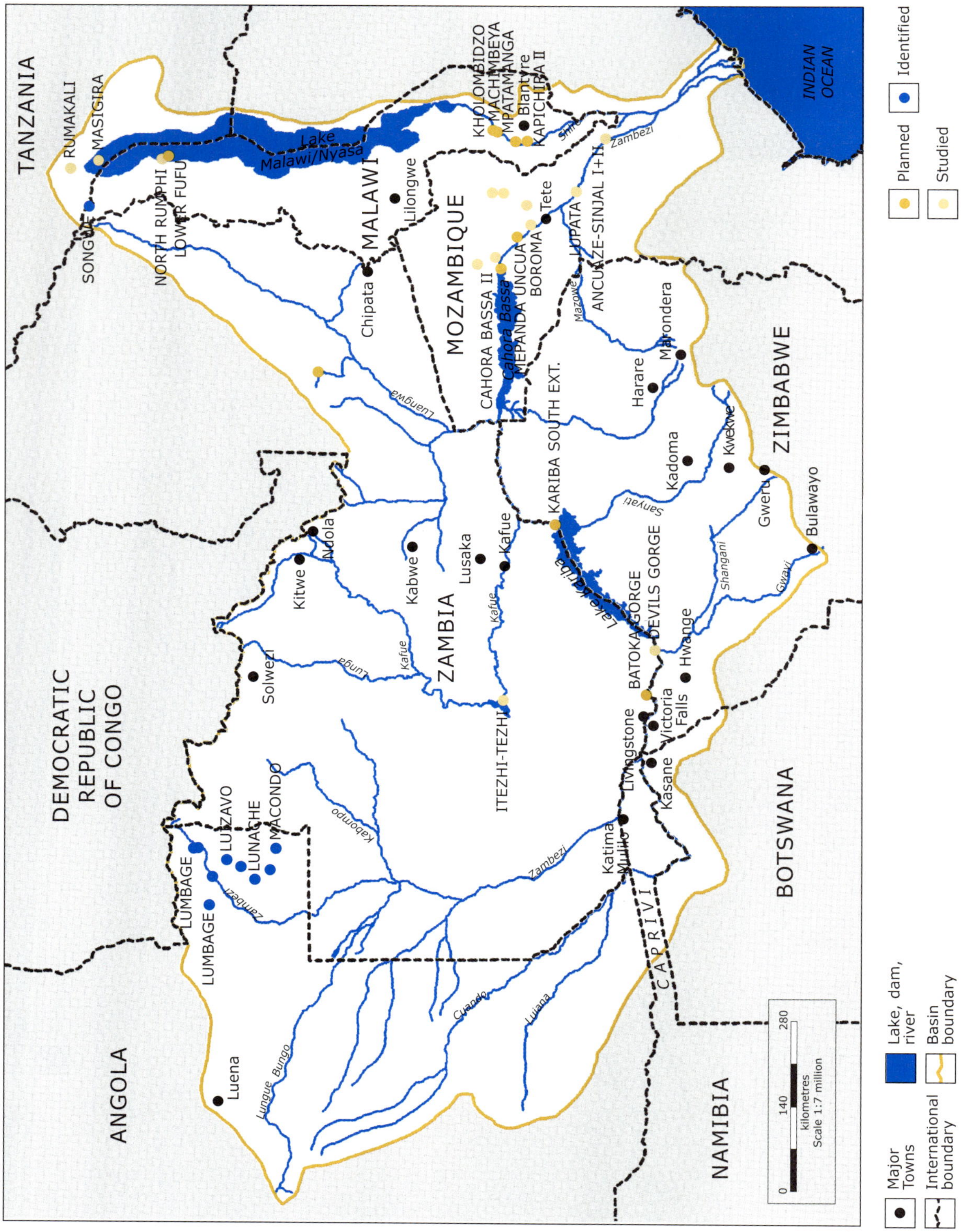

Source: Denconsult/NNR, Zambezi River Basin Atlas, SADC-WSCU, Danida, Lusaka, 1998, p.84.

Table 7.4: Potential hydropower developments

Country	Sub-basin	Facility	River	Capacity (MW)	Completed
Angola	Upper Zambezi	1	Lumbage	1	
Angola	Upper Zambezi	2	Zambezi	4	
Angola	Upper Zambezi	3	Zambezi	2	
Angola	Upper Zambezi	4	Luvua	1	
Angola	Upper Zambezi	5	Luizavo	11	
Angola	Upper Zambezi	6	Ludéu	3	
Angola	Upper Zambezi	7	Lunache	1	
Angola	Upper Zambezi	8	Lufuige	2	
Angola	Upper Zambezi	9	Macondo	3	
Malawi	Shire River-L. Malawi/Nyasa	Kapichira II	Shire	64	2002
Malawi	Shire River-L. Malawi/Nyasa	Kholombidzo	Shire	*	
Malawi	Shire River-L. Malawi/Nyasa	Nachimbeya	Shire	*	
Malawi	Shire River-L. Malawi/Nyasa	Mpatamanga	Shire	*	
Malawi	Shire River-L. Malawi/Nyasa	Lower Fufu	S. Rukuru/N. Rumphi	90	2006/2010
Malawi	Shire River-L. Malawi/Nyasa	Songwe	Songwe	150	
Mozambique	Tete	Cahora Bassa II	Zambezi	1,200	
Mozambique	Tete	Mepanda Uncua	Zambezi	2,000	2005
Mozambique	Tete	Boroma	Zambezi	444	
Mozambique	Tete	Luapata	Zambezi	654	
Mozambique	Tete	Ancuaze-Sinjal I	Zambezi	300	
Mozambique	Tete	Ancuaze-Sinjal II	Zambezi	600	
Mozambique	Tete	Chemba	Zambezi	1,040	
Mozambique	Tete	5.8	Revubue	36	
Mozambique	Tete	5.9	Revubue	110	
Mozambique	Tete	5.13	Revubue	85	
Mozambique	Tete	7.6	Luia	267	
Mozambique	Tete	7.11	Capoche	60	
Zambia	Kafue River	Kafue Gorge Lower	Kafue	600	2008
Zambia	Luangwa River	Itezhi-Tezhi	Kafue	80	2004
Zambia	Kariba	Lusiwasi Ext.	Lusiwasi	40	
Zambia	Kariba	Victoria Falls Ext.	Zambezi	390	
Zambia	Kariba	Katombora	Zambezi	*	
Zambia	Kariba	Kariba North Ext.	Zambezi	300	
Zambia/Zimbabwe	Mupata	Mupata Gorge	Zambezi	640	
Zambia/Zimbabwe	Kariba	Batoka Gorge	Zambezi	1,600	
Zambia/Zimbabwe	Kariba	Devil's Gorge	Zambezi	1,600	
Zimbabwe	Kariba	Kariba South Ext.	Zambezi	300	
Zimbabwe	Kariba	Victoria Falls	Zambezi	300	

* Not included in total

Source: Denconsult, Sector Studies under ZACPLAN, Sector Study 5, Final Report, Lusaka 1998, p.24.

In Mozambique, seven of the possible sites are on the Zambezi River itself, while the remaining five are on its tributaries. Cahora Bassa II, Mepanda Uncua, Luapata and Boroma are all large compared to the existing domestic power market in Mozambique, with a present peak load of about 200 MW. The large investments connected to such plants will normally not be economically and financially viable without an export agreement to sell surplus power, or development might be in connection with a sharp increase in domestic demand from a specific project. A South African aluminium company, for example, now plans to build an aluminium smelter in Maputo.[22]

Zambia has the largest potential for hydropower generation in the Zambezi Basin since the greater length of the Zambezi River is in this country and a number of other rivers such as the Luangwa and Kafue also offer potential for hydro-power. However, at present only 1,660 MW have been developed. Some of this potential is shared with Zimbabwe.

The development potential in the Zimbabwean part of the Zambezi Basin is limited to the Zambezi River itself, where it forms the border with Zambia. The possible sites for new hydropower plants — Batoka Gorge, Devil's Gorge and Mpata Gorge — are therefore joint with Zambia.

Solar energy

Since the Zambezi Basin lies between latitudes 8°–20° South, solar radiation is available for most of the year. Radiation levels are close to the average between December–May, dropping below average between June–August, then rising above average from September–November before the rainy season begins. This means that most thermal and photovoltaic applications of solar energy can be successfully used throughout the Zambezi Basin.[23]

Solar energy provides clean energy and can improve the standard of living of those communities that otherwise have no access to grid electricity. Thanks to solar energy these communities can, for example, engage in commercial activities and use radio and television. The cost of solar technologies has been hampering faster dissemination of the technology, but over the last few decades, the prices have been falling. International efforts to make solar products affordable have also assisted in bringing solar energy to poor communities. An example is the Global Environment Facility (GEF) Solar project, which ran in Zimbabwe from 1992–97, that has installed over 10,000 systems since 1992.[24]

Wind energy

The exploitation of wind energy in the Zambezi Basin is mainly limited to water pumping due to limited wind speeds that tend to be low, averaging between 1–3 m/s. Very few places experience wind speeds in excess of 3 m/s.[25]

However, a pilot project in Zimbabwe by ZERO Regional Environment Organisation has shown that it is possible to generate electricity even at low wind speeds, using specially designed turbines.[26]

Animal power

Animal power is an asset in the Zambezi Basin countries. Cattle and donkeys offer tremendous opportunities for use as draught power for agricultural and transportation tasks. In many areas, small-scale farmers use animal power instead of tractors because of its lower costs and cheaper mechanical equipment.

Use of animal driven pumps in Botswana, for example, is an attempt to allow draught animals to operate a mono pump by pulling bars in a circular path. This technology relies on donkeys whose physical condition is sometimes poor. Botswana's draft Agricultural Policy paper of 1990 had a proposal that "animal draught power be encouraged by improved methods and by the withdrawal of grants for tractors".[27]

Photo: W Nyamucbengwa

Solar energy provides clean energy and can improve the standard of living of rural communities and reduce energy costs in urban areas.

Photo: D Martin

The use of animal drawn implements is preferred in rural communities because of relatively greater flexibility, lower cost and greater linkages to other sectors of the rural economy.

In Malawi, draught animal power refers to the use of animals for ridging, weeding, transporting and planting. The common animals used in Malawi are oxen and donkeys. About 95 percent of the work animals are found in the Northern Region and most of the donkeys are presently in the Central Region. There are few draught animals in the Southern Region. Donkeys are generally used only for transport and then in pairs, as they are generally too light for ploughing and tire more easily than oxen.[28]

Oxen and donkeys have been used for many years in Mozambique for draught power, mainly for pulling wagons and ploughs. In the Zambezi Basin, animal power is used in the Manica Province.

In Zimbabwe, the majority of communal farmers use oxen as a basic source of traction for clearing land, ploughing, cultivating and in-farm transportation. Donkeys are also used in some dry marginal lands, usually to supplement ox power. The use of animal drawn implements is preferred in rural communities because of its relatively greater flexibility, lower cost and greater linkages to other sectors of the rural economy.[29]

NON-RENEWABLE ENERGY AND SELECTED ENVIRONMENTAL IMPACTS

The mining, processing, storage and usage of fossil fuels has detrimental effects on the environment. Thermal power generation, for example, has many negative environmental impacts. These range from coal discards at the collieries to gas emissions from coal and diesel combustion, and groundwater pollution. It affects air quality, surface and groundwater, and landuse.

Mining and processing

When coal is mined by open cast methods large pieces of land are stripped to remove the overburden in order to access the coal seams. This disrupts the natural environment around that area, removes the vegetation, and takes away land that could have otherwise been used for other purposes. The removal of the overburden results in large quantities of spoil heaps that can spontaneously combust or present serious wind and water erosion problems. Spontaneous combustion can cause air pollution and poses a danger to the mining communities.[30]

When the coal is processed, a lot of coal fines are generated. The processing involves washing and grading the coal for the various industrial and domestic applications. These fines are also dumped into heaps at the mine site where they are blown away by wind and swept into water bodies by rain.

Photo: IUCN

One of the major impacts of mining on the environment is the physical alteration of land.

When coal is transformed into coke, further environmental damage occurs. The coke making process results in byproducts such as tar, which are dumped and pollute

Table 7.5: Main mining emissions and effluents from coal in the SADC region

Commodity	Mines	Principal emissions	Areas of concern
Coal	Morupule (Botswana)	Siliceous dust	Spontaneous combustion of stockpiles
	Maamba (Zambia)	Gases and smoke from spontaneous combustion, Smoke from coal burning, Acid drainage from the open pit	Spontaneous combustion of coal faces and stockpiles, Effect of acid drainage to Lake Kariba
	Wankie (Zimbabwe)	Gases and smoke from spontaneous combustion, Smoke from coal burning	Spontaneous combustion of coal faces and stockpiles

Source: Zulu, E.H., Environmental Impacts of Mining in the Zambezi Basin, for SARDC, 1998.

water bodies. Several poisonous gases, which pose both local air pollution problems and global warming, are also produced. Among these gases are carbon monoxide, methane and carbon dioxide.

During the combustion of coal, to generate electricity, solid, liquid and gaseous wastes are also produced. Gaseous emissions include sulphur dioxide, the oxides of nitrogen and other greenhouse gases. Particulates also generated here cause respiratory diseases such as asthma. Sulphur emissions from mining in the SADC region are estimated at one million tonnes per year.[31]

Coal combustion and coke making produces lots of ash, which has to be disposed of before it leaches into groundwater or gets washed into surface water bodies.

Figure 7.3: The greenhouse effect

The atmosphere contains a natural amount of greenhouse gases to sustain all forms of life. Human beings cause the enhanced greenhouse effect. An increased amount of greenhouse gases trap heat energy, contributing to global warming and climate change.

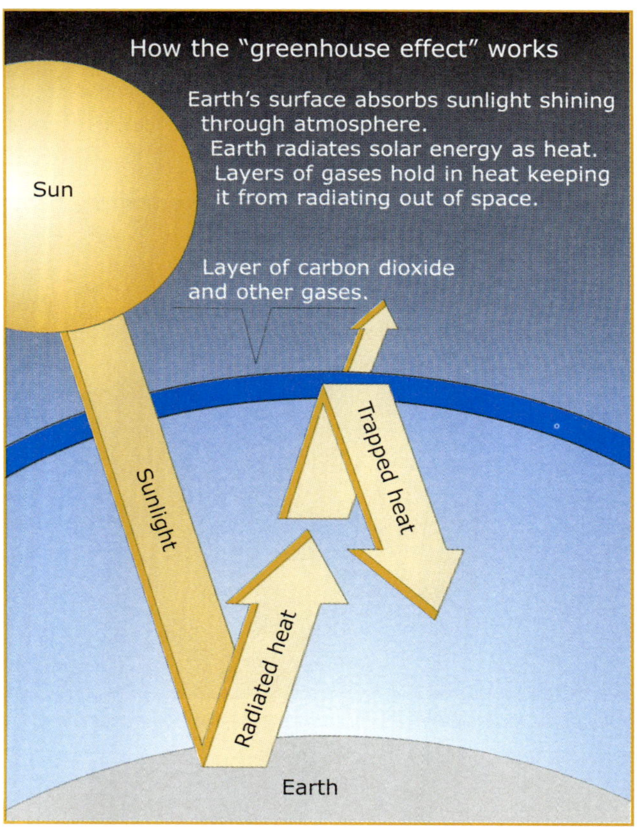

Source: Chenje, M., Sola, L., and Paleczny, D. (eds.), The State of Zimbabwe's Environment 1998, Government of the Republic of Zimbabwe, Ministry of Mines, Environment and Tourism, Harare, Zimbabwe, 1998, p.125.

Usage of non-renewable energy resources

When burned for thermal power production and transportation, fossil fuels release greenhouse gases such as carbon dioxide and acidic gases such as sulphur-dioxide. It is generally believed that global warming due to an increase of greenhouse gases in the atmosphere will alter natural vegetation, wildlife habitats, crop growing seasons, and the distribution of pests and diseases worldwide, the Zambezi Basin included.

The energy sector is responsible for about a third of the greenhouse gas emissions in Africa as a whole. Estimated emissions of the greenhouse gas carbon dioxide for the Zambezi Basin states in 1995, were about 23.14 million tonnes (information was not available for Namibia). This can be compared to South Africa, which emitted 305.81 million tonnes in 1995 and with the USA, having emitted 5,468.56 million tonnes the same year (see also Chapter 9: Pollution).[32]

While the magnitude of air pollution hazards to the regional environment is largely unknown, it is, however, believed that vegetation, soils and water are being impacted from direct effects of gaseous pollutants and acid rain, which is formed when sulphur dioxide combines with rainwater. Data collection on emissions and impacts is not well

Figure 7.4: Greenhouse gas emissions in Africa

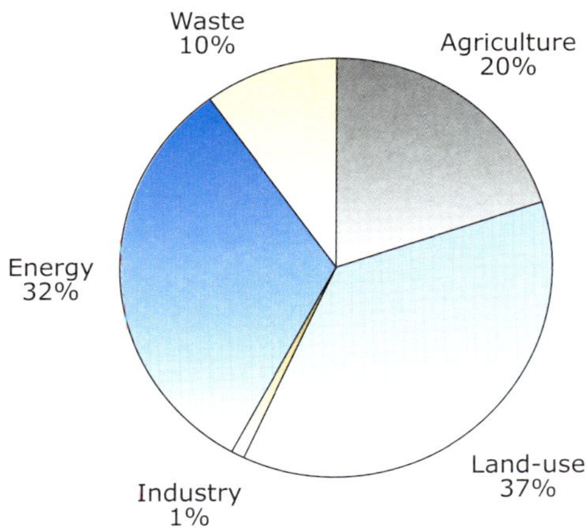

Source: United National Environment Programme (UNEP), Climate Change Mitigation in Africa, Proceedings of International Conference, Victoria Falls, Zimbabwe, 18-20 May, 1998. Southern Centre/Riso, Denmark, 1998.

developed across the region but 89 percent of electricity generation in SADC is from coal,[33] contributing thousands of tonnes of sulphates to the atmosphere annually, which in turn might lead to acid rain in the basin countries.

Another problem related to the use of non-renewable energy sources is exhaust fumes from vehicles driven by fossil fuels. Economic development has impacted on lifestyle, with many people, particularly in urban areas aspiring to be part of the driving crowd. Zimbabwe, for example, has more than 500,000 vehicles on its roads.[34] The World Resources Institute reports that the growing use of motor vehicles is

Table 7.6: Major impacts of common pollutants associated with motor vehicles

Emission	Health impacts	Environmental impacts
Nitrogen oxides (NO_x) include nitric oxide (NO) and nitrogen dioxides (NO_2), which is formed from the oxidation of NO.	• NO_2 is a lung irritant at high concentrations. • NO_2 may lead to depression of the immune system, with children and the elderly at risk.	• NO_2 reacts with water to form nitrate (NO_3^-), a source of acid rain. • NO_2 contributes to the formation of ground level ozone. • NO_2 is associated with suppressed vegetation growth. • NO_2 contributes to corrosion of metals and degradation of textiles, rubber and polurethane.
Carbon monoxide (CO) is a gas produced through the incomplete combustion of organic materials.	• CO reduces the ability of the blood to carry oxygen, with smokers, persons with heart disease, and those with anaemia being especially sensitive.	• CO may contribute to the formation of ground level ozone by depleting the atmosphere's supply of hydroxyl radical (OH).
The combustion or decay of organic material releases carbon dioxide (CO_2).		• CO_2 is an important greenhouse gas, contributing to global warming.
Volatile organic compounds (VOCs) are a chemically diverse group of compounds that have at least one carbon atom and are highly volatile.	• Many individual VOCs such as benzene, are known to have or are suspected of having human health effects ranging from carcinogenicity to neurotoxicity.	• VOCs contribute to the formation of ground level ozone.
Ozone (O_3) is formed from the reaction of NOx, VOCs and sunlight.	• Exposure to O_3 is associated with changes in lung function, decreased immune function, and possibly the development of chronic lung disease.	• O_3 reduces agricultural productivity and the growth of trees. • Ground level ozone is a global warming agent.

Source: Chenje, M., Sola, L., and Paleczny, D. (eds.), The State of Zimbabwe's Environment 1998, Government of the Republic of Zimbabwe, Ministry of Mines, Environment and Tourism, Harare, Zimbabwe, 1998, p.119.

increasing emissions of greenhouse gases such as methane, ozone, carbon monoxide, nitrous oxide and most important, carbon dioxide. Exhaust fumes are also impacting on the health of the people, especially those living in large cities, and ecosystems. Heavy metals and nitrogen oxides are examples of pollutants released by car engines.

RENEWABLE ENERGY AND SELECTED ENVIRONMENTAL IMPACTS

The environmental impacts associated with renewable energy sources such as biomass and hydropower include deforestation and hydrological changes. Threats associated with solar power include disposal of solar panels and batteries.

Biomass

Deforestation, that leads to soil erosion and ultimately land degradation and siltation, is the main current problem associated with biomass energy utilisation in the basin.

The rate of woodfuel use is, in some places, greater than the rate of afforestation. This is for example the case in parts of Mozambique where the effects of war, which ended in 1992, tended to concentrate rural communities around large settlements for better security. This increased population densities, resulting in increased consumption of fuelwood in those areas.[35]

Two aspects of Zambia's woodfuel supply situation are of concern: localised deforestation, due to regional imbalances between woodfuel supply and demand; and agricultural land clearing, which is depleting wood resources more rapidly than energy demand. In Zambia, deforestation is mainly due to agricultural clearing as opposed to charcoal burning. Nationally, there is no shortage of woodfuel. However, around the principal towns in the Copperbelt, Lusaka Province, and Southern Province, tree-cutting for charcoal production, coupled with agricultural land clearing and woodfuel demand, have created worsening problems of local deforestation.

Regarding ethanol, there are both environmental and socio-economic debates that remain to be explored. One of these is the question of food versus fuel, where sometimes debate is on whether to produce sugar or fuel.

The distillation process in ethanol production leaves stillage, which is an environmental eyesore. If the stillage overflows from ponds or burst pipes into waterways, then serious pollution can result. However, solutions can be found to utilise the stillage in a more useful and economic way such as road surfacing and cattle feed.

Hydropower

Depending on location and size, damming rivers for hydropower plants has widespread negative environmental impacts. These include the displacement of people and animals; alteration of river flow regimes; burial of heritage sites, and loss of some species of plants and animals. Other impacts include depletion of wetland habitats and aquatic life; increases in water-borne diseases; sedimentation, siltation and siltation-induced flooding; and loss of irrigation water downstream. Seismicity or reservoir-induced earthquakes, are also possible if the reservoir is large.

During the construction of a dam, the ecosystem along the river is disrupted. Vegetation is destroyed, people and wildlife are displaced. Land that was previously available for other uses is taken up by the reservoir created. Roads have to be built to access the construction site and bush has to be cleared to make way for roads, transmission lines and settlements. The building operations introduce hydrological changes as water flow is altered. This affects aquatic life and their breeding patterns. Water pollution also results from the construction process as

Photo: M Chenje

These petrified trees are a constant reminder of how dams destroy vegetation.

Map 7.3: Environmental impacts of hydropower in 1995

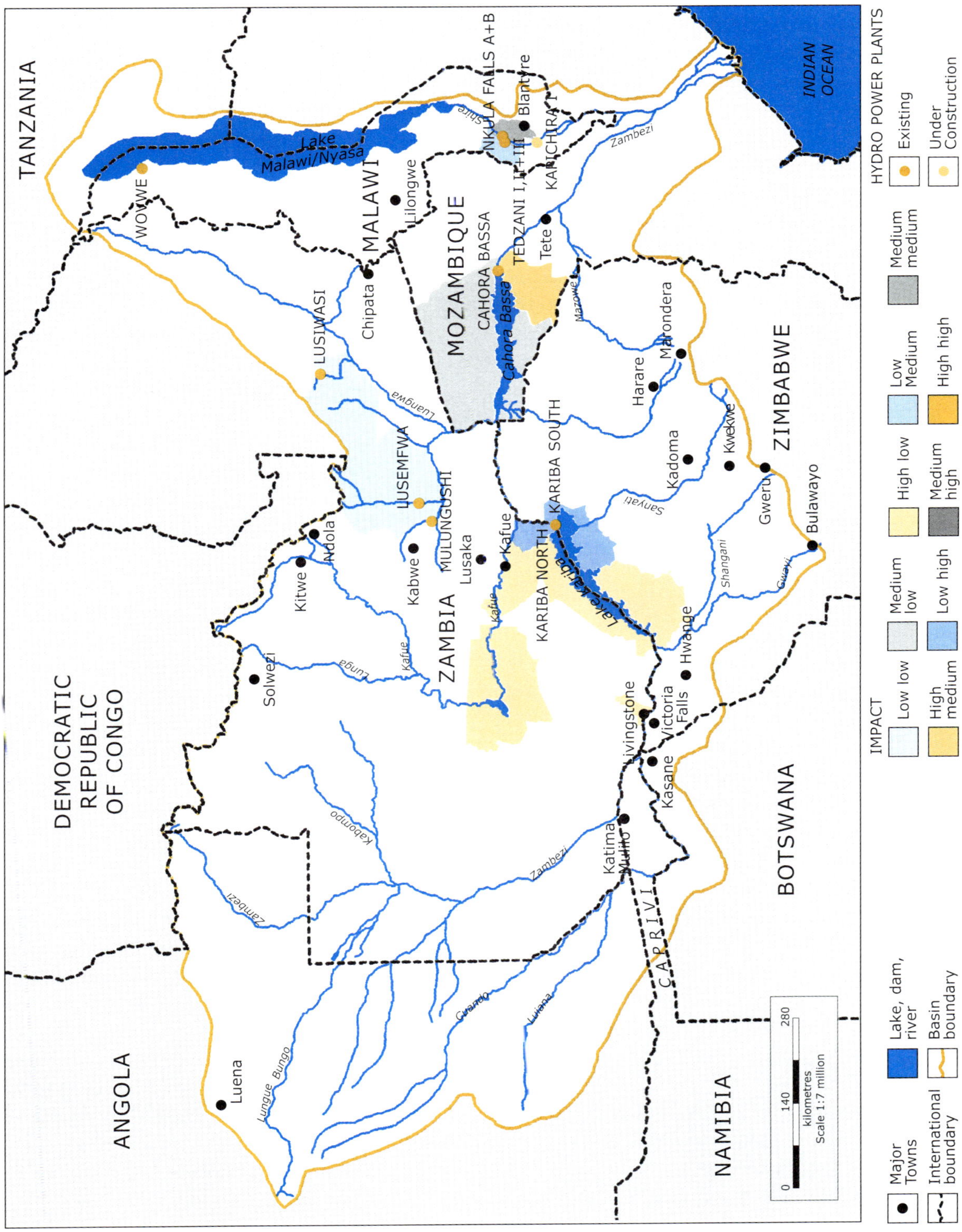

Source: Denconsult/NNR, Zambezi River Basin Atlas, SADC-WSCU, Danida, Lusaka, 1998, p.85.

materials and fuels, especially oils, find their way into the watercourse.

As water accumulates in a reservoir, the incidence of water-borne diseases increases, resulting in health risks to workers and the surrounding communities. Tropical water-borne diseases such as malaria, schistosomiasis (bilharzia), onchocerciasis and encephalitis can easily spread. The filling of the reservoir can also result in flooding of adjoining basins, affecting some upstream activities. Downstream activities are likely to be affected, as water impounded upstream is no longer available.

The Kariba Dam, shared by Zimbabwe and Zambia is an example. The construction of the Kariba Dam forced about 57,000 people, dismantling a nucleated community with a common culture and socio-political economy.

On the positive side, hydropower schemes can open up an area to investment and links with the rest of the world. Before Kariba Dam was built, the Zambezi Valley was an inaccessible disease-prone area. There was very little development in the area; no roads, hospitals, schools or telecommunication facilities. Since the construction of Kariba Dam, the area has been transformed into a hive of recreational and commercial (especially fishing) activities.

Legal/institutional framework

Due to the sudden energy price rise in the early 1970s all the countries in the Zambezi Basin awoke to the need to have concerted energy policies. In the case of Botswana, for instance, the country's rapid economic growth meant that energy demand also rose. The country, therefore, took steps to promote the use of less expensive domestic energy sources and safeguard essential imported supplies. The government's policies centred on substituting imported oil with alternative fuels; building up strategic oil reserves and developing indigenous energy sources. Government established an energy unit, commissioned a National Energy Assessment study and developed an energy master plan.

The framework for energy planning and development in Mozambique aims to:[36]

- Intensify the survey of existing energy resources, giving priority to hydropower and to prospecting for oil, coal and gas.
- Profit from large existing hydropotential in order to meet energy requirements for industrial and agro-industrial development by developing larger hydropower schemes.
- Develop a rural energy policy in areas far from the national electricity network to meet increasing energy demands and avoid environmental degradation and natural resource depletion.
- Optimise and rationalise the use of oil through –
  - Improvements in maintenance and efficient equipment
  - Putting controls on oil consumption
  - Promoting coastal sea, river and rail transport
  - Promoting public transport in the rural areas
  - Controlling imports of equipment driven by oil
  - Introducing speed limits.
- Provide attention to technology evolution.

In Tanzania, different ministries are responsible for different energy sources. For instance, fuelwood is administered by the ministry responsible for natural resources while coal is under the ministry of mines. These ministries formulate policy for the exploitation of resources for which they are responsible and undertake development planning. Specific entities are mandated to exploit and distribute energy resources within the country.

Management of demand

Botswana's energy requirements for the rural population is more than half the total national supply and yet the per capita energy supply for the rural areas is only about one-fifth to one-quarter of that for the urban areas.[37] Fuelwood is the major source of fuel in the rural areas of Botswana as is the case in all of the basin countries. Electricity and petroleum products are mainly used by the mining industry, transport and in urban areas.

Oil-derived products constitute 74 percent of Malawi's commercial energy needs. By 1999, the consumption of petroleum products in the country was about 160,000 tonnes per year. A study by the World Bank forecasts an overall 4.5 percent growth rate in consumption of petroleum products to the year 2000 with gasoline, kerosene and diesel, consumption to grow by 4.2, 4.1 and 4.8 percent, respectively.[38]

To cut down on demand for fuelwood, the government of Malawi has intensified its initiatives to develop new alternative energy sources by, for example, constructing more hydropower schemes and mining coal in the Northern Region. Ethanol is blended with petrol to compensate for added demand on the product.[39]

Statistics for energy demand in Mozambique have been difficult to obtain as there is no central body that has compiled data on the total energy balance for the country. The Electricidade de Mozambique (EDM) has collected data on the energy sector outlining the overall situation on energy consumption. However, information on energy demand within the basin is not readily available.[40]

Pricing

The price of fuel in Botswana depends on location and the amounts purchased. Prices for fuelwood have been increasing rapidly, reflecting growing demand. The main sources of fuelwood are woodlots and natural forests. Prices for fuelwood from natural forests compare favourably with the price for charcoal. Different tariffs are applied for electricity for the southern and northern areas of the country. The tariffs for the southern division are higher than those of the northern division that generates electricity using oil or coal.[41]

All of Malawi's petroleum products are imported in refined form. BP Africa is the main supplier. Imports are made through Tanzania, Zambia and South Africa. Imports and distribution of petroleum products are mainly by road and rail systems. There are 17 products storage depots in the country.

Malawi has a single pricing formula for all products, which determines a maximum price that can be charged. The tax on petroleum products is about 22 percent. Annual petroleum tax revenue is approximately US$33 million and represents 22 percent of government indirect taxes.

Because of the high cost of oil imports mainly from the Middle East, the government of Tanzania has had to introduce conservation measures from time to time, which have included rationing of petroleum products and pricing. Overall, pricing of oil and petroleum products seems to reflect the rise in demand in all the countries of the Zambezi Basin.[42]

Institutions

The Botswana Power Corporation (BPC), a parastatal, is responsible for the supply of electricity to the main urban centres and is accountable to Parliament through the Minister responsible for Energy and governed by the Botswana Power Corporation Act.

In the case of Malawi, government established the energy department with a view to centralising energy planning and management. The main functions of the department are to:

- Ensure that existing energy resources are conserved.
- Identify energy resources, and increase available knowledge as to the country's resource potential and how it can be developed.
- Identify, and give priority to those energy projects, which are likely to produce energy most cheaply and efficiently.
- Control and coordinate energy pricing.
- Examine and coordinate such other functions related to energy as could be necessary from time to time.

Photo: IUCN

The road to energy conservation not only involves the various institutions but individuals as well.

The Electricity Supply Commission of Malawi (ESCOM) is responsible for the generation, distribution and supply of electricity throughout Malawi although, lately smaller private institutions are also engaged in this task. Electricity generation in the country is mainly from hydropower sources centred on the Shire River in the south and Wovwe River in the north.

Energy in Mozambique is the mandate of the National Energy Administration, which is composed of a number of sectors. The EDM is a state enterprise responsible for the planning and coordination of electricity in the country with two major companies, the Cahora Bassa Hydropower Company (HCB) and the Revue Hydroelectric Society (SHER) responsible for generation and distribution. In addition, there are also other enterprises such as local authorities that are responsible for electricity supply to sugar estates, for example.[43]

The exploitation and development of oil and natural gas in Tanzania is the responsibility of the Tanzania Petroleum Development Corporation (TPDC), which is owned by the government. It is also its responsibility to import and refine oil, which is distributed by oil companies such as BP, AGIP, ESSO, CALTEX and TOTAL.

The Tanzania Electric Supply Company (TANESCO) is responsible for the generation, distribution and supply of electricity in the country. TANESCO is also in charge of rural electrification.

Energy administration in Zambia falls under several arms of government. These include the Department of Energy, the National Energy Council and the Ministry of Mines. Government emphasis is on reducing dependency on oil, the electrification of rural areas and the reduction of consumption of firewood and charcoal. Programmes and projects have been initiated where woodlots have been developed to supply urban areas such as Lusaka with fuelwood and help reduce the rate of deforestation in natural forests.[44]

Zimbabwe's Ministry of Transport and Energy is the body responsible for energy issues of the country. The National Oil Company of Zimbabwe (NOCZIM) imports petroleum products while several private companies are responsible for distribution. A larger part of the country's population, however, uses fuelwood for its energy needs, especially in the communal areas. Commercial farmers still use wood for curing tobacco. Zimbabwe gets its electricity from Kariba and imports some from the SAPP. The Zimbabwe Electricity Supply Authority (ZESA) is responsible for the generation and distribution of electricity to consumers throughout the country. ZESA spends between US$5 million–US$6 million[45] monthly on electricity imports from neighbouring countries in the SADC region.

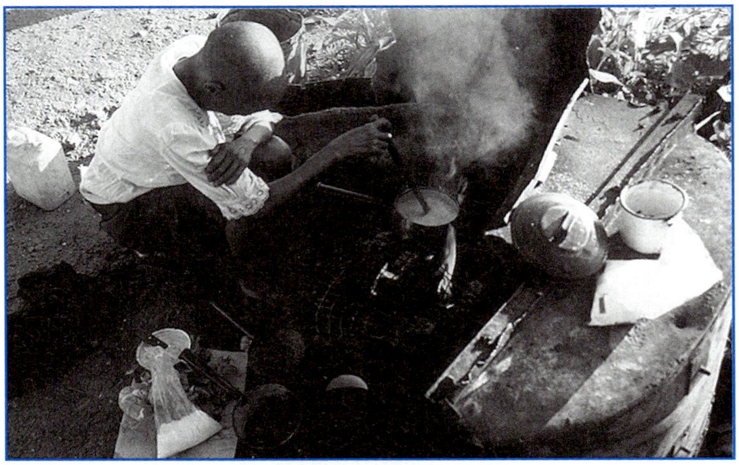
Photo: IUCN

Energy should be considered as a basic requirement and strategies should start with the individual.

REGIONAL INITIATIVES ON ENERGY

In recognition of the importance of energy as a basic requirement for socioeconomic development, SADC policy makers have worked out various proposals and plans on how to develop and utilise the existing vast energy resources to facilitate the process of industrialisation, achieve economic development and raise the living standards of the people. However, the various regional and sub-regional institutions have not been very successful in implementing coordinated and coherent energy policies and strategies.[46]

The SADC Energy Protocol

The SADC Energy Protocol is a regional initiative, which promotes cooperation in the whole energy sector. Since all the countries sharing the Zambezi River Basin are members of the Southern African Development Community (SADC), regional energy initiatives are guided by the protocol.

The SADC Energy Protocol (see also Chapter 12: Environmental Management and Regional Cooperation), which was signed by SADC Heads of State and Government on 24 August 1996 in Maseru, Lesotho, is a legally binding document with seven major objectives, which are to:[47]

- Harmonise national and regional energy policies, strategies and programmes on matters of common interest based on equity, balance and mutual benefit.
- Cooperate in the development of energy and energy pooling to ensure security and reliability of supply at minimum cost.
- Cooperate in the development and utilisation of energy in the sub-sectors of woodfuel, petroleum and natural gas, electricity, coal, energy efficiency and conservation and other crosscutting themes of interest to member states.
- Ensure the provision of reliable, continued and sustainable energy services in the most efficient and cost effective manner.
- Promote joint development of human resources and organisation capacity building in the energy sector.
- Cooperate in research, development, adaptation, dissemination and transfer of low cost energy technologies.

- Strive to achieve standardisation in appropriate energy development and application, including the use of common methods and other techniques.

The protocol, the Policy and Strategy and the SADC Energy Action Plan are the fundamental guidelines for the future development of SADC Energy Sector activities. The SADC Energy Action Plan activities include:

- Development of a regional petroleum training centre.
- Human resources development and training for the coal sub-sector.
- Development of small hydropower plants in Malawi;
- The SADC renewable energy project.
- Development of national woodfuel strategies and plans.

However, some prevailing constraints for the energy development in the SADC region are:

- Concentration of population in limited areas while large areas are left uninhabited.
- Shortage of adequate financial resources to implement the approved projects/programmes and activities, including research.
- Lack of technological experience.
- Shortage of skilled labour.
- Political problems.[48]

The Government of Angola through the Technical and Adminstrative Unit (TAU) coordinates the SADC energy sector.

Southern African Power Pool (SAPP)

The most outstanding and pragmatic cooperation effort in the southern African region is the Southern African Power Pool (SAPP). All SADC countries have agreed to cooperate in electricity projects and trade. Electricity trade in the region, however, has been on-going since the 1950s when interconnections were made between countries such as Zambia, Zimbabwe and later with the DRC.[49]

Trade has since extended to most of the countries in the region with interconnections linking up nearly the entire SADC region. The backbone of the region's interconnected network has been the Kariba Dam. All SADC countries have signed the Intergovernmental Memorandum of Understanding (MoU), Inter-utility MoU, and agreements between operating members and the operating guidelines.

Box 7.4: Some activities under the SADC Energy Action Plan

Regional petroleum training centre
The objective of the project is to train SADC technicians in the petroleum field using the training facilities at the Petroleum Training Centre in Sumbe, Angola.

Human resources development and training for the coal sub-sector
The objective of the project is to carry out a training-needs survey to provide a basis for planning human resources development and training programme for the coal sub-sector.

Development of small hydropower plants in Malawi
The objective of this project is to reduce diesel imports, provide cheap energy for rural centres and increase local self-sufficiency.

SADC renewable energy project
The main objectives of the project are to:

- Strike a balance between development and environmental concerns by using a sustainable model of disseminating basic fuel service to cater for the rural populations lacking access to grid extension in an environmentally benign, more cost effectively and affordable way.
- Build small-scale entrepreneurship capacity to deliver the various NRSE technologies.

Development of national woodfuel strategies and plans
The objectives of the project are to:

- Develop comprehensive woodfuel strategies and implementation plans for each member state.
- Develop projects for woodfuel production integrated with other sectors dealing with biomass management and rural development.
- Provide data to assess national capabilities for the implementation of woodfuel programmes, and means of increasing such capabilities through institutional strengthening.

Source: SADC, Energy, Lusaka, 1999.

SAPP was created by the power utilities in the region to improve reliability and availability without incurring further costs. The main rationale of SAPP is economic development through economies of scale and utilisation of surpluses in some countries to cover deficiencies in others.[50]

Zambezi River Authority (ZRA)

The Zambezi River Authority (ZRA), was established in 1987, succeeding the now defunct Central African Power Corporation (CAPCO). CAPCO used to run the Kariba North and South Bank power stations as a single complex. This was transferred to Zimbabwe Electricity Supply

Photo: M Chenje

In addition to managing the Kariba Dam, the Zambezi River Authority undertakes various research studies on the lake using a well-equipped boat.

Authority (ZESA) and Zambia Electricity Supply Corporation (ZESCO), and the ZRA was given the responsibility of managing the Kariba Dam. The ZRA Act was passed in both Zambia and Zimbabwe parliaments as an agreement to jointly utilise, operate and maintain the existing Zambezi complex as well as additional dams, reservoirs and installations that may be constructed on the Zambezi River. In the 1980s and 1990s, the ZRA has been involved in investigating new dam projects such as the Batoka Gorge Hydropower scheme.[51]

FUTURE CHALLENGES

Energy extraction, conversion and use remain the largest contributor to pollution at the national, regional and global levels. It also remains one of the most important tools for human development. The challenge for the Zambezi Basin therefore, is to extract and use energy in such a way that the other components of the environment are minimally harmed and that the future supplies of other derivatives of the environment are sustained.

The use of the energy resources within the basin cannot be seen in isolation from the other socio-economic activities in the region. The use of the water resources of the basin, for example, impacts greatly on the electricity generation capacities of hydro plants along rivers, while the environmental impacts of electricity generating plants on the hydrology, flow and quality of water in the basin affects other sectors of the economy. It is, therefore, imperative that exploitation of energy resources in the region is properly managed, and cross-sectoral consultations and linkages are established. The political implications of energy resource management, and sometimes ownership and bilateral relations have often far-reaching consequences on energy use.

Using treatment equipment of different types, changing to cleaner fuels, and using energy in a more efficient manner can solve many of the problems within the energy sector.

There are clear indications that substantial opportunities exist to conserve energy through better housekeeping and more efficient equipment in all sectors. By avoiding the wasteful energy practises of previously industrialised countries, the Zambezi Basin countries could reap both economic and environmental benefits. This will, however, require carefully designed and comprehensive programmes to promote cost-effective conservation and efficiency.[52]

LINKAGES TO OTHER CHAPTERS

Chapter 1: Regional Overview: People and Environment
Increasing populations require more energy to power various activities. Imbalance between population growth and energy demand/supply leads to environmental degradation. The fuelwood sector already manifests indicators of stress due to increasing populations.

Chapter 2: Physical Features and Climate
Energy production and use lead to climatic change. Energy-related activities have to be well planned to avoid climate change.

Chapter 3: Water and Wetland Resources
Energy and economic issues in the region are dominated by availability of water resources. Energy production in the hydro sector has impact on the magnitude of competing uses of water.

Chapter 4: Biological Resources and Diversity
Activities in the energy sector have major implication for biological diversity in the basin. Energy development activities change, fragment and eliminate habitats, leading to reduction and/or loss of biological diversity among plants and animals.

Chapter 5: Agriculture
Land clearance, in the form of deforestation, for arable agriculture is increasing in many of the basin states.

Chapter 6: Industry
Energy is required for industrial purposes and needed for economic development. The question is how to provide energy without degrading the environment.

Chapter 8: Tourism
Some aspects of tourism and energy depend on water as the principal resource. Energy development activities, therefore, impact on tourism related ones.

Chapter 9: Pollution
Pollution in the form of greenhouse gases contributes to climate change. In-door pollution of woodfuel stoves reduces people's quality of life as it negatively affects their health.

Chapter 10: Poverty
One of the indicators of poverty in the Zambezi Basin is high level of fuelwood consumed by the majority of households. Poor people do not have resources to invest in better forms of energy. Dependence on traditional energy sources impacts negatively on the environment.

Chapter 11: Gender and Women's Roles
Gender issues have a major effect on the traditional energy sector. Women and children face great burdens related to collection of fuelwood. With reducing woody resources, women and children have to devote more of their time to collect fuelwood.

Chapter 12: Environmental Management and Regional Cooperation
The energy sector contributes to regional cooperation in the form of trade, information sharing, shared infrastructure and institutions. Cooperation mechanism are already in place among countries, within the framework of SADC and the Common Market for Eastern and Southern Africa (COMESA).

Chapter 13: Trends and Scenarios
The energy sector provides one of the greatest opportunities for socioeconomic development in the region. Energy trade, among others, will continue to foster cooperation among states based on a common goal of purpose.

ENDNOTES

1 Denconsult, Sector Studies under ZACPLAN, *Sector Study 5, Final Report*, Lusaka 1998
2 ibid
3 SADC, *Energy*, Lusaka, 1999
4 ibid
5 SADC, *Mining*, Lusaka, 1999, p.6
6 ZESA Corporate Planning Department, Unpublished file material, 1999
7 ibid
8 Yamba, F.D., Chidumayo, E.N., Jain, P.C., Mbewe, D.N., Mbewe, D.J. and T., Mwale, "Zambia Country Study on Climate Change, Volume 1", Inventories, Ministry of Energy and Water Development GTZ, Lusaka, November 1995.
9 Kalumiana, O.S., "Woodfuel Sub-Programme of the Zambia Forestry Action Programme", Ministry of Environment & Natural Resources, Lusaka, 1998, p12
10 EDG Government of Botswana and GTZ, *Botswana Energy Masterplan - Final Phase, Energy Policy for Botswana*, Energy Affairs Division, July 1996
11 SADC TAU, UNDP and Southern Centre, "FINESSE Activities in the SADC Region, Angola Country Study", SADC Energy Sector TAU, Luanda, May 1997
12 ibid
13 Chimhowu, A., Chinemana, F., Sibanda, A. and D., Tevera, "Namibia", in Sam Moyo, Phil O'Keefe and Michael Sill (eds), *The Southern African Environment, Profiles of the SADC Countries*, Earthscan Publications, London, 1993
14 Ministry of Mines, Environment and Tourism/ Zimbabwe, UNEP and UNDP, "Zimbabwe's initial national communications under the United Nations Framework Convention on climate change", Harare, March 1998
15 Maya, R.S. (eds.), "Options for Greenhouse Gas Mitigation in the power sector in SADC. Phase I Report", Inventory of GHG emissions from the power sector, Mond Books, Harare, 1999
16 Pallet, John (ed.), *Sharing Water in Southern Africa*, Desert Research Foundation of Namibia, Windhoek, 1997
17 SARDC-IMERCSA, "Minister Urges sustainable use of energy in the Zambezi River Basin", *The Zambezi*, Vol. 1 No. 2, Harare, 1998.
18 op. cit 1
19 op. cit 1
20 op. cit 1
21 op. cit 1
22 op. cit 1
23 op. cit 11
24 op. cit 15
25 op. cit 11
26 SARDC-IMERCSA, CEP Factsheet, "Zambezi River Basin Series: No 2, Energy", Harare, 1998
27 Mpotokwane, M., Shaw, S., and R., Segodi, "Botswana". In: Sam Moyo, Phil O'Keefe and Michael Sill (eds.), *The Southern African Environment. Profiles of the SADC Countries*, Earthscan Publications, London, 1993
28 Harris, J, "New and Renewable Sources of Energy (NRSE) in Malawi", SADC TAU, Luanda, September 1989
29 op. cit 11
30 SADC, *Mining*, Lusaka, 1999, p.20
31 UNEP, *Global Environment Outlook 2000*, Earthscan, London, 1999.
32 WRI, *1998-99, World Resources: a Guide to the Global Environment*, Oxford University Press, 1998
33 Chenje, M., "Transboundary air pollution spreading in SADC", *The Zambezi, Vol 2 No. 1*, SARDC-IMERCSA, 1999
34 Chenje, M., *Reporting the Southern African Environment: A Media Handbook*, SADC/IUCN/ SARDC, 1998, p.35
35 Lopes, H., "New and Renewable Sources of Energy (NRSE) in Mozambique", SADC TAU, Luanda, September 1989
36 Beijer Institute/Scandinavian Institute of African Studies., *Energy, Environment and Development in Africa*, Uppsala, 1984
37 ibid
38 WWW.Mbendi.co.za
39 E. Laisi, Personnal Communication
40 op. cit 37
41 ibid
42 ibid
43 ibid
44 ibid
45 Mangwengwende, S.E., ZESA Chief Executive in interview with Zimbabwe Broadcasting Corporation, 2 Feb., 2000
46 op. cit 3
47 SADC, Protocol on Energy in the Southern African Development Community Region, SADC, Secretariat, Gaborone, 1996, p8.
48 op. cit 3
49 SADELEC and MEPC, *Electricity in Southern Africa, Investment Opportunities in an emerging regional market,* Financial Times Energy Publishing, London, 1996
50 Rowlands, I.H., *Climate Cooperation in Southern Africa,* Earthscan Publications/UNEP, London, 1998.
51 op. cit 16
52 Raskin, P., and Lazarus, M., *Regional Energy Development in Southern Africa: Great Potential, Great Constraints,* Stockholm Environment Institute, Stockholm 1991

REFERENCES

BEMP, Botswana Energy Master Plan Report, 1989

Chenje, M. and Johnson, P. (eds.), *Water in Southern Africa*, SADC/IUCN/SARDC, Maseru/Harare, 1996

Dale, A.P., "An energy sector overview of Zambezi Basin developments", in T. Matiza, S. Crafter and P. Dale, (eds.), *Water Resource Use in the Zambezi Basin, Proceedings of a workshop held at Kasane, Botswana, 28 April - 2 May 1993*, IUCN, Gland, Switzerland, 1995

de Cruz Fransisco, J., Lopes, H., Mussa, M., and P., O'Keefe, "Mozambique", in: Sam Moyo, Phil O'Keefe and Michael Sill (eds.), *The Southern African Environment. Profiles of the SADC Countries*, Earthscan Publications, London, 1993

Department of Energy, "Energy Policy for Zambia", Ministry of Energy & Water Development, Lusaka, 1994

Department of Energy, "Energy Policy for Zambia", Ministry of Energy & Water Development, Lusaka, 1998

Department of New and Renewable Sources of Energy (DNRSE), "New and Renewable Sources of Energy (NRSE) in Angola", SADC TAU, Luanda, September 1989

Energy Affairs Division, "1993 Energy Statistics", Ministry of Mineral Resources and Water Affairs, Gaborone, 1993

Energy Affairs Division, "Energy Policy for Botswana", Ministry of Mineral Resources and Water Affairs, Gaborone, 1996

Government of Mozambique, "Country Report on Renewable Energies in Mozambique: Present Status and Future Prospects", Energy Directorate Paper presented at the Forum on Renewable Energy for Africa, 29-31 March 1999, Harare

ESCOM, "How Malawi expects to meet electricity demand in the next millennium", Blantyre, 1997

ESMAP Strategy and Programs Division/World Bank and the Department of Energy Resources and Development/Zimbabwe, *Energy Sector Management Assistance Programme, Zimbabwe Integrated Energy Strategy Evaluation*, World Bank, Washington D.C., January 1992

F.T. Sparrow, Z. Yu, D.G. Nderitu, B.H. Bowen and W. Masters, "Electricity Trade Modeling in Southern Africa", Modeling Seminar, SAPP Meeting, Windhoek, February 13, 1997

FAO, Zambia: "Wood Energy Consumption Resource Survey", FO:DP/ZAM/82/EO8, Document No. 2., November 1986

Goldemberg, J., *Energy for sustainable World*, Wiley Eastern Ltd, New York, 1988

Hamutwe, G.S., "Framework for IRP in the Southern African region: Selected country papers, Tanzania", in *Energy and the Environment. Integrated Resource Planning - Tools and Methods - Volume 6, Proceedings of a workshop held in Cape Town, South Africa 11-15 May 1998*, CDG, Berlin, 1998

Heyns, P., Montgomery, S., Pallet, J. and M., Seely, *Namibia's Water, A Decision Makers' Guide*, Department of Water Affairs, Desert Research Foundation of Namibia, Windhoek, Namibia, May 1998

Karekezi, S., and T., Ranja, *Renewable Energy Technologies in Africa*, ZED Books/AFREPREN/SEI, 1997

Karenzi, P.C., "Introduction to the Biomass Study', in D.O. Hall and Y. S. Mao, (eds.), *Biomass Energy and Coal in Africa*, Zed Books/AFREPREN, 1994

Kgathi, D.L., and C.V., Mlotshwa, "Fuelwood procurement, consumption and substitution in selected areas of Botswana: Implications for Theory and Policy", in D.L. Kgathi, D.O. Hall, A. Hategeka and M.B.M. Sekhwela (eds.), *Biomass Energy Policy in Africa*, ZED Books/AFREPREN/SEI, 1997

Kgathi, D.L., Sekwela, M.B.M., Tietema, T. and Mpokotwane, M.A., "Biomass in Botswana" in D.O. Hall and Y.S. Mao (eds), *Biomass Energy and Coal in Africa*, ZED Books/AFREPREN, 1994

Makoni, I.J., "National Survey of Biomass/Woodfuel Activities in Zimbabwe", SADCC TAU, Luanda, March 1990

Masire, G.M., "New and Renewable Sources of Energy (NRSE) in Botswana", SADC TAU, Luanda, September 1989

Maya, R.S., Muguti, E., Fenhann, J. and P.E., Morthorst, "UNEP Greenhouse Gas Abatement Costing Studies, Zimbabwe Country Study - Phase I", Southern Centre/Department of Energy/RISO/UNEP, May 1992

Maya, R.S., "New and Renewable Sources of Energy (NRSE) in Zimbabwe", SADC TAU, Luanda, September 1989

Nkhata, D., & Kalumiana, O.S., "Energy Needs and Shortfall Assessment of the Barotse Flood Plain", IUCN, Zambia Country Office, Lusaka, 1997

SADC TAU, UNDP and Southern Centre, "FINESSE Activities in the SADC Region, Malawi Country Study", SADC Energy Sector TAU, Luanda, 1998

SADC TAU, UNDP and Southern Centre, "FINESSE Activities in the SADC Region, Zimbabwe Country Study", SADC Energy Sector TAU. Luanda, 1998

SADCC Energy Sector TAU, "New and Renewable Sources of Energy (NRSE) in SADCC Region", SADC TAU, Luanda, August 1989

SADCC Energy Sector TAU, "Wind and Solar Resources in the SADCC Region", Luanda, 1991

Sculork, J., Rosenschein, A., and D.O., Hall, "Fuelling the Future, Power Alcohol in Zimbabwe", Acts Press/BUN, Nairobi/Harare, 1991

Southern African Development Community (SADC), "Management of Shared River Systems". SADC/EU Conference background paper and case studies, Maseru, 20-21 May, 1997

Sparrow, F.T., Yu, Z., Bown, B. H., Bowen, W. M. and P.B., Robinson, "Modelling Electricity Trade in Southern Africa", Prude University, U.K. & ZimconsultWest - Modelling Seminar, SAPP Meeting, Windhoek, February 13, 1997

Sparrow, F.T., Yu, Z., Bown, B. H., Bowen, W. M. and W.A, Masters "Modelling Electricity Trade in Southern Africa, Year 2 Interim Progress Report", Prude University, 1998

United National Environment Programme (UNEP) "Climate Change Mitigation in Africa", Proceedings of International Conference, Victoria Falls, Zimbabwe, 18-20 May 1998, Southern Centre/Riso, 1998

United Nations Environment Programme, "Green House Gas Abatement Costing Studies, Phase Two, Country Summaries", Southern Centre/Riso, 1994

Yamba, F.D, Kalumiana, O.S., Mbewe, D.J and C., Mulenga, "Climate Change in Southern Africa, Zambia Country Study", UNEP Collaborating Centre on Energy & Environment, Riso National Laboratory, 1999

8

TOURISM

The Zambezi Basin is well known for its many tourist destinations. Riparian countries within the basin host hundreds of thousands of tourists annually who visit places such as Victoria Falls, Lake Kariba, Lake Malawi/Nyasa or one of many national or recreational parks. Game-viewing, fishing, whitewater rafting, bird watching, hiking and shopping for crafts are just some examples of the wide range of activities offered.

Photos: M Chenje

Wildlife and water-based activities are some of the major tourist attractions in the Zambezi Basin.

Apart from offering the tourist these attractions, the labour-intensive industry also makes significant contribution to the economies of the southern African countries. Tourism, once considered a byproduct of natural resources conservation, has become the foundation of many economies. In Zimbabwe, for example, the industry is now believed to be the country's second largest foreign currency earner after agriculture.[1]

But tourism, the region's fastest growing and probably the world's largest industry, has serious impacts on the environment. The future threat to the environment lies in the rapid, often uncoordinated growth of the industry and the resultant increasing pressures and demands placed upon the natural resources.

Tourism depends on healthy ecosystems and if well managed, it can be used as a tool for conservation. In addition, with the involvement of the people in the Zambezi Basin, tourism could also be an option that improves the standard of living of the majority of the people, while preventing the depletion of the basin's natural resources. Finding environmentally sustainable forms of tourism is, therefore, a prerequisite for the successful implementation of the industry within the Zambezi Basin.

TOURIST ATTRACTIONS

The wide variety of landforms and vegetation zones has resulted in a remarkably high diversity of ecosystems and species in the Zambezi Basin.[2] The major tourist resources in the region are, therefore, associated with natural resources. By far, the major tourism areas in the Zambezi Basin are national parks and nature reserves, with wildlife and scenery as the major attractions.

Photos: M Chenje

A large proportion of the basin, about 351,000 square kilometres, has been set aside to protect the region's wildlife.

Box 8.1: Defining tourism

Tourism is a generic term, which encompasses a multiplicity of definitions. It is however generally accepted that *tourism is about people being away from their own homes, on short-term, temporary visits, for particular 'tourism' purposes.* This definition captures the essential characteristics of tourism, which include the temporary movement of people to destinations outside their normal places of work and residence, activities undertaken and facilities utilised at the destination, mode of transportation used, length of stay and distance travelled, etc. The term, 'tourism' implies at least a 24-hour stay away from a normal place of residence. Anything less than 24 hours is described as an 'excursion' and the participants are excursionists, not tourists.

The two main purposes associated with travelling, which makes people tourists, are leisure and business. Leisure tourism includes travel for purposes such as holidays, sports (non-professional), cultural events, visiting friends and relatives, etc. Business consists of travel for business meetings, exhibitions and trade fairs, conferences and conventions. Other reasons for tourist travel are for medical purposes, religious reasons and for education.

Source: Nyakunu E.P. "Tourism and its impacts on the environment: Lessons from the Zambezi River Basin", *The State of the Environment in the Zambezi Basin, Proceedings of a regional workshop held at Mazvikadeyi, Zimbabwe, July 20-24, 1998*

Map 8.1: Location of lodges and wildlife estates

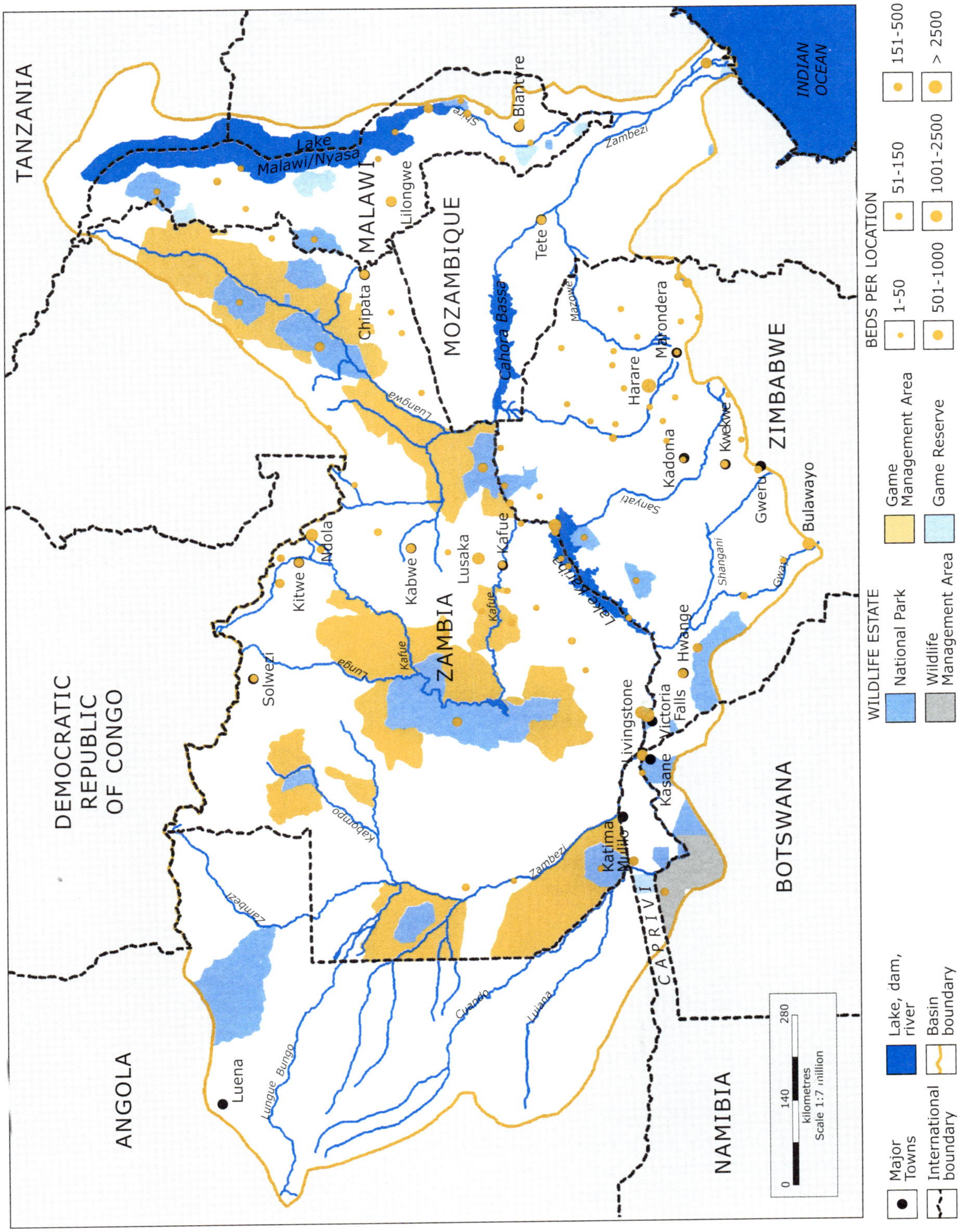

Source: Denconsult/NNR, Zambezi River Basin Atlas, SADC-WSCU, Danida, Lusaka, 1998, p.85.

The Zambezi Basin boasts some of the most important wildlife areas and destinations in southern Africa with great potential for offering exclusive experiences. Some of the national parks and game reserves located within the Zambezi Basin are even protected as internationally important sites. They include sites protected as World Heritage Sites under the World Heritage Convention (1975). Examples include Lake Malawi National Park in Malawi, Victoria Falls in Zambia and Zimbabwe, and Mana Pools National Park in Zimbabwe.

A large proportion of the Zambezi River Basin has been set aside for wildlife and tourism. In general, there is a disproportionately high ratio of protected areas in the basin area of each of the riparian states. Protected areas cover about 27 percent of the basin area. Tanzania is the only exception, with none of its small Zambezi Basin area being protected for wildlife.[3]

Other tourist areas in the Zambezi Basin include the Zambezi River itself with its tributaries and associated wetlands, some of which are either covered by prime wildlife areas or are used for water sports and adventure activities.

The rivers and wetlands of eastern Caprivi make the area quite distinct from the rest of Namibia. While large parts are wooded, about 30 percent of the total land area is floodplain and a major part of this is associated with the Zambezi River system. Caprivi has always boasted the greatest diversity of wildlife in Namibia. However, this has diminished alarmingly because of uncontrolled hunting, competition from livestock and loss of habitat due to low flood levels.[4]

Box 8.2: The Kuomboka ceremony

Barotse life is dominated by the annual flooding of the Zambezi. The rains start in November. By the end of February the floodplain is inundated with water. The Lozi have adapted to this unique situation. For nine months of the year, they live on the plains, their villages built on mounds and ridges. Their cattle graze on rich pastures from November to February when the rains bring a flush of new grass, and from April to July when the receding waters provide fresh green shoots. They plant gardens in the alluvial depressions, fertilised and watered by the flood. Planting is done in July and reaping completed by December. During low water levels, they trawl for fish. As the levels rise and fall, fish traps and reed fences criss-cross the small channels. As the rising flood makes life uncomfortable, the Lozi move to the forest margins to plant their summer gardens.

This yearly migration to the escarpment is the major event in the Lozi calendar. Led by the *Litunga,* the ceremony is known as *Kuomboka,* meaning 'to get out of water'.

Source: Teede, J., and Teede, F., The Zambezi River of the Gods, South Africa, 1990, p. 19-20.

Upper Zambezi

In Angola, there are no tourist areas in the basin because of the civil war. The potential for tourism is reported to be high partly because of the network of important freshwater wetlands and the country's richness in wildlife. A substantial area in the west of the Zambezi headwaters is protected in the Kameia National Park (14,450 sq km). However, there are several population settlements in the park, and the civil unrest has made it difficult to manage the park properly. Poaching, cutting of wood and burning of fields are problems that have negative impact on the environment.[5]

In Zambia, the national parks of West Lunga, Liuwa Plain, and Kafue are all found in this part of the basin. There are nine lodges in Kafue National Park.[6] Except from the protected areas and wildlife attractions, at least three other tourism features are found in this part of the basin: the falls and rapids near Chavuma, Nyamboma and at Sioma; the floodplains at Luena and Luanginga, as well as the Lozi Kuomboka ceremony on the Zambezi Floodplains. These features have not been sufficiently developed for tourism but the Lozi Kuomboka ceremony has the potential to become a major attraction.

The National Parks of Namili and Mudumu together with Caprivi Game Park are the main draw cards in Namibia, and the mainstay of Botswana's tourism industry in the basin is Chobe National Park. The park, established in 1968, covers about 10,570 sq km.[7] A wide variety of different habitats, ranging from deep Kalahari sands to permanent swamp are found here. The park is also the core of the range for the large elephant population of northern Botswana and western Zimbabwe.

Middle Zambezi

This section includes the capital cities of Lusaka and Harare in its catchment and is the most intensively used and industrialised section of the basin. For example, the area between Victoria Falls and Kanyemba draws over 80 percent of Zimbabwe's tourists.[8]

Perhaps the prime tourist attraction in this section of the basin is the Victoria Falls on the Zambia-Zimbabwe border. The Falls comprise one of the wonders of the natural world and were declared a World Heritage Site by UNESCO in 1989.[9]

The development of the tourism industry in the Victoria Falls area has been fourfold over the past 10 years, with the most striking development in the accommodation sector. The number of hotel rooms are more than 1,020 in Zimbabwe and 367 in Zambia.[10] The number of rooms in Livingstone could have even gone down with the closure in 1998-99 of the main hotel in the area. In recent years, there has also been a phenomenal growth of out-of-town lodges and these facilities have been small and exclusive. It is estimated that the average monthly entry into the Zimbabwean side is over 14,500 visitors.[11]

Photo: N Greaves

The development of tourism in Victoria Falls has been fourfold over the past decade.

Next to the Victoria Falls, the most important tourist area in Zimbabwe's share of the basin is Lake Kariba. Since the dam increased water availability in the arid area, it has attracted large herds of wildlife, making it an important tourist destination.[12]

The Zambezi River itself and all of the protected areas are other major attractions in this section of the Zimbabwean part of the basin. Bookings at Mana Pools are generally over-subscribed 12 months in advance and are, therefore, decided in a draw. For this reason, the occupancy rate of accommodation facilities has been 100 percent most of the time.[13]

In Zambia, national parks such as Mosi-oa-Tunya, Blue Lagoon, Lower Zambezi, Lukuzusi and North and South Luangwa are draw cards. In most of the parks, there are lodges and the potential to increase the number exists.[14]

Lower Zambezi and Lake Malawi/Nyasa-Shire River system

Lake Malawi/Nyasa is the most important lake for tourism in this section and so are the wetlands of the Zambezi Delta in Mozambique which are well known for their extensive areas, economic importance and species diversity.

The Gorongoza region, which partly falls within the basin boundaries, occurs in Sofala Province of central coastal Mozambique. The region extends eastwards from the massif of Gorongoza Mountain, across the floodplain and wetlands of the Rift Valley to the forests of Cheringoma plateau, descending to the floodplains and wetlands of the Zambezi Delta, and the caches and mangroves of the Indian Ocean coast. The region has a very high species and ecosystem diversity. It also has a very high carrying capacity for large mammals. Within the basin, the land comprises, among others, the Gorongoza National Park and Marromeu Buffalo Reserve.

Malawi and Tanzania belong to the Lake Malawi/Nyasa-Shire River system. While there are no specific tourist areas developed in the Tanzanian part of the basin, there are many in Malawi, including Lake Malawi National Park. The offshore islands and part of the mainland were first protected in the 1930s but a new concept is that of preserving the actual waters of the lake and the marine life, up to 100 metres from the shore. The Lake Malawi National Park was declared a World Heritage Site in 1984.[15]

TOURISM ACTIVITIES

Wildlife related activities

Most wildlife-based tourism is developed in national parks and game reserves where the big mammals, including lion, leopard, elephant, buffalo and rhino, are the major attraction. Increasingly, however, there is a growing interest in other forms of wildlife such as birds and plants. The location of wildlife areas in the basin is largely an artifact of the presence or recent presence of tsetse fly and poor agricultural potential of designated areas. On a country basis, Botswana has the largest proportion of its basin area under effective wildlife management (65 percent).[16]

Photos: M Chenje

There is increasing interest in various tourism activities, including birding.

Critically important wetlands for birds in the basin include, Lochinvar National Park and Blue Lagoon National Park at Kafue Flats and the intervening game management area which form the most important wetland in the Zambezi Basin both for Wattled Cranes, which reach maximum numbers in the dry season, and for concentrations of waterbirds. The east Caprivi/northeast Botswana wetlands also have a wide diversity of bird species.

Activities related to wildlife include, for example, non-consumptive activities such as walking safaris, game-drives, and bird watching. Sport hunting, which is a consumptive tourism activity, is also a big component of the tourism market in the basin.

Water-based activities

Most of the water-based activities, such as canoeing, boating, whitewater rafting, sport fishing, bungee jumping and scuba diving, are carried out on the Zambezi River and Lake Kariba in the Middle Zambezi and on Lake Malawi/Nyasa.

Canoeing on the Zambezi River was initiated in 1982 and it has since become popular on both the river and Lake Kariba. About 250 km of the river is available for canoeing. The number of canoe trips in Zimbabwe waters in 1994-95 were about 1,900, carrying 22,800 persons during the period.[17]

Sport fishing and boating activities are concentrated on Lake Kariba and Lake Malawi/Nyasa. About 2,080 boats are available for tourist activities, with the majority being on Lake Kariba.[18] Boat cruises are operated from a number of jetties mainly along the Zimbabwean side, but also from two jetties on the Zambian side.

Whitewater rafting and bungee jumping are perhaps the most prominent of the adventure sports on offer. The gorges below the Victoria Falls comprise the finest one-day whitewater rafting run in the world. Bungee jumping off the Victoria Falls bridge, with its 111m drop, is the highest commercial jump in the world and has become an important addition to the adventure sports attractions. An estimated 7,500 people make the jump each year.[19]

Photo: M Chenje

Sport-fishing is one of the water-based activities that attract many visitors into the basin.

In lakes and rivers within the basin, sport fishing is a popular recreational activity. Angling on Lake Kariba peaks during public holidays, long weekends and at angling tournaments. The tigerfish, famous for its spirited fight, is the most sought after by anglers. The most utilised areas are Sanyati Basin (at Nyaodza, the mouth of the Sanyati River and the Charara shoreline), Bumi Hills, and Milibizi.[20]

Lake Malawi/Nyasa is also a popular site for anglers, mainly due to its many fish species. About 500 fish species are found in the lake,[21] making it a popular site for scuba diving and snorkelling.

Cultural activities

This broad category includes activities such as visiting ancient monuments and historic buildings, museums and art galleries. It also includes experiences related to music, dance, food, ceremonies (such as the Kuomboka) and the rural way of life.

Within the basin heritage sites, such as the Khami Ruins in Bulawayo, could be packaged with the most renowned ones such as

Box 8.3: Historical tourism to be promoted

Zambia's tourism industry should move away from traditional tourism based on wildlife to include archaeological and historical sites, to make a meaningful contribution to national development.

Regional Director for the National Heritage and Conservation Commission, Mundemuko S'nvula, said if thousands of people each year visit China because of the Great Wall and Egypt because of the pyramids, then Zambia had an advantage since, in addition to the famous Mosi-o-Tunya and the various types of wildlife, it can add unique archaeological, historical and cultural sites.

During a research tour in the Luangwa Valley, he said his assignment was to locate, identify and document these sites for preservation as well as recommend to National Parks and Wildlife Services which ones in particular had a potential for tourism. S'nvula advised that it was important to start developing other tourism resources so as to expand the revenue base in the tourism sector. However, he also pointed out that it was unfortunate that many of these resources were being vandalised and left to deteriorate since people neither understood the importance of preserving them nor the value of the resources.

Among the examples he cited was the road going to Kasweta Wildlife Camp which passes through petrified forests that are approximately 150–250 million years old and had to be graded away to pave the road. Also included in the list are the Chichele and Mfuwe Lodges which are built on old Stone Age Sites.

He advised that some of the sites need to be rehabilitated while others need quick measures to preserve them before they were destroyed. There is also need to educate the local people about the value of their heritage if they are to value and protect the sites, he added.

Ngambo Makayi, Assistant Information Officer, SLAMU

Photo: S Nkhata

Cultural activities, such as the Kuomboka ceremony in Zambia, help promote ecotourism.

the Great Zimbabwe, even though it is outside the basin. There are also Iron Age sites at Victoria Falls on the Zambian side and fossilised footprints have been found on the Zimbabwean side of the Middle Zambezi valley. Further, rock paintings are found throughout the basin.

Many of the cultural activities in the basin are barely marketed but the economic potential is starting to be recognised.

Other activities

Another important tourist activity is that of business visits and conferences. Nations that share the Zambezi Basin belong to a region, which has high economic potential and is taking steps towards more regional integration to facilitate economic expansion. The proximity of the major cities such as Harare, Lusaka and Lilongwe attracts a large number of business visitors to the Zambezi Basin.

Conferences tend to bring in visitors who may take the opportunity to visit some of these attractions. According to classifications in the Zambia Tourism Policy, the bulk of tourist arrivals to Zambia — 263,986 in 1996 — are visitors to friends and relatives (35 percent) and business visitors (25 percent) as opposed to bona fide holiday tourists (20 percent).

TOURISM AND THE ENVIRONMENT

Tourist behaviour is often motivated by environmental conditions. Even though accommodation and other components such as infrastructure, political stability and social structure are essential to the success of tourism, they are irrelevant if the environment is unattractive. It is recognised that the very existence of the tourism industry within the Zambezi Basin is a result of water and wildlife resources and should be safeguarded to ensure sustainability. The future threat to the environment lies in the uncontrolled growth of tourism and the pressures and demands which will be placed on the natural resources of the Zambezi Basin. Areas within easy reach of large numbers of people are especially under threat.

Photo: M Chenje

The development of infrastructure, including airstrips to facilitate accessibility of remote areas, negatively impacts wildlife habitats.

Infrastructure development

The most significant impact of an increase in the number of people coming to the Zambezi Basin is the expansion of hotels and tourist facilities to cater for this demand as well as an improvement in the road network. Some of the impacts are shown in Table 8.1.

Exploitation of aquatic and wetland resources

The impacts from activities such as rafting, canoeing and, motor-boating directly relate to pollution and waste. The number of boats on Lake Kariba, for example, is very high and oil and fuel pollution is caused by accidental leakages or spillage from engines and/or deliberate waste oil dumping into the lake. Oil is hazardous to animals and people and it reduces the water quality.[22]

Noise pollution is also high and the continued wave action caused by the wake (strip of smooth water left behind a moving boat) damages riverbanks, disturbing the micro-ecology. There is concern that sport angling can overexploit game fish species. The sport has also led to the introduction of alien species in some areas such as trout and bass both of which are favoured by anglers.

The maintenance of good water quantity is as important a habitat requirement for healthy wetlands and river-gorges as it is for water-based tourist activities in the basin. Most water-based attractions, such as rafting, boating and fishing, do not consume water; in fact they require certain minimum flow. Rafting below the Victoria Falls requires a flow of at least 500 cubic metres per second, which is also necessary to maintain the wetlands downstream. One of the major casualties of the planned Batoka Gorge Hydro Electric Project, which will be located below Victoria Falls, will be whitewater rafting, since the river will not provide adequate flows.[23]

Water quality can be impaired by sewage discharges from lodges and towns, siltation due to erosion caused by river-bank developments, overexploitation of wetland resources such as reeds and fish, and pollution by fuels

Table 8.1: Tourism and its impacts on the environment

Biophysical impact	Potential cause
Land take and habitat loss	• Hotel construction, airports, additional roads and access points, residential areas and service sites
Soil erosion (including island and riverbank erosion)	• Roads, boat traffic, bush clearing, deforestation and increase in upgrading and maintenance costs
Wildlife disturbance	• Workers, visitors and traffic around key species and prime view points • Movement constrictions by canoes, hotels and fences, resulting in changes in animal behaviour • Increased human and wildlife conflicts
Loss of biodiversity	• Trampling and picking of flora and disturbance of nesting birds, increased fire hazard
Fire hazards and litter (waste)	• Cigarettes, braais, tins, bottles, plastic bags
Visual impacts and landscape deterioration	• Inappropriate structures
Loss of archaeological sites	• Road infrastructure, purposeful damage, absence of maintenance
Pollution (including noise)	• Motorised crafts, litter, sewage effluent, dumps
Degradation of water quality and quantity	• Domestic water use, effluent discharge and interference with natural water flow regimes for activities such as white water rafting

Source: Nyakunu, E.P., Tourism and its Impacts on the Environment. Paper presented at Regional Workshop, July 20-24, Mazvikadeyi, Zimbabwe, 1998.

Pollution in Kariba harbour is one of the negative impacts of increased boating on the lake.

and pesticides. The flora and fauna, which contribute greatly, if not solely, to the attraction of the Zambezi River Basin require adequate supplies of good quality water.

Wood curio industry

The curio industry has developed as a byproduct of tourism and is a big consumer of indigenous hardwoods.[24] Demand for curio products has increased in many areas with the expansion of the tourism industry. Localised overharvesting of large and increasingly rare indigenous trees is believed to be occuring at an unsustainable rate, and often illegally. A total of about 246 cu m of wood is carved annually.[25]

Photos: D Martin

The curio industry has developed as a byproduct of tourism and is a big consumer of indigenous hardwoods.

Landuse and food security

In commercial farming areas, conservancies and some community-based natural resource management (CBNRM) areas, wildlife management may compete with crop cultivation as cultivated lands are being turned into game ranches and wildlife buffer zones. If the current trend of converting cropland into game ranches continues, food security and self-sufficiency may be threatened. However, in some of these areas, particularly in drier areas, the economics of wildlife management is better than that of crop cultivation. Game farming, coupled with tourism is seen as a more sustainable and productive landuse option than agriculture.[26]

In Botswana, the wild animal species have evolved with, and are better adapted to the harsh and hostile environment. Unlike livestock, wild animals rarely degrade the rangelands they utilise through a variety of behavioural and morphological mechanisms. These adaptive features make wildlife utilisation a better landuse practice, in marginal areas and economically, it is a source of income generated through tourism activities.

Photo: M Chenje

Apart from generating revenue, tourism also contributes to visitor education.

TOURISM AND THE ECONOMY

The tourism sector can make significant contributions to the economies of the basin countries by generating foreign currency. This is, for example, illustrated in Table 8.2, showing the estimated direct value of consumptive wildlife use, of which hunting is the greatest activity. Recent studies under the Zambezi River Action Plan (ZACPLAN) estimate the direct value of wildlife and water-based tourism to the basin at US$148 million annually.[27] Wildlife forms the basis for economically important tourism throughout the southern African region. One of the important justifications for protected areas is, therefore, the realisation that tourism in these areas has the potential to make a significant contribution to the improvement of national economies.

Tourism is also a labour intensive industry, providing employment not only in this sector but also in supporting agricultural, industrial and commercial sectors. The tourism industry has grown considerably in the last few years, and it is expected to continue growing. The period 1993–97 showed an increase in tourist arrivals for all the Zambezi Basin states, except Angola, as shown in Table 8.3.

Table 8.2: Value of consumptive wildlife use in the Zambezi River Basin

Country	Value of wildlife use (US$)
Angola	n/a
Botswana	1,947,749
Malawi	n/a
Mozambique	898,347
Namibia	333,000
Tanzania	n/a
Zambia	3,125,912
Zimbabwe	8,327,813

Source: Zambezi River Authority, *Sector Studies under ZACPLAN, Sector Study No.4, Tourism, Recreation and Wildlife, Draft Report,* Denconsult, 1998.

Table 8.3: Tourist arrivals to the basin states 1993-1997 (Total Year/Arrival/000's)

Country	1993	1994	1995	1996	1997
Angola	21	11	8	8	8
Botswana	607	625	644	700	728
Malawi	153	170	192	232	250
Mozambique	n/a	n/a	n/a	n/a	n/a
Namibia	255	326	399	405	410
Tanzania	230	250	285	310	350
Zambia	157	141	163	264	278
Zimbabwe	951	1,105	1,529	1,743	1,894

Source: Regional Tourism Organisation of Southern Africa (RETOSA)

Botswana, Malawi, Namibia, Tanzania, Zambia and Zimbabwe have all had a steady growth in their tourism industry. Factors such as changes in the political climate and infrastructual development have contributed to the increase in tourist arrivals. Botswana's growth can, for example, be explained by the increased accessibility by road into the country and into Namibia. The proposed Trans-Kalahari Highway (through Botswana) and the Trans-Caprivi Highway linking Namibia to northern Botswana, Zambia and Zimbabwe is certain to increase regional access to the country.

Possible negative economic impacts include the matter of economic leakages, i.e. what proportion of tourist spending benefits the host country. Often only a small percentage of the total amount spent by a tourist reaches the host country. And in return, the host country must divert investment from its social needs to preserving its wildlife and environment, to anti-poaching, roads and airports.[28] For tourism to be a viable conservation strategy, it must lead to economic development while not adversely affecting the natural resources on which it depends.

Rural development

Benefits derived from tourism include the introduction of a broad base upon which new economic opportunities can grow to the benefit of local populations. This can be achieved through employment creation, an injection of money through the wages paid to workers, potentially bigger markets for local produce and opportunities for training and upgrading of local skills. For instance, the curio trade has become one of the most direct ways in which local people can benefit from the tourism industry. Local people dominate the trade as woodcarvers, traders and stallholders.[29]

Apart from contributing to the economy, tourism activities in the basin also provide an opportunity for visitor education and increasing appreciation of the area and natural environment. This in turn works in favour of encouraging tourism investment and providing development opportunities. The Lake Malawi National Park, which covers an area of 140 square kilometres and enjoys a large population and many visitors,[30] has an education complex built to try to ensure that future citizens and visitors appreciate this unique area.

Tourism development also attracts the provision of social amenities and services such as schools, hospitals and recreational facilities and if properly regulated, hunting has proved to be a good caretaker of land. Programmes such as the Communal Areas Management Programme for Indigenous Resources (CAMPFIRE) in Zimbabwe, the Administrative Management Design Programme for Game Areas (ADMADE) in Zambia, and the Communal Wildlife Conservancies being set up in the eastern Caprivi in Namibia as part of the Living in Finite Environment (LIFE) programme are good examples of this. Sustainable wildlife utilisation gives communal residents an alternative landuse option to conventional agriculture.

ACTIONS AND INITIATIVES TO PROMOTE SUSTAINABLE TOURISM

Institutional framework

The Southern African Development Community (SADC) has established a Tourism Coordination Unit (TCU). The prime aim of the TCU is to promote and coordinate tourism development in the SADC region. In September 1998, the SADC Tourism Protocol was signed, providing the legal framework to achieve this. A separate organisation, the Regional Tourism Organisation of Southern Africa (RETOSA), was also established with the full approval of the SADC Council of Ministers.[31] RETOSA's mandate is to market and promote the region in close cooperation with national tourism organisations and the private sector.[32]

Box 8.4: Regional Tourism Organisation of Southern Africa (RETOSA)

RETOSA, located in Johannesburg, is the region's concept of regional tourism cooperation intended to create a concrete destination identity for southern Africa and to develop the combined tourism interests of its member states. For the first time in the region, it combines the full spread of stakeholder interests (public, private and voluntary sectors) in the region into one focused organisation. Among other things, RETOSA seeks to work with conservation and environmental organisations to retain and safeguard the region's heritage into the future.

RETOSA's 1998–2002 Programme aims at increasing southern Africa's market share of the global tourist business.

RETOSA's key objectives are to:

- Create a sound saleable brand of the region.
- Expand and sustain a sound database on the sector.
- Intensify and sustain the profile of southern Africa in the primary and secondary markets.
- Develop and sustain synergy with both new and old tour operators selling the region.
- Increase and improve product knowledge of the region's suppliers.
- Develop appropriate marketing collateral to promote the region.
- Foster and enhance the corporate image of RETOSA to all its stakeholders.

Source: Regional Tourism Organisation of Southern Africa (RETOSA)

At national level, there is more emphasis on the quality of development. Central governments are increasingly controlling the aesthetic aspects while local authorities consider the commercial advantages and weigh these against the 'social and environmental costs' of the proposed or operational tourism projects. Such controls could be through building restrictions, granting of planning approval, control of the scale of development and the selective award of development grants or assistance.

In Malawi, an Endowment Fund Working Group was established under the auspices of the NATURE programme to identify and address issues relating to the creation and operation of a fund to be part of the Environmental Fund established under the Environmental Management Act.[33] The objective of the fund is to provide a sustainable source of financing for conservation and natural resource management.

Policy aspects

The tourism policy for most of the Zambezi Basin countries is to promote "low-volume, high-price/return/value" tourism. In terms of national marketing strategies, this means that the target is the very top end of the market or the high-net-worth discerning traveller. The extent to which the policy is implemented varies but since international visitors provide a considerably higher yield than regional and domestic tourists, this has driven the tourism industry in the basin towards an international market orientation.

This policy of "low-volume, high-price/return/value" has resulted in very few facilities being available for lower income travellers and, in particular, domestic tourism. Travel across borders to immediate country neighbours is motivated primarily for family, cultural and ethnic reasons.

The high prices being charged in the top segment of the tourism industry and the growth of budget tourism are attracting many small operators into the market. The backpacker-lodges, guest houses, lodges and "bed-and-breakfast" facilities are, therefore, growing in number in the main tourism and city areas. For example, Namibia has a well-structured "bed-and-breakfast" type guesthouse system, which is ideally suited to expand into the provision of appropriate services to the middle-income and emerging regional tourism markets.[34]

Privatisation

The process of privatisation of national parks has been launched in a number of the basin countries. Many governments are restructing their tourism and wildlife management agencies away from strict government control to more professional and streamlined needs-driven, semi-government agencies with strong links to the private sector. For instance, in Malawi some of the state lodges and national parks' facil-

Photos: M Chenje

The growth of budget tourism is attracting many small operators into the market.

ities have been or are being privatised. Nyika National Park's facility was recently privatised by way of a 20-year lease agreement.

In Namibia, government-owned resort facilities will ultimately become commercialised to make them more competitive and market-oriented. However, this may take a very long time as the government is against private operations in national parks.[35]

In Zambia, the policy of "privatisation" predominates and it is largely driven by the need to strengthen capital investment for developments in the tourism sector. The Victoria Falls and the Zambezi River are, therefore, seen as under-utilised resources and the government is encouraging investment by releasing various plots of land.

In Zimbabwe, the "indigenisation" programme is driven by the need to spread ownership and benefits of private enterprise to indigenous business people. Developments in Victoria Falls town are dominated by private enterprise, but the bulk of these are not defined as indigenous.

Competition from countries sharing the basin is becoming increasingly intense as countries such as Zambia and Tanzania create an enabling environment by improving investment incentives, entry formalities, infrastructural improvement, personnel training and standards.

Box 8.5: Defining ecotourism

Derived from the two terms ecological and tourism, ecotourism is most commonly used to describe any recreation in natural surroundings. The American Ecotourism Society (one of many organisations involved in ecotourism) adds social responsibilities to define ecotourism as:

"Purposeful travel to natural areas to understand the culture and natural history of the environment, taking care not to alter the integrity of the ecosystem, while producing economic opportunities that make the conservation of natural resources beneficial to local people."

It has been argued that used in this sense, the word refers to a set of processes and underlying ethos, rather than the simple act of travelling to an outdoor, nature-based destination for pleasure or leisure. The CAMPFIRE programme in Zimbabwe is for example not considered by the American Ecotourism Society to offer 'ecotourism' because it's largely based on trophy hunting. This is despite the fact that CAMPFIRE could be regarded as ecologically sound, with minimal environmental disruptions while benefiting local people economically.

Source: Endangered Wildlife Trust, Vision of Wildlife, Ecotourism and the Environment in Southern Africa, Parkview, South Africa, 1995.

Nature-based tourism

The segment of tourism that is undergoing the fastest growth is nature-based tourism or ecotourism. If well managed, this kind of tourism is believed to have many advantages. It:

- Provides a lucrative source of foreign exchange that can be used to finance economic growth and development.
- Suggests a form of industrial growth that is well suited to rural areas. Studies have shown that ecotourism is a better form of landuse than most agricultural pursuits, especially in arid and semi-arid areas.
- Is a form of economic development that can spread revenue to the poorest strata of marginal rural communities.
- Is, by some account, an industry without chimneys, thus can promote the protection of wildlife while promoting social development.
- Has the capacity to promote respect for other people's cultures on the part of visitors from foreign countries.[36]

Yet, nature-based tourism is not necessarily ecologically sound. Some places visited by tourists support fragile ecosystems that cannot endure heavy disturbance.

In addition to the tourism industry, local communities and governments at both national and local level, and organisations concerned with the conservation of nature are now actively involved in the promotion of environmentally responsible tourism.[37]

The Okavango — Upper Zambezi International Tourism project is, for example, a positive initiative found in the Zambezi Basin.

Community Based Natural Resources Management (CBNRM)

As discussed in the chapter on biodiversity, community-based natural resources management systems devolve power and responsibility for resource management to communities. This approach has also introduced a new component crucial to its success - the involvement of local people in the generation and distribution of benefits for community projects.

Box 8.6: Okavango — Upper Zambezi International Tourism (OUZIT)

The initial proponent was the Chobe Wildlife Trust, in 1993, which sought to establish a Southern African Wildlife Sanctuary in wetlands associated with the source of the Zambezi River.

OUZIT is now being sponsored by the South African government, in terms of its Spatial Development Initiative (SDI) part of whose objectives is to stem the tide of "economic refugees" from neighbouring countries. The project seeks to incorporate protected areas in Angola, Namibia, Botswana, Zambia and Zimbabwe, covering a total area of more than 260,000 sq km.

The immediate goals of the initiative have been defined as to:

- Transform the region's comparative advantage in ecological and adventure tourism into a continental and global competitive advantage.
- Assist SADC member states in developing the capacity to command a major share in the revenues generated from tourism.
- Assist in the empowerment of local communities living in the region to participate and share in the rewards of economic growth resulting from the development of the tourism sector.

Case Study: Benefits from natural resources turn Botswana peasants to entrepreneurs

Rural people in the Chobe area of northern Botswana are gradually turning from peasants to entrepreneurs, using thousands of dollars generated from safari operations to establish businesses.

Just 200 km from one of southern Africa's prime destinations, Victoria Falls, the innovative conservation and development activities in the Chobe Enclave are set to significantly improve the standard of living of the residents.

A new partnership is developing between Rann Safaris and the Chobe communities which has encouraged local residents to make ambitious plans to establish a multi-million dollar chalet facility, along with service stations, community shops and hardware stores. The Chobe people may also branch out into lucrative timber industry.

"From 1966 when Botswana attained independence it continued with existing colonial policies in restricting the local communities from utilising their wildlife and other natural resources," said Luckson Masule, Chobe Enclave deputy chief. "All revenues generated from wildlife and other natural resources went to the central government." That is now history to the Chobe villagers as they focus their attention on how best the wildlife industry can improve their standard of living.

Rann Safaris hunts wildlife such as elephant, leopard, lion, buffalo and zebra in the Enclave which contains five villages, each earning an average of US$23,000 annually for a total of about US$116,000.

The joint venture is managed by a community-based organisation, the Chobe Enclave Conservation Trust (CECT) which was established in 1993 through a USAID grant of US$165,000 which covered the costs of building an office complex, radio communications network, motorbike and a house for the programme officer.

The communities pay US$2,000 in annual lease fees to the Botswana government for using state land for hunting. Fifteen percent of the annual revenue generated from Rann Safari's sport hunting is given to the trust to cover administrative costs and to develop joint projects which benefit the five villages.

Each village elects 10 members to represent them on a village committee two of whose members sit on the CECT Board.

More than US$25,000 has been used by the CECT to build campsite facilities, which comprise an administration office, 12 flush toilet stalls and a tuckshop for campers.

Campers pay US$5-a-day to stay and the money generated from this project is shared equally among the villages. The campsite is expected to generate more revenue from tourists who flock to this area to view elephant and other species. Wildlife is so abundant that some tourists say the animals seem to be "falling from the sky and coming out of the ground."

Kachika village has already secured land to establish a filling station and is currently negotiating for a joint venture project with two major fuel companies in Botswana. The scores of tourist vehicles which pass by the area constitute a ready market for fuel, because the nearest service station to this village is about 100 km away.

These projects are suggested by members of the village committee and then presented to villagers who vote to decide whether or not to embark on the business ventures.

One of the major problems experienced by villagers is lack of good business planning skills.

By Emmanuel Koro

To residents living in or near wildlife areas, tourism can offer new opportunities for access and more community control over use of land and natural resources,[38] jobs, enterprises and skills development. It can be a way to earn benefits from parks, wildlife and tourists that in the past have only brought costs. It also means making wildlife management financially viable.

CBNRMs promote biodiversity conservation by creating the necessary conditions for sustainable use. A third key element is that they are also a rural development programme, enabling communities to derive a direct financial benefit from wildlife and tourism activities.

Prior to the introduction of CBNRMs, tourism laws and policies virtually ignored community-based tourism developments and effectively discouraged communities from participating in tourism. Rights to natural resources and tourism development opportunities were vested in government, which in turn channelled most of its development and tourism promotion efforts into the private sector. Post-independence governments have sought to address this inequity. The private sector has been quick to recognise the growing market for nature-based tourism and the possibilities offered by the country's communal areas.

Box 8.7: CAMPFIRE benefits six million people

In just under 10 years, the Communal Areas Management Programme for Indigenous Resources (CAMPFIRE), has expanded from just two pilot districts to 30 full programmes covering nearly half of Zimbabwe and involving six million people. Most of its activities are concentrated in the agriculturally marginal areas of the Zambezi Valley and Lowveld districts of the country.

When the CAMPFIRE programme started in 1990 in the districts of Nyaminyami and Guruve in the Zambezi River Basin, the national average benefit per household was approximately Z$48. As the number of districts participating in the programme increased, the average initially fell to Z$24 in 1991. However, since then the improved marketing of wildlife opportunities in communal areas resulted in the national household average benefits rising to Z$58 in 1993.

A 1997 report by the World Wide Fund for Nature (WWF) estimated household incomes to have increased by 15–25 percent in the Omay communal areas of Zimbabwe, to an average between Z$400–600. Expectations for a further increase are brighter, given the removal in 1997 of the international ban in trade in elephant products, the main output of CAMPFIRE. In 1999, elephant products, including ivory generated Z$124 million in 1999.

In a recent report, the Land Tenure Commission says any reduction of CAMPFIRE will be to the detriment of rural development in Zimbabwe. Wildlife utilisation, through CAMPFIRE, gives communal residents an alternative landuse option to conventional agriculture.

In addition, CAMPFIRE generates employment in the form of game wardenships. In the Chawarara area of Muzarabani, CAMPFIRE activities have also led to the employment of "reporters" whose duties involve compiling reports on crop destruction by animals, and other problems caused by wildlife and people.

Under CAMPFIRE, managerial power over natural resources, particularly wildlife, is devolved to local communities. So successful has this strategy been in promoting both conservation and rural development in Zimbabwe, that some other SADC countries have adapted the concept. Similar programmes include the Living in Finite Environment (LIFE) programme in Namibia and the Administrative Management Design Programme for Game Areas (ADMADE) in Zambia.

CAMPFIRE uses an entrepreneurial approach to development based on wildlife management that uses market forces to achieve economic, ecological and social sustainability.

It depends on changing the perceptions of the people who live adjacent to wildlife areas. Sustainable rural development requires strategies that enable rural people to improve their quality of life while maintaining the natural resource base. CAMPFIRE reconciles these requirements by identifying a range of potential financial and other benefits that rural communities can derive from the careful management of wildlife.

The major aim of the programme is to change the attitude of indifference adopted by local people during the colonial era when authorities usurped their control over wildlife.

Through rural-district councils — communal people's representatives at the local level — government allows communities in CAMPFIRE areas to sustainably utilise their natural resources. Revenue generated is used to finance infrastructure developments such as clinics, schools, roads and water-points. In some cases, households are paid cash dividends.

Partly due to CAMPFIRE, Zimbabwe has seen its elephant population grow to more than 65,000, almost twice the country's environmental carrying capacity of 35,000.

Given its success with the management of wild game, there are strong calls for CAMPFIRE to broaden its scope to include forestry, water and minerals. The concept could also diversify into tourism, other than trophy-hunting and camera safaris. However, such a move requires wide-ranging environmental policy changes.

Source: SARDC-IMERCSA, The Zambezi, Vol. 1 no 1, 1998.

Zambia's approach to wildlife conservation in game management areas is through the ADMADE programme, which is designed to encourage residents to participate in wildlife management schemes and cooperate with land-use planning policies. The origin of the programme stemmed from the experiences gained during the pilot Lupande Development Project.[39] The project demonstrated the effectiveness of local people's participation in wildlife conservation and how revenues accruing from wildlife can economically benefit them. ADMADE is rooted in traditional African values with an appreciation of customary leadership.

In Mozambique, integrated management of the Gorongoza-Marromeu Region Natural Resources Management Area (GMNRMA) has been carried out since 1995 to promote conservation and sustainable use of wildlife and forest resources in the area. This supports national and local economies while ensuring the long-term conservation of the area's biodiversity.

The weaknesses inherent within CBNRMs include over-dependence on donor-supported management, lack of real community involvement and leadership in the management of resources,[40] lack of mechanisms for communities to benefit more effectively from conventional tourism operations[41] and over-dependence on consumptive use of selected wildlife species.[42] It has also been argued that rural communities are being forced to finance infrastructural development which is the responsibility of government.

FUTURE CHALLENGES AND OPPORTUNITIES

Sustainability concerns tourism's effects on the environment and the distribution of its economic and social costs and benefits. One of the future challenges, therefore, for the sustainable development of the tourism industry is to convince stakeholders that tourism cannot continue to grow unless impacts on nature's life-supporting systems are considered.

For tourism to become a strong economic sector in the Zambezi Basin and a force for creating more jobs and combating poverty, it is important that the symbiotic relationship between tourism and healthy ecosystems be maintained. To strike a balance between tourism and ecology, some of the future challenges include:

- efficient planning such as adequate forecasting of tourist pressures and capacities;
- natural resources-oriented tourist marketing; and
- improved cooperation between organisations involved in the industry.

Greater public-private sector cooperation and among governments of the basin states for tourism development is needed.

It is also important that tourism developments and activities in the area develop links with local communities. Mechanisms to allow an equitable sharing of the benefits should be pursued vigorously. This is to avoid potential negative social impacts of tourism such as cultural change and anti-social behaviour, including begging, prostitution and alcohol abuse. There is need, therefore, for planners and national governments to minimise the negative social impacts arising from tourism development in the Zambezi Basin. The tourism business concerns have a responsibility too, to optimise benefits and minimise 'social costs' by promoting a better understanding of the problems.

Overall, tourism as an industry, if properly managed, has the potential to foster sustainable regional growth in the basin countries.

Photo: M Chenje

The future of tourism depends on a symbiotic relationship between the industry and healthy ecosystems.

LINKAGES TO OTHER CHAPTERS

Chapter 1: Regional Overview: People and Environment
Tourism is a fast-growing industry in the basin, providing opportunities to improve the standard of living of the people. The curio trade is one of the most direct ways in which local people benefit from the industry.

Chapter 2: Physical Features and Climate
A combination of climate and drainage has created a wide and spectacular diversity of ecological systems and landforms of international importance. This underpins the tourism industry in the basin.

Chapter 3: Water and Wetland Resources
Given the potential of tourism as a contributor to national economies and water as a limiting factor in ecosystems, any attempts to utilise water and wetland resources in the region should be sensitive to the needs of tourism and the environment.

Chapter 4: Biological Resources and Diversity
An increased appreciation of the state and importance of biodiversity in the Zambezi Basin demonstrates its significance as a tourist attraction, among other values. What is more important is that an understanding of the threats to biodiversity would also shed more light on the impact of tourism on biodiversity.

Chapter 5: Agriculture
With most of the tourism in the Zambezi Basin based on wildlife, its sustainability depends on adequate land being set aside for wildlife with minimal encroachment from agriculture and increasing unplanned settlements.

Chapter 6: Industry
The tourism industry, once considered a byproduct of natural resource conservation, has become the foundation of many economies in the basin.

Chapter 7: Energy
Current energy sources are destructive to the environment through pollution and deforestation. Deforestation could jeopardise the tourism potential of the Zambezi Basin as a combination of poor agriculture, firewood collection and charcoal burning severely degrades the land and encroaches on wildlife habitat.

Chapter 9: Pollution
Eutrophication of water through the release of untreated effluents into water bodies is one of the major causes of the current waterweed infestation that is affecting and threatening wetlands in the Zambezi Basin. Pollution is a threat to the tourism industry.

Chapter 10: Poverty
Tourism contributes to rural development and environmental management, helping poverty reduction efforts. Poverty threatens tourism through degradation of resources upon which tourism depends or through the social unrest, crime and instability that it breeds.

Chapter 11: Gender and Women's Roles
The role of women needs to be enhanced by eliminating barriers, which limit their effective participation in the tourism industry. The industry is dominated by men but provides opportunities for women to invest and become major players in development.

Chapter 12: Environmental Management and Regional Cooperation
Since the environment on which tourism depends extends beyond national boundaries, it is important that basin states consolidate regional cooperation by harmonising policies in various areas, including tourism development.

Chapter 13: Trends and Scenarios
Growing demand for outdoor recreational spaces is fast outstripping existing and proposed facilities as well as the resources available. Appropriate management structures, adequate planning, design and building guidelines for tourism facilities are required to ensure that the facilities enhance rather than detract from the natural setting.

ENDNOTES

1 Chenje, M., Sola, L., and Paleczny, D. (eds.), *The State of Zimbabwe's Environment 1998*, Government of the Republic of Zimbabwe, Ministry of Mines, Environment and Tourism, Harare, Zimbabwe, 1998, p.76-77
2 Endangered Wildlife Trust, *Vision of Wildlife, Ecotourism and the Environment in Southern Africa*, Parkview, South Africa, 1995, p. 6-7
3 Meynell, P. J., Sola L., and N., Nalumino *Strategic Environmental Assessment of Developments Around Victoria Falls*, 1996, p.3
4 Ministry of Environment and Tourism, *Namibia Environment*, Volume 1, January, 1996
5 Chenje, J; "State of the Environment in the Zambezi Basin", Inception Mission Report, Harare, November 1998
6 ibid
7 IUCN, *Protected Areas of the World: A review of national systems,* Volume 3: Afrotropical, Gland, 1992, p.17
8 Nyakunu, E., P., "Tourism Industry in the Zambezi Basin", for SARDC, 1999
9 op. cit 3
10 ibid
11 ibid
12 O'Hagan T., *Wild Places of Southern Africa*, Southern Book Publishers, 1996, p.249
13 Coulson, I. M. and Bowler M., "The Zambezi Valley Complex: Tourism and Economics", an unpublished report for The Zambezi Society, 1992
14 op. cit 5
15 Johnston, F., and V. Garland, *Malawi – Lake of Stars*, Central Africana Limited, Blantyre, 1993, p.149
16 Zambezi River Authority, Sector Studies under ZACPLAN, Sector Study No.4, *Tourism, Recreation and Wildlife*, Draft Report, Denconsult, 1998
17 op. cit 2
18 ibid
19 Nyakunu E.P., "Tourism and its impacts on the environment: Lessons from the Zambezi River Basin", State of the Environment in the Zambezi Basin, Proceedings of a regional workshop held at Mazvikadeyi, July 20-24, Zimbabwe, 1998
20 Sanyanga R. A., "Tourism and Wetlands Management in Zimbabwe: with special reference to the Zambezi River System, T. Matiza and S, A. Crafter, (eds.), in Wetlands and Ecology and Priorities for Conservation in Zimbabwe", Proceedings of a Seminar on Wetlands in Zimbabwe, IUCN, 1994, p.81
21 Chenje, M., and P., Johnson, *Water in Southern Africa*, SADC/IUCN/SARDC, Maseru/Harare 1996, p.63
22 Matiza, T. and Crafter, S., *Wetlands Ecology and Priorities for Conservation in Zimbabwe*, IUCN, 1994, p.84
23 Matiza, T., Crafter, S., and Dale, P., *Water Resources Use in the Zambezi Basin*, IUCN, 1995, p. 144-145
24 op. cit 1
25 op. cit 3
26 op. cit 1
27 op. cit 3
28 Martin, D., "Southern African News Features - Tourism Statistics Misleading", Harare, 29 June, 1998
29 op. cit 2
30 op. cit 15
31 SADC, *Tourism*, Johannesburg, 1996, p.2
32 Regional Tourism Organisation of Southern Africa RETOSA, "Constructive Co-operation in Tourism", (Brochure)
33 Malawi Government, *State of Environment Report for Malawi 1998*, Environmental Affairs Department, July, 1998
34 Imani, "Study on the Pricing Structures of Tourism Products in Zimbabwe", August 1998, p.34
35 ibid
36 op. cit 2
37 op. cit 3
38 Ashley, C., and Roe, D. *Community Involvement in Wildlife-Tourism: Strengths, Weaknesses and Challenges*, IIED, 1997
39 Munyenyembe F. E. C., Mubanga G., "The Integration of Wildlife and Agriculture in Game Management Areas of Zambia: The ADMADE Programme", in Proceedings of the SADCC/GTZ Workshop on Wildlife Based Tourism in the SADCC Region, December, 1990
40 Lewis, D. M., Phiri, A., and Morris, J.(eds.), "Wildlife Snaring - an indicator of community response to a community-based conservation project", *Fauna and Flora International*, Oryx 32 No 2 April, Cambridge, 1998, p.111
41 Chieftainess, C., "Successes and Constraints of ADMADE", in *Kobus*, Newsletter of the Wildlife and Environmental Conservation Society of Zambia, Lusaka, Issue Jul-Sept 1998, p.4
42 Adams, L. A., *Sustainable Use of Wildlife and the International Donor Organisations*, Humane Society of the United States of America/Humane Society International, Washington, 1994, p.7

REFERENCES

Ashley, C., and D. Roe, *Community Involvement in Wildlife-Tourism: Strengths, Weaknesses and Challenges*, IIED, 1997

Chenje, J., "State of the Environment in the Zambezi Basin", Inception Mission Report, Harare, November 1998

Chenje, M., and P. Johnson, *Water in Southern Africa*, SADC/IUCN/SARDC, Maseru/Harare 1996

Chenje, M., Sola, L., and D. Paleczny, (eds.), *The State of Zimbabwe's Environment 1998*, Government of the Republic of Zimbabwe, Ministry of Mines, Environment and Tourism, Harare, Zimbabwe, 1998

Coulson, I. M. and M. Bowler "The Zambezi Valley Complex: Tourism and Economics", an unpublished report for The Zambezi Society, 1992

Endangered Wildlife Trust, *Endangered Wildlife - Ecotourism and Environment – A Vision*, Parkview, South Africa, 1997.

Hiscock E. H. J., Matiza-Chiuta, T., and T.J.F. Lash "Zambezi Basin Wetlands Conservation and Resource Utilisation Project", Inception Mission Report, June 1996

IUCN, *Protected Areas of the World: A review of national systems,* Volume 3: Afrotropical, Gland, 1992

Johnston, F., and V. Garland, *Malawi – Lake of Stars*, Central Africana Limited, Blantyre, 1993

Malawi Government, *State of Environment Report for Malawi 1998*, Environmental Affairs Department, July, 1998

Masundire, H., *The effects of Kariba Dam and its management on the people and ecology of the Zambezi River in Water Management and Wetlands in Sub-Saharan Africa*, (eds.) M.C. Acreman and G. E. Hollis.

Matiza, T., Crafter, S., and P. Dale, *Water Resources Use in the Zambezi Basin*, IUCN, 1995

Matiza, T. and S. Crafter, *Wetlands Ecology and Priorities for Conservation in Zimbabwe*, IUCN, 1994

Meynell, P.J., Sola, L., and N. Nalumino, *Strategic Environmental Assessment of Developments Around Victoria Falls*, Volume I, II & III, IUCN, 1996.

Ministry of Environment and Tourism, *Namibia Environment*, Volume 1, January, 1996

O'Hagan T., *Wild Places of Southern Africa*, Southern Book Publishers, 1996

SADCC Coordinator for Forestry, Fisheries and Wildlife, "Proceedings of the SADCC/GTZ Workshop on Wildlife Based Tourism in the SADCC Region", August 1990

Sanyanga R. A., "Tourism and Wetlands Management in Zimbabwe: with special reference to the Zambezi River System, in Wetlands and Ecology and Priorities for Conservation in Zimbabwe", Proceedings of a Seminar on Wetlands T. Matiza and S A. Crafter,(eds.), in Zimbabwe, IUCN 1994

Zambezi River Authority, Sector Studies under ZACPLAN, Sector Study No.4, *Tourism, Recreation and Wildlife*, Draft Report, Denconsult, 1998

Photo: IUCN

Much of today's environmental pollution problems in the Zambezi Basin stem from development strategies of the 1950–60s, which maximised exploitation of resources regardless of the consequences on the environment.

9 POLLUTION

Pollution of surface and groundwater resources and the atmosphere, as well as improper dumping of both solid and liquid waste are becoming major environmental problems in the Zambezi Basin. Although the severity and impacts of some of the pollution is not well researched

Photos: M Kullberg

High rates of urbanisation in the basin countries are contributing to various pollution problems, impacting on human health and the environment.

and documented, it is believed that the quality of vegetation, soils and water is adversely being affected by the pollution.

The urban areas such as Harare and Lusaka are facing serious problems of air pollution caused mainly by industrial activity and the over-dependence on fossil fuels. Sulphur dioxide, suspended particulate matter, nitrogen oxides, carbon monoxide and lead, are the most common, as well as most harmful air pollutants in the basin. If the growth in demand for energy and vehicular transport is to be met using current technologies, the Zambezi Basin will witness emissions from thermal power stations increasing 11-fold and from vehicles five-fold by the year 2003.[1] The use of mainly leaded fuel in vehicles throughout the basin is a major concern, which coupled with the ageing fleet of cars, is worsening the levels of lead pollution in the basin.

Table 9.1: Carbon dioxide (CO_2) emissions in the Zambezi Basin, 1995

Country	CO_2 emissions (million tonnes)	CO_2 emissions per person (tonnes)
Angola	4.60	0.4
Botswana	2.24	1.5
Malawi	0.73	0.1
Mozambique	0.99	0.1
Tanzania	2.44	0.1
Zambia	2.40	0.3
Zimbabwe	9.74	0.9
South Africa*	305.81	7.4
USA**	5,468.56	20.5

No information is available for Namibia. *For comparison within the SADC Region ** For comparison with a developed country

Source: WRI, 1998–99. World Resources: a Guide to the Global Environment, Oxford University Press, 1998.

The high rates of urbanisation and industrialisation in the basin, which are not being matched with adequate waste treatment facilities have resulted in water pollution in some of these areas. Consequently, gastro-enteric diseases such as cholera and dysentery have become prevalent due to faecal contamination of drinking water supplies.

Much of today's environmental pollution problems in the Zambezi Basin stem from development strategies of the 1950–60s, which maximised exploitation of resources regardless of the consequences on the environment. As a result, no serious or effective measures were taken to control environmental degradation, including pollution. Despite worsening levels, pollution is both preventable and controllable. Options available for pollution control include use of cleaner fuels, purifying emissions and encouraging energy conservation. While pollution control is extremely costly, the benefits outweigh the costs.

SOURCES OF POLLUTION IN THE BASIN

The increase in the amount of discharges and types of pollutants in the Zambezi Basin is due to population growth, intensive urbanisation, as well as increased industrial and agricultural activities. The major sources of pollution are:

- **Point source pollution**
  sewage effluent
  industrial processes
  power generation
  mining activities.

- **Non-point source pollution**
  natural pollution
  storm water runoff
  agricultural activities
  leachate from landfills
  soil erosion
  gold panning.

Point source pollution

Sewage effluent

Domestic effluent is handled using some form of sewage treatment works, which consist of features such as septic tanks, oxidation pond systems, conventional treatment works, and activated sludge types of works. Nearly every town and city in the Zambezi Basin has some form of sewage treatment works. The quality of effluent produced depends on the type of works, but in general pollution problems such as eutrophication of water bodies are being caused by the high concentrations of phosphorous and nitrogen in the effluent.

Urbanisation is probably the biggest threat in terms of pollution in the Zambezi Basin because of the problems of sewage disposal. Most urban centres tend to have localised pollution problems, which hardly get as far as the Zambezi River. However, the ability of some of the rivers in the basin to dilute pollutants varies during the year, since rainfall is seasonal. This results in pollutants travelling over great distances.

The largest and closest settlement to the Zambezi River is Livingstone, in Zambia, with a population of 100,000. In recent years, most of the industries that sustained Livingstone in the past have been on the decline except for tourism. In Livingstone, six wastewater oxidation ponds were commissioned in 1995, prior to which raw sewage was being discharged into the Zambezi River. Despite the ponds, some limited discharge into the Zambezi River, still occurs.[2]

In Zimbabwe, Victoria Falls with a population of more than 30,000 is one of the most popular tourist destination in southern Africa, with a monthly visitor population averaging 32,000. The growth rate of Victoria Falls is estimated at 14 percent annually. Environmentalists, the world over, believe that the Victoria Falls has reached its full infrastructural and population potential and any further development will lead to environmental degradation.

Victoria Falls town discharges 8,000 cu m of wastewater into the Zambezi River. The discharge of effluent by the town into the Zambezi River was condoned under the Water (Effluent and Waste Water Standard) Regulations because the town's population was until 1998 considered small. The town's sewage treatment facilities are also overloaded and subject to frequent breakdowns, forcing the town to discharge raw sewage into the Zambezi River through the Masume River.

Sewage treatment facilities are largely inadequate in all the urban centres and are a major source of water pollution in the Zambezi Basin. Chitungwiza town, south of Harare, has high levels of pollution due to poor disposal of sewage. The town has grown rapidly since 1980, with annual population growth rates of 3.5–7 percent, resulting in serious overload of sewage treatment facilities. The town has more than 500,000 residents making it one of the largest urban areas in Zimbabwe. The town is a significant pollution threat to the Manyame River, which drains into the Zambezi. One particularly serious source of pollution is a set of waste stabilisation ponds on the banks of the Nyatsime River, which receive raw sewage from the Tilcor Industrial Area of Chitungwiza. The sewers conveying the sewage to the ponds fail occasionally, resulting in the discharge of raw sewage into the Nyatsime and eventually the Manyame rivers.[3] Up until recently, the Zengeza Sewage Works were discharging 15 megalitres of partially treated effluent into a tributary of Nyatsime River, while Firle Sewage Works in Harare discharged a fair amount into the Mukuvisi River, which also feeds the Manyame River.[4]

Khami Dam on Khami River, which feeds into the Gwayi and eventually the Zambezi, used to supply the City of Bulawayo with drinking water, but is now too polluted to be used for that purpose. This is due to the sewage effluent, which is discharged from the Southern Areas Sewage Treatment plant. The plant is failing to cope with the 700 cu m of effluent per day, which it receives, resulting in partially treated effluent being discharged directly into Khami Dam.[5]

The major consequence of discharging partially treated sewage effluent into any aquatic ecosystem is eutrophication due to high levels of nitrates and phosphates. A recent survey showed that total phosphates and total nitrogen in the final effluent at Victoria Falls, Livingstone, Kariba and Kasane exceeded permissible levels by Zimbabwean standards.[6] The total coliform count was also high. Nevertheless, the discharges do not seem to have significant effects on the nutrient levels of the Zambezi River, whose mean concentrations of phosphates and nitrates average 2μg/l and 10μg/l, respectively.[7] The phosphate and nitrate levels have not changed much since the 1970s,[8] and the Zambezi River basically remains poor in nutrients.

Despite the insignificant effect in altering the Zambezi River's nutrient levels, the persistent discharge of poor quality effluent into the river affects the microbiological quality of the water of the Zambezi and associated reservoirs.[9] For example, a 1996 study

Photo: IUCN

Sewage treatment facilities are largely inadequate in all the urban centres and are a major source of water pollution in the Zambezi Basin.

showed a high incidence of coliform bacteria in the inshore water of Lake Kariba.[10] The main source of the coliform bacteria is poorly managed private and municipal wastewater disposal systems, as well as large tourist boats. The large coliform bacterial count partly explains the high incidences of cholera and dysentery experienced by Kariba and Siavonga throughout the 1990s.[11]

Faecal coliform is still the primary water quality issue in rivers within the Zambezi Basin, particularly in areas where animal and human waste is not adequately collected and treated. The water quality situation is more critical in fast growing urban areas such as Chitungwiza, Victoria Falls and Livingstone where the population growth rates still far exceed the rates of development of wastewater collection and treatment facilities.[12] Coliform counts for the Zambezi River between Victoria Falls and the Victoria Falls town outfall, exceed those found in clean water by as much as 100-fold, confirming that Livingstone discharges raw sewage or effluent into the river. Below the outfall of Victoria Falls town, the river is grossly polluted, with faecal coliform counts of as much as 350 times that found in unpolluted water.[13]

Industrial pollution

Sources of industrial effluent range from pulp-and-paper mills to fertiliser factories and granulation plants; abattoirs; textile manufacturing, using dyes and noxious cleaning chemicals; cloth manufacturers; various chemical industries; water treatment plants; steel industries and others. These industrial processes produce large quantities of different kinds of pollutants.

Industrial pollutants from poorly designed solid waste dumpsites find their way into the environment through direct discharge and leaching into aquifers. Studies along the Zambezi Valley found that metal and batteries dominated the industrial solid waste in Kariba, contributing 60 percent and 10 percent of the total waste, respectively. Livingstone's solid waste was predominantly food remains (73 percent) and metal (nine percent), while in Victoria Falls, paper (34 percent) and plastic (22 percent) dominated the solid waste.[14]

Major pollution effects due to industry occur in the highly urbanised parts of the Zambezi catchment area, which includes the Manyame, Kafue, and the Kwekwe/Sanyati river systems.

The Kafue River, for example, is the lifeline for 40 percent of Zambia's population, in addition to being the hub of the country's industrial activity. Industrial activities include chemical, fertiliser and textile manufacturing. More than 93,000 tonnes of industrial waste are produced annually, and most of it finds its way into the Kafue and eventually the Zambezi River.[15]

Photo: IUCN

Industrial processes produce different types of pollutants, contributing to air, land and water pollution.

Table 9.2: Waste generation by different industrial sectors in Zambia

Waste Type	Tonnes per year
Soap, stock and spent bleaching earth	15,060
Asbestos	570
Broken glass	336
Wood splints	70
Molasses and bagass	2,800
Lime sludge	490
Caked bitumen and sludge	10
Paint sludge	100
Packaging material	30
Petroleum sludge	4
Chromium sludge	350
Boiler ash	27,000
Acid slurry	46,000

Source: Banda, P.M., Pollution in the Zambezi Basin, for SARDC, 1999.

In Harare, the Mukuvisi River, which drains into the Manyame, is generally considered the most heavily polluted river system in Zimbabwe. The river flows through both industrial and residential areas of Harare, receiving waste discharges at various points. In its upper reaches, the river receives effluents from a phosphate fertiliser manufacturing plant. Most of the pollutants that find their way into the river are soluble, including pyrites cinder, aluminium sludge and fluoride-rich scrubbing alcohol.[16] During the rainy season, there is always a carryover of finely divided gypsum into the river.[17] The groundwater of the Mukuvisi catchment area is also heavily contaminated with calcium, manganese and copper, whose concentrations average 30 percent, 195 mg/kg and 133 mg/kg, respectively.[18] High levels of water contamination pose a serious threat to people and wildlife.

Sulphate concentrations have been found in excess of 1 000 mg/l in some groundwater sources in parts of Dowa District, Malawi.[19] Fluoride concentrations in excess of 3 mg/l have also been found in groundwater in some parts of Karonga and Lilongwe districts.[20]

A recent analysis of the quality of water in the Zambezi River shows that pollution in the urbanised parts of the Zambezi catchment area does not cause a significant build-up of trace metals in the river. For example, a 1974 study estimated the conductivity value of the Zambezi River during the months of April and October at 40μS per cu m and 121μS per cu m, respectively,[21] and in 1998 another study observed conductivity of 80μS per cu m in April and 100μS per cu m in October.[22] The two studies did not differ significantly in the concentrations of potassium, sodium, calcium and magnesium.

However, localised inorganic pollution is potentially a serious problem in some parts of the Zambezi Basin. For example, a study conducted in 1995 showed unacceptably high levels of lead and cadmium within the drainage basin of Lake Kariba, as well as in aquatic organisms.[23] There is circumstantial evidence to indicate that trunk paralysis in elephants at Lake Kariba could be due to lead poisoning.[24] Cadmium levels in the lake's fish have also been observed to be 10 times higher than those found in fish from temperate waters. The Cadmium, whose levels are below the World Health Organisation's permissible limits, is said to emanate from some steel manufacturing industries in the Kwekwe/Redcliff areas of Zimbabwe.

While it is undesirable to have high concentrations of trace metals in the environment, it is important to note that the same metals, in right proportions, are necessary to support both terrestrial and aquatic life. Metals such as manganese, zinc and copper are important for the physiological functions of living tissue, as well as regulating many biochemical processes.

Power generation

Power in the Zambezi Basin is currently produced through a combination of hydroelectric schemes and coal-fired conventional thermal power plants. Each of such schemes has its environmental impacts, which are closely related to the means required for generating the power. There is reason to anticipate an increase of power requirements in the basin as urban centres and industrial processes expand. A number of both thermal and hydroelectric schemes are planned to meet increasing demand due to population growth and economic development. These include the Batoka hydropower scheme and the Gokwe North thermal power project in Zimbabwe.

The generation of thermal power is a major source of air pollution within the Zambezi Basin. Major thermal power stations found within the basin include Hwange, Munyati and Harare, all located in Zimbabwe, as well as two stand-by thermal power plants located on the Copperbelt of Zambia. Tanzania also has a thermal power plant, which lies within the basin. The gases emitted by

Photo: IUCN

Thermal power plants are a major source of pollution.

the thermal power plants, which include carbon dioxide, nitrogen dioxide and sulphur dioxide, have not been quantified. However, in Harare, sulphur dioxide levels have been observed to exceed acceptable WHO limits by as much as 70 percent during certain times of the year.[25]

Outside the basin, South Africa provides some lessons to the basin countries on the pollution impacts of thermal power generation. The country's Gauteng Highveld produces the worst air pollution in southern Africa due to the high concentration of thermal power plants and other coal-burning industries. The effects of acid rain resulting from emissions from the Gauteng Highveld have not been well studied, but it is known that some of the potentially vulnerable ecosystems and areas are found within the basin.[26]

While most thermal power plants produce limited volumes of effluent, and therefore have limited impact on the aquatic environment, the high temperatures of water due to cooling of machinery used for power generation may induce low oxygen solubility, causing physiological distress for aquatic organisms.

On the other hand, air pollution affects areas outside national boundaries, threatening global environmental conditions. Some pollutants, once in the atmosphere, are converted to acidic compounds that destroy forests, acidify lakes and rivers, corrode metals and cause human health problems.

The impact of air pollution on human health in the Zambezi Basin has not been thoroughly investigated, as it is generally not considered as a major threat as water pollution.

Mining activities

Mining is a major cause of pollution in the Zambezi Basin, causing water acidification and releasing a variety of highly toxic metals such as arsenic, mercury, cadmium and lead. The disposal of solid mine waste also presents serious problems.

There is a considerable amount of mining in the Zambezi Basin, including copper mining in the Kafue subbasin in Zambia; and tin, gold and coal mining in Zimbabwe and Mozambique. Coal and copper are, however, the major minerals mined within the Zambezi Basin. Both open and underground mining methods are used in the extraction process, wasting vast tracks of land through excavations of the ground and build up of mine dumps. The Kafue River has over 90,000 ha of land under mining, 11 percent of which is occupied by solid waste dumps. The solid waste, which in 1996-97 amounted to 17 million tonnes, comprises of overburden, waste rock, tailings and smelting slag.[27]

Photo: IUCN

The disposal of solid mine waste presents serious environmental problems.

Mining causes serious environmental problems for local communities in the Zambezi Basin. The problems range from the perpetual water pollution caused by acid mine drainage to cyanide spills in gold mining and heavy metals contamination. Such contamination presents a growing threat to groundwater resources. It also poses a threat to wildlife habitat, causing among other effects, fish kills. Another environmental problem caused by mining is the destruction of the landscape by abandoned hardrock in mine sites. These abandoned hardrock mines can pose serious environmental threats to public health and safety. The extent to which mine effluents have polluted the Zambezi Basin needs to be established and appropriate mitigatory policies put in place.

Mine wastes form the largest part of the pollution problem and they fall predominantly into three categories:

- Acidic waste from groundwater seepage into the mine.
- Mining and processing waste.
- Seepage and leachates from mine tailings and slime dumps.

Examples of this type of acidification have been noted in a number of localities in Zimbabwe in the vicinity of iron mines and processing plants such as the Zimbabwe Iron and Steel Company in Redcliff.[28]

Copper mine dumps are alkaline and stable, while those of coal tend to be acidic and susceptible to spontaneous combustion. The main pollutants associated with mining include suspended particulate matter, heavy metals and chemicals used in the processing of the mineral. A recent study found as much as two percent calcium, 1.6 percent copper and 929 parts per million (ppm) cobalt in water samples from the Kafue River.[29]

The extent of mine dumps, which result in visual pollution, has not been quantified in the Zambezi Basin. Elsewhere, in South Africa, mine dumps number 450, cover some 10,000 hectares and hold 20 billion tonnes of rocky waste.[30] The smelting of the minerals also releases some gas pollutants, chief of which is sulphur dioxide. Copper mines around the Kafue River discharge about 246,000 tonnes of sulphur dioxide into the atmosphere annually.[31]

Photo: IUCN

The extent of mine dumps, which result in visual pollution, has not been quantified.

Mining also results in some noise and dust pollution. The noise emanates from the use of heavy plant and equipment, while dust pollution is common in some open cast types of mining such as quarrying. The dust pollution from quarrying can be quite serious but localised such as the case at a quarry near Chegutu in Zimbabwe. Similarly, cement factories discharge a lot of dust into the atmosphere, discolouring landscapes within the vicinity of the cement plants. This is the situation in areas surrounding two of Zimbabwe's major cement plants located in Harare and Bulawayo. Chapter 6 discusses other environmental problems due to mining.

Non-point source pollution

Agricultural and forestry activities

Pollutants from agriculture and forestry activities include nitrogen, phosphorous, insecticides, herbicides, as well as poisonous residues contained in irrigation seepage water and livestock feedlots. Such sources of pollution are difficult to control due to their diffuse nature, as well as the fact that agriculture is the lifeline for millions of people in the Zambezi Basin.

The agricultural industry takes up the greatest proportion of activities in the Zambezi Basin, requiring the extensive use of both organic and inorganic fertiliser and pesticides. DDT has been extensively used in the basin, not only for agricultural purposes, but also for human health as well through the campaign to eradicate tsetse fly during the period 1961–91. For example, 127 tonnes of DDT were used for tsetse control in the Binga, Gokwe and Hurungwe areas of the Zambezi operational front on the Zimbabwe side in 1983 alone.[32]

The tsetse control operations are probably the main reason for the current levels of DDT in the aquatic fauna of Lake Kariba.[33] An analysis of DDT and its metabolites in fish, mussels and birds in Lake Kariba showed that the pesticide bioaccumulates in living organisms.[34] Levels of the pesticide were found to be highest in bottom living species such as the mussels and benthic feeding fish, such as Manyame (or *Hunyani labeo*). High DDT levels have also been found in the breast milk of nursing mothers living around Lake Kariba, particularly those residing in the Nyamhunga Township.

One symptom of DDT poisoning is eggshell thinning in birds, leading to reduced hatching. The Fish Eagle is one such species, which is seriously affected by DDT in Lake Kariba.[35] Besides causing eggshell thinning, the following impacts were also observed in woodland birds:[36]

- Progressive population declines in some bird species in sprayed areas.
- Mortality among some bird species following spray operations.
- Fewer bird species in sprayed areas (only six in sprayed areas compared to 23 species in unsprayed areas).
- Eighty-eight percent decline in black chat population over 33 months after cessation of spraying in the sprayed area compared to 13 percent decline in unsprayed area.

Story: DDT in the Kariba District of Zimbabwe

Mary Chileshe is one of many Zambezi Basin residents on whose life DDT (dichloro-diphenyl-tricholoroethane) has left an indelible imprint.

A resident of Kariba district in Zimbabwe, she has lost two infants to the deadly chemical used since the 1970s to control tsetse fly and mosquitoes. She is now uncertain about the survival of future children and fears a third pregnancy.

"These deaths are again straining my marriage," she says.

Prof. Chris Magadza, director of the University of Zimbabwe's Lake Kariba Research Station, wonders why medical doctors mislead government by not certifying the causes of these high mortality rates as DDT. He says research has established that DDT effects were causing population declines in people and animals such as birds, bats and lizards.

"Independent studies have also shown that DDT use has resulted in a very high rate of human infant deaths in the area. This is because the mothers contain very high levels of DDT in their milk."

Medical officers at the Kariba District Hospital confirm the presence of DDT in breast milk of some women in Nyamhunga high-density suburb. Other consequences of DDT poisoning are reproductive failure, weakened resistance to illness, loss of body weight and sex imbalances in some fauna.

Dr. Rudo Sanyanga, a former researcher at the Kariba Research Station, describes the chemical as a threat to both people and animal health because it penetrates the food chain. The aquatic life gets DDT residue from the surrounding water or food which results in *kapenta* fish passing it on to its predators, among them people, crocodiles, mallard ducks, African goshawk and tiger fish.

"On entry into the aquatic environment through various pathways, these fairly persistent substances may remain within the water body unchanged for a long time, or undergo degradation to simpler compounds that may be more toxic or more persistent than the parent compounds," Sanyanga explains.

Dr. Ngoni Moyo, a biologist with the UZ, says DDT use has adverse effects on the ecosystem.

"DDT is a long-lasting chemical as its residue, if consumed by animals or human beings, accumulates in human fat, causing cancer in old age."

Tobacco farmers who grow the country's highest foreign currency-earning crop, have protested against continuing import and spraying of DDT in the tobacco growing districts of Mount Darwin, Bindura, Shamva and Centenary following deteriorating yields.

Environmental and crop organisations such as Environment 2000, the Zambezi Society and Tobacco Research Board have lobbied for the total ban of DDT.

But the cash-strapped government insists DDT is the only cheap method it can apply to kill the two insects, which cause nagana in animals and sleeping sickness and malaria among people.

"The alternative to DDT is pyrethroid which is 20 times more expensive and beyond the reach of the ministry," says Dr. Timothy Stamps, Minister of Health and Child Welfare. DDT use was banned in the West two decades ago because of its long-lasting toxicity but Zimbabwe defied the ban, arguing that the hazardous substance was the only affordable and effective means to eradicate the two insects.

By Zunde Makhuza

Photo: M Chenje

DDT, which is used to control tsetse fly and mosquitoes, is a threat to people.

- Decline in population and high nest desertion in the African Goshawk in sprayed areas. One study found DDT levels in excess of 130 ppm in the African Goshawk.[37]

A study in which DDT has been found to damage the brain leading to death in large mammals[38] has helped dispel the misconception that the pesticide does not harm large mammals. Local studies have shown negative impacts of DDT on the Ground Lizard, in which the populations of the reptile declined progressively with spray intensity.[39]

The impacts of DDT on the fish population in the Zambezi Basin have not been well studied. However, circumstantial evidence indicates the following major impacts:[40]

- Higher DDT levels in fish from sprayed areas.
- Growth performance is better in fish from unsprayed areas.
- Fish mortality is higher in sprayed areas than unsprayed areas.
- Fish live longer in unsprayed than in sprayed areas.

A curious imbalance in the ratio of males and females of the squiker was noted by another study.[41] This sex imbalance was attributed to high DDT levels, and it has since been well established that some chlorinated hydrocarbons mimic sex hormones, causing imbalances.

Since 1982, the use of DDT and other organochlorine pesticides in the Zambezi Basin has been gradually phased out, with the goal of a total ban, in line with global trends. The biodegradable pyrethroid insecticides are replacing the organochlorines, which are persistent in the environment.

Other pesticides used in the basin include organophosphates and carbamates. However, the impact of these pesticide groups on flora and fauna is not well understood. Herbicides and fungicides are also extensively used in the Zambezi catchment area. Commonly used fungicides include dusting sulphur, captan, benomyl, triforine and copper oxychloride. However, no study has been conducted to determine the levels of the herbicides and fungicides in the basin. There is, however, concern over the continued use of copper and sulphur-based compounds, which are suspected to have long-term effects on soil fertility.

Great usage of inorganic fertilisers is common throughout the Zambezi Basin. While judicious applications increase crop yields, as well as boost soil nutrient reserves, concerns have been raised that in the long run, continued use of fertilisers may cause pollution, mainly through the accumulation of nitrogen salts, which eventually may lead to eutrophication of lakes and reservoirs. A country-wide study conducted in 1985 in Zimbabwe, showed that on average, groundwater contained more than one milligramme of nitrogen per litre.[42] Although the average results were within WHO permissible limits, results from other sites, particularly disused mineshafts and areas around septic tanks, pit latrines and cattle pens, showed high levels of contamination.

The most serious effect of both agriculture and forestry is soil erosion. In Zimbabwe, soil loss from agricultural land ranges from 1–20 tonnes per hectare per year.[43] Soil losses due to erosion contribute to the sediment load carried by the rivers, increasing the siltation of reservoirs. Another side effect of agriculture and forestry that is related to soil erosion is the leaching of plant nutrients and other chemicals from the catchment areas. It has been observed elsewhere that nutrient loading through leaching from catchment areas may exceed nutrient loading from the more obvious point sources in urban and industrial areas.[44]

Other deleterious effects stemming from agricultural activities can arise as a result of the use of municipal wastewater to irrigate pastureland. The practice is becoming widespread in Zimbabwe as a means of tertiary treatment of sewage effluent. There is a possible accumulation of trace metals and toxic substances through the process. However, the build-up of such metals and toxins may not attain serious proportions in the near future, but it is necessary to constantly monitor the levels of the substances in flora and fauna if the quality of agricultural produce is to be maintained.

Table 9.3: Pesticide levels in sediments from selected areas (mg/g)

	Charara	Sanyati	Ume	Sengwa	Ruziruhuru	Gachekache	Nyaodza
DDT	117.7	12.3	-	64.7	16.6	20.4	30.3
Dieldrin	1.67	-	-	51.5	32.7	20.5	63.7
Endosulfan	24.4	-	1.91	16.1	25.5	12.0	167.8
Heptachlor	-	0.88	4.90	40.02	2.66	-	3.59

Source: Zaranyika, M.F., Mambo, E. and Makhubalo, J.M., *Organochlorine Pesticide Residues in the Sediments of Selected River Bays in Lake Kariba, Zimbabwe*, University of Zimbabwe, Harare, undated.

Photo: M Chenje

The run off from urban areas contribute to high concentrations of lead from vehicle exhaust fumes.

Effluent discharge and run off from forestry and agro-based industries can also have deleterious pollution effects on the aquatic environment. The effects of pulp-and-paper processing on water quality have been well documented on the Manyame River where there has been a complete destruction of benthic habitats downstream of Lake Chivero.[45]

Natural toxicity

Nutrient loading may occur naturally from the vegetation washed by rainwater. Phosphorous tends to be adsorbed onto silt particles, and as such phosphorous particles tend to be locked in the sediments except if anaerobic conditions occur and the formation of acidic compounds redissolves the phosphorous.

Natural toxicity also stems from natural phenomena such as vulcanicity. Although the Zambezi Basin has not witnessed any form of acute vulcanicity in recent years, there are some dormant volcanic features such as the Great Dyke in Zimbabwe where some mineral nutrients are found in high levels.

Storm water run off

Urban run off can easily pollute water sources if towns are not well planned to contain drainage. This drainage water is difficult to manage since it emanates from non-point sources.

The run off from towns and cities contributes high concentrations of lead (from vehicle exhaust fumes), phosphorous and nitrogen (from decaying matter), and spillages from sewage works due to stormwater ingress into sewers. Such pollution could be high for Harare, Chitungwiza and Norton, which sit on the Manyame catchment area, as well as the Copperbelt that lies on the Kafue sub-basin. Eutrophication is further worsened by the street run off from built-up areas that washes into water bodies. In a basin that is predominantly rural, the activities still have a profound effect on the quality of water in impoundments.

Leachate from landfills

A landfill can be any area of land used to deposit mainly solid wastes. Landfills constitute important potential sources of groundwater pollution since they may contain organic micro-pollutants, heavy metals and salts. Leachate from such landfills requires large volumes of water, which can come from precipitation or even moisture in the refuse.

Most municipal wastes contain little hazardous material, but solid waste of industrial origin may contain toxic materials such as metals and organic pollutants. Methane is also present in groundwater close to waste disposal sites such as the Golden Quarry in Harare where oils and grease are dumped.

Gold panning

The proliferation of riverbed and alluvial gold panning has resulted in serious damage to the basin's rivers, exposing water supplies to serious pollution. About 57 percent of the gold panners, numbering hundreds of thousands throughout the Zambezi Basin use the most destructive horizontal tunneling method of extracting the mineral.[46] During heavy rains, the riverbanks often collapse, resulting in loss of vegetation, erosion and siltation. Chemicals such as mercury are often used adjacent to the water bodies, contaminating rivers and streams. Water bodies contaminated with chemicals are not only a health hazard to people who drink the water, but can also be consumed through the food chain system.

Siltation of rivers from gold panning can reduce the storage capacity, as well as the operational life of water bodies. A 1992 report suggests that half of Zimbabwe's dams are 50 percent silted and that for some rivers, it was no longer economically viable to construct new dams.[47]

Transport

Road, air and water transport causes various forms of pollution, including noise, air and water pollution. Noise pollution from aircraft is mainly restricted to airport termini, which in most areas of the Zambezi Basin tend to be located far away from residential areas. Noise from road transport is worsening due to the growth in the number of vehicles and the absence of regulations that limit the level of noise generated by cars. The regulations are also silent on the permissible levels and types of fumes, which cars should emit.

EFFECTS OF POLLUTION

Eutrophication and water weeds

One of the major consequences of water pollution is the worsening of eutrophication of lakes and river impoundments.

Eutrophication of water bodies causes the proliferation of aquatic weeds. Until recently, the Kariba weed, was the most widespread within the Zambezi Basin. The water hyacinth has since become the most problematic weed in the basin, and it is found in nearly every water body within the Zambezi catchment area. The water hyacinth weed invaded Lake Kariba five years ago. Other aquatic weeds found within the basin include the water fern and water lettuce.

The water hyacinth is a deceptively attractive plant, with shiny green leaves and delicately transparent lavender flowers. The weed can cause enormous ecological and economic damage when it is not controlled.

Due to its high reproductive capacity, the water hyacinth must be removed considerably faster than its rate of multiplication. Various options for controlling the weed have been tried in the basin, including manual, chemical and biological control methods. All methods have met with limited success.

Box 9.1: Characteristics of water bodies experiencing cultural eutrophication

Biological factors

a Primary productivity: usually much higher than in unpolluted water and is manifest as extensive algal blooms.

b Diversity of primary producers: initially green algae increase, but blue-green algae rapidly become dominant and produce toxins. Similarly, macrophytes (e.g. reed maces) respond well initially but due to increased turbidity decline in diversity as eutrophication proceeds.

c Higher trophic level productivity decreases.

d Higher trophic level diversity decreases. The species of invertebrates, which tolerate more extreme conditions, increase in numbers. Fish are also adversely affected and populations are dominated by surface dwelling coarse fish such as pike and perch.

Chemical factors

a Anoxic conditions develop, especially at night when algae are not photosynthesising. Thus seasonal and diurnal patterns of oxygen availability occur. The decay of algal blooms also produces anoxic.

b Salt content of water increases, further restricting floral and faunal diversity.

Physical factors

a Mean depth of water body decreases.

b Turbidity increases, as sediment input increases, and restricts the depth of light penetration, which can become a limiting factor for photosynthesis.

Water uses

a Water quality for domestic and industrial uses deteriorates.

b Amenity use of water is severely impaired due to the production of odours and loss of floral and faunal attractions.

Source: Hubert Meena, Draft chapter contribution for SARDC, 1999.

Photo: M Chenje

The water hyacinth can cause enormous ecological and economic damage when it is not controlled, but it also provides cover for some animals, for example, crocodiles.

Health effects

Among the organisms that are found within water bodies, pathogens pose the greatest threat to public health, especially when the receiving water is used for domestic, recreational or agricultural purposes.

The presence of faecal indicator organisms in a sample of water shows that intestinal pathogens could also be present and that the water supply may pose a risk to human health.[48] Common water-borne diseases in the basin include dysentery, cholera, malaria and bilharzia. Cholera and dysentery causing pathogens, like for other water-borne diseases, are often found in water contaminated with poorly treated human waste and sewage.

Air pollution also negatively affects human health. Most of the gaseous pollutants that are released into the air are poisonous to human health. Ozone depletion, which results from the introduction of ozone depleting substances into the atmosphere, puts people at the risk of skin cancer diseases and eye cataracts.

Ecological effects

Pollutants degrade water quality and harm aquatic life by interfering with photosynthesis, respiration, growth, and reproduction. For example, pollution of water bodies can impair reproduction in fish, retard their growth or even kill them. The late March to early April 1996 fish deaths in

Case study: Kafue's toxic sugar cane

Kafue is a small industrial town near Lusaka. It is home to several industries, such as Nitrogen Chemicals of Zambia (NCZ), a fertiliser manufacturing plant. Everyday NCZ discharges huge quantities of toxic residue into its treatment lagoons.

The lagoons are now the hub of a thriving sugar cane growing business for scores of poor residents of outlying shanty compounds. The nitrate-rich waste nourishes the sugarcane, making it grow much faster. The sugar cane grows in tufts of up to 10 canes and each cane fetches on average K100 (about 14 US cents).

Investigations by health experts and chemists indicate that the lagoon sugar cane is badly contaminated and is a threat to human health. A principal chemist at the NCZ, Wart Jude, says toxic materials in the cane accumulate in the body over time. Gasper Sichalwe, district health inspector for Kafue, says laboratory tests show traces of toxic substances harmful to human health in the lagoon sugar.

John Malunda (20), a resident of Soloboni Compound, about one kilometre from the lagoons, is one of those affected. Says Malunda: "It all started... after I ate a lot of sugarcane harvested from the lagoons."

Malunda has since given up selling the cane after discovering the "evils" associated with it. But for others, like Jacob Banda, a retired civil servant, it's business as usual. He has been growing sugar cane for the past five years and has no plans to retire from this business. "Survival comes first," he says.

A father of five, Banda confesses that he is aware that the sugarcane is not safe for human consumption. "But times are hard. Being close to the lagoons, I joined scores of other people in growing sugarcane for sale. It grows fast and customers come from as far as Lusaka to buy the cane in bulk. I make good money which helps me to look after my family."

One of Mulanda's regular customers is Mary Chuma, a 35-year-old mother of five, who travels to Kafue from Lusaka twice a week to buy sugar cane for resale at Lusaka's Soweto market. "It's good business for me. I have heard that the lagoons are full of poisons, but my interest is to make money."

But like so many other men and women who make a living out of the cane, she finds it difficult to believe that the sugarcane is not fit for human consumption. "It tastes just like any other sugarcane," she said matter-of-factly with a bundle of sugarcane delicately balanced on her head.

Efforts to fence off the lagoon from the public have failed. As soon as the barbed wire fencing and warning signs are erected, they are pulled down.

The vast size of the lagoons also makes it difficult to monitor, especially for the handful of security guards, some of whom have been accused of being in league with sugarcane growers and hawkers.

So fierce is the battle for survival that each time discharge pipes from the factory are closed to dry out the lagoons, people from the surrounding compounds puncture the pipes to let the waste water out and nourish the cane.

While most sugarcane growers and hawkers might be aware of the dangers, the same cannot be said of the consumers many of whom either don't care or don't seem to accept how cane that tastes so sweet can be said to be poisonous. "I eat sugar cane once in a while. I don't suspect or feel that there's anything wrong with it," said one man who did not want to be named. Nevertheless, the dangers are real.

For as long as Zambia's social and economic problems remain, and unemployment and poverty persist, Kafue cane will continue to endanger the lives of thousands of Zambians.

Source: Energy and Environment, ZERO, Harare, Zimbabwe March 1995

Lake Chivero illustrate the effect of pollution on aquatic organisms. The fish, mainly the Greenhead Bream, died from high levels of ammonia compounded by low levels of oxygen.[49]

Oils and grease released into water bodies can also suffocate aquatic organisms. Heavy metals, on the other hand, are toxic to aquatic life and can also contaminate groundwater.

Pollution also threatens fishery grounds, recreation sites and avenues of transport, in addition to many other practical and aesthetic values. The water hyacinth for example, threatens to choke the river system, making it unusable for transport, fishing, recreation and other activities.

Nutrient levels in rivers have been showing a steady increase over the years due to high rates of urbanisation, as well as growth in the agricultural sector. For example, nitrate levels in the Sebakwe River, which flows into the Zambezi through the industrial region of Kwekwe/ Redcliff, has been experiencing a steady rise in the levels of nitrates.

Acid rain, which results from air pollution, causes problems in soils in various ways. The acidity leads to leaching of some nutrients such as calcium, magnesium and potassium; increases solubility of others such as aluminium, iron and manganese; and mobilises toxic metal ions such as lead and nickel. Acidity also inhibits decomposition of organic matter by bacteria and fungi.

Climate change

One of the greatest impacts of atmospheric pollution is climate change, which is threatening not only the Zambezi Basin but the whole world. Anticipated changes in climate will mainly be manifested through a rise in temperature, as well as an increase in the incidence of droughts and floods.

The rise in the global average temperature is mainly attributed to the increase in the level of greenhouse gases, which include carbon dioxide, methane, dinitrogen oxide, ozone and chlorofluorocarbons (CFCs), all of which are emitted by some industries found in the Zambezi Basin.

The impact of global warming so far is not well known, although it is believed that the density of stomata in plant cells has increased by 40 percent during the past 200 years in response to high carbon dioxide levels.[50] Related to climate change is the depletion of the ozone layer by atmospheric pollutants such as CFCs and halons. Major ozone depleting substances found in the basin include methyl bromide, which is widely used as a fumigant in tobacco farming in Zimbabwe, and the CFCs, which are used in the foam, rubber and refrigerant manufacturing industries.

Box 9.2: Global warming crisis

Failure to control the emission of greenhouse gases like carbon dioxide (CO_2) will contribute to a temperature rise of 1.0–3.5°C by the year 2100.

According to the recently launched United Nations Environment Programme (UNEP) Global Environment Outlook 2000 (GEO-2000), the most authoritative assessment ever of the environmental crisis facing humanity in the new millennium, the grave consequences of this rise have been well documented.

These include a rise in sea levels which will displace millions of people in low-lying delta areas and on island states, reduced agricultural production in the tropics and sub-tropics, unpredictability in the supply of freshwater, the reintroduction of serious diseases like malaria to Europe and the wholesale loss of important ecosystems and biodiversity.

In addition, the report states that there is little doubt that climatic change is a major contributory factor in the increase in natural disasters.

The UN estimates that three million people perished as a result of natural disasters in the last three decades. According to the Munich Reinsurance Company, global economic losses during 1997 and 1998 reached $120,000 million — eight times higher than the 1960s.

As a result of the UN-sponsored Kyoto Protocol, efforts are underway to slow down CO_2 emissions though there is little sign that these are sufficiently far-reaching to have any major impact on the problem. In 1996, CO_2 emissions hit a record high of 23,900 million tonnes.

Source: UNEP., Global Environment Outlook, Nairobi, 1999

MANAGEMENT OF POLLUTION

The Zambezi River Basin is witnessing worsening pollution problems, including increased levels of uncontrolled waste dumping, mainly in urban and tourist areas. While the basin countries have made great strides in ensuring access to safe drinking water to many, incessant droughts and high population growth rates are threatening to reverse such gains. Indoor air pollution from heating and cooking activities is worsening, particularly in the rural areas, due to lack of cheap alternatives to firewood and cow-dung.

The basin is also witnessing growing pollution problems resulting from industrial and energy-related pollution, deforestation and overuse of natural resources. Other issues of concern with respect to pollution and waste management are:

- Proper management of waste is not a priority by both local and central governments of the basin when compared to issues of economic development, housing and employment.
- There is a general lack of awareness regarding the impact of waste on the environment and human health in the basin.
- Legislation and regulations relevant to pollution and waste management are in most cases inappropriate, incomplete or diffuse.
- There is inadequate enforcement of existing regulations, and this is worsened by limited monitoring capacity.
- Existing industrial processes are inefficient, resulting in excessive waste generation and inordinate resource and energy consumption.
- Small to medium-scale enterprises, which do not have adequate finances and technical know-how to manage their pollution and waste, are set to worsen pollution and waste management problems in the Zambezi Basin.

Actions and challenges to control pollution

Point source pollution can easily be measured and controlled while non-point sources are difficult to monitor and control. It is crucial that the quality of the effluent discharged, directly or indirectly, into any water body be strictly controlled by regulations, which prescribe a comparatively high standard of purity. In addition, a system of issuing permits through governments or basin water boards or authorities may help in keeping the levels of pollution of both water and air under control.

A number of instruments can be put in place in most of the riparian states to control the levels of land, water and air pollution. Among such instruments are measures to monitor the levels of pollution, public education and awareness campaigns, environmental production standards, legal policies, waste treatment requirements, and provision of disincentives to pollute.

Pollution monitoring

Monitoring of levels of pollution requires a close check up of the following:

- Substances entering the environment, noting their quantity and quality, sources and distribution.
- Effects of the substances on the environment.
- Variation in concentration and effects of the pollutants, paying attention to the causes of the changes.
- What can be done and at what cost to ameliorate the environmental effects of the released substances.

Pollution control and prevention consists of systems of measurement, criteria, standards, laws and regulations, which are directed at the sources and causes of the various forms of pollution. The systems also include water quality monitoring, which is a pattern of health care delivery designed to control all factors in the human physical environment that may cause deleterious effects on physical, mental and social human wellbeing. As such water quality monitoring is undertaken in order to check if the quality of the water meets the required regulations with respect to human health and the environment.[51]

Pollution monitoring is increasingly gaining prominence in the basin with countries such as Zambia and Zimbabwe undertaking legal reforms to strengthen monitoring. Both countries have set up units responsible for the physical inspection of all sources of pollution.

Photo: M Kullberg

The increase in motor vehicles has exacerbated air pollution problems, particularly in urban areas.

Figure 9.1: A generalised pollution monitoring pathway

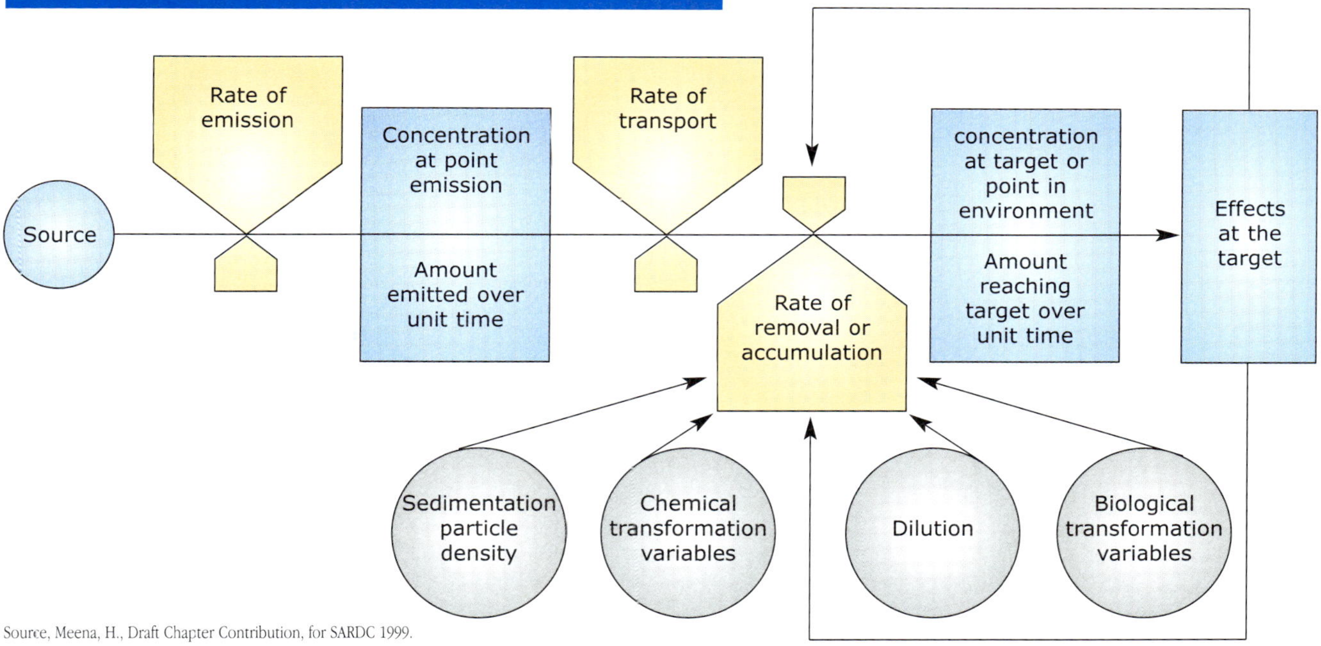

Source, Meena, H., Draft Chapter Contribution, for SARDC 1999.

Through the Environmental Council of Zambia, inspectors physically check all sources of pollution. Figure 9.2 shows a pathway that can be adopted for effective environmental pollution monitoring in the basin. Generally, water pollution is detected by either a complainant or from results of the routine monitoring and surveillance programmes.

Education and awareness campaigns

The level of environmental awareness is strongly gaining momentum in the basin courtesy of global initiatives such as the World Water Day and World Environment Day, as well as national initiatives such as the National Tree Planting Days in Zimbabwe and Malawi. Both individuals and corporate organisations are taking heed of such campaigns. For example, in commemorating the 1997 World Water Day, the Zambia Sugar Company announced its plans to put in place mitigation measures against pollution effects that could result from factory effluent, solid waste and gaseous emissions discharged into the environment, particularly the Kafue River.

Non-Governmental Organisations (NGOs) and private citizens are central in spearheading measures to reduce the levels of pollution. The private citizens, for example, has dual roles to play:

- Control their own pollution.
- Deal with the effects of pollution on their wellbeing.

Legislation

Several pieces of legislation covering water and air pollution are in place throughout the Zambezi Basin, despite problems of poor enforcement and weak institutional set-up. Other laws are so outdated that they have now become obsolete. Zimbabwe's Atmospheric Pollution Act (1971), for example, allows industries that were in existence before the promulgation of the act to continue using outdated equipment even if the industries are major polluters.

This legal loophole has seen Wankie, which mines coal, pollute the air over many years without prosecution.[52] Elsewhere, Zambia's Air Emission Regulation Statutory Instrument No.141 of 1996 sets out limits for mainly sulphur dioxide, dust and heavy metals, ignoring other gaseous pollutants, particularly greenhouse gases. The effectiveness of legislation against atmospheric pollution is further weakened by the lack of monitoring equipment, as well as the paltry fines imposed, making it cheaper for companies and individuals to pollute and pay fines.

Water pollution control is fairly covered by national legislation, including Zimbabwe's Water Act (1998) and Zambia's Effluent or Wastewater Regulation (1993). The Zambian legislation sets out limits for effluent quality based on bacteriological, chemical and physical parameters.

Other important pieces of legislation as far as pollution is concerned, include those relating to pesticides and

Table 9.4: Environmental laws in the basin states

COUNTRY	SECTOR/AREA	LAWS
Angola[1]	Water Resources and pollution	Decree of 1 December 1892
		Decree Law No. 39193 of 2 May 1953
	Fisheries and Wildlife	Decree Law No. 40040 of 20 January 1955
	Water Resources	Decree Law No. 46186 of 11 February 1965
	Land and Water Resources	Legislative Edict No. 3574 of 28 August 1965
Botswana[2]	Agriculture	Agricultural Resources (Conservation) Act
		• *Agricultural Resources (Grapple Plant) Regulations*
	Water Resources	Water Act
		• *Water Regulations*
		Aquatic Weeds (Control) Act
		• *Boat (Import, Registration and Movement) Regulations*
		• *Declaration of Infested Waters Order*
	Atmosphere	Atmospheric Pollution (Prevention) Act
		• *Atmospheric Pollution (Prevention) (Declaration of Controlled Areas) Order*
		• *Declaration of Controlled Areas (Selebi-Phikwe) Order*
		• *Declaration of Controlled Areas Order*
	Health	Factories Act
		• *Factories Act (Building Operations and Works of Engineering) (Safety and Health) Regulations*
		• *Notification of Industrial Diseases Regulations*
		• *Sanitary Accommodation Regulations*
		Public Health Act
		• *Public Health Regulations*
	Mines, Quarries, Works	Mines, Quarries, Works and Machinery Act
		• *Mines, Quarries, Works and Machinery Regulations*
	Tourism	Tourism Act
		• *Tourism Regulations*
Malawi[3]	Environment management	Environment Management Act (EMA), 1996
	Land use and management	The Land Act, 1965
	Fisheries	Fisheries Management and Conservation Act, 1997
	Forestry	The Forestry Act, 1997
	Wildlife	The National Parks and Wildlife Act, 1992
	Water	The Water Resources Act, 1969
	Waste Management and Sanitation	There is no principal statute but the EMA (1996) gives power to promulgate regulations on hazardous waste to the Minister
	Air and Noise pollution	There is no principal statute
	Toxic substances and chemicals	No principal statute. But there is a proliferation of institutions claiming responsibility over this sector
	Transboundary Water Resources	No statute, but there is a SADC Protocol on Shared Watercourse Systems
Mozambique[4]	Land	Land Act, 1979
	Water Resources	Water Act, 1991
		• *Foreign Fishing Boats Regulations, 1978*
	Mining	Mining Act, 1986
	Health/Pollution	• *Pesticides Regulations, 1987*

COUNTRY	SECTOR/AREA	LAWS
Namibia	Land/Water	Water Act, 1956 Minerals (Prospecting and Mining) Act, 1992 Soil Conservation Act, 1969 Agriculture (Commercial) Land Reform Act, 1995
	Pollution	Public Health Act, 1919 Agricultural Pests Act, 1973 Atmospheric Pollution Prevention Ordinance, 1976
Tanzania[5]	Fisheries	Fisheries Act, 1970 • *Fisheries (Explosives, Poison, and Water Pollution) Regulations, 1982*
	Water Resources	Water Utilisation (Control and Regulation) Act, 1974 Water Utilisation (Control and Regulation) (Amendment) Act, 1981 Urban Water Supply Act, 1981
	Mining	Mining Act, 1979
	Petroleum	Petroleum (Exploration and Production) Act, 1980 • *Petroleum (Exploration and Production) Transitional Rules, 1982*
	Health	Public Health (Sewerage and Drainage) Ordinance CAP 336 Merchant Shipping Act, 1967 Protection from Radiation Act, 1983
Zambia[6]	Water Resources	Water Act, Chapter 312
	Health	Public Health Act, 1930 Local Government Act No. 22 of 1991 • *Local Administration (Trade Effluent) Regulations No. 161 Statutory Instrument of 1985* Factories Act, 1967 • *Construction (Safety and Health) Regulations*
	Land	Action for Smoke Damage (Prohibition) Act, 1959
	Land/Water Resources	Noxious Weeds Act No. 59 of 1993
	Environment	Environmental Protection and Pollution Control Act No. 12 of 1990 Petroleum (Exploration and Production) Act of 1985 The Explosives Act, 1974
Zimbabwe[7]	Natural Resources	Natural Resources Act
	Pollution/Health	Mines and Minerals Act
	Atmosphere	Atmospheric Pollution Prevention Act
	Water Resources	Water Act (1998)
	Land/Water	Fertiliser, Farm Feeds and Remedies Act Hazardous Substances and Articles Act Plant Pests and Diseases Act Noxious Weeds Act

Note: For sources of this information, see footnotes. It will be apparent that laws and regulations on pollution in particular and the environment in general, fall under different sectors in each of the countries and there is no central authority to govern environmental issues alone. As such these laws are repetitive.

1 IUCN., *Angola: Environment Status Quo Assessment Report-Main Report,* IUCN ROSA, Harare. 1992 Note that these consititute environmental legislation enforced during the colonial period.
2 Government of Botswana., *Executive Summary of the National Conservation Strategy* (NCS) Action Plan Consultancy Report, NCS/A, Gaborone, March 1998
3 Malawi Government., *State of Environment Report for Malawi 1998,* Environmental Affairs Department, Lilongwe, 1998
4 Government of Mozambique., *Country Report for UNCED 1992,* Environment Division, Maputo, 1991
5 Khan, M.R. and H.J. Gijzen (eds.), *Environmental Pollution and its Management in East Africa,* University of Dar es Salaam, 1989
6 CIDA/NORAD, *State of Environment Report 1994: Zambia,* Environmental Council of Zambia, Lusaka, 1994
7 Chenje, M., L. Sola, and D. Paleczny (eds.), *The State of Zimbabwe's Environment 1998,* Government of the Republic of Zimbabwe, Ministry of Mines, Environment and Tourism, Harare, 1998

Box 9.3: Provisions from Zimbabwe's new water law on pollution

Zimbabwe's new water law carries stringent penalties for pollution, with fines of up to US$13,000. The 1998 Water Act also carries the following principles:

- The release of any matter into natural watercourses should be controlled through the issue of permits.
- The permit will specify the monitoring and reporting requirements of the discharges.
- A polluter may be required to remedy the effects of the pollution.
- Levies raised from the discharge of poor quality effluent will be used for environmental restoration and research.
- Local authorities have delegated powers to control pollution.

Source: Government of Zimbabwe., Water Act, Harare, Zimbabwe 1998

toxic substances, as well as waste management. Such laws are in place throughout the basin states. It is important that these states collectively harmonise existing national legislation governing pollution.

Waste management

A general way of dealing with waste is waste reduction, which should have elimination as its first goal and reuse as its second. Waste reduction involves the following practices:

- Composting the waste at points of origin.
- Re-using or repairing rather than throwing things away.
- Recycling of waste, which can contribute to a much larger segment of the waste stream being diverted from traditional disposal means.

Photo: M Chenje

Solid waste management has become a mounting problem in the urban areas of the basin.

Waste reduction requires shifts in business culture. Manufacturers, at each stage of the design and manufacturing chain must review their processes and consider the implications that those processes have in as far as waste generation is concerned. This should be done while focusing on ways to reduce waste which often bring a further benefit by exposing hidden inefficiencies in manufacturing processes as well as reducing cost of waste disposal.

There is also need to adopt new and appropriate compaction technologies. For example, it is possible to compact 30 cu m of loose waste to one cubic metre of dense waste, depending on the waste's contents. Such small amounts of waste are easy to dispose of at landfills.

Disincentives to pollute

A number of disincentives to pollute have been introduced in the basin countries to discourage pollution by mainly industrial concerns. The disincentives include the polluter pays principle (PPP), which has not been very effective due to low penalties that have not been periodically reviewed and updated. As a result, industries have continued to pollute because they can afford the paltry penalties.

Strategies to abate pollution effects have often been derailed by the need to balance economic development with adequate provision of social services. For example, calls to install scrubbers to reduce pollution from the Hwange thermal plant were rejected on the grounds that the same money could be used to electrify surrounding areas without electricity. The idea to reject the scrubbers was supported by the argument that the cost for scrubbers for Hwange was enough to provide electricity to a settlement as big as Soweto in South Africa.

As far as cement factories are concerned, some investment in equipment to prevent dust pollution has been made, particularly for Zimbabwe's two major cement plants in Harare and Bulawayo. The problem with the equipment is that it consumes a lot of electricity, making the cost of cement very expensive.

LINKAGES TO OTHER CHAPTERS

Chapter 1: Regional Overview: People and Environment
Being the most industrialised region in the Southern African Development Community outside South Africa, pollution in the Zambezi Basin poses one of the major threats to human and environmental health. Stringent pollution laws and enforcement mechanisms are therefore required to curb environmental dearth.

Chapter 2: Physical Features and Climate
Waste dumps constitute the biggest contribution of mining to pollution, resulting in significant changes to the landscape. The dumps, which in essence are some form of visual pollution, preclude the affected land from any other form of landuse as they are expensive to reclaim.

Chapter 3: Water and Wetland Resources
A major consequence of water pollution is the worsening eutrophication of lakes and other river impoundments. Eutrophication causes the proliferation of aquatic weeds. Until recently, Kariba weed was the most widespread weed within the basin but the water hyacinth has since become the most problematic weed and is found in nearly every water body in the catchment.

Chapter 4: Biological Resources and Diversity
Pollution is major threat to the richness of flora and fauna. The sensitivity of such plant and animal life to pollutants is variable. As such, pollution tends to selectively kill certain species, causing an imbalance in the food chain, resulting in the malfunctioning of ecosystems.

Chapter 5: Agriculture
Agriculture is the main source of pollutants in the form of herbicides, fertilisers, pesticides and other chemicals that find their way into river systems and other water bodies. Pollution from agriculture poses threats to human health and aquatic life.

Chapter 6: Industry
Industrial pollution is varied, depending upon the types and amounts that are discharged. Different types of industrial pollution include organic, bacterial and pollution due to the accumulation of nutritive salts.

Chapter 7: Energy
The generation of thermal energy is a major source of pollution within the Zambezi Basin. Gases that are emitted by thermal power generation include carbon dioxide, nitrogen dioxide and sulphur dioxide. The emitted gases often combine with falling rain to form acid rain.

Chapter 8: Tourism
The main tourist attractions in the Zambezi Basin are wildlife and scenic areas such as the Victoria Falls. The aesthetic value of such scenic sites is often degraded by pollution. Dust pollution, for example, causes the discolouration of vegetation, buildings and scenic sites, causing them to lose their appeal to tourists.

Chapter 10: Poverty
It is the poor who usually suffer most from pollution. The poor are more likely to work in polluted environments, and are also more likely to suffer from inadequate nutrition, which exacerbates the toxic effects of pollution. Industrial workers and those in the mining sector are often exposed to noxious gases and poisonous chemicals, while workers on commercial farms handle poisonous chemicals.

Chapter 11: Gender and Women's Roles
While pollution does not affect humanity on the grounds of their gender, it is common knowledge that certain members of society tend to be more exposed to pollution. By virtue of their traditional role of cooking food for their families, women and children tend to be more affected by indoor pollution than men. In addition, women tend to pass on bio-accumulated pollutants to their children.

Chapter 12: Environmental Management and Regional Cooperation
Being transboundary in nature, pollution tends to create opportunities for both regional conflict and cooperation. Basin states need to harmonise pollution control legislation as pollutants respect no boundaries.

Chapter 13: Trends and Scenarios
Pollution is a worsening problem in the Zambezi Basin, which can lead to disastrous consequences if the business as usual approach is maintained. There is therefore, need for timely interventions through policies, strategies and laws to reduce the levels of pollution.

ENDNOTES

1 World Bank, *Development and Environment: World Development Report,* Oxford University Press, Oxford and New York, 1992
2 Sanyanga, R.A. and H. Masundire, *Waste Management in Major Population Centres in the Zambezi Valley*, Unpublished Project Report CIDA-ECEP 32PP, 1998
3 Bagg, W.K. "River Pollution in Harare", *Geographical Education Magazine*, Vol.15, No.1, 7-15, 1992
4 Mtetwa, S., personal communication, 1999
5 ibid
6 op cit 2
7 ibid
8 ibid
9 Magadza, C.H.D. and Dhlomo, "Wet Season Incidence of Coliform Bacteria in Lake Kariba Inshore Waters in the Kariba Town Area", *Lakes and Reservoirs Research and Management,* 2:89-91, 1996
10 ibid
11 Moyo, N.A.G. Mazvimavi, D. Mtetwa, S. Dube, B.N. Mhlanga, A.T. and W. Mhlanga, "Water Resources", In: *The State of Zimbabwe's Environment – 1998,* Harare, 1998
12 UNESCO/UNEP/WHO, *Water Quality Assessments*, University Press, Cambridge, 1991
13 Feresu, S.B. and J. van Sickle., "Coliform as a Measure of Sewage Contamination on the River Zambezi", *Journal of Applied Bacteriology*, Vol. 68, 97-403, 1989
14 Moyo, N.A.G. "Pollution in the Zambezi Basin", For SARDC, 1999
15 Banda, Pollution in the Zambezi Basin, for SARDC, 1999
16 Mathuthu, A.S. Zaranyika, F.M. and Jonnalagadda, "Monitoring of Water Quality in Upper Mukuvisi River in Harare, Zimbabwe", *Environmental International*, Vol.19, p.51-61, 1993
17 Bagg, W.K. "River Pollution in Harare", In: Heath, R.A. (ed.), *Geographical Education Magazine*, Vol.15, No.1, 7-15, 1992
18 Nyamangara, J. and J. Mzezewa, "Maize Growth and Nutrient Uptake in a Zimbabwean Red Clay Soil amended with Anaerobically Digested Sewage Sludge", *Journal of Applied Science in Southern Africa*, Vol.2, No.2, 1996
19 Geake, A.K. et al; "Groundwater Recharge and Pollution Mechanisms in Urban Aquifers of Arid Regions", Hydrogeological Report", 86/11, BGS, Wallingford, 1986
20 ibid
21 op cit 2
22 ibid
23 Berg, H. Kiibus, H. and N. Kautsky, "Water, Air and Soil Pollution", *Ambio,* 83:237-252, 1995
24 ibid
25 Ngara, T. and A. Rukobo; Environmental Impacts of the 1991/92 Drought in Zimbabwe: An Extreme Event. Radix Consultants (Pvt) Limited, Harare, Zimbabwe, 1999
26 *State of the Environment in Southern Africa,* SADC/IUCN/SARDC, Maseru/Harare, 1994
27 Masinja, J.H., *Proceedings of a Conference on Man and Biodiversity*, Ministry of Science, Technology and Vocational Training/UNESCO, 1998
28 Mtetwa, S., Pollution in the Zambezi Basin, for SARDC, 1999
29 Isakson, S and A. Safuestad, *Environmental Effects from the Mining Industry on the Sediment Quality in the Kafue River, Zambia,* Lulea University of Technology, Sweden, 1996
30 Durning, A.B. "Apartheid's Environmental Toll", *Worldwatch Paper*, No.5, Worldwatch Institute, Washington DC, 1990
31 CEEZ/ECZ, *Enabling Activities for the Preparation of Initial Communications Under the Framework Convention on Climate Change: Inventories of Anthropogenic Greenhouse Gas Emission and Removals in Zambia*, Lusaka, 1998
32 Zaranyika, M.F. Mambo, E. and J.M. Makhubalo, *Organochlorine Pesticide Residues in the Sediments of Selected River Bays in Lake Kariba, Zimbabwe*, University of Zimbabwe's Chemistry Department, Harare, undated
33 Berg et al; DDT and other Insecticides in the Lake Kariba Ecosystem, Zimbabwe, *Ambio*, 21 (7):444-450, 1992
34 ibid
35 Douthwaite, B. Article in: *Zimbabwe Wildlife*, Dick Pitman, (ed.), p.13, 1991
36 Magadza, C.H.D., "DDT in the Tropics", *A Review of the NRI Report on Impacts of DDT in the Zambezi Valley, Zimbabwe,* 1996
37 Hartely, R.R. and T.J. Douthwaite, "Effects of DDT Treatments Applied for Tsetse-fly Control on the African Goshawk in Northwest Zimbabwe", *African Journal of Ecology*, 32:265-272, 1994
38 Anjum, F.J. and M.K. Siddiqui, "In Vitro Inhibition of Fish Brain Ca2+-atpase by Monocrotophos, Dimethoate, Diazinon and DDT", *Indian Journal of Experimental Biology,* 28(5):488-489, 1990
39 op. cit 36
40 ibid
41 ibid
42 Interconsult S/A and Norad, *National Master Plan for Rural Water Supply and Sanitation*, Draft Report, Vol.2, 1985
43 Ellwell, H.A. and M.A. Stocking, "Loss of Soil Nutrients by Sheet Erosion", *The Zimbabwe Science News*, Vol.22, No.7/8, 1988
44 Toerien, D.F., "A Review of Eutrophication and Guidelines for its Control in South Africa", *National for Water Research Spec. Rpt*, No.WAT.48, *CSIR,* 1977
45 Marshall, B.E., "Some Effects of Organic Pollution on Benthic Fauna", *Rhod. Sci. News*, Vol.6:142, 1972
46 Milne, G.R. and D.A. Marongwe, *Small-scale Gold Panning in Zimbabwe*, Ministry of Environment and Tourism, Harare, 1993
47 op. cit 2
48 Anonymous, "The bacteriological Examination of Drinking Water Supplies 1982: Reports on Public Health and Medical Subjects, No.71, *Methods for the Examination of Waters and Associated Materials*, p.1-91, London: HMSO, 1982
49 Moyo, N.A.G., "The Poisoned Lake", *The Zimbabwe Review*, 2(iv): 3-5, 1996
50 Alma, P; *Environmental Concerns*, Cambridge University Press, Cambridge, 1993
51 Mtetwa, S., "No Room for Complacency – Water Pollution in Zimbabwe", *Water Lines*, Vol.14, No.1, p7-10
52 Moyo, N.A.G., "Pollution in the Zambezi Basin", For SARDC, 1999

REFERENCES

Chenje, M. and P. Johnson, (eds.), *Water in Southern Africa*, SADC/IUCN/SARDC, Harare/Maseru, 1996

Chenje, M. and P. Johnson, (eds.), *State of the Environment in Southern Africa*, SADC/IUCN/SARDC, Harare/Maseru, 1994

Clark, R.B., *Marine Pollution*, Oxford University Press, Oxford, 1986

Crick, H., *Poisoned Prey in the Heart of Africa: DDT and Dieldrin may be Killing African Birds of Prey*, British Trust of Ornithology, London, 1990

Greenhalgh, T., *Environment Today*, Longman Group Ltd., Essex, 1994

Harrison, R.M., *Pollution: Causes, Effects and Control,* The Royal Society of Chemistry, Cambridge, 1992

Haslam, S.M., *River Pollution: An Ecological Perspective*, Belhaven Press, London, 1992

Laurie, F. *The Fragile Environment*, The Press Syndicate of the University of Cambridge, Cambridge, 1991

Lewis, W.J., Foster, S.S.D. and B.S. Drasar, *The Risk of Groundwater Pollution by on-site Sanitation in Developing Countries*, IRCWD, Duebendorf, 1982

Lopes, H., Jergman, K. and U. Aswathanarayana, *Proceedings of the International Africa Conference on Environment, Technology and Sustainable Development*, Eduardo Mondlane University, Maputo, 1991

Mason, C.F, *Biology of Fresh Water Pollution*, Harlow Longman Scientific, Essex, 1993

Matengu, C., *Investigation on the Spillage of Rotenone in the Nabwizu Lake*, Environmental Health Services, Katima Mulilo, 1999

Matthews, R.B., *Agroforestry in Zambia*, Unpublished, 1998

Partanen, T., Kurppa, K., V.F. Ngowi, *Occupational Pesticide Hazards in Developing Countries: Epidemiological Considerations*, Institute of Occupational Health, Helsinki, 1991

Simpson, D.E., *Measurement and Modelling of Stormwater Runoff Pollution from an Urban Catchment*, Rhodes University, Grahamstown, 1987

Simpson, D.E. and P.H. Kemp, *Quality and Quantity of Stormwater Runoff from a Commercial Landuse Catchment in Natal, South Africa*, Council for Scientific and Industrial Research, 1982

Zaranyika, M.F., Mambo, E. and J.M. Makhubalo, *Organochlorine Pesticide Residues in the Sediments of Selected River Bays in Lake Kariba, Zimbabwe*, University of Zimbabwe, Harare, undated

10

POVERTY

Poverty is both the cause and result of environmental degradation. Rapid population growth in a situation of limited livelihood options is a major factor in growing poverty and environmental degradation in the basin. Once a community is subjected to poverty, there is a vicious poverty-environment circle, which is difficult to break. Efforts to reduce the poverty-environment circle in the Zambezi Basin states are hampered by slow economic growth and development, and a fragile natural resource base.

Photos: M Chenje

Poverty is both the cause and result of environmental degradation. The exploitation of marginal land, including hill slopes, has become a major issue of concern in the basin.

Marginal areas are sometimes cultivated to meet the food requirements of the poor. The exploitation of marginal land, including wetlands, streambanks and hill slopes, contradicts one of the SADC objectives, which emphasises that the utilisation of natural resources "requires good management and conservation, to ensure that development does not reduce or impair the diversity and richness of the region's natural resource base and environment."[1]

POVERTY IN THE BASIN

Causes

The main causes of poverty in the Zambezi Basin states and in Africa in general, include:[2]

- Inadequate access to employment opportunities.
- Inadequate access to land and capital, and limited access to credit by the poor.
- Inadequate access to the means of supporting rural development in poor regions.
- Inadequate access to markets where the poor can sell goods and services.
- Low endowment of human capital.
- Limited choices to livelihood options.
- Overdependence on one form of economic activity, in particular agriculture, whose performance is governed by water availability.
- Natural disasters such as drought and floods, for example, the February 2000 floods in Mozambique destroyed the livelihoods of over a million people, leaving them with no food or shelter.
- Destruction of natural resources, leading to environmental degradation and reduced productivity.
- Inadequate assistance for those living at the margins and those victimised by transitory poverty.
- Lack of participation by and failure to draw the poor in the design of development programmes.

Limited land available has seen increases in population densities due to rapid population growth. The average population density is slightly more than 28 people per sq km, but this masks serious overpopulation in some areas. However, a number of points should be borne in mind:

- The present density is relatively higher if the natural resource endowment and economic possibilities for intensification are considered.
- Present distribution of land-rights means that some areas are highly utilised and others under-utilised in relation to potentials.
- If present growth-rates of population are sustained, the population will double within the next generation.[3]

The persistence of poverty and deprivation are pervasive causes of high birth rates throughout the basin states. One of the primary reasons for having large numbers of children per family is economic: Children provide labour for the survival of families lacking capital and technology, as well as insurance in old age. Children in poor households assume responsibilities for fetching water, herding livestock and collecting fuelwood, in addition to caring for their elderly parents. In societies where infant mortality rates are high, this creates a powerful rationale for high birth rates. For example, young people dominate the pop-

Figure 10.1: Human poverty indices in the Zambezi Basin (1995)

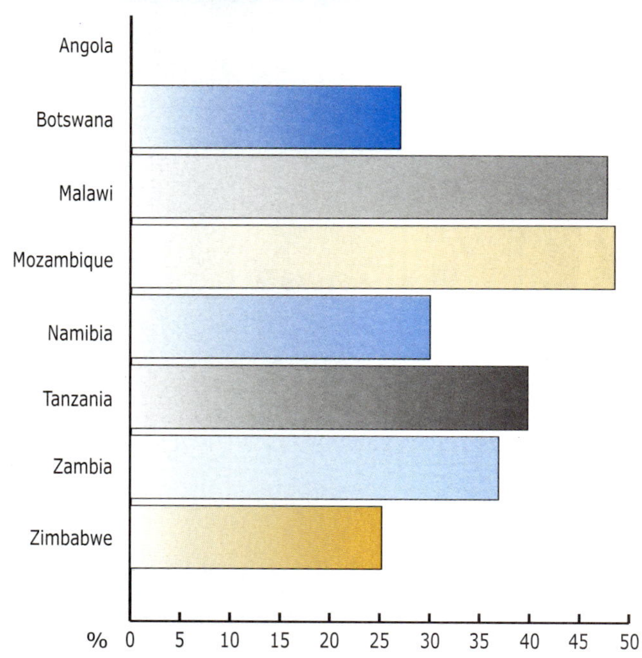

Note: No information is available for Angola.

Source: Statistics obtained from the SADC Human Development Report, 1998

Box 10.1: Poverty

Poverty may be defined in several ways depending on the point of focus. It relates to income, social and economic relativity and social indicators. From an income point of view, people are poor when they are in a state where, "their income (or consumption) is less than that required to meet certain defined needs." This definition however does not adequately address society's capacity to achieve a socially acceptable standard of living that may entail having some education, information, legal rights and infrastructure.

Poverty can also be absolute or relative. Absolute poverty often assumes that needs are fixed at a level where the basics are food, habitat, water, sanitation, health, education and road infrastructure. However, this definition requires periodic review because absolute levels of poverty respond to growth and technological advancement. Relative poverty is when there is an involuntary withdrawal by a society from active participation in social activities because they are deprived. Other definitions consider social indicators such as life expectancy, infant mortality, school enrolment ratios, access to health facilities and access to safe water and sanitation.

Five factors — poverty, physical weakness, isolation, vulnerability and powerlessness — are closely linked to each other and impact on people in a state of deprivation to the extent that they cannot somehow get away from it.

Poverty

People or a country can be poor in relation to the level of development of the immediate environment. Poverty relates to lack of resources for production to afford a decent standard of living. Inability to access basic, but essential goods and services leads to both physical and mental dearth. Female-headed households are especially disadvantaged due to the lack of access to productive resources or because they are denied their rights to own resources such as land.

Physical weakness

Lack of food and cash, results in physical weakness in people. A general lack of strength, malnutrition, poor health, physical (and mental) disability, make more people dependent either on other family members or the government.

The dependency syndrome is followed by latent laziness and the government has to feed its people almost every year. As the country may not have produced and stocked enough food, the handouts to its citizens are usually imported food aid.

Isolation

Due to inadequate infrastructure, especially in rural areas, the majority of the people are isolated by physical barriers. In such conditions, access to agricultural markets, schools, clinics, information and other services is difficult. Ignorance, disease, illiteracy and poverty abound. Because essential needs are difficult to obtain, such communities become vulnerable to 'decay' and become poorer and deprived.

Vulnerability

When a people are weak and cannot afford the basic needs that sustain their life they reach a stage of desperation. The result is that the rural (and sometimes urban) poor reach a point when they are powerless and cannot cope with any socio-economic transformations for they are weak in body and spirit, and the powerful exploit them. Thus, poverty cannot be considered in the microcosmic sense alone. It has been noted that, "the processes that concentrate wealth and power operate at three levels: internationally, the richer countries have made and kept the poorer countries relatively poor through colonial exploitation and post-colonial unequal exchange and benefit from the investment of capital and the expatriation of profits. Internally, within the poorer countries, urban middle class interests gain at the cost of the rural interests through shifts in the rural-urban terms of trade (cheap food for the towns, dear goods for the countryside), and through investment in urban industries and services. Within the rural areas themselves, the local elite, land owners, merchants, moneylenders and bureaucrats consolidate their power and wealth." This condition does not only apply to people per se, but also to countries, where disparities in income between the rich North and the poor South have a bearing on the relationship between these countries.

Sources: Wratten, Ellen, "Conceptualising Urban Poverty", in *Environment and Development*, IIED, London, 1995
Holmberg, J., "Poverty", in *Environment and Development*, IIED, London, 1991

ulation in Caprivi with the average number of children born to a woman being six or seven.[4] According to a survey carried out in Mozambique, the poor households have about twice as many dependents (mainly children) as the non-poor households.

The discrepancies in the distribution of land, concentration of a large population on poor communal land, and population growth have resulted in increasing population pressure on the land in the Zambezi Basin. While agriculture is the mainstay of most of the economies of the basin and key to reducing poverty, it is also a potential threat as it cannot absorb the increasing number of people dependent on the shrinking land and water resources.[5] For example, Zimbabwe's grossly unequal agrarian structure has been described as the 'food security paradox': the co-existence of widespread malnutrition in the midst of a food system, which produces a surplus of food for export.[6] In 1994, almost one-third of children had low height-for-age, indicating chronic malnutrition.[7] Low wages, insecurity and the seasonality of employment in the midst of vast agricultural wealth are the main causes of poverty in the country.

Photos: IUCN

Children assume responsibilities for herding cattle and other chores to support their families.

Economic performance

A mono-economy; – where countries such as Malawi, Mozambique and Tanzania depend on agriculture; Angola on oil and diamonds; Botswana on beef and diamonds; and Zambia on copper; – limits options available for sustainable economic growth. The dependence on a mono-economy means that countries in the basin suffer in times of poor commodity prices, and more often than not, the balance of trade is not in their favour. This has in the past constituted a perpetual burden of debt servicing, depleting the very resources that would otherwise have been used for development.

Foreign debt continues to rise in the basin states. For example, according to the World Bank, Zimbabwe's external debt increased more than five times from US$786 million in 1980 to US$4.168 million in 1993. Ruinously low commodity prices, crushing external debt burdens, anarchic capital markets, failure to prevent conflicts, and misallocated aid, all restrict the prospects for poverty reduction[8] in the basin.

External debt (see Table 10.1) is one of the factors that have forced the countries of the basin to rely on official development assistance. Zambia's development assistance, for example, rose from about US$55 per capita in 1980 to US$221 in 1995.[9] Other countries, Angola, Botswana, Malawi, Mozambique, Tanzania, Namibia and Zimbabwe registered ODA levels of US$38, US$61, US$43, US$66, US$117, US$29 and US$44 per capita in 1995, respectively.[10] Only Botswana and Tanzania were able to reduce this dependence by 1995.

The introduction of economic structural adjustment programmes (ESAPs) in the Zambezi Basin states has seen pressure on urban scarce resources increase. ESAPs have led to the liquidation of some companies and massive job losses. For example, in Zimbabwe, seven textile companies have been liquidated since January 1997.[11] The result has been increased poverty in urban areas.

Poverty and low human development in the region are also tied to colonialism in all the basin states. Other contributing factors are migration and uneven development

Table 10.1: External debt of basin states, 1974-95

Basin States Country	Total External Debt in million US Dollars 1974	1981-83	1984	1983-85	1989	1991-93	1993-95
Angola*				1,707		9,233	11,087
Botswana	128	205	261	286	513	641	678
Malawi	661	851	876	926	1,394	1,733	1,987
Mozambique				1,568	4,737	5,055	5,547
Namibia*	n.a.	n.a.	n.a.	n.a.	n.a.	n.a.	n.a.
Tanzania	2,197	3,341	3,431	3,677	4,917	7,334	7,028
Zambia	3,046	3,697	3,824	4,054	6,873	7,006	6,761
Zimbabwe	559	1,774	2,252	2,262	3,088	3,870	4,502
Basin States	6,591	9,868	10,644	14,480	21,522	34,872	37,590

* Not included in the debt amount of the basin total.
Data source: World Resources: *World Resources Report* (1990-91; 1992-93; 1994-95; 1996-97 and 1998-99). World Resources Institute/United Nations Environment Programme/United Nations Development Programme/World Bank, Oxford University Press, Oxford/New York.

Figure 10.2: The cycle of poverty

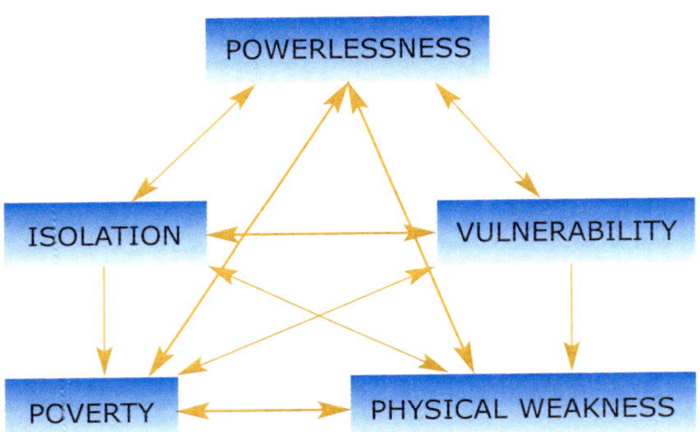

Adapted from: Robert Chambers, *Rural Development: Putting the Last First*, John Wiley and Sons Inc., New York, 1983.

Photo: IUCN

The basin's poorest people are insecure and vulnerable, and are most affected by drought and/or floods.

between centres of growth and wealth, and peripheral areas, which supply labour to those centres and register stunted growth.[12]

Impacts of poverty

The poorest families in some rural areas, most of them headed by single parents, mostly women, typically have less than two hectares of land, and no irrigation.[13] They often do not have cattle for draught power. Instead, they plough the soil with hoes, limiting the area of land they are able to plough, and the variety of crops they are able to produce. Their maize stalks are usually smaller and paler than those of richer farmers because they cannot afford fertiliser. Few have savings or other assets, except a few goats, which they sell to get through times of stress. In a good year, the poorest families grow enough maize to feed their families for three or four months.

The basin's poorest people are insecure and vulnerable, and are most affected by drought and/or floods. During the 1992 drought, which affected many countries in the region, unskilled workers were highly vulnerable to the rising unemployment caused by the economic recession, which accompanied the drought and the introduction of the structural adjustment programmes.[14] Typically, the poorest families lack the capacity to cope with stress, and suffer acute difficulties, such as sickness, physical weakness, and economic impoverishment. Such disadvantages become self-perpetuating; inadequate nutrition and ill health reduce the productivity of the poor, and lower productivity means less income and less food.[15] Vulnerability leads to a downward spiral, as events that suddenly make people poorer also reduce their assets.

Natural disasters and economic crises claim their victims from among the most vulnerable sections of society. In every society the poor live shorter, less healthy lives than those who are better off. Limited state provision of health care, education, clean water, and sanitation exposes the poor to health risks, reducing their productivity and opportunities.

Photo: IUCN

Poor economic performance has the greatest impact on remote areas, which often have inadequate transport and communications systems.

The economic downturn in Zambia in the 1970s and 1980s dealt a severe blow to social services nationwide. The hardest hit were the remote areas, which were also poorly serviced by transport and communication systems.

In Mozambique, poverty affects the overwhelming majority of the people. About 60 percent of the people have a monthly income equal to or lower than US$20.[16] The national poverty line, adjusted for differences in the cost of living in various parts of the country based on the national average prices prevailing in April 1997, was half a dollar (US$0.5) per person.[17] The incidence of poverty (headcount index) was 69.4 percent, indicating that in 1996–97 more than two-thirds of the Mozambican population, or 10.9 million people, were living in absolute poverty. The poverty gap index and the square poverty index were also high, at 29.3 and 15.6 percent respectively.[18]

In some parts of the basin, structural adjustment has become a euphemism for suffering. Zimbabweans redefined the acronym for the Economic Structural Adjustment Programme (ESAP) as "enhanced suffering for African people." Simultaneous reduction of food subsidies and decline in wages, which accompanied structural adjustment dramatically, reduced household incomes in the basin. Between 1990–92, price decontrol increased by half the price of an average food basket for low-income urban families in Zimbabwe. At the same time, average wage settlements in 1992 were 25 percent lower than the rate of inflation, implying a further erosion in purchasing power,[19] and unemployment increased sharply as structural adjustment and drought combined to cause a deep recession.

A study carried out in Kambuzuma, a low-income urban settlement in Harare, showed the proportion of families living below the poverty line doubling to 43 percent. Most households also changed their diets, eating less high protein foods and more carbohydrates. The poorest families cut their number of meals from two to one per day. Women bore the brunt of this household-level adjustment, foregoing meals themselves in order to maintain the food intake of children. Single-parent households were particularly badly affected, reducing their spending on food by one-fifth.[20]

Increasing unemployment, declining real wages, and reduced social welfare provision, have been almost universal features of structural adjustment. In Tanzania, the minimum wage covered only 14 percent of the cost of the

Photo: IUCN

The increased carrying capacity on small areas results in excessive pressure from domesticated animals, causing further land degradation due to overgrazing.

most basic food requirements for a one-person household at the end of the 1980s.[21]

IMPACT OF POVERTY ON THE ENVIRONMENT

Rural and peri-urban areas

The rural economy of the basin countries is principally dependent upon subsistence agriculture. In the face of increasing population, peasant farmers have resorted to cultivating marginal lands. These fragile ecosystems include steep slopes, streambanks, *vleis* and other ecologically sensitive areas. Tilled land now extends its borders onto areas meant for grazing, squeezing the commons to even smaller units than before. The increased carrying capacity on these small areas results in excessive pressure from domesticated animals, causing further land degradation due to overgrazing.[22]

In Tanzania, all indices of poverty show rural people to be more disadvantaged than those in urban areas. Over 50 percent of the people living in rural areas (including farmers) are poor, whereas only 39 percent of urban dwellers are poor.[23] The World Bank estimates based on income for Malawi, show that 55 percent of the people live in poverty, 98 percent of whom are in rural areas. Food insecurity has become common, especially among smallholder households with less than one hectare.[24]

The energy needs of rural and peri-urban communities are also of great relevance in resource use and management in the basin. As the number of people grows so does the demand for energy. People invade forests, mainly near urban areas, to collect fuelwood to meet their energy needs. While the poor cannot easily afford the energy sources that are readily available to the middle and upper class, they often have no alternative but to log forested areas for firewood and charcoal. Malawi, for instance, has a deforestation rate of about three percent annually although the use of the harvest is not entirely for firewood but also for curing tobacco.

Due to land degradation, there is continued and accelerating soil erosion in many parts of the basin, leading to siltation of water sources. This condition further leads to diminishing water supply in the face of rising demand. Together, soil erosion, diminishing water supply and poor farming methods lead to a significant reduction in the productivity of land, exacerbating rural poverty.

In Malawi, poverty has contributed to the increase in informal sector activities, tenancy, estate employment and encroachment on estate and public land.[25] The informal marketing of firewood and charcoal has also increased over the years. The government has linked dependence on more expensive woodfuel to poverty and high population growth. The government has acknowledged that the removal of subsidies and rising tariffs on paraffin and electricity have contributed to the increased dependence on fuelwood in urban areas.[26]

Education levels are also important in explaining rural poverty. According to a World Bank study: "If access to wage employment is regarded as an asset, then it is a more significant contributor to inequality than landholding." Access to high wage employment is closely linked to level of education. The study points out that the poor are

Photo: M Chenje

Photo: IUCN

In a situation of scarce fuelwood or rising electricity costs, rural people turn to cheapest energy sources, using dung or cutting down trees to make charcoal.

Story: The Zambezi, a great river of potential and lost opportunities

Rich as it is in flora and fauna, the Zambezi River Basin plays host to some of the poorest people in the region, an unfortunate paradox.

From the Kalene hills to its delta in Mozambique where it empties into the Indian Ocean, the great Zambezi has made limited economic contribution to the lives of millions of Africans who live on its banks. Only subsistence survival from wetland and floodplain food resources, and building materials offer a meagre existence.

The basin communities depend heavily on natural resources for a living, the result of which is a high rate of environmental degradation. Firewood is a major source of energy for the communities, contributing to increasing levels of deforestation.

The Kariba Hydroelectric Power Station, one of the major reasons for the birth of the lake which bears its name, generates hundreds of megawatts of electricity annually but none of this power finds its way into the homes of the local communities. Other electricity generating centres in the basin are situated at Hwange and Munyati, Kafue Gorge and Cahora Bassa.

Even the few social amenities like schools and clinics, which are thinly spread in the basin, do not benefit from the hydroelectric power schemes through either piped water or electricity.

Up to today, the Nyaminyami community does not have a high school or a decent hospital. The people do not have a good gravel road, let alone a tarred one, or a fixed telephone network.

Lake Kariba is the largest water body in Zimbabwe but the Zambezi Valley communities face severe water shortages each year owing to poor rainfall patterns in the valley area. There has never been an attempt to draw water from this massive body for the local community's consumption.

"Nothing good has come from the project for us. There are more deaths now in this area and malnourishment is a serious problem. As Tonga people, we deserve to be compensated for the forceful eviction from our ancestral lands," says Jonah Habwalo, the Lusitu Village headman in Southern Zambia.

It was a frustrating experience for the local community when they found out that there are long-standing plans to draw water from their doorstep to supply urban populations in Bulawayo, which is more than 400 km away, when they don't have access to the same water.

The major economic activity from the Zambezi Basin is tourism, which is not fully exploited to reduce the poverty. Tourism generates substantial income for most of the countries and is the economic mainstay of the river basin. Top attractions include the river basin itself, a unique ecosystem that supports large mammals, including elephant, hippo, lion, giraffe and hundreds of bird species and beautiful flora.

Communities in the Zambezi Basin are rated as poor in terms of monetary income but social scientists say they could raise their economic status if they fully exploited tourist dollars through purchases of local arts and crafts, and game viewing in the massive conservation areas dotted along the river.

Unsustainable game populations also contribute to poverty when wildlife decimates crops in villages. With the basin's huge wildlife populations, it is no wonder that Zimbabwe's CAMPFIRE programme was first pilot-tested and rated as a success in the two valley districts of Nyaminyami and Guruve.

To some, who lived through forced removals when they were forcibly resettled many kilometres away during the controversial construction of dams creating Lakes Kariba and Cahora Bassa, the community management schemes coming since the mid-1980s are a classic show of "too little, too late".

Kariba Town has become a "new" settlement, attracting "new" ethnic groups from inland who work there at the expense of the dispossessed Tonga who were evicted to make way for the dam.

The environs of Kariba, like those of Binga and Victoria Falls up-river and Kanyemba downstream give an overwhelming impression of lost opportunities.

The beautiful gorges and impressive mixes of biodiversity are, in places, dotted with shanty dwellings which are skillfully and shamefully tucked away from hilltop mansions, posh hotels and beautiful, but environmentally insensitive, leisure boats.

"The large amounts of money that have been spent studying the effect of the Kariba Dam project on the Tonga have not yielded much for the ordinary person who remains buried in deep poverty," says University of Zambia academic, Emmanuel Chidumayo.

Zambia's dilapidated lakeshore town of Siavonga, situated just across the dam a stone's throw from Kariba Town, also exudes a sense of lost opportunities.

The Zambezi slumbers with its mighty potential, impressive geography and endless variety. The river basin can best be described as a mirror image of the rest of the region where many resource-endowed communities wallow in poverty while there is plenty on their doorstep.

By Tawona Tavengwa

poor because of the low returns they receive for their labour. If poor rural people had more education and more livestock, they would be better off.

Urban areas

Movement of people from rural to the urban areas puts pressure on services such as water supply and sanitation, health and education as well as on accommodation. The lag between economic growth and labour is wide and more often than not, job seekers find themselves destitute due to shortage of jobs. Although some city or town bylaws may prohibit marketing of goods within their area of jurisdiction, the informal sector has become vibrant in the basin states, an "indicator" of survival for those who have no access to formal employment. In Zimbabwe, for example, 29 percent of the labour force is employed in the informal sector, but the majority of them live in poverty.[27]

Box 10.2: Findings on poverty levels in Zimbabwe

- Poverty is more prevalent in rural areas with up to 75 percent of households in the total poor category compared with 39 percent of urban households.
- The highest incidence of poverty is in the communal lands (84 percent of households), followed by the resettlement areas and small-scale commercial farms (70 percent), large-scale commercial farms (57 percent), and urban areas (39 percent).
- Households headed by females (31 percent of the total) have a greater incidence of poverty than those headed by males. About 57 percent of female-headed households are very poor compared to 40 percent of male-headed households. About 72 percent of female-headed households fall into the combined poor and very poor category compared to 58 percent of male-headed households.

Source: UNDP/PRF/IDS, *Zimbabwe Human Development Report 1998*, UNDP/PRF/IDS, Harare, 1998

Urbanisation in the basin states is occurring at a very fast pace, faster than the continental average of 4.3 percent. The highest urbanised countries of the basin in 1995 were Zambia (more than 50 percent), Namibia (37 percent), Mozambique (34 percent) and Zimbabwe (32 percent) with urban growth rates of 3.3, 5.3, 7.1 and 4.4 percent respectively.[28] The challenge for the basin countries is to come up with effective measures that can control environmental problems arising from population pressure.

Most of the urban poor live in unplanned squatter settlements on the periphery of urban centres, where their lack of legal status and inadequate service provision make them extremely vulnerable. That vulnerability is made worse by insecure, low-wage employment. The vast majority of the urban poor work in the informal sector in a variety of activities, including petty-trading and casual labour. There has also been an increase in numbers of "street kids" who run away from poor homes to fend for themselves on the streets of major centres such as Lusaka and Harare.

Photo: M Chenje

The sale of curio pieces in urban and tourist areas has become one of the major activities in which people are engaged to earn income. However, such activities are in the majority of cases subsistence with very few breaking the vicious environment-poverty circle.

Urban residents generally consume more renewable resources than rural people, placing heavy demands on the environment. Large concentrations of people also generate huge volumes of waste, resulting in pollution and health problems. Cholera and other water-borne diseases are often found in water contaminated with untreated human waste and sewage. In the SADC region, more than 171,000 cases of dysentery were reported in 1994 for Malawi, Mozambique and Zimbabwe, with at least 600 lives lost.

Due to industrialisation and urbanisation, industrial and consumer waste is increasing in many urban areas. For example, in Harare, people generate 124,000 tonnes

Photo: IUCN

One of the challenges of urbanisation is waste handling.

of solid waste annually.[29] Based on the Harare statistics, Zimbabwe's urban population, estimated at about 3.66 million in 1997, generates a total of about 732,000 tonnes of solid waste annually. The problem of solid waste disposal is further compounded by the growing production of plastics and chemicals, sometimes toxic, which are difficult to recycle.

Some of the challenges of urbanisation in the basin today include, among other things, proper urban planning, provision of safe drinking water, infrastructure, waste handling and other services. Many of these are often sacrificed due to unplanned settlements.

ACCESS TO RESOURCES

Land

The links between poverty and environmental degradation are related to poor people's rights to use and control of natural resources. Land tenure and property rights in the Zambezi Basin are a blend of the traditional systems, the colonial system and the post-independence government policies. In most basin countries, land is classified as state or public land, free or leasehold land and customary communal land. State land consists of national parks, gazetted forest and protected land and state farms.

In Zimbabwe, the state holds customary land in trust and the local communities have rights to reside, till and graze their livestock on it. The land is inequitably owned with most of it still belonging to only about 4,000 white commercial farmers.

Box 10.3: Land — Zimbabwe a case study

In Zimbabwe, the problems of access to resources illustrate the strong relationship between poverty and the environment. This relationship is contained in the country's history dating back to over a century from the time the country was first colonised by European settlers. To protect their interests, successive colonial governments enacted several laws. These included the 1932 Land Apportionment Act conceived as the genesis of the current problems of overcrowding, poverty and land degradation faced by the country's communal areas today.

The Act made way for the white settlers to subdivide the country into state, European and African land. State land constitutes national parks, gazetted forestland and state farms. African and European land was almost equal in size, in spite of the fact that Africans constituted well over 90 percent of the country's population by that time. Africans did not get and still do not have title over their land, as it was held in trust by the State. All they got were rights to reside, till and graze their livestock on the land.

Over time, this situation was made worse when the Africans were thrown onto land that is in agro-ecological regions IV and V, which are characterised by semi-arid conditions suitable for extensive agriculture and ranching. The low rainfall of between 400-800 mm annually is punctuated by seasonal droughts. The soils are generally very poor as they contain very little mineral clays, being sand-derived from granite and quartzite.

The communal areas are also characterised by extensive rock outcrops that reduce the amount of productive land available. Estimates suggest that up to one-third of the land in some localities is comprised of rock outcrops. In contrast, much of European land is located in regions I, II and III - areas suitable for intensive agriculture as there is both high rainfall and generally good soils.

Land distribution in Zimbabwe, remains largely the same today. For instance, communal areas today carry about 70 percent of the national population. In 1992, about 74 percent of the communal land area in Zimbabwe contained 34 percent of the population (over three million people) that was located in regions IV and V. While land in the communal areas has not increased over the years, the population increased at about 3.1 percent annually between 1970-95. To compound the situation, some of the land previously owned by the local communities was annexed to that of settlers or declared public and private land and this has not changed even after the settlers lost power to the nationalists.

Source: Mashingaidze I.C., "Poverty and the Environment: The Case of the Communal Areas in Zimbabwe", Presented at a Regional Workshop on the State of the Environment in the Zambezi Basin, Chinhoyi, 20-24 July, 1998

In Namibia, the poor and the non-poor have about the same amount of land per household: an average of 0.4 ha per capita for the former and an average of 0.6 ha per capita for the latter. Maize is the most common crop grown by the poor and non-poor alike. It is cultivated by about 75 percent of all households.

In Botswana, resources are under pressure because of increasing population. Development needs have raised concern about the capacity of the environment to sustain the needs of future generations in terms of fuelwood.[30]

In Tanzania, access to land does not appear to be linked to poverty. However, a review of regional characteristics, indicates that access to land with good agricultural potential may in fact determine expenditure levels.[31] More of the poor have land than do the non-poor. Over 50 percent of the difference in income between the poor and non-poor is due to the difference in endowments such as cattle and education available to each family. About 21 percent of the total income differential between the two groups is the result of livestock.[32]

More and more people are being forced to settle on fragile lands, exacerbating environmental problems in various parts of the basin.

In cases where for instance, agriculture is unable to offer all their needs, communities find alternative sources of income. Logging for fuelwood and charcoal production are common in Malawi, Mozambique, Namibia and Zambia.[33] Indigenous forests and their biodiversity are lost every year as people encroach on them for a living.[34] This situation is aggravated by disparities in land distribution.

Studies in Tanzania have shown that household requirements are about 42 million cu m of wood annually while forestry plantation programmes can only replace about 20,000 ha/year.[35] In rural areas, wood apart from being used for cooking is also utilised to support local activities such as beer brewing, fish-smoking, and charcoal making.

Photos: M Chenje

In cases where agriculture is unable to support needs, the disadvantaged turn to other natural resources, logging and making bricks for either immediate needs or to sell.

In Angola, the Moxico Province in the Zambezi Basin remains quite isolated from the rest of the country because of the war. Electricity supply is undeveloped, forcing the majority of people to rely on wood and charcoal for most of their energy needs. This is the main cause of deforestation in the more densely populated areas in the basin, such as Cazombo, Lumbala and Luena.

Soil erosion and land degradation are likely to continue at alarming rates as the population pressures increase throughout the basin. The Omay area in the Kariba subbasin is a classic example where a scheme to eradicate tsetse fly has opened up a marginal, ecologically fragile area to settlement. The peasants who moved into the region brought with them their cattle and goats, leading to large-scale degradation of the habitat. An associated problem, and one that may have long-term repercussions, is the rapidly declining fertility of the soil. The soils in this area are very fragile even in their pristine state.[36] They become even more susceptible to erosion, and decrease in fertility rapidly when cultivated. Any subsequent erosion of these soils will lead to an increase in siltation of the Zambezi River.

Photo: M Chenje

Soil erosion and land degradation are likely to continue at alarming rates as the population pressures increase throughout the basin.

To a far greater extent than the wealthy, the poor people depend for their survival on the use of common resources. They fish, hunt, produce goods for sale from leaves and fibres, make medicine from forest sources, and gather nuts and berries for food. During the drought of 1991–92, villagers in Zambia and Zimbabwe drew upon local knowledge of roots, leaves, and other forest products, to survive chronic food shortages. Without that knowledge, which has been passed down over generations, the human costs of the drought would have been much higher.

The war in Angola has had a catastrophic effect on the country's economic and social infrastructure. Thousands of Angolans have been internally displaced or are refugees in neighbouring countries, marginalising them to a life of poverty. Displaced people and refugees often have to overexploit resources to survive.

Water

Increasing water demand is a concern in the Zambezi Basin because of the increasing human population and associated demand for resources, especially food. Water resources in the basin are not equally distributed in space or time. Some parts of the basin receive more rain than others.

In addition to surface water, groundwater is very important in the basin during the dry season and year-round, particularly in the rural areas. Groundwater is the main source of water for many rural people throughout the basin, catering for the majority of the 69.4 percent of the basin's rural population. Increasing population in communal areas is straining groundwater resources.

In the Caprivi town of Katima Mulilo, almost all water is piped, either to communal or indoor taps. In rural areas, however, 32 percent of households obtain their water from rivers and other open, surface sources. Twenty-six percent of households depend on wells, 24 percent on piped water, and 16 percent on boreholes.[37] While Caprivi has more rivers and open water than any other region in Namibia, people in the area perhaps have greater problems with the supply of water than anyone else. Sources of water may be far from their homes, so women have to walk long distances and spend much time collecting domestic water.

Photo: M Chenje

Very few communities have access to piped water, with the majority of the rural people and some in urban areas depending on groundwater resources.

In the areas where the Tonga people were resettled when Kariba Dam was built, water supplies and sanitation facilities were inadequate because water sources were difficult to find. Some boreholes and wells either dried up or the yields were too low to sustain the communities or they simply broke down due to maintenance problems.[38]

In urban areas, the factor most affecting water use for domestic purposes is the economic status of the individual. This will often dictate whether the water supply is reticulated through the residence and is always accessible, or available only through a communal standpipe some distance from the residence. It will also dictate affordability of water, and whether water is required for garden use and topping up a swimming pool, for example. The extremes in per capita use will range from around 20 litres per capita per day (l/c/d) among economically constrained users to 300 or more l/c/d among the rich.[39]

ACCESS TO SERVICES

Social

Human development has been substantially reduced in the region, especially considering the levels of investment in essential services such as education, health, housing and water provision, impacting on the quality of the services delivered.[40]

Geographic isolation, low income, and lack of productive assets often reinforce the effects of poverty in the basin. In terms of geographical isolation, the majority of the rural homes are located far away from social and commercial services.

There is a close correlation between distance from markets and poverty. About one-third of the poorest families in Zambia live over 20 km away from the nearest food market, and cannot afford transport to these markets. Inability to obtain credit is another major problem for poor farmers, preventing them from adopting new technologies.

Katima Mulilo is Caprivi's capital and administrative centre. It serves as the seat for the regional government and ministerial offices and it is the only commercial centre in the region. Small retail outlets can be found in other centers, but Katima Mulilo is the only centre in Caprivi which provides banking, postal and other commercial services.[41]

Only a small area in the Caprivi has access to telephone and electricity supply services, and very few households are connected. Expansion of these services is constrained both by the unavailability of distribution lines and the inability of households to pay user fees. There were only about 1,000 subscribers in the region in 1997 and only five percent of all households had electricity supply in 1991.[42]

Photo: IUCN

The majority of the rural homes in the basin are located far away from social and commercial services.

Safe water and sanitation

Although countries such as Malawi and Zimbabwe have tried to improve the water supply situation for rural communities, these efforts seem to have only gained momentum during the United Nations International Water and Sanitation Decade of the 1980s. The situation regarding access to water and sanitation in the basin states needs more attention. According to a recent report,[43] percentage of the population with access to safe water and sanitation is not very high for Angola, Malawi, Tanzania (except for sanitation) and Zambia. These countries have less than 40 percent of the population with access to safe water. There is, therefore, need for additional attention, especially in the form of renewed effort and investment in these areas. Table 10.2 gives a picture of the basin states in respect of water supply and sanitation.

More often governments, through their water supply institutions, have taken giant steps in improving water supply in particular supply zones but have paid lesser attention to improving sanitation facilities. The situation has put pressure on health services, especially when liquid and solid waste removal facilities are insufficient, and diseases abound. The absence of sufficient water supply and sanitation facilities is in itself poverty. Due to this, there are very few means of improving the situation, resulting in more poverty.

Water contamination in the basin is widespread due to industrial pollution, as well as through discharges of untreated sewage into surface and groundwater. Loaded

Table 10.2: Population with access to safe water and sanitation in the basin

Country	Percent of population with access to safe water 1990-96			Percent of population with access to adequate sanitation 1990-96		
	Total	Urban	Rural	Total	Urban	Rural
Angola	32	69	15	16	34	8
Botswana	93*	100*	91*	55	91	41
Malawi	37	80	32	16	22	4
Mozambique	63	n/a	n/a	54	n/a	n/a
Namibia	57	87	42	34	77	12
Tanzania	38	73	29	86	96	84
Zambia	27	50	17	64	89	43
Zimbabwe	79	99	69	52	96	32

Note: * indicates data that refer to years or periods other than those specified in the column heading, differ from the standard definition, or refer to only part of a country.

Source: UNICEF, *State of the World's Children 1998*, Oxford University Press, New York, USA, 1998

with bacteria and viruses, these waters can be a major threat to human health in the downstream localities, and water treatment is commonly inadequate or unreliable.

For example in Tanzania, 62 percent of water supplies are unsafe for drinking. Throughout the region, water-borne diseases such as typhoid, cholera, dysentery gastroenteritis, and hepatitis and water-based diseases, including schistosomiasis and guinea-worm. are common.

In some areas, the groundwater is unpalatable, being salty and unfit for consumption, while in other areas people have to use dirty water from wells and backwaters. Water from these unhygienic sources contributes to the high incidence of diarrhoea, especially among children. In the case of Caprivi for instance, a major pipeline runs along the road from Kongola to Katima Mulilo, carrying water pumped from both the Kwando and Zambezi rivers. While many people depend on this pipeline, the water is not treated, and the supply is inadequate to the extent that little water reaches the middle section of the line.[44]

The Botswana Water Utilities Corporation (WUC) has made that country one of the few in Africa with a safe urban water supply supported by a semi-autonomous public agency. It has done so through strict recognition of water as a scarce economic good, while still considering equity concerns. The monthly tariffs are high, but progressive rate structure does not exceed ability-to-pay of poorest customers. The urban areas have 24-hour supply and high-quality water treatment.[45]

Education

The low earnings of poor people are partly a consequence of their low levels of skill and literacy. Studies for agriculture and industry have shown that better-educated people adapt more easily to new technologies, and have higher rates of productivity.[46] This increases their earning capability and employment prospects, bringing wider benefits for society.

Children from poor households are more likely to work, primarily as unpaid family labourers, restricting their educational opportunities, which may in any case be unaffordable for the poor, creating an inter-generational poverty trap.

As seen from Table 10.3, the levels of illiteracy in the basin states are highest in Angola and Mozambique where more than 70 percent of the women and more than 40 percent of the men are illiterate. This situation is reflected in the school enrolment ratios where less than 32 percent of both the women and men enroll for education.

One of the major reasons for low enrolment, however, is that guardians cannot afford tuition for their children in the face of escalating costs of services. Malawi has taken a step in addressing this situation by providing free primary school education for eight years and facilitating entry to post-primary educational institutions, especially for girls. The enrolment ratios of 71.4 percent for females and 79.7 percent for males as shown in Table 10.4 are some of the highest of the basin states. Both Botswana and Namibia have comparable school enrolment ratios.

Generally, girls tend to receive less schooling than boys. Regarded as more important to household subsistence than their brothers, girls are often kept at home to perform domestic work or to care for younger siblings, and they are likely to be the first to be withdrawn from school if the family faces a crisis. Such practices have kept women trapped in circles of poverty and illiteracy, which are passed down generations.[47]

The basin states must recognise the diversity of languages and review their language policies to ensure that:

- Language use promotes full participation in development rather than be a divisive factor.
- Future censuses include questions on home language to provide planners with accurate data.
- Children are given basic education in a language they understand so that they have equal learning opportunities.

Table 10.3: Education and child health in the basin states, 1970-93

Country	Adult female literacy (%)		Adult male literacy (%)		Gross primary school enrolment as a percentage of age group: Female		Male		Births attended by trained personnel (%) in 1990	ORT use (%) in 1993	Low birth weight in infants (%) in 1990	Percentage of one-year-olds fully immunised against the following diseases in 1993			
	1970	1990	1970	1990	1960	1993	1960	1993				A	B	C	D
Angola	7	29	16	56	14	79	30	86	15	48	19	53	30	28	47
Botswana	31	55	60	78	43	118	30	112	78	64	8	50	57	57	60
Malawi	20	37	58	69	26	78	50	86	55	50	20	96	92	92	92
Mozambique	6	18	26	52	43	51	71	69	25	60	20	66	49	49	62
Namibia	n.a.	n.a.	n.a.	n.a.	48	141	43	138	68	75	12	92	73	79	71
Tanzania	21	49	55	75	16	69	33	71	53	83	14	92	82	81	79
Zambia	32	65	64	82	40	81	61	87	51	90	13	88	64	62	62
Zimbabwe	58	77	75	88	65	114	82	123	70	82	14	79	69	69	73

A = TB B = DPT C = Polio D = Measles

Note: n.a. stands for data not available.

Source: Adapted from: WRI/UNEP/UNDP/ The World Bank., World Resources 1996-97. Oxford University Press, Oxford, 1996

Photo: IUCN

Rural pupils often have to go with inferior educational services, impacting on their ability to derive the best out of their years at school.

Table 10.4: Adult illiteracy rates and school (primary and post-primary) enrolment ratio (%) in 1995

COUNTRY	Adult illiteracy rate: Female	Male	School enrolment ratio: Female	Male
Angola	71.0	44.0	27.5	31.8
Botswana	40.1	19.5	71.6	69.0
Malawi	58.2	28.0	71.4	79.7
Mozambique	76.7	42.3	20.5	29.0
Namibia	26.0	22.0	84.9	78.7
Tanzania	43.2	20.6	32.1	33.4
Zambia	28.7	14.4	48.5	55.0
Zimbabwe	20.1	09.6	65.3	69.7

Source: Adapted from: UNDP/SADC/SAPES, *SADC Human Development Report 1998*, SAPES Books, Harare, 1998

Health

Human health concerns are widespread because of lack of safe drinking water and inadequate sanitation/sewage collection systems, and pollution of many rivers, lakes, and groundwater. The primary causes of disease and poor health in the basin are water-related. The lack of safe water for drinking, the shortage or lack of accessible water for hygiene, and the scarcity of sanitation services all contribute to the human health problems.

In Zambia, 30 percent of the rural population lives 10 km or more from the nearest health facility,[48] making services inaccessible. In Caprivi, there are two hospitals and three health centres. Clinics and primary health care facilities are found at 34 other places throughout Caprivi. In the 1991 census, low life spans were largely due to the high rates of disease, which affected Caprivians at birth. The infant and child mortality rates were 50 percent higher than the national average. Diarrhoea and measles caused a high proportion of infant deaths, while 55 percent of school children suffered from goiter and 42 percent were physically stunted. Among the older people, malaria, tuberculosis and HIV took an increasing toll.

COPING MECHANISMS

Indigenous knowledge systems

Indigenous knowledge systems (IKS) were not only critical in resource management and sustainability, but also in cushioning communities, particularly the poor, against natural disasters. Families dried some foodstuffs and also had granaries to stock their harvests until the next agricultural season. Due to the abrupt decay of the traditional authority over resource use and management as a result of colonial rule, and in some instances continued even after independence, the traditional rulers and their people lost control of their rights over resources and survival techniques proven over generations. The management of resources was centralised, vesting ownership and responsibility over the natural resources in central government. Resources became inaccessible, limiting the people's livelihood options.

In Botswana, the powers of the chiefs have been reduced in favour of the land boards.[49] This transfer of power of land management alienated people from managing the resource. Since the introduction of these new regulations, adequate adherence to indigenous conservation methods is limited. People now cut trees or collect firewood from any locality, even if they are not residents of that area, a practice not accepted in the past. The depletion of forest resources in Botswana is encouraged by lack of alternative affordable fuels.

In an attempt to seek solutions to various problems associated with resource depletion, the government and other institutions, for example, the Forestry Association of Botswana, are working with various communities to investigate the traditional knowledge systems, which were used to maintain resources in the past.

In Malawi, the indigenous people's knowledge about the following resources was particularly valuable:

- Soil characteristics and capabilities.
- Local climatic factors, especially rainfall, drought incidence and frost or hail occurrences.
- Local plant and animal diseases.
- Key dates for the most appropriate planting.
- Farming technologies such as cassava production on mounds, using potash-rich heaps for potato production and water-harvesting basins in dry areas in south Malawi, together with the protection of trees with goat dung, and the use of *Faidherbia albida* as part of traditional agroforestry.

In Mozambique, due to the assimilation policies of the Portuguese colonisers, many African systems were marginalised, despised or destroyed. After independence, the new government viewed ethnic loyalty as an obstruction to nation building. However in 1994, the Mozambican Parliament approved a law on decentralisation and local government reform. This resurrected traditional authorities, bypassed since independence in 1975, and provides an opportunity for local people to be responsible for their resource management and use of indigenous knowledge systems.

Traditional societies in Zimbabwe like elsewhere in the basin, practised and enforced wildlife conservation through timely hunting of animals and birds, avoiding indiscriminate killing, and fostering selectivity. Societies believed wanton killing was punishable by the spirits and, as a result, control mechanisms are found in traditional taboos, totems and customs. For example, the custom of totems forbade people to eat certain animals – scavengers such as vultures and hyenas. Taboos forbade the killing of young animals and females in gestation. Hunting in sacred places was also prohibited. The killing of rare species such as python and the pangolin could only be done with permission of the chief.

There is growing realisation by governments and civil society that indigenous knowledge systems have much to offer natural resource management in the region. The

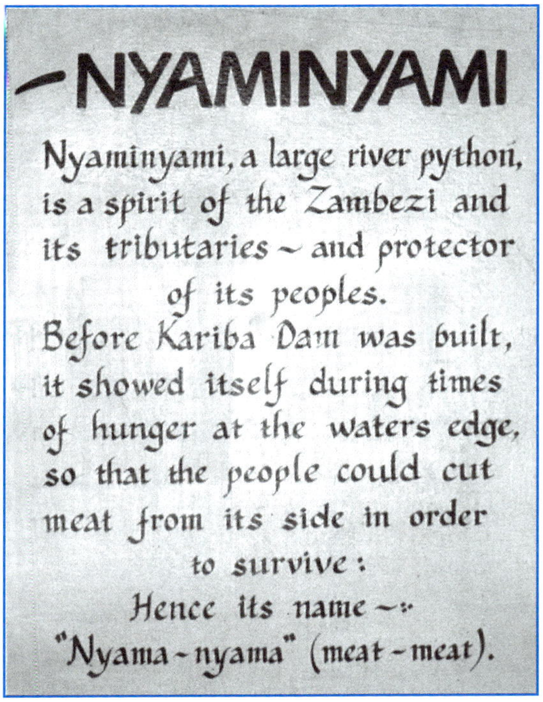

Photos: M Chenje

The people of the basin have a rich cultural history and indigenous knowledge which over generations has been key in the sustainable management of wildlife and other resources.

Box 10.4: Land and forest resources in Zambia

In Zambia, the legislation since independence up to 1974, provided for the protection of forests through the declaration of areas of state lands as forest reserves and the declaration of protected forest areas on any land, including reserve land. In forest reserves and protected forest areas, a license was necessary in order to cut trees, remove firewood, fruits, grass, honey, plants, rubber, seeds, timber and other prescribed forest produce. People needed licenses to settle in such reserves.

The most telling factor of the policy on the declaration of these areas is that central government did not even consult communities directly affected by the declaration of protected forest areas. The communities were forcibly resettled elsewhere.

In the case of areas within rural councils, the local inhabitants, subject to any bylaws of the local council, were permitted to fell, take, burn or injure or remove any of the species of trees in a protected forest area without a license. The protection of species in the Zambian Forests Act was limited to conserving trees on state lands outside forest reserves by listing them as reserved trees in the Schedule to the Act, without specifying the process or the criteria for the listing of such species.

The laws and regulations governing resource use and management were also tailored in such a fashion that those who chose to exploit them would pay dearly for their actions. Income gained from these resources did not, and still does not in some countries, directly benefit those that live adjacent to such resources. They are, therefore, forced to look for alternatives to support themselves. The result has been dependency on the government for employment, habitat and food. Over time, people started encroaching on marginal lands for subsistence agriculture, including areas that would have best suited wildlife conservation parks, wetlands and mountain slopes.

Societies and communities have moved from long- to short-term management options and a culture of poverty has crept in as people struggle to survive. This trend, has led to accelerated environmental degradation in the basin. Due to the high concentrations of people in generally unstable ecosystems, widespread land degradation has occurred in Zambia's communal areas.

Land issues in Zambia, as elsewhere in the basin, have a role to play in terms of environmental sustainability. It has been noted that, "with the possible exception of climate whose vagaries are manifest in the periodic droughts that intermittently plague the country, all other processes of desertification are basically the consequences of this primary factor: the unequal distribution of land."

Sources: Government of Zambia, The Forest Act, Government Printer, Lusaka, 1941
Murphree, M., "The Lessons from Mahenye: Rural Poverty, Democracy and Wildlife Conservation", Wildlife and Development, Series No.1, IIED, London, 1995
Darkoh, M. B. K., "Desertification in Africa", Journal of Eastern African Research and Development, Vol.19, Nairobi, 1989

Photos: M Chenje

The custom of taboos, which is respected in some parts of the basin, forbids people from killing some animals during particular seasons. Such customs have played an important role in wildlife conservation even before modern conservation became in vogue.

array of public and other institutions that have indigenous knowledge systems on their agenda in each of the countries bears witness to this.

Colonial legacies, especially the imposition of national boundaries that were insensitive to the cultural homogeneity of indigenous peoples, also contributed significantly to the process of degradation of indigenous knowledge systems. The poverty of knowledge on indigenous knowledge systems is reinforced by the fact that these systems are rarely reflected in the curricula of schools, professional colleges and universities.

Deliberate political, social, cultural and economic empowerment of people and the recognition of their indigenous knowledge systems are essential. Legitimising indigenous knowledge systems is necessary. It entails proper understanding of all facets of indigenous knowledge systems, including comprehending how its inherent qualities provide specific solutions to problems, giving indigenous systems an appropriate status, using scientific methods emanating from indigenous knowledge systems and giving due recognition to information from indigenous knowledge systems.

Displacement of the Tonga

Some development projects not only destroy traditional knowledge systems but ironically cause more misery and poverty among people than raise their standard of living. An example is the Kariba Dam, which forced the resettlement of thousands of the Tonga people on the north and south banks. They had to build their homes from scratch. Intact communities on either side of the river were cut off from each other by the dam and an imaginary border between Zambia and Zimbabwe.

Many of the people lost highly productive alluvial fields on the edge of the Zambezi and forced into dryland farming in the rugged foothills of the escarpment.

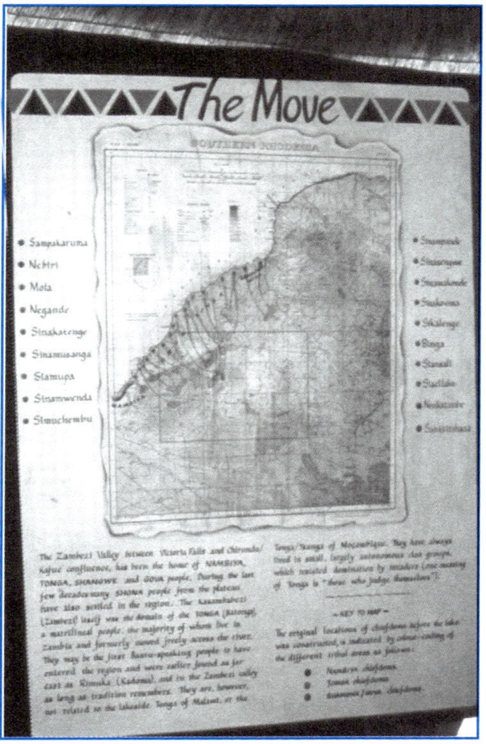

Photo: M Chenje

Memories are still fresh in the Tonga people whose forefathers were forcibly removed from the Zambezi Valley about four decades ago to make way for the Kariba Dam from which they have benefited little.

Before flooding, the life of the displaced people depended on the Zambezi River in many aspects of their livelihoods, consisting mainly of gathering wild fruits, fishing, hunting and subsistence farming on small fields and gardens in the valley. They lived with very little intervention from colonial rule and could engage in as many occupations as they so wished.[50]

Cultivation was done throughout the year, on the floodplains when the Zambezi had receded and in small gardens and fields during the rainy seasons. They caught fish using different methods like hooks, spears and baskets, providing them with a constant source of protein. They hunted freely and could kill dangerous animals by setting up traps. Food was adequate for their needs.

Today the Tongas on both sides of the Zambezi live in abject poverty without even benefiting from the electricity generated at Kariba for which they were displaced.

Box 10.5: Impacts of the displacement of the Tonga

The poorly planned resettlement programme and the subsequent harsh displacement from the Zambezi Valley brought more misery and poverty than good in the lives of the victims. A once-upon-a-time unified, self-sufficient community, the displaced people today live in a negative socio-economic climate caused by the forced displacement from their natural habitat.

Social

It has been reported that some families broke up during the displacement because the wives preferred to follow their kinsmen. On the contrary, polygamy increased in some homes in Zambia. The life of the displaced people was characterised by dire poverty, which in turn leads to a socio-cultural crisis and comparative illiteracy.

The psychological impact was seen in increased levels of alcoholism, which subsequently leads to household, village and at inter-village conflicts. Strained relationships led to cases of witchcraft and crime. People lost contact with their relatives and relationships broke up because people could no longer move freely across the river as before.

Economic

The displaced people lost their fertile alluvial land. They were forced to open new fields in poorer, drought prone Karroo uplands, which affected food security. In Zimbabwe, they could no longer hunt as freely as they used to, as the wildlife areas were designated national parks. Similarly, fishing practices could no longer be undertaken to supplement food resources. As a result of continually falling yields due to drought, poor farming practices, and infertile fields, they resorted to selling off their cattle for survival. As if that were not enough, bovine *trypanosomiasis* took its toll on the very herds on which they depended.

Water resources

"... We went away with our bodies but we left our water behind. We would like the water to follow where we are today so we can plough throughout the year." – Simpongo Munsaka, a Tonga.

It appears that among all the complaints relating to the displacement, inadequate supplies of water was the biggest. None of the resettlement areas were provided with as much water as the people were used to in the valley. Moreover, they became susceptible to erratic rainfall patterns whose effect was less pronounced in the valley by virtue of proximity to the Zambezi River. This greatly affected their crop productivity.

The quality of water found in the resettlement areas varied to different degrees for different uses. In Lusitu, for example, the piped water supply scheme delivered not only inadequate, but untreated water as well to the community from the Zambezi River downstream of the dam in Zambia. By comparison, the Siamupa area in Binga suffered from hard borehole water.

Demographic

It is reported that mortality rates generally increased among the displaced people. Reasons advanced for this were shortage of food, poor sanitation and inadequate health facilities. All these are deemed to be direct results of economic and social impacts.

Environmental

The inadequate resettlement planning caused land pressure, resulting in deforestation, soil erosion and overgrazing. The people were resettled among host communities, forcing them to share the limited resources available, such as land for cultivation, firewood, building materials and pastures. This in turn accelerated the degradation of the environment as they grossly exceeded the carrying capacity.

Source: Zambezi River Authority, *Kariba Dam's Operation Noah Re-launched*, ZRA, Lusaka, 1997

Rural areas

Mechanisms of coping with rural poverty include gold panning, informal sale of agricultural produce and craft production. Along the major roads that link the urban areas, rural people are on the road sides selling agricultural produce, for example potatoes, mangoes, tomatoes and wild fruits, as well as stone sculptures, wooden crafts and baskets woven from reeds.

Photo: M Chenje

Rural people make a living on road sides selling various things such as mushrooms.

Gold panning has become a major problem in some areas because of the localised environmental degradation associated with it. The degradation caused includes:

- Removal of trees to use for shelter and mining.
- Erosion caused by the many paths in and around the riverine vegetation caused by the panners' movements.
- Excavations into riverbanks in search of gold, causing massive erosion and sedimentation.
- Organic and inorganic pollution resulting from lack of proper sanitation and uncontrolled use of mercury to extract gold.

A 4,500km stretch of riverbank is affected by gold panning in Zimbabwe.[51] The majority of gold panning in the country occurs in the rivers draining into the Zambezi, such as the Mazowe, Manyame, Angwa and Ruenya. One of the reasons that started the large number of people (between 150,000-300,000) to engage in gold panning was the very large increase in fertiliser price in 1985.[52] This resulted in communal farmers moving off the land to find other sources of income to survive the harsh economic climate.

In the early 1980s, poverty, to some extent, fuelled poaching of high-value species, encroachments into protected areas and loss of habitat in many countries, reaching crisis levels in some areas. For example, many Zambian nationals accused of poaching in the Zambezi Valley were killed by Zimbabwean game wardens, leading to an acrimonious debate on whether wildlife was more important than human life. It is important to note that once development needs and poverty begin to compete with conservation, the latter is certain to lose.[53]

Urban areas

Mechanisms of coping with the increasing poverty in the urban areas include urban agriculture and informal markets, for example, flea markets and the marketing of agricultural produce and sweets at street corners. Tuckshops are now mushrooming in the suburbs due to the economic hardships.

The rise in urban agriculture in the basin states is also largely attributed to economic hardships. It is considered as a spontaneous haphazard activity, which is not planned for and hence not supported. Thus, the practice is widely viewed as illegal because it is not supported by any statutory instrument.[54]

Urban agriculture in the basin states is for producing food for family consumption and the urban market. This activity is largely practised on a small-scale on land around the homesteads and open spaces and infrastructural servitudes within the cities and towns. The harsh economic conditions, due to structural adjustment policies, prominently feature as the main driving force behind the activity.

The economic structural adjustment programmes worsened the plight of the urban households, mostly the poor. Removal of subsidies, the consequent increase in prices of basic foodstuffs and retrenchments, meant that the urban poor became poorer. The primary benefits of urban agriculture are household food security, employment, income generation and savings on purchases. Between 1996-97 households in Zimbabwe involved in the informal marketing of agricultural products accrued an average income of Z$5,951.40 (about US$400 at that time) within a period of eight months. Households producing and selling chickens within the same period had a net average income of Z$4,982.76.[55]

Urban agriculture has become one of the major activities across the basin with thousands of people encroaching on private property and hills to supplement their incomes. Open spaces for residential areas have been converted into grain fields.

In Gweru, in Zimbabwe's Midlands Province, aerial surveys conducted in 1996, showed that a total of 2,257 ha of land was under crop cultivation, representing 8.5 percent of the total land area in Gweru.[56] Open space cultivation is common in the high density areas where most of the low income urban residents live.

Orthodox planning principles view open spaces in urban areas as green wedges or 'ecological lungs' of the city but the urban farmers view such lands as 'idle' and therefore, should be put to productive use. The usual response to such practices by city authorities has been the slashing of the semi-mature crops as a deterrent measure to curb the activity. However, most urban farmers view such responses as a manifestation of colonial practices aimed at promoting the interests of a minority. As a result, the farmers have been unyielding and continue with their farming.

POLICY RESPONSES

Agenda 21 realises that while managing resources sustainably, an environmental policy that focuses mainly on the conservation and protection of resources must take account of those who depend on the resources for their livelihoods. Otherwise it could have adverse impact both on poverty and on chances for long-term success in resource and environmental conservation. Equally, a development policy that focuses mainly on increasing the production of goods without addressing the sustainability of the resources on which production is based will sooner or later run into declining productivity, which could also have an adverse impact on poverty.

An effective strategy for tackling the problems of poverty, development and environment simultaneously should begin by focusing on resources, production and people, and should cover demographic issues, enhanced health care and education, the rights of women, the role of youth and of local communities and a democratic participation process in association with improved governance.

Poverty alleviation

Almost all the countries in the basin are addressing poverty and undertaking various initiatives. The Poverty Alleviation Programme in Malawi and the Poverty Alleviation Action Plan in Zimbabwe are clear examples of such initiatives.

In Malawi, the government has recognised the dire poverty situation in the country and has pursued a vigorous development initiative that directly impacts on the poor. Rural communities have often requested facilities such as boreholes, clinics, roads, primary and secondary schools and farm inputs such as fertiliser and seed. Their needs are promptly addressed through the Malawi Social Action Fund (MASAF) and other sources. The Malawi case demonstrates that the needs of communities are basic: clean water to safeguard health, clinics to treat diseases, roads to facilitate access to markets and education to reduce illiteracy.

In Zambia, only 1.74 percent of public spending has been allocated to improve water services compared to education and health, for example, which have received 18.5 and 13.3 percent respectively.[57]

The needs of the rural people are basic and include grinding mills, which should be accessible to the majority of the people to lessen the burden of travelling long distances.

The government notes:[58] "The alleviation of poverty centres on empowering the people to enable them to improve their standard of living. The background of low levels of investment as well as deferred maintenance in water and sanitation infrastructure has culminated into a serious lack of safe water for both domestic consumption and industrial use. This situation poses one of the most immediate threats to community health."

The Zambian government adds that policies and initiatives to alleviate poverty and improve the population living standards are being carried out through activities such as:

- Liberalisation of the economy to attract investment to build the industrial sector.
- Increasing agricultural production, particularly by irrigation and promotion of high value crops through investment programmes like ASIP.
- Improved urban water supply and sanitation through reorganisation of the water sector and rehabilitation of schemes.
- Environmental management programmes such as NEAP and ESP.
- Pursuance of increased hydropower generation through proposed schemes such as the Kafue Gorge Lower III.

Zimbabwe has made several initiatives in addressing basic needs of the poor. Zimbabwe's Social Dimension Fund, an arm of ESAP, was instituted in 1991 to buffer vulnerable groups and poor communities from the impacts of economic structural adjustment.[59] This initiative consisted of several major programmes:

- The Social Welfare Programme "intended to compensate the poor for new or increased user charges in education and health, and for deregulating maize prices. Payments were designed to finance exemptions from health fees, school and examination fees for households with incomes below US$87 per month together with a per capita cash food benefit of US$1 per month for urban households earning less than US$44 per month."[60]

- The Employment and Training Programme "provided an introductory one-week training course in starting up a new business, along with loans for small enterprises."[61] This programme currently caters for retrenched persons in both private and public sectors, victims of the country's economic reform programme, and persons with disabilities. The objective of the programme is to train these people so that they can be able to generate self-employment in the non-formal and small-scale enterprise sectors. The programme finances soft loans up to about US$26,100. The loan attracts an interest rate of 15 percent annually, payable over a maximum of 10 years, with a maximum grace period of up to a year.

- Social Safety Nets Programme is designed to alleviate the suffering of the vulnerable low-income groups and the poor with incomes below US$87. The scope of the programme includes: education assistance, involving payment of school fees; health assistance, involving payment of health fees; and food security, involving direct payment of food money to the urban poor.

- Micro-Enterprise Development Programme (MEDP) has a broadened target group which covers not only ESAP-induced poverty, but addresses transitory and structural poverty as well through income generation and employment creation. The programme targets all poor people engaged in income generating activities wishing to expand their options through credit, and who can absorb as well as repay credit under market conditions. The MEDP uses existing financial intermediaries as the delivery mechanism.

- Community Action Project (CAP) is the latest component of the PAAP. It was prepared as a result of the findings of the PASS that showed the prevalence of poverty in the country's rural areas. CAP seeks to contribute to poverty alleviation in Zimbabwe's poorest communal and resettlement areas, especially of the most vulnerable and disadvantaged groups. CAP is to achieve its goal through two major objectives, which are to:

  - Strengthen the local capacity to identify and prioritise needs, plan and implement solutions, manage and maintain investments, and apply the experience and lessons learned to future needs.
  - Promote participatory and collaborative planning and implementation of projects through community partnerships with rural district councils, government agencies, NGOs and the private sector.

FUTURE CHALLENGES

Agriculture

Working the soil with simple farm implements has been a long tradition of the people and governments seem to have done little in revolutionising the agrarian economy. Farm inputs such as fertiliser, pesticides, and loans to purchase seed are hard to get for the smallholder while this facility is generously extended to the commercial farmer.

A clear paradox exists: in an economy that is wholly dependent on agriculture, prices of farm inputs are very high. Inputs such as fertiliser, pesticides and herbicides are expensive. While the application of such inputs helps boost production, environmental pollution results if the inputs are abused. This is often a reflection of inadequate land husbandry and extension methods extended to poor rural farmers.

Food aid to the Zambezi Basin states is common as seen from Figure 10.3. Yet, it should be possible to do away with imports of food, especially when there is potential to do so from its abundant land and water resources.

Figure 10.3: Average annual food aid (cereals) for the basin states 1975-95 ('000 metric tonnes)

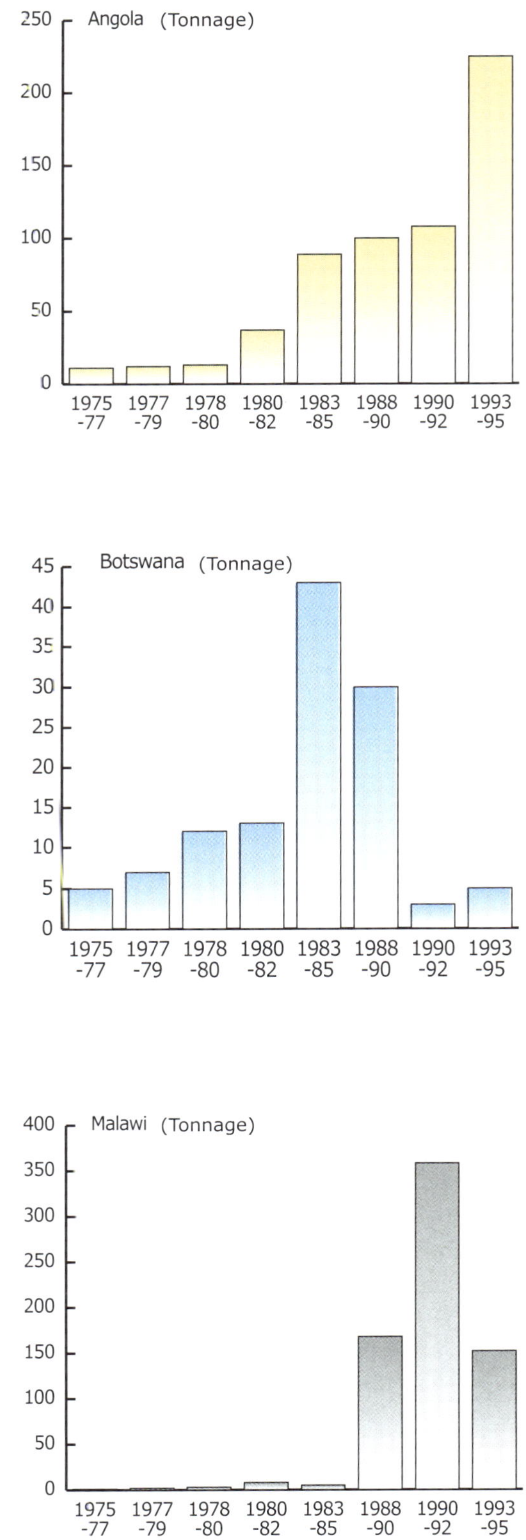

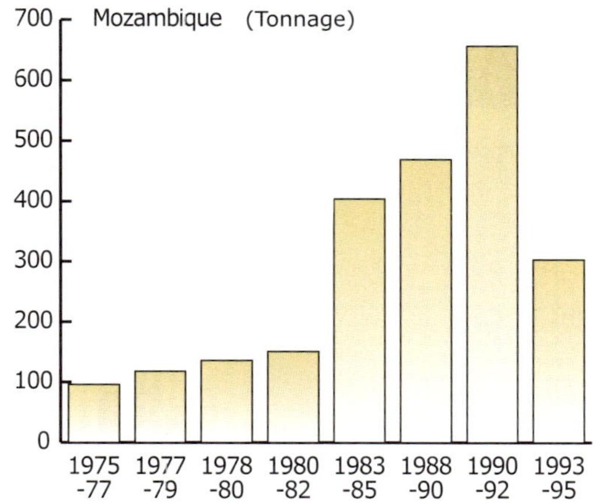
Mozambique (Tonnage)
0
100
200
300
400
500
600
700
1975 -77
1977 -79
1978 -80
1980 -82
1983 -85
1988 -90
1990 -92
1993 -95

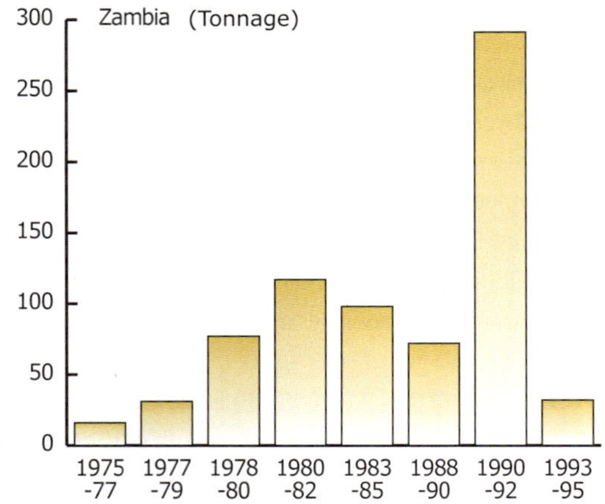
Zambia (Tonnage)
0
50
100
150
200
250
300
1975 -77
1977 -79
1978 -80
1980 -82
1983 -85
1988 -90
1990 -92
1993 -95

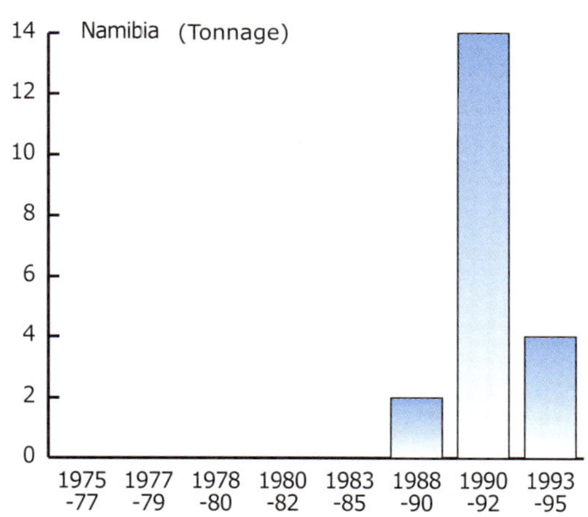
Namibia (Tonnage)
0
2
4
6
8
10
12
14
1975 -77
1977 -79
1978 -80
1980 -82
1983 -85
1988 -90
1990 -92
1993 -95

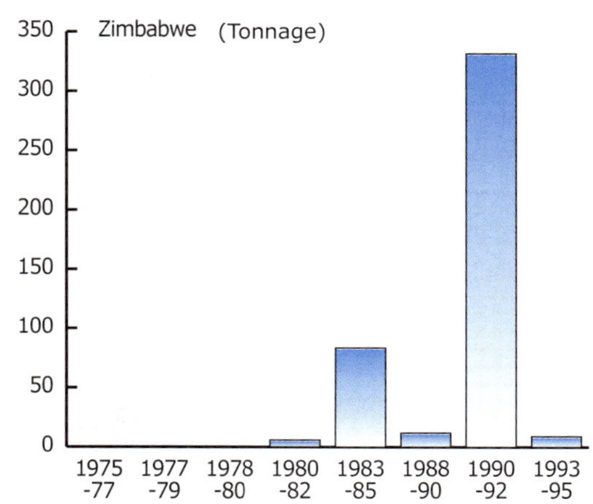
Zimbabwe (Tonnage)
0
50
100
150
200
250
300
350
1975 -77
1977 -79
1978 -80
1980 -82
1983 -85
1988 -90
1990 -92
1993 -95

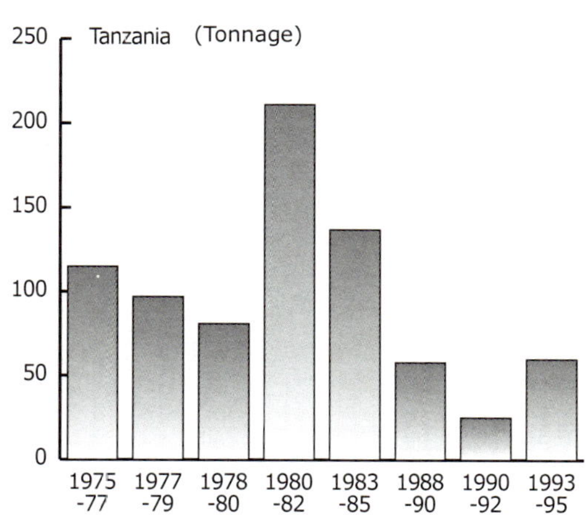
Tanzania (Tonnage)
0
50
100
150
200
250
1975 -77
1977 -79
1978 -80
1980 -82
1983 -85
1988 -90
1990 -92
1993 -95

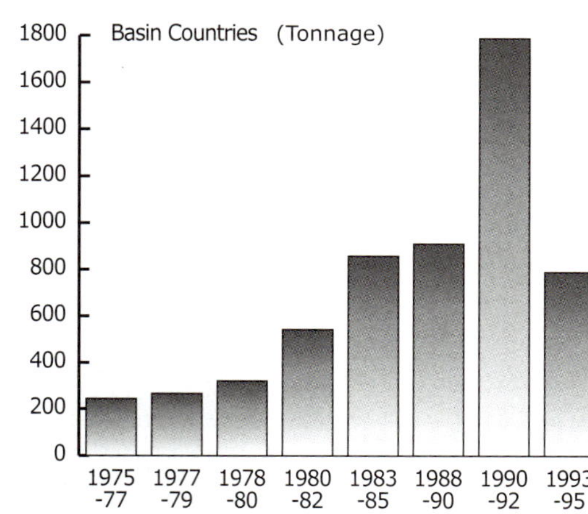
Basin Countries (Tonnage)
0
200
400
600
800
1000
1200
1400
1600
1800
1975 -77
1977 -79
1978 -80
1980 -82
1983 -85
1988 -90
1990 -92
1993 -95

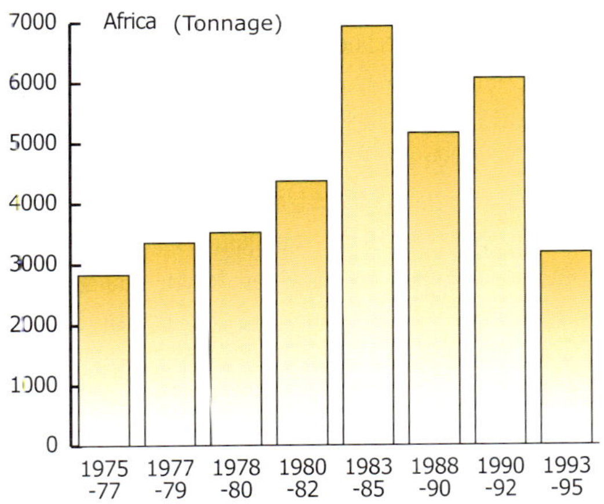

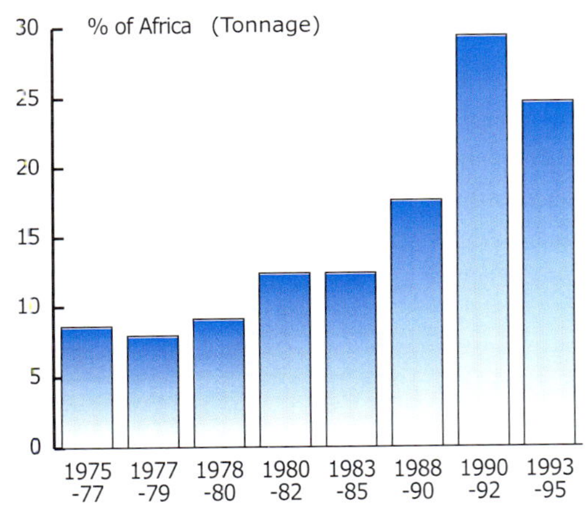

Source: World Resources Institute: *World Resources Report (1990-91; 1992-93; 1994-95; 1996-97 and 1998-99)*, WRI/UNEP/UNDP/World Bank, Oxford University Press, Oxford/New York.

There are several reasons that can be attributed to the shortfalls in the yields of food and non-food crops in the basin states. These include among others, poor rainfall in some years, reduced soil fertility, little access to credit facilities, less or no subsidies to agricultural inputs, exorbitant prices for farm inputs that the majority of the small holders are unable to afford, and poverty.

In spite of extension services, free seed and fertiliser that the basin states provide to the rural farmers, the dependency syndrome can be entrenched and will conflict with poverty alleviation measures. The real need of the rural and peri-urban poor must be recognised and attention given. This can only be done if further comprehensive research on the poverty situation in the basin countries is undertaken to give the poor a chance to indicate how their lives can be improved.

Tourism

In terms of the tourism industry, foreign currency is generated in the riparian countries but little is being done to inject a substantial part of the earnings into programmes that take the poor into consideration. Exceptions and perhaps what may now be considered as models are the Chobe Enclave Conservation Trust in Botswana, Tchuma Tchato in Mozambique, ADMADE in Zambia and CAMPFIRE in Zimbabwe. Under these programmes, communities are directly involved in planning, design and management, and have a say in the distribution and use of the revenue generated from the sale of the resources, mainly wildlife.

However, contrary to popular belief that tourism will directly benefit the rural communities, the resources from tourism are part of the central government revenue account and are used for diversified programmes.

Poverty reduction strategies

To achieve poverty reduction and steer towards prosperity, there are several action items that countries need to consider. According to the World Bank, these include:[62]

- Fostering efficient macro-economic and sectoral policies for sustained growth and poverty reduction.
- Establish a forum for poverty reduction at which stakeholders discuss, evaluate, and coordinate efforts to reduce poverty. In particular, establish opportunities for listening to the poor.
- Carry out regular reviews of public expenditures as the basis for a pro-poor public investment programme that can be supported by donors.
- Shift actual expenditures on social services from urban to rural areas.
- Decentralise government decision-making, especially on public expenditures and promote community participation in design, implementation and monitoring of programmes that are essential to successfully reduce poverty.
- Monitor poverty through, for example, collection of household data to assess the difficulties faced by both men and women and to evaluate the progress being made.
- Governments and donors should build a consensus on a strong vision for and commitment to poverty reduction.

- Assistance, except for clearly humanitarian aid, should be related to the governments' commitment to poverty reduction.
- Donors must demonstrate greater flexibility and openness with each other in coordinating assistance programmes better for a concerted attack on poverty.
- Donors should reduce the bias in assistance programmes towards urban and high-potential agricultural areas and seek a balance of programmes that are more pro-poor.

While the above principles and actions are not a panacea for poverty alleviation, countries and governments in the Zambezi Basin states must blend these with the requirements of their resources and the needs and aspirations of their people.

The web of poverty is linked to all the sectors of the economy and all groups of society. There are inter-linkages between the major factors that create the poverty-environment circle. As such, there cannot be one intervention to address poverty in the basin. Often than not, poverty alleviation programmes may fail especially in situations where governments have conditioned their citizens to receiving handouts each time there is crop failure. Self-reliance must, therefore, be an important goal of macro-economic policies of the countries of the basin. What is required is a multi-faceted approach by government, non-governmental organisations, the private sector and communities that considers the different factors.

The capacity of poor people to improve their standard of living has always been underestimated. Because of this, people's needs have been determined externally. The selection of the type of intervention and technology employed has been based on "affordability" rather than "willingness to pay." Decisions have been made through the extensive involvement of outsiders, excluding the local decision-makers. Such supply-driven intervention strategies have failed to improve the well being of the people and the way they utilise natural resources in a sustainable manner. Interventions should empower communities to initiate, select, implement, manage, and maintain systems that they are willing and able to maintain.

The benefit of spreading growth to the rural communities is that there is cumulative appreciation of output and growth as fewer resources are likely to be allocated for handouts to other essential services such as health. This also has a secondary benefit of reducing resentment, social tensions, crime and political instability. Deliberate policies that provide for the direct injection of resources for programmes are, therefore, needed to alleviate poverty.

The prime objective of poverty alleviation and eradication programmes centres on the desire to create economic growth that permits social and economic conditions acceptable to and desirable by all, and leads to a lifestyle that harmonises different cultures and fulfils the aspirations of the people. Education, health, sanitation, water, mortality rates are some of the factors that determine human development as a fulfilment of peoples' cultural and economic aspirations and expectations.

One of the most fundamental requirements towards achieving positive results in poverty alleviation and environmental management programmes is to take into account the rights of the individual as well as devolving power and authority to those that are close to and in direct contact with the environment and the natural resources. An interesting Zimbabwean axiom can literally be translated thus:

"When you are favoured by the spirits of your ancestors who are guardians of the land and the resources on it, and they bring within the scope of your bow and arrow with wild meat, be thankful but share it. Do not sit in the bush and eat it by yourself. Take it back to the village and share it with your neighbour."[63]

One of the most fundamental requirements towards achieving positive results in poverty alleviation and environmental management programmes is to take into account the rights of the individual.

LINKAGES TO OTHER CHAPTERS

Chapter 1: Regional Overview: People and Environment
High levels of rural and urban poverty and high rates of population growth are adversely affecting the environment, leading to increasing depletion of the natural resource base of the basin.

Chapter 2: Physical Features and Climate
Natural resources such as forests, grasslands, hills, knolls and soil provide the ground for the people's food, clothing and shelter. When these resources are difficult to access, poverty creeps upon them, forcing people to overexploit the resources leading to degradation.

Chapter 3: Water and Wetland Resources
Food production for consumption and for trade depends on water resources, and droughts have had a marked impact on all sectors of the economy. With poor water supplies, people are unable to get access to adequate nutrition, they are weak and cannot produce for themselves. As a result, poverty dominates and people are forced to migrate to wetlands. Pressure on these ecosystems further degrades the environment.

Chapter 4: Biological Resources and Diversity
The flora and fauna of the basin provide rural communities with food, medicines and other basic requirements. The sustainable use of the biodiversity of the basin states is therefore important in order to provide the basic material required for poverty alleviation programmes.

Chapter 5: Agriculture
When agricultural produce does not meet the daily requirements of the community, due to inadequate access to land, drought and inappropriate land practices, poverty prevails.

Chapter 6: Industry
The dependence, in many of the basin states, on agriculture continues to put many people at a disadvantage especially when commodity prices fluctuate or fall, resulting in lower incomes per family. Agro-based and manufacturing industries would alleviate the present poverty crisis under proper management to avert the negative impacts that industry brings on the environment.

Chapter 7: Energy
Poor communities mainly collect fuelwood to meet their energy needs because they cannot afford other energy sources. As fuelwood becomes scarce, people turn to crop residues and cow dung, depriving the soil of nutrients and resulting in decreased productivity.

Chapter 8: Tourism
Tourism could accelerate poverty alleviation through direct employment of the rural poor. Ploughing back the revenue received from tourism into community projects and programmes that provide basic needs of the poor would go a long way in reducing the levels of poverty in the basin states.

Chapter 9: Pollution
Due to the shortages in or total absence of waste collection and disposal facilities, waste is disposed of close to habitats, putting the residents at great risk from diseases. Pollution contributes to the suffering of the urban poor and is the major cause of disease in these areas.

Chapter 11: Gender and Women's Roles
Alleviation of poverty in the basin states requires the full participation and education of men and women, through better access to, use and management of natural resources.

Chapter 12: Environmental Management and Regional Cooperation
Regional collaboration is crucial in improving the social and economic status of the people of the basin. The diversity of the resources in the basin provides much chance for development of industry to ensure employment of a large sector of the people, reducing the levels of poverty.

Chapter 13: Trends and Scenarios
Unless the basin states take drastic steps towards poverty alleviation, the situation is likely to get worse as the population increases in relation to the finite resources. The end result is that sustainable development will remain a chimera.

ENDNOTES

1 Southern African Development Community, *Declaration Treaty and Protocol of Southern African Development Community*, SADC, Gaborone, Botswana, 1993
2 The World Bank, *Taking Action to Reduce Poverty in Sub-Saharan Africa*, IBRD/The World Bank, Washington, D.C., 1996
3 Denconsult, *Water Use and Effluent Discharge in the Domestic and Industrial Sectors*, Sector Studies under ZACPLAN, Sector Study No.1, Zambezi River Authority, August 1998
4 CSO, "1991 Population and Housing Census: Basic Analysis with Highlights", CSO Report, National Planning Commission, 1995
5 The World Bank, *African Water Resources, Challenges and Opportunities for Sustainable Development*, World Bank Technical Paper No. 331, African Technical Department Series, The World Bank, Washington, D.C., 1996
6 Stoneman, C. and G. Thompson., "Background Paper on Trade and Food Security in Zimbabwe", Harare, 1994
7 UNICEF, *Children and Women in Zimbabwe*, UNICEF, Harare, 1994
8 Watkins, K., *The Oxfam Poverty Report*, Oxfam, 1995
9 UNDP/SADC/SAPES, *SADC Regional Human Development Report 1998*, SAPES, Harare, 1998
10 ibid
11 ENDA, "Urban Agriculture in Gweru", Household Nutrition, Economic Costs and Benefits, Research, Development and Consultancy Division, ENDA, August 1997
12 Op. cit 9
13 Op. cit 8
14 ibid
15 ibid
16 INE, *National Household Inquiry into Living Conditions 1996-97*, Planning and Finance Ministry, Poverty in Mozambique: Profile Determinants and policy Implications, INE, 1997
17 ibid
18 ibid
19 IMF, "Zimbabwe Background Paper", Paper for Members of Executive Board, 10 May, 1993
20 Kanji, N. and N. Jazdowska., "Structural Adjustment and Women in Zimbabwe", *Review of African Political Economy*, Vol. 56, 1993
21 Jamal, V., "Changing Poverty and Employment Patterns under Crisis in Africa", International Labour Organisation, July 1994
22 Mashingaidze I. C., "Poverty and the Environment: The Case of the Communal Areas in Zimbabwe", Paper Presented at the Regional Workshop on the State of the Environment in the Zambezi Basin, held at Mazvikadei, Chinhoyi, 20-24 July 1998
23 Ndissi, E., "Poverty in Tanzania is Rural", *Business Times*, 5-11 April, 1996
24 Chenje, J., "State of the Environment Reporting Programme for the Zambezi Basin", Inception Mission Report, IUCN/ERCSA/Sida, Harare, 1998
25 Malawi Government, *State of Environment Report for Malawi 1998*, Department of Environmental Affairs, Lilongwe, July 1998
26 ibid
27 UNDP/PRF/IDS, *Zimbabwe Human Development Report 1998*, UNDP/PRF/IDS, Harare, December 1998
28 UNFPA, *The State of World Population 1996*, UNFPA, New York, 1996
29 Chenje, M., Sola, L., and D. Paleczny., (eds.) *The State of Zimbabwe's Environment 1998*, Ministry of Mines, Environment and Tourism, Harare, 1998
30 ECA/SADC/SA, "Report of the Ad-hoc Expert Group Meeting and the Regional Policy Framework on Managing Issues in the Nexus of Population", *Food Security and Environment in Southern Africa*, ECA/SADC/SA, 1998
31 Op. cit 23
32 ibid
33 SADC/IUCN/SARDC, *The Biodiversity of Indigenous Forests and Woodlands in Southern Africa*, unpublished report, SADC/IUCN/SARDC, Maseru/Harare, 1997
34 Matiza-Chiuta, T., "Water and Wetland Resources of the Zambezi Basin", For SARDC, Harare, 1998
35 REPOA Secretariat, "Poverty and the Environment", *Business Times*, February, 1996
36 Op. cit 3
37 SIAPAC, "Integrated Summary Report: Community Consultation and Survey Programme", Department of Water Affairs, Ministry of Agriculture, Water and Rural Development, 1997
38 Zambezi River Authority, *Kariba Dam's Operation Noah Re-launched*, Zambezi River Authority, Harare/Lusaka, 1997
39 Op. cit 3
40 Op. cit 27
41 Op. cit 34
42 ibid
43 Op. cit 27
44 Government of Namibia, "Feasibility Study for the Development of Water Supply for the Katima Mulilo and Kongola in the Eastern Caprivi", Report No. 2300/6/1/2/P1, Ministry of Agriculture, Water and Rural Development, 1994
45 Op. cit 5
46 Mooch, P., "Education and Agricultural Productivity", in *International Encyclopedia of Education*, Pergamon, 1994
47 Op. cit 8
48 ibid
49 Matowanyika, J., "The Region Develops its Institutions on Indigenous Knowledge Systems: Experiences from Countries in Southern Africa", *Hearing the Crab's Cough*, SADC-ELMS, IUCN, March, 1996
50 Op. cit 42
51 Chamber of Mines, *Chamber of Mines Journal*, Chamber of Mines, October 1996
52 Op. cit 3
53 Op. cit 24
54 Op. cit 11
55 ibid
56 ibid
57 Government of Zambia, "Zambia Country Situation Report on Water", Paper prepared for the SADC Round Table Conference, Ministry of Energy and Water Development, Lusaka, December 1997
58 ibid
59 Op. cit 27
60 ibid
61 ibid
62 The World Bank, *World Development Report 1992 - Development and the Environment*, The World Bank, Washington, D.C., 1992
63 Murphree, M., "The Lessons from Mahenye: Rural Poverty, Democracy and Wildlife Conservation", *Wildlife and Development*, Series No. 1, IIED, London, 1995

REFERENCES

Amanor-Wilks, Dede Esi; *In Search of Hope for Zimbabwe's Farm Workers*, Panos Institute, London, 1995

Association For Rural Advancement; *Drought, Relief and Rural Communities*, AFRA, Pietermaritzburg, 1993

Chambers, R., *Rural Development: Putting the Last Things First*, John Wiley & Sons Inc., New York, 1983

Dalal-Clayton, B., "Southern Africa Beyond the Millenium: Environmental Trends and Scenarios to 2015", *Environmental Planning Issues*, No. 13, IIED, London, 1997

Darkoh, M. B. K., "Desertification in Africa", *Journal of Eastern African Research and Development*, Vol.19, Nairobi, 1989

Drakakis-Smith, D., "And the Cupboard was Bare: Food Security and Food Policy for the Urban Poor", *Geographical Journal of Zimbabwe*, No. 23, Harare, 1992

Foundation for Research Development, *Rural Poverty Alert: Challenging Science and Technology*, FRD, Pretoria, 1992

Government of Zambia, *The Forest Act*, Government Printer, Lusaka, 1941

Herbold Green Reginald; "Ecology, Poverty and Sustainability: Environmental Portents and Prospects in Rural Namibia", IAAE/AGRECONA Inter-Conference Symposium, Swakopmund, 1990

Holmberg, J., *Poverty, Environment and Development*, IIED, London, 1991

IIED; *Environment and Urbanisation-Urban Poverty: Characteristics, Causes and Consequences*, IIED, London, 1995

Jones, D., Akester, S. and A. Palfreman, "Evaluation of ODA Institutional Support to Malawi Fisheries Department", Evaluation Report EV 446, Overseas Development Administration, London, 1990

Lindberg, C., "Poverty and Land Degradation", Working Paper No. 22, School of Geography, Stockholm University, Stockholm, 1993

MacGarry, B. (ed.); "Waste or Want?" *Environment and Poverty Seminar Papers*, Mambo Press, Gweru, 1995

Mitchell Beazley; *Caring for the Earth: A Strategy for Survival*, IUCN/UNEP/WWF, Reed Consumer Books Limited, London, 1993

Moyo, S., *The Land Acquisition Process in Zimbabwe (1997/8)*, UNDP, Harare, 1998

Munasinghe, M. and Wilfrido Cruz; "Economy-wide Policies and the Environment: Lessons from Experience", *World Bank Environment Paper*, Number 10, IBRD/The World Bank, Washington, D.C., 1995

Murphree, M., "The Lessons from Mahenye: Rural Poverty, Democracy and Wildlife Conservation", *Wildlife and Development*, Series No.1, IIED, London, 1995

South African Panorama; *Special Environment Issue*, March/April 1992, ABC Press (Pty) Limited, Cape Town, 1992

Sudhanshu Handa; "Structural Adjustment Policies, Income Distribution and Poverty: A Review of the Jamaican Experience", *International Food Policy Research Institute Research Perspectives*, IFPRI, Reprint No. 374, Volume 25, No. 6, Washington, 1997

Swatuk, Larry, A., "Power and Water: The Coming Order in Southern Africa", *Southern African Perspectives*, No. 58, Centre for Southern African Studies, University of Western Cape, Bellville, 1996

The World Bank; *Striking a Balance: The Environmental Challenge of Development*, The World Bank, Washington, D.C., 1989

The World Bank; *Taking Action to Reduce Poverty in Sub-Saharan Africa*, IBRD/The World Bank, Washington, D.C., 1996

The World Bank; *Water Supply and Waste Disposal: Poverty and Basic Needs Series*, The World Bank, Washington, D.C., 1983

The Zimbabwe Trust/Department of National Parks and Wildlife Management/The Campfire Association; *People, Wildlife and Natural Resources - The Campfire Approach to Rural Development in Zimbabwe*, The Zimbabwe Trust, Harare, 1990

Timberlake, L., *Africa in Crisis: The Causes, the Cures of Environmental Bankruptcy*, Earthscan Publications Limited, London, 1988

United Nations; *Agenda 21: Adoption of Agreements on Environment and Development*, United Nations Conference on Environment and Development, Rio de Janeiro, 1992

United Nations Development Programme; *Human Development Report 1994*, Oxford University Press, Oxford, 1994

United Nations Development Programme; *Human Development Report 1997*, Oxford University Press, Oxford, 1997

WRI/UNEP/UNDP; *World Resources 1992-93: A Guide to the Global Environment*, WRI/UNEP/UNDP, Oxford University Press, Oxford, 1992

WRI/UNEP/UNDP; *World Resources 1998-99: A Guide to the Global Environment*, WRI/UNEP/UNDP, Oxford University Press, Oxford, 1998

11

GENDER AND WOMEN'S ROLES

The understanding of the linkage between gender and the environment in the basin is still at an early stage, even though it is becoming more obvious that there is a clear connection based on women's close relationship with natural resources.[1] Women, particularly those living in the rural areas, play a major role in managing natural resources such as soil, water and forests. The recognition of the relationship between gender and the environment is still far from being generally accepted. Often there is an implicit assumption that women and men perceive the environment and environmental issues in the same way. Given the context of women's and men's different experiences with the environment and the divergent needs that they have of the natural environment, this assumption may be ill-founded.[2]

If the Zambezi Basin is to achieve sustainable development, one necessary step is to reduce gender inequalities and increase women's participation in issues concerning the environment and natural resource management.

GENDER ROLES

Women and men interact with the environment within specific gender relations and processes concerning the use of the resources. Society has allocated different roles, responsibilities and activities to women and men according to what is considered appropriate. Although not static, each society has gender prescribed roles and responsibilities. The gender specific roles for women and men are a

Photo: M Chenje

Women and men interact with the environment within specific gender relations and processes concerning the use of the resources.

Box 11.1: Defining gender

Gender is defined as socially constructed and culturally variable roles that women and men play in their daily lives. Gender structures social, political and economic relations and shapes values and attitudes through society. While sex is biological, gender is acquired and constructed by society. Gender is reinforced by custom, law and specific development policies.

The concept of gender can be explained by distinguishing it from sex. One's sex is biologically determined: male or female. Gender, on the other hand, recognises that girls and boys are brought up differently, or are socialised to play different roles in accordance with the norms, attitudes, beliefs, practices and expectations of their societies. Unlike sex roles, which are biologically determined, gender roles are taught or learnt; they are the result of socialisation and are, therefore, changeable. For example, while the biological fact of having a child does not in itself make it impossible for a woman to have a job, she may be prevented by a number of factors determined by gender, among them cultural norms, from taking up the job. Inspite of the different degrees and forms of gender inequality, the system is universal.

Sources: SAPES Trust *Gender in Southern Africa – Conceptual and Theoretical Issues,* Harare, Zimbabwe, 1992.
Okeyo, T., M., "Traditional Behaviour and AIDS: Analysis and Change Through Community Participation," November 1990-1991.
SADC, "Report on the SADC Gender Strategy Workshop," South Africa, 1997.

reflection of society's historical, cultural, religious, socio-economic, legal and development conditions. The roles of women and men in society are not fixed but are in constant change due to dynamic interactions.

Cultural attributes

The cultural attributes of a community have a very strong influence on societal gender relations. Culture and tradition have fashioned out society in such a way that there is a very strong relation between the type of work, rights, access and gender. In the Zambezi Basin, there are several types of African cultures, ranging from the Lunda, Luchaze and Quioca cultures in Angola; the Tswana culture in Botswana; the Chewa, Tumbuka, Yao, Ngoni, Mang'anja and Sena cultures in Malawi; the Sena culture in Mozambique; the Lozi, Mbunda, Tonga and Chewa cultures in Zambia; and Ndebele, Shona and Tonga cultures in Zimbabwe. With regard to the way gender issues are treated, there is not much cultural diversity among the tribes who reside in the basin. All cultures within the Zambezi Basin place women in a subordinate position.[3]

Roles of women and men in the basin

Within the Zambezi Basin, there are gender specific roles played by women and men in all socioeconomic activities such as water resources development and management, agriculture, mining, fishing, hunting and gathering, forestry, tourism, recreation, crafts, transport, and environmental management.

In general, men's traditional role in society is to be the breadwinner and defender of the family against any

Figure 11.1: Factors that determine gender roles in society

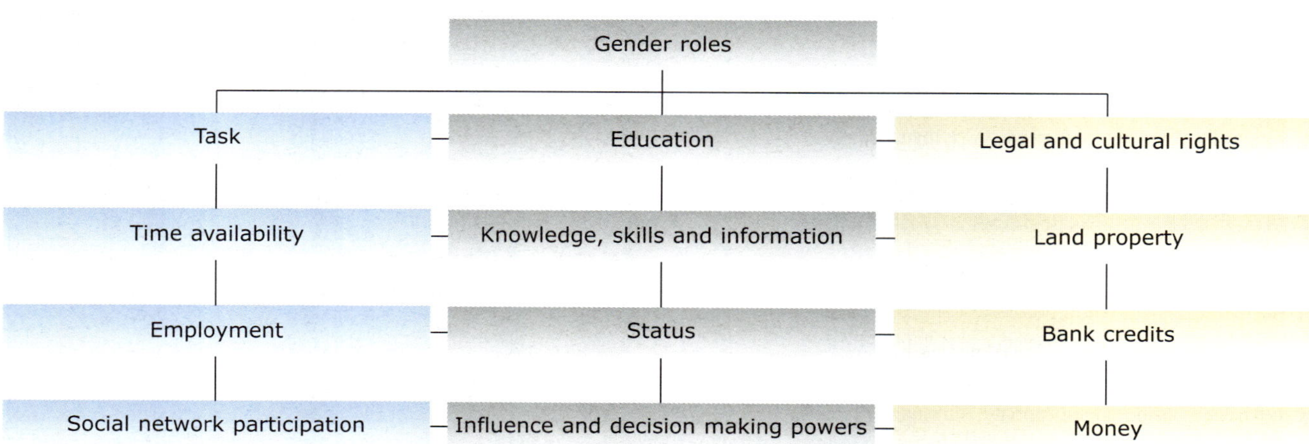

Source: Matiza-Chiuta, T., "Gender and the environment," For SARDC, 1998.

danger. Men attend social and political meetings, are responsible for animal husbandry, hunting and fishing. Income generation for the family is also a primary responsibility of men, and this makes them the main provider of income for the family. Men make key decisions on what crops to be grown, what land preparation procedures to be used, when to harvest, and how much crop to be sold. Men usually also clear and burn bush, fell trees, fence farms against animal pests. Water transportation is another major activity in the basin that is largely dominated by men. In the Zambezi Floodplains, for example, the building of canoes and transporting the family in and out of the floodplain is the role of men.

Women too, attend social and political meetings, including traditional ones, for example, *Kgotla* in Botswana. They cook, maintain and till gardens, fetch fuelwood, water, and keep small stock, for example, goats.

Regarding common tasks, both women and men plant but women are usually responsible for household food crops while men focus on cash crops. Within the fishing industry, both women and men fish but the methods and fishing grounds vary by gender. Deep water fishing is usually done by men and young boys while women are involved in the post-harvest activities such as processing and marketing of fish and fish byproducts.[4] Together with men, women also perform cash production to earn a living.

Women usually harvest renewable resources such as thatching-grass, reeds and grasses to make mats, bowls and other crafts that are sold to hotel owners and tourists. Women, therefore, play an important role in the craft industry that contributes to the tourism industry.[5]

While both men and women participate in the economic spheres, women are assigned additional obligations, which fall within their domestic domain such as food preparation and

Photos: M Chenje

Women and men play specific roles in all socio-economic activities, but increasingly they share responsibilities.

reproductive role, including childcare. As society develops, women's roles shift to suit both the household social needs, but still within the traditional division of labour. Regarding women's tasks, most of them are performed in and around the home, making them homebound.

Photo: M Chenje

Women undertake most of the reproductive tasks, including child-rearing.

At the household level, women perform most of the domestic tasks – reproductive and productive. These refer to child bearing and rearing, food processing, care of the sick and spouses, as well as most of the duties needed for the production of food.

In the area of environmental management, women are active and knowledgeable managers and caretakers of the environment. In many rural settings, women carry out natural resources conservation work, such as soil conservation and tree planting.

In urban areas, women take primary responsibility for the maintenance of adequate living conditions for their families. They usually are responsible for overseeing the family's water and sanitation system. Both in rural and urban areas, women are usually the first level health care providers and cope regularly with disease caused by unhealthy environments.[6]

Due to factors such as urbanisation, gender roles and responsibilities are changing. More women are, for example, now being found in decision-making positions that traditionally have been totally dominated by men.

Box 11.2: Women's workload

"But women already work hard. They already grow most of the developing world's food, market most of its crops, fetch most of its water, collect most of its fuel, feed most of its animals, weed most of its fields.

"And when their work outside the home is done, they light the Third World's fires, cook its meals, clean its compounds, wash its clothes, shop for its needs, and look after its old and its ill. And they bear and care for its children. The multiple burdens of womanhood are too much."

Source: UNICEF, WHO, UNESCO, "Facts for Life: A Communication Challenge," Undated.

Table 11.1: Activities by gender in the Zambezi Basin

Gender group	Activities
WOMEN	Basket making, beer brewing, pottery-making, brick-making, baby care, cooking for the family, cultivating fields, weeding fields, harvesting crops, collecting veld products and firewood (including grass cutting for thatch), collecting water, some fishing and processing, and other household chores.
MEN	Working as farm labourers, brick-making, livestock rearing and tending, ploughing with oxen, hunting, deep water fishing, land clearing, defending the family against physical danger, house-building, community manual work, and making decisions for the community.

Source: Chenje, J., *State of the Environment Reporting Programme for the Zambezi Basin, Inception Mission Report,* Harare, November 1998.

Division of labour

Substantial evidence is available documenting the existing unequal division of labour between women and men in the basin. In rural Tanzania, for example, women work more than 14 hours a day compared with men's 10 hours.

Declining resource availability and very high rates of male out-migration from the rural areas in search of employment have

Table 11.2: Percentage of total labour in working hours between men and women

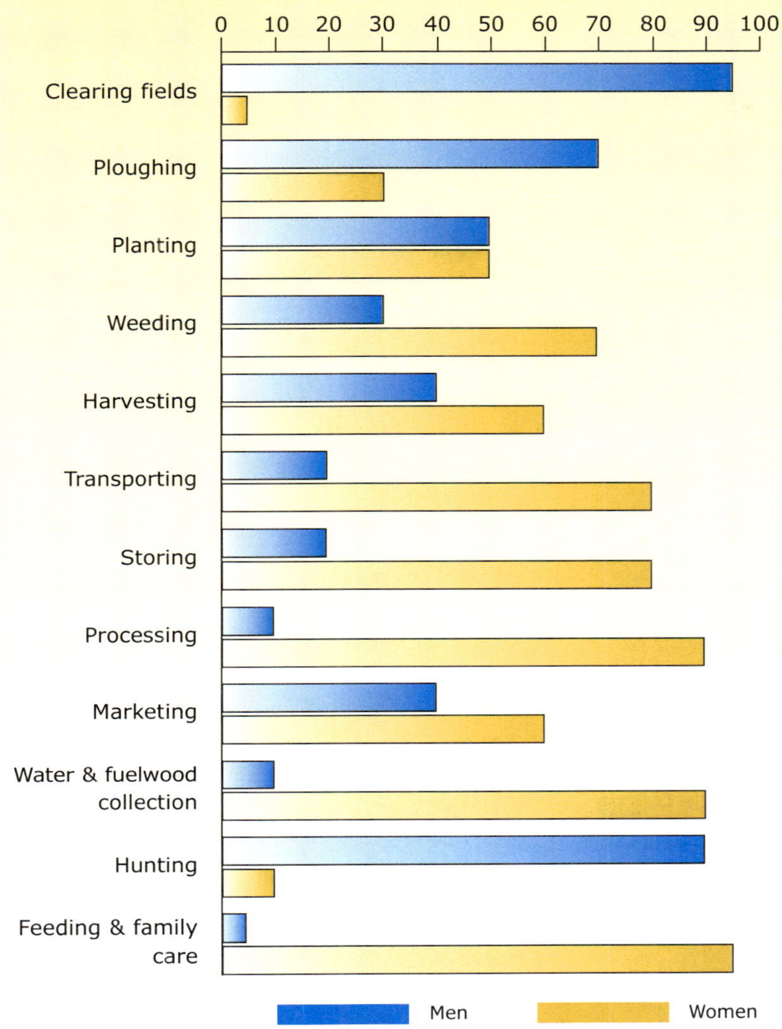

Source: Rodda, A., *"Women and the Environment," Women and World Development series,* London: Zed Books and the UN-NGO Group on Women and Development, 1991.

increased labour demands on women. This is to such a point that unless women are provided with extra resources under their control, there will be a spiralling deterioration in resource management practices.[7]

In Zimbabwe, for example, a comparison of men's tasks in rural areas reveals important differences.[8] Men's work usually has seasonal peaks (ploughing and cattle herding) or are done once in many years (constructing housing and granaries). This gives men an ability to take seasonal or permanent employment elsewhere. By contrast, some of the women's tasks may be seasonal, for example, weeding and harvesting, but social reproductive tasks are daily chores. These are not transferable to men due to cultural practice.

However, women's roles in natural resources management are often portrayed as a major asset that should be harnessed in initiatives to conserve natural resources. This has a negative impact of overburdening women and treating them as a source of cheap labour.

GENDER ISSUES AND INEQUALITIES

A gender issue is realised when there is inequality between women and men, commonly referred to as a gender gap. This may be inequality in the provision of welfare benefits such as education or in the access to production factors such as land. It may also be inequality of participation in the decision-making process at work or at home, and therefore, inequality in control over the distribution of resources and rewards.[9]

In general, the balance sheet of the Zambezi Basin reveals that women do not have an equal share of political power and income compared to men. This is despite the fact that women comprise more than half of the total population in the basin.

The gender statistics in the Zambezi Basin also reveal that:

- Women constitute a majority of the population, up to 65 percent, as most men have migrated to urban centres outside the basin in search of employment.
- Women constitute a larger number of the poor.
- Women are rural based to a larger extent than men, and remain on the periphery of human development and governance.
- Women have access to a lower educational level than men.
- Women work longer hours than men.

Table 11.3: Gender gaps in power sharing and income

Country	Seats held in Parliament (%)		Administrators and managers (%)		Professional and technical workers (%)		Earned income share (%)	
	Women	Men	Women	Men	Women	Men	Women	Men
Angola*	9.6	90.4	4.3	95.7	27.7	72.3	31.5	68.5
Botswana	8.5	91.5	36.1	63.9	61.4	38.6	39	61
Malawi	5.6	94.4	4.8	95.2	34.7	65.3	42	58
Mozambique	25.2	74.8	11.3	88.7	20.4	79.6	42	58
Namibia*	6.9	93.1	20.8	79.2	40.9	59.1	18.9	81.1
Tanzania	n/a	n/a	n/a	n/a	n/a	n/a	n/a	n/a
Zambia	9.7	90.3	6.1	93.9	31.9	68.1	39	61
Zimbabwe	14.7	85.3	15.4	84.6	40	60	38	62

Source: *UNDP, Human Development Report, 1998.*
* Source: *UNDP, Human Development Report, 1995.*

Figure 11.2: The percentage of women who reside in rural areas

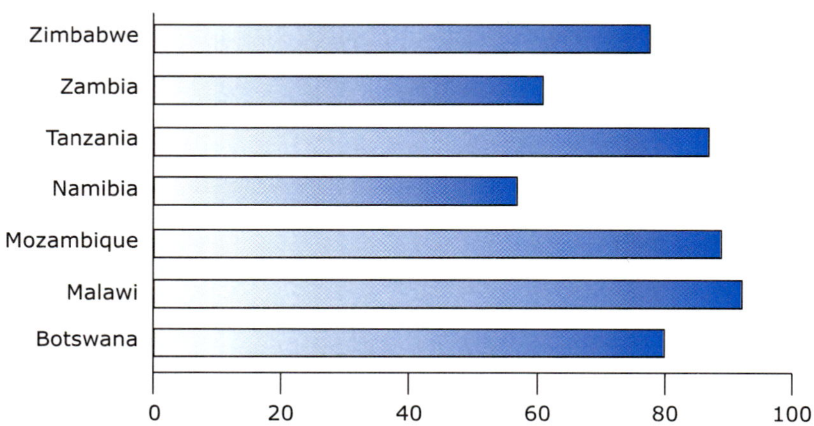

Source: UNDP/SADC/SAPES Trust, *SADC Regional Human Development Report 1998,* SAPES Books, Harare, 1998, p.11.

Photo: M Chenje

The majority of women are rural based and remain on the periphery of human development.

- Most of women's work is unpaid, and their share of household income and decisions are not in proportion to their labour.

All these gender differences and gaps have implications on the environment and the environmental degradation, in turn, has implications for both women and men. Gender and the natural environment are crosscutting issues that cannot be seen in isolation from one another; nor can they be seen in isolation from broader social implications such as health, economics or legal rights.[10]

Key gender issues that impact on the environment are related to access to and ownership of resources.

Access to and ownership of resources

Studies indicate that while women carry a heavier load in both production and reproduction, they do not enjoy similar rights in terms of access to resources.[11] Within a community, rights and access to resources are also often organised through kinship, sexual division of labour and type of production. Men commonly have rights to resources such as land, water, and fisheries. They even have rights to control and allocate certain amounts of women's and children's labour to crops which men control and for which they retain revenue. Women generally do not have rights to most of the natural resources and do not have such rights over men's labour. In natural resources management, men commonly have full disposal rights while women have user rights.

Means of production

Throughout the region, the pattern is that women have less access to the means of production such as land, and support services such as credit.

While women constitute the majority of the agricultural labour force in the region, and are mainly responsible for food production, their land rights are limited in all the countries in the basin.[12] In Zambia, for example, 90 percent of the land available for agriculture falls under traditional land controlled by chiefs who follow patriarchal principles in its allocation. Women have no direct access to land in this situation.

Women are normally entitled to cultivate land controlled by their husbands. Under Tswana Customary law, women's access to land in Botswana is secondary, through men. Although everyone should have equal access to land, in reality, access is differentiated by gender. Unmarried and divorced women are among the poorest households without access to productive land because of lack of means to use the land. In the rural villages, women are often refused allocation in their own right and advised to apply for land through their husbands, brothers or fathers for the unmarried, who have marital power over household assets.

In the Lozi, Tonga and Ngoni cultures of Zambia, a widow cannot inherit her husband's land. In that culture, a woman can get land only if she returns to her village of birth. Even in that case, the land rights held by women usually weakens with time. In Zimbabwe, land-use rights are biased against women. It is assumed that the primary land-use rights to males on behalf of the family will meet the normal needs of women. Customary law vests land-use rights in men, denying women access to land. This has resulted in the uneven distribution of land between men and women. In some parts of Malawi where matriarchal principles are followed, women own land but men decide how the land should be used.

Food insecurity and malnutrition in most of rural Africa can partly be attributed to lack of access and control over land by women.[13]

Closely related to the issue of land is livestock ownership. Men generally own more cattle than women throughout the Zambezi Basin. In Botswana, for instance, of the total national herd, women own only 14 percent. This marginalisation from cattle ownership of women not only reflects female poverty, but also means that those

Table 11.4: Distribution of landholders in Zimbabwe by sex and farm sector

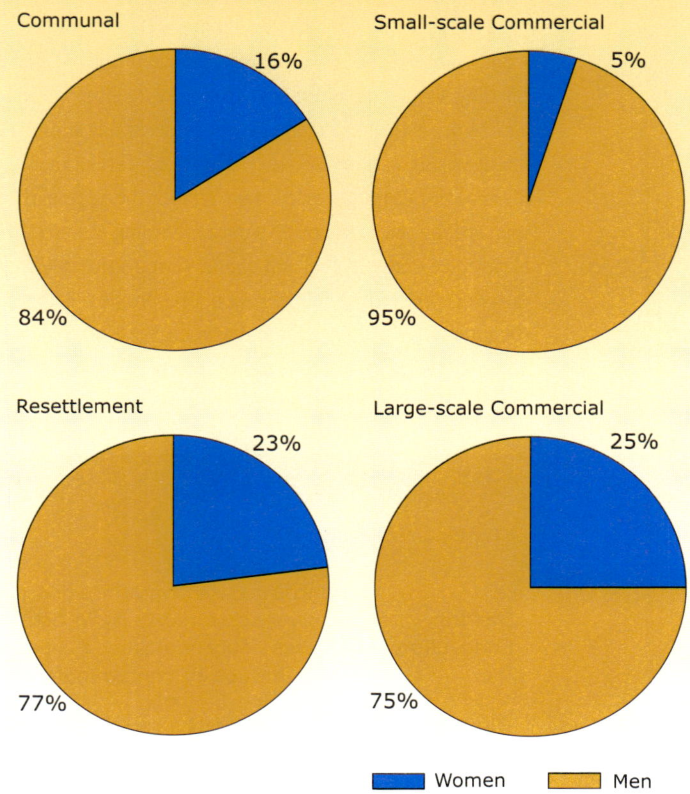

Source: ZERO, *Enhancing Land Reform in Southern Africa,* ZERO, Harare, 1998.

Box 11.3: Development programmes are not targeting women

In Zambia, former Ministry of Agriculture Food and Fisheries (MAFF) permanent secretary, Namukolo Mukutu admitted that many well-intended agricultural development programmes were focused on the wrong target group — men, when the producers of food for household food security are women. He noted that government's agricultural activities such as agricultural research and extension, agricultural credit, land allocation, agricultural education and training, and agricultural mechanism were all focused on the needs and requirements of men. The lack of involvement of women in the mainstream of agricultural development strategy had left a legacy of economic defranchisement of a group that constitutes more than 50 percent of the national population.

Source: "Mulasikwanda Liswaniso," *Zambia Daily Mail,* Lusaka, 4 March 1999.

that have land but no access to cattle have no access to draught power for ploughing. They end up ploughing late, ploughing fewer hectares and getting lower yields.[14] This has a bearing on household food security.

Further, in technology development, the use and maintenance of the resource base and the problems identified by women are often neglected. New technology in agriculture is developed and targeted specifically at men, marginalising women farmers. Technology development is mainly for crops grown by men, and for the large part, men are the ones who receive master farming training. In Malawi, for example, fuel and energy saving stoves were introduced to reduce deforestation. The problem is that the intended users, mainly rural women, have not accepted the technology, because they were not consulted at the design stage of the stove.

The Zambezi Basin also presents dilemmas such as land mines, especially in present and former war zones in Angola, Mozambique and Zimbabwe. These mines make it difficult for women to access raw materials for crafts, for example.

Chemical spraying against tsetse fly and mosquitoes in the Zambezi Basin is a threat to human health. Many of the chemicals used in agricultural production have similar health impacts.

Water and sanitation

As domestic managers, women not only do most of the work regarding collection of water, they also decide which water sources to use for what purposes in what season and how much water to collect, and how to dispose of the various types of waste water. Women also

Table 11.5: Some side effects of residual applications of chemicals to control tsetse fly

Chemical	Some side effects
DDT	Contributes to contamination of human breast milk and chicken.
Dieldrin	Causes acute mortality in fish and carries the risk of contamination of human food.
Endosulfan	Causes acute mortality in bees and fish, carries risk for human health through consumption of contaminated fish.

Adapted from Nagel, P., *Environmental Monitoring Handbook for Tsetse Control Operations,* Margraf Verlag, 1995.

Figure 11.3: Linkages between gender and water related issues

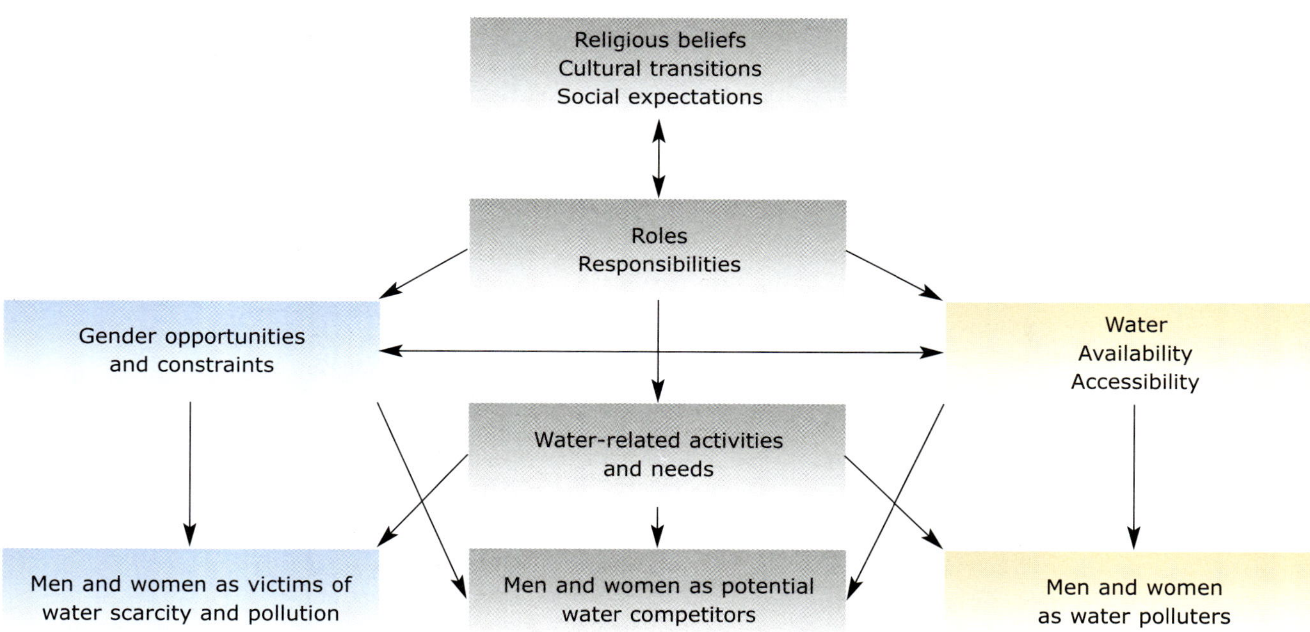

Source: Brismar, A., *Freshwater and Gender, A Policy Assessment, Comprehensive Assessment of the Freshwater Resources of the World,* Stockholm, Sweden, 1997.

direct children's water collection and use, and guide them on sanitation, waste disposal and general hygiene.

An average household of about five members needs 80,000 litres of water annually for household use. Women also need access to water to undertake work for commercial purposes.[15]

Access to safe water supplies and sanitation in the region is generally limited. Over 50 percent of the people in the SADC-region are without access to sanitation and over 40 percent are without access to safe drinking water.[16] With the scarcity of water in many riparian countries, women, particularly those in rural areas, are forced to walk long distances to fetch water. According to a World Bank estimate, some African women use 40 percent of their daily nutritional intake travelling to collect water.[17]

Another survey, carried out by UNICEF in rural areas of Namibia, found that female-headed households are 20 minutes farther away from water sources in the dry season compared to the wet season. The daughters in these households are responsible for fetching household water from distant sources, an activity, which can take over three hours, and compete directly with school attendance.[18]

While estimates vary in the various countries, in Zambia, for example, access to protected water sources in the rural and peri-urban areas range between 25–60 percent and access to sanitation ranges between 40–50 percent.[19]

Due to environmental degradation and deforestation, women in the basin now spend more time looking for fuelwood, water and food, and have less time to cook, forcing them to cut back on the number of cooked meals and their nutritional value. In Mozambique, women spend 15.3 hours during the dry season and 2.9 hours during the wet season each week drawing and carrying water.[20] In Malawi, a study indicates that each household spends five percent of all days and all nights in a year collecting water.[21]

In addition, as a result of their tasks as water suppliers, women are more exposed to water-related diseases. In many urban areas, for instance, poor women living in squatter camps and makeshift homes with no water supplies, often resort to washing clothes and bathing in small polluted rivers flowing through the cities. This is especially the case with the Mukuvisi River in Harare, which flows through both residential and industrial areas.[22] A chemical fertiliser producing company in Harare has for many years been responsible for the discharge of chemical pollutants into the river.

The concept of productive water (based on the premise of relative benefits that could be obtained by allocating water to alternative productive uses and the constraints that may limit such production) has been adopted by the water management reform programmes in the

Daughters are not only responsible for washing dishes and fetching water but also in taking care of their young siblings.

Photos: M Chenje

basin. This principle is determined by gross margins and employment per unit of water used, giving comparative indicators for alternative use. However, there is a danger of supplying water where economic benefits are obvious, substantial and easily quantifiable to the neglect of those cases where the benefits are less clear.

The principle of productive water, first, may have negative impacts on women, whose water-related activities might not be classified as productive. Men in the region predominate in the use of productive water, while women are largely involved in domestic water use. The allocation of water under the principle of productive water may negatively affect women. A study in Nkayi, Zimbabwe, revealed that, at boreholes where water is used for both cattle watering and domestic water needs, conflicts often arose over priorities of different uses with cattle watering generally taking precedence over domestic uses.

Regarding water pricing, most governments have opted for the user pays/cost recovery option, where the users are required to pay for services received and possible additional measures to make the water sector self-financing. If applied without social measures, the option can easily deny women access to water resources. Lack of ability to pay may deny them access to adequate and safe water.

Closely linked to the allocation of property rights over water, is the idea of using water markets as a system for improved allocative efficiency ultimately leading to reduced demand for water. The rationale behind such markets is that they would allow alternative uses for water to emerge through the operation of rational choices induced by a better determination of water price, and that this in turn will help improve its allocation. Considering the gender disparities in resource allocation and property rights in the Zambezi Basin, the use of water markets to improve allocative efficiency would adversely affect women.

Energy

Fuelwood is used for domestic purposes such as cooking and heating. It is also used in crop processing, beer brewing, making pottery, fish-smoking, brick-burning and in other small-scale rural based industries.

Most of the total energy consumed in the Zambezi Basin comes from biomass. In Malawi, for example, 90 percent of the country's net energy supplies come from fuelwood and main consumers are rural households who use fuelwood for cooking and heating.

The intensive use of fuelwood is contributing to serious environmental degradation, due to deforestation and ultimately soil erosion and poor soil fertility. But contrary to some beliefs, women are more of victims than perpetrators of deforestation because generally, they obtain fuelwood from branches and deadwood, not from cutting live trees.

The rate of deforestation was about 0.7 percent during the period 1990–95 in the SADC-region. At present rates of deforestation, one-third of the SADC forests will disappear by the middle of 21st century.[23]

Photo: D Minty

Women walk long distances to collect fuelwood for heating and cooking.

Since collecting fuelwood is mainly a woman's task, she is most affected by fuel scarcity. In Malawi, for example, were forests are disappearing at the rate of 3.5 percent per year, women spend six to nine hours per week collecting fuelwood for household consumption.[24]

Emissions from biomass fuels are dangerous sources of air pollution in the home, where women are usually responsible for cooking and, therefore, also more exposed to the emissions. Wood fuels are capable of producing pollution concentrations higher than fossil fuels under slow-burning conditions.[25]

When fuelwood is in short supply, women shift to alternative fuels such as biogas. This comes from the fermentation of animal dung, producing methane which can be used for lighting and cooking. The major constraint is lack of livestock to produce the dung, water to mix with the dung and sometimes, the labour required in collecting both.

Forestry

Women's gathering activities are very important to household income and nutrition. The traditional uses of forests

Table 11.6: Distance to fuelwood supplies (percent households by Agricultural Developments Divisions)

	<2 km*	2-8 km	>8 km
Karonga	69	31	<1
Mzuzu	89	10	<1
Kasungu	84	13	3
Salima	62	36	2
Lilongwe	68	26	6
Liwonde	56	33	11
Blantyre	63	30	8
Ngabu	52	38	10

* < 2 km = up to one return trip
2-8 km = up to three hours
> 8 km = minimum three hours

Source: Engberg, L., "A Comparison of Rural Women's Time Use in Two Villages in Malawi, 1985"; cited in World Bank, *Women and Development in Malawi: Constraints and Actions*, Washington, D.C., 27 August 1991.

and woodlands are influenced by the economic trends in which communities are subjected. There is widespread occurrence of a variety of edible worms and wild fruits in the basin, which women collect. However, once economic markets for these products are established, women are usually displaced by men who often plunder the resources for economic gain. This has been the case with mopane worms and wildfruits and, to some extent, grasses. In the Caprivi area, the marketing of thatching-grass has seen a displacement of women from their traditional roles of harvesting.

Wetland resources

Communities all over the basin have been exploiting wetland products for centuries and many of the wetlands are centres of large population concentrations. In these communities, women and men have specific roles in the management and utilisation of the wetland resources.[26]

In communities where women fishers play an important role in harvesting fish, gender blind extension and education can cause irreversible damage to the fishery, since women, who often fish in shallow waters where fish breed, may not be aware of conservation and wise use.[27] Lack of alternative sources of income and increasing destruction of aquatic habitats often leads to overfishing to satisfy subsistence needs and generate income.[28]

The good habitat provided by wetlands has attracted a lot of wildlife, ranging from the big mammals such as elephants to waterfowls such as geese. Protected areas have been gazetted and developed around many wetland ecosystems and many more are being planned. In areas where protected areas have been developed on and around wetland ecosystems, gender issues such as women's access to the resources provided by these wetlands are often overlooked. Developers have argued that affected communities can still get limited access to natural resources through licences. But the hustle of obtaining a licence is reported to be unconquerable to many people, who are in the end totally denied access.[29]

Photo: IUCN

Women's access to even basic resources such as thatching grass is limited in some areas of the basin.

ACTIONS AND INITIATIVES TO REDUCE GENDER INEQUALITIES

Legal aspects

All countries in the Zambezi Basin practice plural legal systems that comprise general laws and indigenous customary law together with people's practices and religion. General law includes common law and modern statutes. This plurality is the basis for highly complex legal relationships processes and attitudes. It has profound implications for gender and the changing socio-legal relations within the basin. However, efforts have over the years been made to advance women's status as well as to protect and safeguard human rights in the basin. At the regional level, heads of state and government have committed themselves to protecting and promoting human rights of women and children by adopting the Gender and Development Declaration in September 1997. Among others, the declaration commits SADC governments and states to:

- Repeal and reform all laws, amend constitutions and change social practices which still subject women to discrimination, and enact empowering gender sensitive laws.
- Encourage the mass media to disseminate information and materials in respect of the human rights of women and children.
- Promote women's full access to, and control over productive resources such as land, livestock, markets, credit, modern technology, formal employment, and a good quality of life in order to reduce the level of poverty among women.[30]

At the national level, all countries in the basin have ratified the Convention on the Elimination of All Forms of Discrimination Against Women (CEDAW). The CEDAW, dubbed as the global constitution for women, provides a basis for realising equality between women and men. This is aimed at ensuring women's equal access to means of production, and equal opportunities in, political and public life, including the right to vote and to stand for election as well as to education and employment.

National initiatives

At national policy level, a number of responses have been identified regarding environmental policies. These include institutional strengthening, legislative and policy review, command-and-control strategies using standards, regulations and penalties and those using macro-economic incentives. Other response measures have been far-reaching, seeking to change production and consumption systems and encouraging popular participation in the development of environmental policies.

An even more far-reaching effort has been observed in various countries where environmental rights have been entrenched into the constitution. In Mozambique, the constitution guarantees, all citizens the right to live in a balanced natural environment and the duty to defend it.[31] The Mozambican government also passed an Environmental Law, in July 1997 that aims to achieve the country's sustainable development, and places special emphasis on the role of women in environmental management.[32] The government further integrated gender issues in activities of various departments namely, training, human resources, and management of natural resources. Other measures adopted to meet the environ-

Box 11.4: Malawi integrates gender in its National Environmental Policy

Objective:
To integrate gender, youth and children concerns in environmental planning decisions at all levels to ensure sustainable social and economic development.

Guiding Principles:

a Both women and men should play a key role in the sustainable utilisation of renewable natural resources and other development programmes.

b Basic training in environmental and natural resource management will include gender analysis methodologies and tools.

c Training, research, and programme implementation should incorporate gender issues.

d Developing environmental awareness among youth and children is critical to environmental sustainability now and in the future.

Strategies:

- Ensure that public awareness campaigns on environment include women.
- Facilitate women's participation in population and environment decision-making, resource ownership and management.
- Collect and disaggregate by gender information related to the environment.
- Recognise the importance of gender roles and gender analysis in environmental management in all training programmes at all levels.
- Tailor public awareness campaigns, formal and informal education, and extension services in environmental and natural resource management to the different roles and priorities of men and women.
- Develop and implement an environmental education programme for the youth and children, including their involvement in pilot projects in natural resources and environment.

Source: Government of Malawi, National Environmental Policy, Ministry of Research and Environmental Affairs, Lilongwe, Montfort Press, Limbe, Malawi, 1996.

mental challenges include the commissioning of gender studies on access to usage, control and management of natural resources in several districts of the country.[33]

Similarly, Malawi's constitution mandates the state to actively promote the welfare and development of the people, by progressively adopting and implementing policies and legislation aimed at, for example, managing the environment responsibly. This is in order to accord full recognition to the rights of future generations, by protecting the environmental and sustainable development of natural resources.[34] A section of the Malawi National Environmental Policy calls for integration of gender, youth and children concerns in environmental planning and decisions at all levels to ensure sustainable social and economic development.[35]

In Namibia, the government has defined in its national gender policy strategies that will involve and integrate women into natural resource and environmental issues.[36]

Regional initiatives

At the regional level, SADC has defined achieving sustainable utilisation of natural resources and effective protection of the environment as among its eight main objectives. To this effect, the region has outlined an environmental policy whose main goals are to:

- Protect and improve the health, environment and livelihoods of the people of southern Africa with priority to the poor majority.
- Preserve the natural heritage, biodiversity and life-supporting ecosystems in southern Africa.
- Support regional economic development on an equitable and sustainable basis for the benefit of present and future generations.

Gender has been incorporated into the SADC agenda and a gender unit has been established within the SADC Secretariat in Gaborone.

Some SADC sectors have also developed sectoral gender policies. The SADC Environment and Land Management Sector (ELMS) has an elaborate gender policy and the SADC Water Sector Coordination Unit (WSCU) is working on a programme to empower women and improve their participation in the sector.

Some environmental organisations in the region have also been designing projects that integrate gender from

Box 11.5: Namibia integrates women in natural resources and environmental issues

Namibia has specified in its national gender policy, strategies that seek to involve women in environmental issues. The strategies mandate the government to establish mechanisms that:

- Provide opportunities to ensure that women have access to relevant information and education that will enhance their knowledge and skills, to enable them to effectively participate fully in decision-making and management of the environment.
- Encourage and support, without contravening legislation, indigenous women's innovations and practices relating to additional technologies for food preservation, prevention, relief, and treatment of all types of diseases, as well as the protection of the environment.
- Encourage recycling of wastes by giving economic incentives to both women and men who engage in recycling activities. This will, in particular, help the disadvantaged, low skilled households.
- Enact legislation to protect and safeguard copyright, without exploitation of women and men who hold indigenous knowledge and skills with respect to use and misuse of natural resources and the environment as a whole, for the purpose of national development and economic advancement of the individuals concerned.
- Protect women and men by taking appropriate measures, for example, through strict legislation and compulsory medical care, to reduce identified environmental health hazards originating from industrial development.
- Ensure, in collaboration with other interested parties, that a gender perspective is integrated in the design and implementation of all viable and sustainable environmental and natural resource management programmes at all levels.
- Design research programmes to assess the extent to which environmental hazards affect women in poverty and those belonging to marginalised groups.
- Integrate women's traditional knowledge and practices of sustainable resource use and management in developing and designing gender sensitive research programmes for environmental development.
- Sponsor and promote research to generate information on the role of indigenous women in food production and preservation, soil conservation, irrigation, sanitation, pest management and control, water storage, sources and preservation of energy.

Source: Republic of Namibia, *National Gender Policy*, Office of the President, Department of Women's Affairs, Windhoek, 1997.

planning to implementation levels. One such project is the Zambezi Basin Wetlands Conservation and Resource Utilisation Project (ZBWCRUP) managed by IUCN's Regional Office for Southern Africa.[37]

Global initiatives

At the global level, governments have expressed their commitment to creating a new development paradigm that integrates environmental sustainability with gender equality and justice within and between generations as contained in Chapter 24 of Agenda 21. "As consumers and producers, caretakers of their families and educators, women play an important role in promoting sustainable development through their concern for the quality and sustainability of life for present and future generations."[38]

The Rio Principles, adopted in 1992 in Rio de Janeiro, state: "Women have a vital role in environmental management and development. Their full participation is therefore essential to achieve sustainable development."[39]

Women's role in water management was especially highlighted in a pre-UNCED conference held in Dublin 1992. The Dublin Conference stated that: "Women play a central part in the provision, management and safeguarding of water. This pivotal role of women as providers and users of water and guardians of the living environment has seldom been reflected in institutional arrangements for the development and management of water (and other natural) resources."[40]

The efforts to address gender and mainstream it in all aspects of development received a boost by the endorsement of gender equality at the 1995 Fourth World Conference on Women, held in Beijing, commonly dubbed the Beijing Conference. This endorsement is seen as one of the most radical changes that have taken place in recent years. It is transformational in character, and it is also inclusive; it is beneficial for all, for women as well as for men.

The rationale for mainstreaming a gender equality perspective is twofold: a human right perspective to ensure that women and men have equal access and control over productive resources and information; and an effective development perspective. The latter recognises the different interests, roles, responsibilities and needs of women and men in order to meet and support them, to achieve sustainable development. A mainstreaming strategy has two major aspects:

- It seeks to enable women and men to benefit equitably from the resources, opportunities and benefits of the mainstream development process. This requires integration of equality concerns into the analyses and formulation of policies, programmes and projects, with the objective of ensuring that these have a positive impact on women and reduce gender disparities.

Box 11.6: Integrating gender in the Zambezi wetlands conservation project

The Zambezi Basin Wetlands Conservation and Resource Utilisation Project (ZBWCRUP) has devised an action plan to integrate gender in all its activities at the overall project management, coordination and monitoring, field project implementation and management, and community levels. The action plan however, cautions that a mainstreaming approach should not preclude initiatives specifically directed towards women or towards equality between women and men. In order to achieve effective gender mainstreaming in the implementation of activities, the following requirements need to be met:

- Adequate expertise on gender issues in project management, coordination, implementation, monitoring and evaluation systems.
- Attention to gender issues needs to be incorporated in all terms of reference for all studies under the project.
- Statistics should be disaggregated by sex.
- Identification of areas requiring more data to highlight important gender disparities.
- Reporting on the ZBWCRUP project progress should include attention being given to gender issues in the area.
- The project should take gender issues as an integral part of all its dialogue discussions, programming and planning process, and in all quarterly and annual reviews.
- Gender issues should be raised with implementation partners at all levels of project implementation.
- An ongoing process of dialogue and consultation on gender issues integration should be established with field officers and networks.
- Need for adequate capacity within partner institutions to work with gender.

Source: Lopi, B., "Gender and the Environment in the Zambezi Basin" For SARDC, 1998.

Box 11.7: Mainstreaming gender

In addressing the lack of adequate recognition and support for women's contribution to conservation and management of natural resources and safeguarding the environment, governments and other actors should promote an active and visible policy of mainstreaming a gender perspective in all policies and programmes, including, as appropriate, an analysis of the effects on women and men, respectively, before decisions are taken.

Source: UN, *Women: Platform for Action and Beijing Declaration*, New York, 1996.

Photo: M Chenje

The rationale for mainstreaming gender equality ensures that women and men have equal access to resources.

- It also seeks to include the interests, needs, experiences and visions of women in the definition of development approaches, policies and programmes and in determining the overall development agenda. This requires women's empowerment through strategies to enable them to formulate and express their views and participate in decision-making across all development issues.

The Global Women, Environment and Development programme, designed by the United Nations Development Fund for Women (UNIFEM) in 1990, is being used by some NGOs in developing countries as a guide to promote the integration of women and development concerns into global environmental policy. The programme, referred to as "A Vision for the 21st Century," has four goals to:

- Provide women with the means to farm more productively.
- Provide women with knowledge and technologies to use and manage resources more effectively.
- Provide the financial support necessary for women to act in their own self-interest and in the interest of environmental sustainability.
- Stimulate a more effective voice in environmental decision-making and the issues that impact on sustainable development.[41]

FUTURE CHALLENGES

Legal aspects

Despite the legal reforms that are in place, women in the basin still suffer gender discrimination and violations of their rights. It is usually assumed by legislators that the law affects women and men equally, but the reality is different. Many women in the basin do not exercise the rights that the laws specifically guarantee them due to many factors. These include:

- Ignorance of the law and its administration.
- Economic hardships that make it difficult for them to pursue their legal rights.
- Insufficient information on existing rights.
- Lack of awareness on the existence of the laws.
- Lack of assertiveness and confidence on the part of women and fear of breaking valued relations with the family.[42]

For the basin population therefore, law reform must be complemented with the adoption of targeted interventions aimed at correcting gender inequalities, and equalising opportunities for women and men. A real and critical challenge remains in that even where laws providing for gender equality have been put in place, mechanisms for enforcement of these laws are either weak or non-existent. Customary law and beliefs have a great impact on the legal system that is practised in the basin. Women, however, tend to lose out under both systems of the law. On the one hand, they may not be able to implement effectively the rights they formally have under general law; while on the other, they are deprived of their rights under customary law because of the way society and the courts interpret it.

Policy and decision-making

Although countries in the basin have a number of policies that, directly and indirectly, protect the environment, not many of these incorporate gender. Many of the policies tend to be blind to the fact that women relate to the environment differently from men. As a result of this gender blind approach, some environmental policies and legislation in basin countries do not effectively take care of the different groups of people who are differentiated by gender.

For example, only Angola and Namibia out of the eight riparian states have identified the issue of women and the environment as a national priority. And at SADC level, women and the environment is not among the top four regional gender priorities. Environmental related issues are, however, mentioned in the SADC Gender and Development Declaration.

Women remain largely absent at all levels of policy formulation and decision-making in natural resource and environmental management, conservation, protection and rehabilitation. Their experience and skills in advocacy for and monitoring of natural resource management too often remain marginalised in policy-making and decision-making bodies as well as in educational institutions and environment-related agencies at the management level.[43]

Within the basin, there are a number of developments and environmental programmes such as ZACPLAN, the Zambezi Basin Wetlands Conservation and Resource Utilisation Project, wildlife management and tourism in national parks and other protected areas, major fisheries development in Kariba, Lake Malawi/Nyasa and Zambezi Delta. Despite their long list of roles and responsibilities, women in the Zambezi Basin have not been actively involved in the planning, implementation, management and evaluation of these projects and programmes.

This is, for example, due to the fact that women are rarely trained as professional natural resource managers with policy-making capacities, such as land-use planners, agriculturists, foresters, marine scientists and environmental lawyers. Even in cases where women are trained as professional natural resource managers, they are often under-represented in formal institutions with policy-making capacities at the national, regional and international levels. Women are also not equal participants in the management of financial and corporate institutions whose

Photo: IUCN

Many of the policies in the basin countries tend to be blind to the fact that women relate to the environment differently from men.

decision-making most significantly affects environmental quality.

Furthermore, there are institutional weakness in coordination between women's non-governmental organisations (NGOs) and national institutions dealing with environmental issues despite the recent rapid growth and visibility of women's NGOs working on these issues at different levels.

Despite that the Rio and Dublin principles on water resources call for the effective participation of women in water resources development and management, these principles as they are applied today in the Zambezi Basin have not as yet contributed to the improvement of women's access to water resources. A number of countries in the basin have undertaken water sector reform programmes whose main objectives are to improve the management and development of water resources as well as revise the water rights that were entrenched within minority groups. Principles applied in these reform programmes include the commoditisation of water or treating water as an economic good, allocating water on the productive water principle, user pays principle and demand management through economic instruments. The application of the principles have serious gender implications that need to be considered, if women are not to be denied access to water resources.

Gender specific research

Information on the extent and severity of the various environmental problems and their impact on women as compared to men in the Zambezi Basin is very scanty. This is due to lack of gender specific research in the basin. There is need for more data to highlight the important gender disparities in the Zambezi Basin.

Understanding gender issues is a logical part of the resource-sensitive approach to development. Therefore, gender differences and inequalities must be taken into account if development interventions are to be effective in serving the needs of women as well as men, now and into the future.

LINKAGES TO OTHER CHAPTERS

Chapter 1: Regional Overview: People and Environment
Women in the Zambezi Basin are poorly represented in sectors related to environment and development, despite the fact that they comprise more than half of the total population. It is, therefore, crucial to engage both women and men in identifying and analysing problems related to environmental as well as socio-economic and political aspects that are critical to achieving sustainable development.

Chapter 2: Physical Features and Climate
Global climate change is a limiting factor in the Zambezi Basin states' efforts to eradicate poverty, with the most immediate problem being drought with consequences such as land degradation and food insecurity. Since women are at the centre of the regions' food production the analysis of the implications of climate change must have a gender perspective.

Chapter 3: Water and Wetland Resources
Water constitutes the most important component of a wetland ecosystem. In rural communities, women are usually the main managers of water resources. They fetch and supply drinking water and for household use, and are heavily dependent on rain-fed agriculture and, in some areas of the region, fisheries.

Chapter 4: Biological Resources and Diversity
With primary responsibility for providing their families with food, water, fuel, medicines, fibre, fodder and other products, women rely on healthy and diverse ecosystems. Women's important role in the management of biodiversity and biological resources must be recognised, and their participation in decision-making must be ensured at all levels of resource management.

Chapter 5: Agriculture
While women constitute the majority of the agricultural labour force in the region, and are mainly responsible for food production, issues related to food insecurity and malnutrition must be tackled with a gender perspective.

Chapter 6: Industry
Due to factors such as urbanisation and industrialisation, gender roles and responsibilities are changing. More women are, for example, now being found as professional or technical workers within industry.

Chapter 7: Energy
Since collecting fuelwood is mainly a woman's task she is most affected by fuel scarcity. To improve women's standard of living and move towards food security, there is need to decrease the rate of deforestation and the growing scarcity of fuelwood.

Chapter 8: Tourism
Women in the Zambezi Basin contribute substantially to the family budget through income generating activities such as food processing, trading of agricultural products and production of handicrafts, which are sold to hotel owners and tourists. Women play a very important role in the craft industry that is contributing to the tourism industry.

Chapter 9: Pollution
Household burning of fuelwood, crop residues, manure and coal for heating and cooking leads to localised air pollution. Since women usually are responsible for cooking, they are more exposed to the emissions from biomass fuels.

Chapter 10: Poverty
The links between poverty, gender and the state of the environment are becoming more and more obvious. Poor people, the majority of whom are women, have no choice but to exploit natural resources in order to survive. This way of living may in turn induce further environmental degradation.

Chapter 12: Environmental Management and Regional Cooperation
Women remain largely absent at all levels of policy formulation and decision-making in natural resource and environmental management, conservation, protection and rehabilitation. Instruments and arrangements put in place at the regional level in the monitoring and management of the environment in the Zambezi Basin should include gender concerns and perspectives.

Chapter 13: Trends and Scenarios
Gender, as it applies to the people of the Zambezi Basin states needs to focus on changing the quality of life in the riparian states, especially that of women, through better access to, use and management of natural resources and setting a solid foundation for development and sustenance of human rights.

ENDNOTES

1 Rathgeber, E., M., *Bringing Gender and Environment into the Curriculum: Some African Examples*, International Development Research Centre, Nairobi, Kenya, 1994, p.4-5
2 ibid
3 Matiza-Chiuta, T., "Gender and the environment", for SARDC, 1999, p.3
4 IUCN, "Zambezi Basin Wetlands Conservation and Resource Utilisation Project: Gender Integration Strategy", unpublished document, IUCN, Harare, 1997
5 ibid
6 op. cit 1
7 SAPES Trust, *Gender in Southern Africa*, Harare, 1992
8 ZWRCN, SARDC, *Beyond Inequalities Women in Zimbabwe*, Harare, 1998
9 TAMWA/UNICEF, *How to plan for women's empowerment*, 1992, p.3
10 SADC-ELMS, "Gender Strategy", Draft, Maseru, 1997, p.5
11 SARDC, *Beyond Inequalities, Women in Southern Africa*, Harare, 1999
12 Ruth, M., "Gender in Southern Africa, Conceptual and Theoretical Issues", Undated, p.14
13 ibid
14 UNDP/SADC/SAPES Trust, *SADC Regional Human Development Report 1998*, SAPES Books, Harare, 1998, p.21
15 SIWI, "Water - The Key to Socio-economic Development and Quality of Life", Proceedings Stockholm Water Symposium, 1998, p.221
16 op. cit 14 p. 39.
17 Chenje, M. and P. Johnson; *Water in Southern Africa*, Maseru/Harare, 1996, p.12
18 Gender Net, http://www.worldbank.org/gender/info/namibi.htm
19 UNICEF/GRZ, "UNICEF/GRZ programme of Cooperation for Women and Children 1997-2001", UN Building, Lusaka, Zambia p.3
20 Gender Net, http://www.worldbank.org/gender/info/mozamb.htm
21 Nankhuni, F., "The Rural Transport Burden on Women and the Children: Implications on Economic and Social Development of Malawi and Suggestions for Policy Reform, 1995". Paper presented at the Seminar on Women and the Democratisation Process in Malawi, Ryalls Hotel, Blantyre, 29-31 May, 1995
22 Royal Netherlands Embassy, "Women, Water and Sanitation", Paper by the Royal Netherlands Embassy, Harare, 1988
23 UNDP/SADC/SAPES Trust, *SADC Regional Human Development Report 1998*, SAPES Books, Harare, 1998, p.14
24 UNIMA, SARDC, *Beyond Inequalities, Women in Malawi*, Harare, 1997
25 Dankelman, I., and J. Davidsson, *Women and Environment in the Third World: Alliance for the Future*, Earthscan Publications Limited, London, 1988
26 op. cit 4
27 ibid
28 op. cit 17
29 op. cit 4
30 SADC, *Gender Mainstreaming at SADC, Policies, Plans and Activities*, Gaborone, 1999
31 UEM, SARDC, *Beyond Inequalities Women in Mozambique*, Maputo and Harare, 1999
32 ibid
33 ibid
34 Chenje, M., "Environmental Rights are Human Rights", Paper presented at a seminar on Human Rights and Justice, 1-2 August 1995, Harare, 1995
35 Government of Malawi, *National Environmental Policy*, Ministry of Research and Environmental Affairs, Lilongwe, Malawi, 1996
36 Republic of Namibia, *National Gender Policy*, Office of the president, Department of Women's Affairs, November, Windhoek, 1997
37 op. cit 4
38 United Nations, "UN Beijing Declaration and Platform for Action", 1996, pp. 138-139
39 United Nations, *Earth Summit: Agenda 21 - The United Nations Programme of Action from Rio*, UN Department of Public Information, New York, 1992
40 Sida, "A Gender Perspective in the Water Resources Management Sector", Department of Natural Resources and the Environment, Stockholm, Undated, p.9
41 Brasileiro, A. M., *Gender and Sustainable Development: A New Paradigm*, UNIFEM, New York, USA, 1996
42 University of Western Cape, SARDC, *Beyond Inequalities Women in South Africa*, Cape Town and Harare, 1997
43 op. cit 38

REFERENCES

Athukorala, K., "The Need for Gender Analysis in Strategic Planning for Effective Water and Management in Sri Lanka", *Water Resources Development*, Volume 12, Number 4, 1996

Bhatt, M., R., "*Women in Water Management:* The Need for Local Planning" *Development In Practice*, Volume 5, *Number 3,* August 1995.

Brismar, A., *Freshwater and Gender, A Policy Assessment; Comprehensive Assessment of the Freshwater Resources of the World*, SEI, Stockholm, Sweden, 1997

Chenje, J., "State of the Environment Reporting Programme for the Zambezi Basin", Inception Mission Report, Harare, November 1998

Cleaver F., C., *Water as a Weapon: The History of Water Supply Development in Nkayi District*, *Zimbabwe,* Review, Vol. XVII, Harare, 1994

de Sherbinin, A., (ed.), *Water and Population Dynamics, Local Approaches to a Global Change*, IUCN The World Conservation Union & Population Reference Bureau, Gland, 1997

FAO, *Gender; Key to Sustainability and Food Security*, FAO Plan of Action for Women in Development, 1996 - 2001, Rome, 1995

FAO, *Integrating Gender Considerations Into FAO Forestry Projects*, Guidelines, FAO, Rome, 1993

FAO, *Guidelines for the Improvement of Statistics on Women; Obtaining Statistics from National Agricultural surveys in selected countries of the Near East*, FAO, Rome, 1996

Guijt, I., *Gender and the Environment in Development Cooperation; An Assessment of Agenda 21 and the Platform for Action*, Department for Policy and Legal Services and Department for Natural Resources and the Environment, 1997

Hannan-Andersson, C., "Gender Perspectives on Water Resources Management: Domestic Water Supply and Environmental Sanitation", 1995

Hiscock, E.H.J., Matiza-Chiuta, T., and T.J.F. Lash, *Zambezi Basin Wetlands Conservation and Resource Utilisation Project, Inception Mission Report,* Harare, 1996

Kwaule, F., "Gender and Peri-Urban Water Supplies in Malawi", Lilongwe, 1993

Kullberg, M., "The State of the Environment in the Zambezi Basin", Proceedings of a Regional Workshop held at Mazvikadeyi, Zimbabwe, 1998

Leach, M., "Gender and the Environment: traps and opportunities", *Development in Practice*, Volume 2, No.1, 1992

Longwe, S.H., "The Evaporation of Gender Policies in the Patriarchal Cooking Pot", *Development and Practice, Volume 7, Number 2,* May 1997

Maya, R.,S., "Energy and Environment Project Study on Energy - Environmental Interactions Zimbabwe", Harare, 1994

Mbewe, A., "Organisation of Institutions Related to Energy and Environment in Zambia", Lusaka, 1995

Ministry of Environment and Tourism, *Zimbabwe Environmental Impact Assessment Policy*, Harare, 1994

Ministry of Environment and Tourism, "Proceedings of the First National Workshop on State of the Environment Reporting in Zimbabwe, held at Kadoma, Zimbabwe, 21-23 January 1997", 1997

Governent of Namibia, *Wildlife Management, Utilisation and Tourism in Communal Areas*, Ministry of Environment and Tourism, Windhoek, 1995

Muzungu, E. and P. van der Zaag (eds.), *The Practice of Smallholder Irrigation, Case Studies from Zimbabwe*, University of Zimbabwe, Harare, 1996

Ress, P., *Women's Success in Environmental Management*, *Volume 4, Number 1,* 1992

SADC, *Marine Fisheries Policy and Strategy*, Marine Fisheries Co-ordinating Unit, Windhoek, 1994

SADC, "Food, Agriculture and Natural Resources", SADC Annual Consultative Conference, 1-4 February, Johannesburg 1996 (a)

SADC/SARDC-WIDSAA, *SADC Gender Monitor*, *Issue 1,* Harare, February 1999

SAPES Trust, *Gender in Southern Africa, Conceptual and Theoretical issues*, Harare, Zimbabwe, 1992

SARDC-WIDSAA and National Partners, series: *Beyond Inequalities: Women in Angola, Botswana, Lesotho, Malawi, Mauritius, Mozambique, Namibia, South Africa, Swaziland, Tanzania, Zambia and Zimbabwe*

Woroniuk, B., "Gender Equality, Experience and Results Analysis Exercise", Synthesis Report, 1996

Zimbabwe Women's Resource Centre and Network, *Gender and Water*; Bibliography No: 25, 1997

ZERO-Regional Environmental Organisation, *Enhancing Land Reform in Southern Africa, Reviews on Land Reform Strategies and Community Based Natural Resources Management,* Harare, 1998

12

ENVIRONMENTAL MANAGEMENT AND REGIONAL COOPERATION

The Zambezi Basin states face many environmental challenges, ranging from the physical and ecological to social and economic. The problems vary and include land degradation, deforestation, air and water pollution, and habitat conversion due to overexploitation of resources. Growing poverty, rapid population and economic growth are causing increased demand on resources, particularly at a time when many of the existing policies and institutions are too weak for effective management of such demand. The problems also include increasing discharge of both solid and liquid wastes, some of which are not properly managed due to inappropriate laws or antiquated technology. All these add up to the mammoth challenge to achieve effective environmental management and sustainable regional cooperation.

Photo: M Kullberg

The construction of the Kariba Dam has facilitated cooperation between Zambia and Zimbabwe and laid the foundation for expanded cooperation in the whole basin.

Perhaps the greatest symbol of regional cooperation in the Zambezi Basin, is the Kariba Dam. With its completion, Lake Kariba has spawned a successful inland *kapenta* fisheries industry, a thriving tourism industry, and hydroelectricity generation, which has helped some countries in the region to meet part of their power requirements.

Reviewing regional cooperation in the Zambezi Basin, requires acknowledging that cooperation does not necessarily always yield positive results. For example, the environmental and social impacts due to the construction of the Kariba Dam were devastating. Thousands of local people on both sides of the Zambezi River were displaced and more than 5,000 animals were rescued as the impounded waters of the river rose. The majority of the displaced Tonga people are still living in abject poverty on both sides of the river, still depending on government handouts.[1]

For regional cooperation to be successful, a multi-sectoral approach is necessary, taking into account not only water and environmental management, but also social and economic aspects such as poverty and economic development.

SOME AREAS STRENGTHENING REGIONAL COOPERATION

Water and environmental management

The sharing of water across national boundaries has over the years become one of the world's most critical foreign policy issues. This is also true of the 14-member Southern African Development Community (SADC) region in which the countries share at least 15 large and small river basins, covering about 70 percent of the region. Sharing water resources is just as complex for a small water basin such as the Pungwe between Mozambique and Zimbabwe as it is for large ones such as the Zambezi in which eight countries have common interests.

Shared watercourses are a potential source of conflict between or among countries, usually over water usage. A finite amount of freshwater is available to satisfy the needs of the growing population, the production of food, industrial and energy production, cities and conservation of nature. Poor environmental management in the basin such as through pollution and deforestation causes problems, including the spread of diseases, soil erosion, and siltation downstream.

Photos: M Kullberg, insert: D Minty

People overexploit the environment for survival, worsening environmental degradation and reducing their own ability to derive a living from the same environment.

But shared watercourses can also be instrumental in creating opportunities for strengthening cooperation. The fact that the SADC countries share 15 river basins means that cooperation should be as natural as the watercourses themselves, which are used, in some cases, to demarcate national boundaries. The common river basins have also since time immemorial fostered close ties between different communities who today find themselves behind imaginary boundaries drawn on a map and enforced through the strict adherence of longitudinal and latitudinal coordinates.

Another example justifying regional cooperation is the issue of natural resource management. Environmental problems and issues such as deforestation, migration of alien species such as the water hyacinth, and movement of wildlife, know no boundaries. Ecosystems transcend both national and political boundaries and often do occur in two or more countries and are subject to different and sometimes opposing management practices.

Poverty

Cooperation in the Zambezi Basin largely depends on its large pool of human resources. With a total population of about 38,4 million people living in the basin, the region's human resources are central to any effective management of the environmental resources in the region.

Poverty is a major problem in the Zambezi Basin, forcing hundreds of thousands of people to overexploit natural resources for survival. The Human Poverty Index, introduced by the United Nations Development Programme (UNDP) in 1997 to measure poverty in developing countries, shows that 40 percent of the people in SADC live in poverty. Of the eight basin states, Mozambique has the highest level of human poverty with two-thirds of the people affected. Malawi and Tanzania are second and third with 47.7 percent and 39.8 percent respectively.[2]

People overexploit the environment for survival, worsening environmental degradation and reducing their own ability to derive a living from the same environment. The result is increased vulnerability at the household level, and increased poverty generally. Poverty in resources due to overexploitation translates into human poverty. While SADC countries have for more than two decades invested a lot of human, financial and infrastructural resources into food security programmes and have identified poverty alleviation as a priority, increasing poverty remains an indictment on the region. Its initiatives seem to have so far failed to break the poverty and environment circle.[3]

It is generally accepted that structural adjustment programmes, which have been adopted by some of the basin states, have also contributed to increased poverty in the basin. In some cases, structural adjustment programmes have not led to expected economic growth and improved quality of life. They have instead, contributed to the increased commercial exploitation of natural resources for export. Often these resources are not processed but exported as raw materials, limiting foreign currency earnings for these countries.

Conflicts, for example the war in Angola between the government and Unita rebels, continue to be a major factor in the impoverishment of the people. The resumption of war in Angola has increased the internal displacement of people, leaving about 1.5 million homeless.[4] It has also forced thousands of refugees into neighbouring countries such as Zambia and Namibia. Often these refugees have no choice but to overexploit resources in their new settlements, increasing conflicts with local communities. Poverty alleviation and conflict resolution are, therefore, major activities for regional cooperation, and should not been seen only as national activities.

Economic development

Economic growth in the region has resulted in problems such as waste production and management, air and water pollution, among others. It has also led to heavy water and energy demands to satisfy both a growing population and industry. Several energy sharing projects are currently either ongoing or planned within the SADC region to meet demand. Among other reasons, limited natural loca-

Photo: M Chenje

The use of coal in tobacco curing produces carbon dioxide, one of the greenhouse gases.

Story: Energy sharing initiatives in the basin

In a bid to alleviate environmental degradation, Tanzania has signed an agreement to link that country's power grid to that of Zambia. The project, estimated to cost US$153 million, will enable Tanzania to receive 200 megawatts of electricity from its neighbour to augment its own supply.

The agreement will entail construction of a 700 km transmission line from Serenje via Kasama in Northern Zambia to Mbeya in southern Tanzania.

Baruany Luhanga, managing director of state-owned Tanzania Electricity Supply Company, (TANESCO) believes the inter-linkage will go a long way in solving power problems in southern Tanzania.

Except for regional centres, most of the southern parts of the country have been without electricity since independence 38 years ago.

Residents rely on fuelwood as their main source of energy, resulting in systematic tree-felling and desertification.

According to Edward Lowassa, the minister of state responsible for environmental conservation, about 60 percent of the country's land mass is threatened by desertification, much of which he blames on the cutting down of trees. This same indiscriminate tree-cutting has destroyed water sources in Tanzania, affecting rainfall patterns and fuelling drought.

"At least seven regions out of 25 face water shortages because their water sources have been destroyed. Consequently, only between 28–43 percent of the people in those regions have access to potable water," he says.

Lowassa says the destruction of water sources has also caused reduced agricultural production. Farmers who once produced 30 bags of maize on 0.45 ha, are now harvesting six bags using the same inputs.

Data from the National Environment Management Council (NEMC) shows that 400,000 ha of forest are destroyed every year, to produce charcoal and firewood for cooking. Only 10 percent of the country's 30 million people use electricity, the rest depend on firewood.

Luhanga believes that the inter-linkage of power from Zambia will significantly help the country fight deforestation.

He says the regions most likely to benefit are Mbeya, Rukwa, Iringa, Ruvuma and Mtwara which border Malawi, Mozambique and Zambia. TANESCO and the Zambia Electricity Supply Corporation (ZESCO) have completed technical, economic and financial studies, which were initiated four years ago. The studies confirm the project is viable.

Baldwin Msowoya, TANESCO's deputy managing director, says the technological and economic studies involved selection of the line route and the design of the transmission line and substations.

An environmental impact assessment (EIA) study of the line route has been carried out to identify any adverse impacts from the project.

The linkage to the Zambian power grid will enable Tanzania to be part of the Southern African Power Pool (SAPP) which aims at creating a common market for electricity in the SADC region.

SAPP is already operational in Botswana, Lesotho, Mozambique, Namibia, South Africa, Swaziland, Zambia and Zimbabwe.

Tanzania's installed electricity generation capacity is about 642 megawatts of which 62 percent is from hydro stations, and the rest from 44 diesel generators scattered throughout the country.

However, only about 400 megawatts is actually generated which is not nearly enough to meet the nation's demand of more than 1,000 megawatts.

Unreliable supply frequently forces Tanzania to resort to power shedding, affecting the country's economy.

Basil Mramba, the Mbeya Regional Commissioner, believes that the power inter-linkage will hasten rural electrification. Apart from slowing down the rate of deforestation, the project will also help to alleviate poverty in villages.

"If villages were electrified, they could set up small-scale industries like carpentry, fruit canning factories and small-scale mining," he said.

The electricity could also be used to cure tobacco. Currently, the crop is cured by firewood, which has led to depletion of forests in southern and central Tanzania.

At the moment for every 20 trees cut, only one tree is replanted. To address this imbalance, the government has started a campaign to plant 100 million trees by June, 2000.

By Alfred Mbogora

tions to generate hydroelectricity have seen some governments turn to thermal power generation, resulting in serious air pollution.

Economic development has also impacted on lifestyle with many people, particularly in urban areas, aspiring to be motorists. One of the results has been a massive growth in motor vehicles in the SADC countries in general and the Zambezi Basin states, in particular. Zimbabwe, for example, has more than 500,000 vehicles on the road and most of them are found in cities within the basin. Linked to economic development is also growing consumption with solid waste management problems as a consequence.

Land, water and air pollution, both local and transboundary, have a negative impact on sustainable development, influencing social, economic and ecological aspects, all of which contribute to sustainable development. The issues above and many others, provide many opportunities for increased regional cooperation.

EVOLUTION OF REGIONAL COOPERATION

The Zambezi Basin has been managed as disparate parts within eight national boundaries defined during colonialism. Until recently, those eight national parts, together totalling about 1,321,900 square kilometres, were not seen as part of a whole but independent components able to survive outside the natural unit that is represented by the basin, the third largest in Africa. The result was that the Zambezi River and/or its tributaries were seen as beginning and ending within national boundaries.

Historically, development projects in the basin were first and foremost undertaken within the national context rather than regional. This was a result of the existing political situation then. Regional cooperation in the Zambezi Basin only started in earnest as the political winds of change blew across the region. The 1980s saw the independence of Zimbabwe and the start of the process, which led to Namibian independence in 1990.

Table 12.1: Energy projects within the Zambezi Basin 1996-98

Title	Site
Regional Rural Electrification Programme	SADC
Regional Hydroelectric Hydrological Assistance Programme	Upper and Middle Zambezi and Kafue sub-basins
Kafue Gorge Regional Training Centre	Kafue Gorge
Small Hydropower Plants in Malawi	Malawi
Malawi/Zambia Power Cooperation in the Border Region	Malawi/Zambia Border
Mozambique-Malawi Interconnection of Electricity Supplies	Malawi/Mozambique
Cahora Bassa Power to SADC	SADC
Upgrading of the Kafue Gorge Power Plant	Kafue, Zambia
Refurbishment of National Control Centre	Zambia
Refurbishment of Victoria Falls Power Station	Victoria Falls, Zambia
132kV Tie-line Zambia-Malawi	Zambia-Malawi
330/220 kV Tie-line Zambia-Tanzania	Zambia-Tanzania
Power cooperation between Zambia and Namibia	Victoria Falls-Katima Mulilo
SADC Renewable Energy Project	SADC
SADC Capacity Building on Utilisation of Renewable Technologies	SADC
Development of National Woodfuel Strategies and Plans	SADC
Identification and Support of Non-Governmental Organisations and women's groups dealing with Woodfuel	SADC
Assessment of Environmental and Socio-economic Impact of Woodfuel Scarcity	SADC
Improvement of Woodfuel End-use Efficiency in Rural Industries of the SADC Region	Angola, Tanzania, Mozambique Zimbabwe
Rural Energy Planning and Environmental Management Training Programme	SADC
Strengthening the Coverage of Woodfuel and Environmental Protection in relevant SADC Training Institutions	SADC
SADC Industrial Energy Management	SADC
Demand Side Management Opportunities for SADC	SADC
Energy Efficiency Improvements in SADC Heavy Industry	SADC

Source: SADC Energy Sector Reports, Maputo 29 – 30 January, 1998 & Lusaka 10-12 February, 1999.

Before these changes, countries in the region were in conflict, supporting liberation movements in both Zimbabwe and Namibia in their struggle against colonialism and illegal occupation. This was particularly the case during the 1970s.

Impact of war

Even though efforts to manage the Zambezi Basin as a unit started in the 1980s before Namibian independence, significant progress was only realised after the end of civil war in Mozambique in 1992, following the signing of a peace accord between the Frelimo government and Renamo rebels. This was consolidated with the all-party elections in 1994, which not only retained the Frelimo government but also confirmed Renamo as the official opposition party.

Photo: M Kullberg

Increased understanding of some ecosystems has reshaped thinking on how the Zambezi Basin should be managed.

Hopes for peace were also raised in Angola in 1992, but the country is now back at war following the violation by Unita of the 1994 Lusaka Peace Agreement. The Angolan war has diverted the government's full attention and resources from some regional issues. War has made it difficult for meaningful work to be carried out in parts of the basin within Angola. Insecurity has been reported in some parts of northern Zambia bordering Angola as a result of this conflict. The Angolan war has also recently spilled into Namibia with the Unita rebels launching raids into that country for its support of the Angolan government. A number of lives have been lost in Namibia since the beginning of 2000.

Where regional cooperation existed in the basin, it was defined within the railway-track parameters: harnessing water for either hydroelectric generation or irrigation. Engineering rather than social and/or ecological considerations were given priority. The impacts of dam construction on displaced people or flooded habitats were secondary. The Kariba and Cahora Bassa dams are good examples.

However, in the 40 years that the Kariba Dam has been in existence and the adoption of the 1966 Helsinki Rules governing management of shared watercourses, increased understanding of environmental management systems and ecology has helped reshape thinking on how the Zambezi Basin should be managed. The separate parts within national boundaries are now being seen as part of a whole, where the effective management of the whole is critical to the sustainability of the independent parts. The Zambezi Basin is now being seen as a whole ecosystem rather than the sum of its parts.

Under the auspices of the United Nations Environment Programme (UNEP), the basin states adopted in 1987 the Zambezi River Action Plan (ZACPLAN), a blueprint for effective management of the basin. ZACPLAN has 19 projects covering various aspects of sustainable environmental management. ZACPLAN is not a hard law, i.e. it is not binding on the parties. However, it is the first comprehensive attempt to coordinate basin activities and provides practical implementation of the SADC Treaty, the OAU Common Position, Agenda 21, the UN Conventions, Dublin/ Rio Principles and the Global Water Partnership initiatives.

Regional programmes and cooperation

Environmental law in the region has been enhanced to protect threatened species of some animals and plants. Zimbabwe, like other countries in the region, is currently

Box 12.1: ZACPLAN Objectives

The following are some of the objectives:

- Prepare an inventory of existing and potential development, evaluate the environmental impact of major projects, and initiate the basin-wide exchange of information.
- Develop regional legislation necessary for management of the Zambezi and the minimum national legislation required by riparian states for enforcement.
- Develop human resources, administrative and institutional structures and technical capabilities in riparian states to enable the aims of ZACPLAN to be achieved.
- Develop a basin-wide unified monitoring system related to water quality and quantity.
- Develop an integrated water development and management plan for the Zambezi.
- Promote environmental education and public participation in ZACPLAN.
- Establish minimum standards for drinking water and wastewater disposal.

Source: SADC, SADC Water, Lusaka, February 1999.

drafting an environmental management bill that is expected to harmonise environment-related legislation. To curb illegal traffic of wildlife products and trophies, the SADC Wildlife Sector Technical Coordination Unit, and TRAFFIC, are enhancing effective law enforcement by identifying 'hot spots' and illegal wildlife trade routes in the region. The project is in part fulfilment of the objectives of the Convention on International Trade of Endangered Species in Wild Flora and Fauna (CITES).

A wetlands conservation programme being implemented in the region is helping to promote awareness of the role, value and appropriate uses of wetlands among policymakers, extension workers and users of wetlands. Phase I of the programme reviewed wetland systems in the SADC region in 1992 while Phase II gained support of regional partners in developing recommendations for implementation.

The SADC Forestry Sector developed six programmes in line with the regional forestry policy aimed at improving forest resource management through training and education. Under the special facility for SADC forestry project development, Malawi's capability in coordinating forestry development in the region was enhanced. Training has also been offered by some cooperating partners in strengthening national tree seed centres in order to achieve sustainable supply of quality tree seeds.

Community forestry development for fuelwood production and environmental conservation projects have been implemented at Nacala (Mozambique) and Mount Mulanje's montane forest (Malawi). Under the management of indigenous forests some pilot projects have been implemented in Botswana, Malawi, Mozambique and Namibia.[5]

The level of cooperation among different stakeholders at the national level has a bearing on cooperation at a regional level within the Zambezi Basin. At the national level, environmental management is complex with many ministries and departments involved in one way or another.

Table 9.4 shows that virtually all the Zambezi Basin states have many institutions dealing with water and environmental management. Such a state of affairs may be necessary, as it is generally believed that there is strength in numbers. However, it may also be counterproductive, particularly if there is conflict among such disparate government players in the same area.

While the SADC countries have committed themselves to foster regional cooperation in many different sectors, by and large, this has not been translated into practical implementation at the national level. Cooperation is mostly at the government level, and in some cases, quasi-government institutions such as universities, but lacking at the community level. It is only recently that communities involved in natural resources management programmes have started sharing experiences.

Cooperation among the eight Zambezi Basin states is largely defined in the context of the Southern African Development Community (SADC). Such cooperation is articulated in the 1992 SADC Treaty and other relevant protocols, policies and treaties.

The success of regional cooperation depends on common political values, systems and institutions; complementarity between national and regional strategies and programmes; and the sustainable utilisation of natural resources and effective protection of the environment.[6]

In addition to the SADC Secretariat headquartered in Gaborone, Botswana, SADC has decentralised its programmes, entrusting each SADC member state to coordinate a particular sector or sectors. Lesotho manages the

Box 12.2: SADC Treaty

The objectives of SADC shall be to:

- Achieve development and economic growth, alleviate poverty, enhance the standard and quality of life of the peoples of southern Africa and support the socially disadvantaged through regional integration.
- Evolve common political values, systems and institutions.
- Promote and defend peace and security.
- Promote self-sustaining development on the basis of collective self-reliance, and the interdependence of member states.
- Achieve complementarity between employment and utilisation of resources of the region.
- Promote and maximise productive employment and utilisation of resources of the region.
- Achieve sustainable utilisation of natural resources and effective protection of the environment.
- Strengthen and consolidate the long-standing historical, social and cultural affinities and links among the peoples of the region.

Source: SADC, Declaration Treaty and Protocol of the Southern African Development Community, Gaborone, 1993.

Box 12.3: Water Sector Coordination Unit Vision

The vision of the Coordination Unit is: "to attain the sustainable, integrated planning, development, utilisation and management of water resources that contribute to the attainment of SADC's overall objective of an integrated regional economy on the basis of balance, equity and mutual benefit for all member states."

Source: SADC, Water, February 10-12, Lusaka, 1999

Other SADC sectors/institutions whose mandates have a bearing on cooperation in the Zambezi Basin, include:

- Energy Sector Coordination Unit, Luanda, Angola.
- Drought Monitoring Centre, Harare, Zimbabwe.
- Food Security, Harare, Zimbabwe.
- Inland Fisheries, Forestry and Wildlife, Lilongwe, Malawi.
- Southern African Transport and Communications Commission, Maputo, Mozambique.
- Marine Fisheries, Windhoek, Namibia.
- Tourism, Port Louis, Mauritius.
- Mining, Lusaka, Zambia.

Photo: M Kullberg

The basin's human resources have a strategic role to play in the management of the environment.

environment and water sectors under separate coordination units: Environment and Land Management Sector (ELMS) under which ZACPLAN was initiated, and Water Sector Coordination Unit (WSCU).

SADC Policy and Strategy for Environment and Sustainable Development

In August 1996, the *SADC Policy and Strategy for Environment and Sustainable Development* was approved following a three-year consultation process coordinated by SADC ELMS and involving various stakeholders in different SADC countries. The main goals for a regional environment policy and strategy are to:

- Protect and improve the health, environment and livelihoods of the people in southern Africa with priority to the poor majority.
- Preserve the natural heritage, biodiversity and life-supporting ecosystems in southern Africa.
- Support regional economic development on an equitable and sustainable basis for the benefit of present and future generations.[7]

Three complementary but more functional goals are to:

- Strengthen the analytical, decision-making, legal, institutional and technological capacities for achieving sustainable development in southern Africa.
- Increase public information, education and participation on environment and development issues in southern Africa.
- Expand regional integration and global cooperation on environmental and natural resources management for sustainable development.[8]

Under Strategic Priorities for Action, the SADC Policy and Strategy identifies the following issues and projects as falling within the region's Environment and Sustainable Development Programme:

- Major problems, which are common to two or more countries, for example, land degradation, deforestation, water and sanitation, agrochemicals, hazardous wastes.
- Resources and ecosystems shared by two or more countries, for example, the Zambezi River, migratory wildlife, international fisheries, and the Kalahari-Namib Desert.
- Problems with transboundary impacts in two or more countries, for example, transboundary air pollution and siltation of international rivers.[9]

Photo: M Kullberg

The use of arms is sometimes inevitable in trying to resolve areas of conflict, but it's important that such measures be discouraged and dialogue be promoted.

SADC protocols
In addition to the Policy and Strategy, a number of SADC protocols have been adopted to deal with various issues and all of them have a bearing on regional cooperation in the Zambezi Basin. Perhaps the most critical is the SADC Protocol on Shared Watercourse Systems.

Other SADC initiatives
Conflict resolution
In 1996, SADC launched the Organ on Politics, Defence and Security to help resolve disputes involving member states. The organ became a source of acrimony among the SADC member states in terms of its operations. However, by the beginning of 2000, the tensions appeared to have been eased.

Poverty
It's only recently that concerted efforts are being made at the regional level to address growing poverty in the basin states. Through the SADC Food Security Unit based in Harare, the region has embarked on a major poverty alleviation project. One of its aims is to build national capacity to effectively tackle poverty issues.[10]

Wildlife management
Four Zambezi Basin states, Botswana, Namibia, Zambia and Zimbabwe are taking part in a USAID-funded project called the SADC Community Based Natural Resources Management Programme (CBNRMP), which among other objectives, devolves authority to respective local communities to manage wildlife.

In Malawi, the SADC Wildlife Sector Coordination Unit programme of action has gathered momentum following approval by the SADC Council of the Regional Wildlife Policy and Strategy in 1998. The coordination unit promotes sustainable utilisation of wildlife resources by integrating wildlife conservation with economic development and assists member states in combating poaching. A wildlife management protocol was signed in 1999.

Box 12.4: SADC Protocols

SADC Protocol on Shared Watercourse Systems
The objectives of this protocol are:

- To develop close cooperation for judicious and coordinated utilisation of the resources of the shared watercourse systems in the SADC region and coordinate environmentally sound development of the shared watercourse systems in order to support sustainable socio-economic development.
- Have regional conventions on equitable utilisation and management of the resources of shared watercourse systems in the SADC region and consolidate other agreements in the SADC region regarding the common utilisation of certain watercourses.
- Promote SADC integration process in accordance with Article 22 of the Treaty establishing SADC.

The protocol, which was signed in 1995, came into force in 1998 following its ratification by two-thirds of the SADC members. Efforts to amend some of the provisions are underway. Angola, Botswana, Lesotho, Malawi, Mozambique, South Africa and Zambia have proposed amendments.

Mozambique, for example, is concerned with wording such as "drainage basin". It is also opposed to the wording, "water flowing into a common terminus". It has suggested that this be amended to "water flowing into a natural lake or the sea".

By end of 1998, Botswana, Lesotho, Malawi, Namibia, South Africa, Swaziland, Tanzania, Zambia and Zimbabwe had ratified the protocol while Mauritius had acceded to it.

SADC Protocol on Transport, Communications and Meteorology
The protocol, which was signed in August 1996 and entered into force by September 1998, commits member states to, among other things, adopt and implement road traffic measures, which enhance their capacity to assess and control the impact of road transport on the environment. These measures include:

- Common standards to enhance vehicle pollution control.
- Measures relating to the transboundary movements of hazardous substances.
- Developing and implementing incident management systems for environmental incidents impacting on road transport.

The protocol, coordinated by Mozambique, provides for cooperation of SADC states in areas such as meteorology and transport, including road traffic and inland waterways of which the Zambezi River is part. While the region's basic meteorological network is in place, many member countries are facing maintenance problems.

In addition, Article 6.7 of the protocol encourages SADC states which have not acceded to the 1989 Basel Convention on the Transboundary Movements of Hazardous Substances and their disposal, to consider doing so.

Of the eight Zambezi Basin states, only Zambia had not ratified the protocol by the end of 1998. SADC countries are now developing annexes for incorporation into the protocol and these include:

- Model legislation.
- Model agreements.
- Technical standardisation and harmonisation.
- Memoranda of understanding.

SADC Protocol on Energy
The protocol deals with the region's energy issues, including renewable and non-renewable sources of energy. Of particular significance is the region's commitment to reduce the dependence of the majority of the people on biomass to meet their energy needs. Under general principles, the protocol commits member states to "ensure that the development and use of energy is environmentally sound".

Under Annex 1 on Guidelines for Cooperation, the agreement commits states to: "Develop and utilise electricity in an environmentally sound manner, and subject electricity projects to environmental impact assessments in conformity with agreed basic environmental standards".

On the use of coal (which produces carbon dioxide) the Protocol states: "More deliberate efforts should be implemented to substitute coal for woodfuel. To this end, the Commission shall develop a strategy for expanded coal utilisation considering such issues as cost, efficient coal stove development, coal briquetting, coal gasification and the use of coalbed methane."

The protocol is, however, silent on the issue of coal's contribution to global warming through the release of carbon dioxide, and also acid rain due to sulphur dioxide. Research has also shown that increased use of coal will increase metal exposures because coal contains many toxic metals and can be breathed deeply into the lungs. Some of the associated health problems are lung cancer and cardiopulmonary disease.

The SADC Energy Protocol was approved in 1996 but has yet to be ratified by two-thirds of the member states for it to come into force. The SADC Council of Ministers has approved both the Policy and Strategy for Regional Energy Cooperation Development in SADC and the SADC Energy Action Plan, developed under the energy protocol. These provide the framework for regional cooperation in the energy sector.

SADC Protocol on Mining

Under Article 8, which deals with environmental protection, the Protocol states that SADC member states shall:

- Promote sustainable development by ensuring that a balance between mineral development and environmental protection is attained.
- Encourage a regional approach in conducting environmental impact assessments especially in relation to shared systems and cross border environmental effects.
- Collaborate in the development of programmes to train environmental scientists in fields related to the mining sector.
- Undertake to share information on environmental protection and environmental rehabilitation.

The Protocol, adopted in Lilongwe, Malawi, in August 1997, is not yet in force.

SADC Protocol on Trade

The SADC Trade Protocol was finalised in 1996. By 1999 five basin states, Botswana, Malawi, Namibia, Tanzania and Zimbabwe had ratified the Trade Protocol. According to the SADC Industry and Trade Coordination Division, in 1996, SADC states imported goods and services worth US$41 billion from the rest of the world and US$8 billion from within and among the member states. In the same year exports to the rest of the world amounted to US$46.2 billion while intra SADC exports amounted to US$8.8 billion. Thus intra-SADC business accounts for only about 16 percent of annual business activity as measured by values of imports and exports.

The SADC countries cooperate with several partners, including the European Union (EU), Nordic countries, the Commonwealth, the US, Japan, Germany and Asian countries. The SADC member states have an economic forum that allows them to form SADC common positions where this is necessary. For example, at the World Trade Organisation Ministerial Conference held in Geneva in 1998, SADC member states presented a common position through Tanzania, the sector coordinator for Trade and Industry.

Protocol on Wildlife Conservation and Wildlife Law Enforcement

A SADC Protocol on Wildlife Conservation and Wildlife Law Enforcement has been developed and was signed in August 1999. Its objectives are to:

- Harmonise relevant sections of national "Wildlife Acts" to create an enabling environment for more effective law enforcement procedures.
- Standardise penalties for poaching of, and trafficking in, wildlife and wildlife products; establishment of extradition procedures for poachers and traffickers.
- Establish wildlife agencies for improved cooperation in law enforcement activities.
- Find ways of sharing information on the movements and whereabouts of poachers and identification of illegal trafficking routes.
- Establish a database on illegal traffickers, their arrests and convictions.
- Undertake joint operations in the field to apprehend poachers, recover and dispose of illegally obtained wildlife products.

Box 12.4: SADC Protocols (continued)

SADC Protocol on Tourism
The SADC Protocol on Tourism has been signed by all member states except Angola.

The objectives of the Protocol, among others are to:

- Use tourism as a vehicle to achieve sustainable social and economic development through the full realisation of its potential for the region.
- Ensure equitable, balanced and complementary development of the tourism industry within the region.
- Optimise resource usage and increase competitive advantage in the region vis-à-vis other destinations through collective efforts and cooperation in an environmentally sustainable manner.
- Ensure the involvement of small and micro-enterprises, local communities, women and youth in the development of tourism throughout the region.
- Contribute towards the human resource development of the region through job creation and the development of skills at all levels in the tourism industry
- Improve tourism service and infrastructure in order to foster a vibrant tourism industry.

Sources: SADC, Water, Lusaka, 1999.
SADC, Industry and Trade, Lusaka, 1999.
SADC, Energy, Lusaka, 1999.
SADC, Tourism, Lusaka, 1999.
SADC, Protocol on Transport, Communications and Meteorology in the Southern African Development Community (SADC) Region, Gaborone, 1996.
SADC, Transport, Communications & Meteorology, Lusaka, 1999.
SADC, Protocol on Energy in the Southern African Development Community (SADC) Region, Gaborone, 1996.
WRI, 1998-99 World Resources: A Guide to the Global Environment, World Resources Institute, New York, 1998.

Photo: M Chenje

One of the major successes of regional cooperation involving some basin states, was the successful application in 1997 by Botswana, Namibia and Zimbabwe for CITES to downlist the African elephant from Appendix I to II.

One of the major successes of regional cooperation involving some of the Zambezi Basin states, was the successful application by Botswana, Namibia and Zimbabwe at the 1997 10th conference of the parties to CITES to have the African elephant downlisted from Appendix I to II. The downlisting enabled the three countries to have limited trade in ivory with Japan. The three SADC countries successfully auctioned their ivory stocks to Japanese ivory dealers in early 1999, closing off one of the most emotive issues ever to confront parties to CITES. The controversy is certain to be reopened ahead of the 11th COP in April 2000 in Kenya.

Gender initiatives

Governments in the basin have, under SADC, made a commitment to articulate and implement gender policies. One of the achievements of regional cooperation is the adoption of the SADC Gender Policy.

While significant steps have been made since 1997 in strengthening regional cooperation in terms of gender, some of the results are however, yet to translate themselves into reality in SADC countries of the Zambezi Basin. Some of the achievements include the recognition of women as major players in environmental management but

Box 12.5: Declaration on Gender by SADC leaders

In the Declaration adopted in 1997, the Heads of State and Governments of SADC have committed themselves and their countries to, inter alia:

- Placing gender firmly on the agenda of the SADC Programme of Action and Community Building Initiative.
- Ensuring that there is equal representation of women and men in the decision-making of member states and SADC structures at all levels, and the achievement of at least 30 percent target of women in political and decision-making structures by year 2005.
- Promoting women's full access to, and control over productive resources such as land, livestock, markets, credit, modern technology, formal employment, and a good quality of life in order to reduce the level of poverty among women.
- Repealing and reforming all laws, amending constitutions and changing social practices which will still subject women to discrimination, and enacting empowering gender-sensitive laws.
- Enhancing access to quality education by women and men, and removing gender stereotyping in the curriculum, career choices and professions.
- Taking urgent measures to prevent and deal with the increasing levels of violence against women and children.

Source: SARDC, Beyond Inequalities: Women in southern Africa, 1998.

Box 12.6: NGO role in SADC regional programmes

Article 23 of the SADC Treaty states:

"1. In pursuance of the objectives of this Treaty, SADC shall seek to involve fully, the people of the region and non-governmental organisations in the process of regional integration.

2. SADC shall cooperate with, and support the initiatives of the peoples of the region and non-governmental organisations, contributing to the objectives of this Treaty in the areas of cooperation in order to foster closer relations among the communities, associations and peoples of the region."

Source: SADC, Declaration Treaty and Protocol of the Southern African Development Community, Gaborone, 1993.

this has not yet translated into project and programme development. However, women's input is increasingly being sought in decision-making.

Involvement of NGOs

The decision by SADC governments to involve non-governmental organisations (NGOs) and other stakeholders in SADC activities has also seen the establishment of regional NGOs or refocusing of national NGOs to more regional issues. NGOs in the SADC region are cooperating in areas such as environmental management, gender, as well as democracy and governance issues. Examples of national NGOs cooperating at the regional level include members of the World Conservation Union (IUCN).

In addition, at least 14 NGOs are working together with the Southern African Research and Documentation Centre (SARDC) in raising awareness on gender issues in the Zambezi Basin.

Water management

In terms of water management, many initiatives involving both government and NGOs are being undertaken by the Zambezi Basin states as part of the wider SADC programmes. Examples include the SADC Roundtable on Integrated Water Resources Development and Management, which is coordinated by SADC WSCU and assisted financially and technically by the United Nations Development Programme (UNDP).

In December 1998, Zambezi Basin and other SADC states as well as their partners held a major conference on this initiative, in Geneva. The conference provided input for further work in integrated water resources management in the region.

Other activities include programmes such as:

- Water demand management in southern Africa.
- Global Water Partnership Southern Africa Programme.
- Environmental Monitoring Programme for Lake Kariba.

Water demand management in southern Africa

The objective of the Water Demand Management (WDM) project is to promote approaches that encourage water conservation by various users. It involves the application of selective economic incentives, to promote efficient and equitable use of water as well as a number of water conservation measures aimed at raising awareness on the scarcity and finite nature of the resource. IUCN-ROSA undertook an 18-month research study to investigate and establish WDM applications in the SADC region. The research was based on five case studies involving four basin states: Botswana, Mozambique, Namibia, and Zimbabwe, as well as South Africa.

The project identified and evaluated current policies and strategies for WDM in order to establish the benefits and impacts, as well as effectiveness in meeting social and environmental needs in the southern African region. It also sought to improve understanding and knowledge of water demand management among decision-makers, and to influence national and regional water policies and strategies. Results indicate that the project has led to:

- A better understanding of water demand management in the SADC region.
- Development of a network of WDM experts and practitioners.
- Development of cutting edge documents that articulate very clearly positions of relevance to the region.[11]

Box 12.7: Roundtable conference communiqué

Following a two-day roundtable conference in Geneva on integrated water resources management, SADC and its cooperating partners issued a communiqué which, inter alia, stated:

- The conference agreed that the large number of shared water bodies among member states clearly demonstrates that commonly agreed approaches and solutions must be found for water resources if SADC is to attain its goal of an "integrated regional economy on the basis of balance, equity and mutual benefit" It was also agreed that such a common approach would not only foster the socio-economic development of SADC, but would also promote stability, peace and security.

- In order to lay a solid foundation for an integrated approach to water resources development and management in the SADC region, the conference agreed that, as a first step, emphasis should be on enhancing the institutional and managerial capacity for improved water resources management and development. It was noted that the role of the Water Sector Coordination Unit is that of coordination, facilitation, harmonisation, mobilisation and monitoring.

- The conference observed that the Regional Strategic Action Plan was in line with international consensus on integrated water resources development and management, and therefore, endorsed the priority intervention areas identified in the plan. These are the legal and regulatory framework; institutional strengthening; sustainable development policies; information acquisition; management and dissemination; awareness building, education and training; public participation; infrastructure; and public-private partnership.

- The conference underscored the need for an effective legal and regulatory framework in the SADC region. It was noted that national water legislation in most SADC countries is inadequate and weakly enforced and that national laws are not consistent with widely accepted international water principles.

Source: SADC, SADC Roundtable Conference on Water Resources Development and Management, 1998.

Photos: M Kullberg

Water Demand Management (WDM) tries to promote approaches that encourage water conservation by various users.

Global Water Partnership Southern Africa Programme

The GWP is a network established by Sida, the World Bank and UNDP in August 1996 to support countries in the sustainable management of their water resources. The GWP supports integrated water resources management by collaborating with governments and existing networks, and by forging new collaborative agreements.

The network encourages governments, aid agencies and other stakeholders to adopt consistent and complementary integrated water resources management policies and programmes.

The Southern Africa Technical Advisory Committee (SATAC) supports the implementation of integrated water resources management practices and cross-sectoral dialogue at all levels in southern Africa.

SATAC's goal is to encourage sustainable water resources management by promoting the application of the Dublin/Rio principles in southern Africa and encouraging dialogue among the various stakeholders by creating an information network and interactive fora at national and regional levels.

The Environmental Monitoring Programme for Lake Kariba

The three-year programme, being implemented by the Zambezi River Authority (ZRA) together with the Stockholm Environment Institute (SEI), focuses on the development of an internal corporate environmental policy for ZRA. It also looks at the development of a water quality monitoring system for Lake Kariba and the development of a long-term control programme for the water hyacinth.

COOPERATION AT GLOBAL LEVEL

The Zambezi Basin states are signatories to several multilateral environmental agreements (MEAs) which influence national programmes and regional initiatives that are outside the SADC protocols.

OAU initiatives

States in the basin are members of the Organisation of African Unity (OAU) and the United Nations. Under the ambit of the OAU, these countries participate in fora such as African Ministerial Conference on the Environment (AMCEN). The involvement of AMCEN in promoting regional cooperation in shared basins is due to the recognition that even though Africa has more than 50 interna-

Box 12.8: List of some regional institutions

Biodiversity Foundation for Africa (BFA)
The Biodiversity Foundation for Africa is an NGO established by a group of scientists and environmentalists. Its principle is to improve scientific knowledge of African biodiversity to support its sustainable management at the landscape and ecosystem scale. The foundation is currently active in carrying out biological inventories and monitoring of miombo woodland in the catchment area of the Zambezi River.

Common Market for Eastern and Southern Africa (COMESA)
In addition to the SADC Treaty and its protocols, other regional economic fora include Common Market for Eastern and Southern Africa (COMESA) which groups together more than 20 states in eastern and southern Africa.

Desert Research Foundation of Namibia (DRFN)
Desert Research Foundation of Namibia is an NGO dedicated to appropriate research, training and environmental education in Namibia and elsewhere in the arid areas of Africa. It has been identified by SADC-ELMS as a focal point for training in desertification as part of SADC-ELMS' support to national initiatives to address the Convention to Combat Desertification.

Musokotwane Environment Resource Centre for Southern Africa (IMERCSA)
IMERCSA provides researchers, the media, NGOs and governments with up-to-date information on the environment in southern Africa.

The World Conservation Union–Regional Office for Southern Africa (IUCN-ROSA)
IUCN-ROSA's mission is to facilitate and strengthen an integrated approach for the sustainable and equitable use of natural resources and the conservation of biological diversity in southern Africa. Wetlands of the Zambezi Basin are currently being addressed by a project called "Zambezi Basin Wetlands Conservation and Resource Utilisation Project". The goal is to conserve the critical wetlands of the Zambezi River Basin, and the project focuses on four pilot sites each with its own theme.

Zambezi River Authority (ZRA)
The Zambezi River Authority is a binational organisation responsible for water management along with environment and pollution monitoring of Lake Kariba and the portion of the Zambezi River common to both Zambia and Zimbabwe. It also maintains the Kariba Dam wall and the telemetry system located upstream of the Dam which forms part of the Kariba complex.

Source: Various reports

tional river basins, most of these basins do not have international agreements.[12] The Zambezi Basin states endorsed, as part of the OAU, the African Common Position on the African Environment and Development Agenda. The document highlights the need for African cooperation in areas such as:

- Food self-sufficiency and food security.
- Efficient and equitable use of water resources.
- Management of shared waters and basins.
- Managing demographic change and population pressures.
- Optimising industrial production and preventing pollution.
- Management of biodiversity, rational management of forest resources.
- Drought monitoring.
- Development of environmental legislation.
- Poverty eradication.
- Capacity building, environmental education, training and public awareness.[13]

It still remains to be seen whether these agreements will produce any tangible results for cooperation and development for the basin or remain shelf documents.

Global initiatives

In almost three decades since the 1972 Stockholm Conference on the Human Environment, various legal instruments have been adopted by the international community to facilitate a global approach to tackle many environmental problems. While half of the Zambezi Basin states (Angola, Mozambique, Namibia and Zimbabwe) were not independent and, therefore, did not participate in the 1972 conference, two decades later all the basin states participated in the 1992 Earth Summit in Rio de Janeiro. The Earth Summit spawned its own share of international agreements. These include the Convention on Biodiversity, the UN Framework Convention on Climate Change, the Convention to Combat Desertification, Forest Principles and Agenda 21.

Implementation of Agenda 21

Unsustainable development, largely driven by economic and sectoral policies that are narrowly focused and ignore adverse impacts on the environment, is a topical issue in the basin states. A global action plan for environment and development, Agenda 21, adopted in 1992 provides an integrated policy framework for national and regional action towards sustainable development. Environmental concerns are increasingly being incorporated as an integral part of development policy and decision-making, guiding sectoral ministries and other activities impacting on the environment in the region.

Agenda 21 remains the overall framework within which multilateral environmental agreements; national programmes and community-based activities are premised. It provides a range of regulatory measures, economic incentives, policy goals and priority programmes that ensure ecologically and economically sustainable development. Based on these and other convention-specific provisions, the basin states are implementing various programmes under the auspices of SADC and their respective line ministries. National governments in partnership with development partners are implementing programmes in fulfilment of some environmental agreements. For example, local authorities are also implementing Agenda 21 at their level to address their socio-economic and environmental concerns.

Box 12.9: Management of shared river basins in Africa

One of the objectives of the African Common Position on Environment and Development is to:

"Promote the environmentally sound management of shared water resources through international agreements so as to give governments a solid basis for distributing water resources."

AMCEN outlines a number of activities, including international/regional cooperation and coordination. It states:

"National governments, in collaboration with other governments and in cooperation with appropriate regional and international organisations should establish machineries to promote technical cooperation in research and transfer of technology in relation to the environmental management of shared waters and basins.

"African governments, in collaboration with the national and international scientific community and with the cooperation of relevant national and international organisations should promote and support research and development relating to shared water resources and basin management."

Source: UNECA, African Common Position on the African Environment and Development Agenda: Annex IV – Programme Briefs, Addis Ababa, 1992.

Selected environmental conventions

In implementing the Convention to Combat Desertification (CCD), some of the basin countries are participating in the Sub-regional Action Programme (SRAP) through their respective institutions that were identified to take a lead in different aspects. For Botswana, the Department of Environmental Science at the national university, Kalahari Conservation Society and Thusano Lefatsheng were identified while the Desert Research Foundation of Namibia represents that country. In Mozambique, the National Institute of Agricultural Research and National Remote Sensing and Cartography Centre were also assessed; the Faculty of Forestry and Institute of Resource Assessment in Tanzania and the Drought Monitoring Centre and ZERO-Regional Environmental Organisation in Zimbabwe, among others, were also assessed to participate in implementing the conventions programmes.

Implementation of the Convention on Biological Diversity (CBD) is quite advanced with a regional biodiversity strategic action plan that was translated to national level action plans for different countries. Initiatives such as the Southern African Botanical Diversity Network (SABONET) are helping to conserve plant species in botanic gardens. Other strategies covering biodiversity, wildlife, forests and parks are enunciated in more comprehensive policies such as National Conservation Strategies or National Environmental Action Plans. Outside parks, wildlife management is growing as some community-based natural resources management programmes, private safari operators and commercial farmers take part in conservation. For example in Zimbabwe, the total area under wildlife is now about 30 percent.[14]

These new conventions, including the 1997 Convention on the Law of the Non-navigational Uses of International Watercourses, as well as those adopted prior to the 1990s, provide a challenge for regional cooperation in the Zambezi Basin. The basin states have ratified the majority of them. These agreements provide a basis for cooperation and harmonisation of laws and regulations needed for the effective management of the basin.

Given the complex situation at the national level, regional cooperation within the Zambezi Basin is a challenge. Countries have to navigate bilateral agreements, SADC level structures, Africa region initiatives as well as keep abreast of global trends and development.

FUTURE CHALLENGES

Management of the Zambezi Basin should extend beyond just the river and its immediate environs. Cooperation should not only be restricted to the immediate mandate of institutions such as the ZRA, but others even if they may be about 400 kilometres away from Kariba. What hap-

Photo: M Kullberg

Vultures are often viewed negatively by people but have an important role to play in nature.

Table 12.2: Basin states and international conventions

CONVENTION/TREATY	Angola	Botswana	Malawi	Mozambique	Namibia	Tanzania	Zambia	Zimbabwe
1963: Banning Nuclear Weapon Tests	-	R	R	-	-	R	R	-
1968: African Nature Conservation	-	-	R	-	R	R	R	-
1971: Wetlands of International Importance (RAMSAR)	-	R	R	-	R	-	R	-
1972: Bacteriological and Toxic Weapons	-	R	S	-	-	S	-	R
1972: World Cultural and Natural Heritage	R	-	R	R	-	R	R	R
1973: Convention on International Trade in Endangered Species	-	R	R	R	R	R	R	R
1982: United Nations Convention on the Law of the Sea	R	R	-	-	R	R	R	R
1985: Protection of the Ozone Layer	-	R	R	R	R	R	R	R
1987: Montreal Protocol on Ozone Layer	-	R	R	R	R	R	R	R
1989: Transboundary Movement of Hazardous Wastes (BASEL)	S	-	R	S	R	R	R	S
1991: Ban on import of Hazardous Wastes into Africa and on Control of their Transboundary Movements (Bamako)	-	-	-	-	-	R	-	R
1992: Biological Diversity	R	R	R	R	R	R	R	R
1992: Climate Change	S	R	R	R	R	-	R	R
1994: Desertification	R	R	R	R	R	R	R	R
1995: Protocol on Shared Watercourse Systems in the SADC Region	-	R	R	S	R	R	R	R

Note: S= Signed, R= Ratified

Source: SADC, SADC Policy and Strategy for Environment and Sustainable Development, ELMS, Lesotho, 1996

UNEP, Register of International Treaties and other Agreements in the Field of Environment, Nairobi, 1996

pens in Luena, the largest urban area in the Angola part of the basin should be of interest to decision-makers as far as Tete and Chinde, in Mozambique.

There is, therefore, need for a basin-wide coordinating structure, for example an expanded ZRA or ZAMCOM as recommended under the Protocol on Shared Watercourse Systems. Such a structure would foster multi-sectoralism, harmonious policies, integrated strategies, conflict resolution and definition of priorities. Cooperation should be multi-sectoral, taking into consideration not only water and environmental management, but also social and economic policies and programmes. Policies on population, human settlements and distribution, human resources development, poverty alleviation, education, urbanisation, industrialisation and energy are just as critical. Since poverty is a critical factor in resource overexploitation, lack of regional cooperation in addressing the causes of poverty limits successes in other areas of cooperation.

Generally, regional cooperation is restricted to the more charismatic activities along the Zambezi River, focusing on areas such as hydroelectricity generation, water resources management, water quality and supply, climate and weather monitoring, and tourism. However, other equally important issues such as the spread of the water hyacinth have not received much regional attention, although there are proposals under the UNEP and SADC WSCU groundwater programme to control infestation and translocation of aquatic weeds. To date, countries have largely dealt with the problem on their own.

Photo: M Chenje

Regional cooperation means abandoning individual interests but striving to accommodate the needs of various stakeholders.

Despite the steps that have been made in terms of regional cooperation, both national and regional policies are still sectoral. Despite efforts aimed at an integrated water resources management strategy, institutions at different levels are still pursuing narrow sectoral interests. The danger is that the management of the Zambezi Basin will still be business-as-usual: one step forward and two steps back.

One of the major challenges to regional cooperation is conflict among regional institutions, both government and non-government, some of which may have competing interests. Overlaps and such competition among different institutions are common at the national level, and threaten strengthened cooperation at the regional level. In terms of water management, the integrated water resources management strategy is designed to encourage cooperation among stakeholders who may otherwise not be inclined to work together. Integrated management should not only be relevant at the regional level, but also at the national and sub-national levels.

Lack of clearly defined priorities can lead to over-management at the regional level. For example, the Kariba area is over-managed due to the existence of several institutions, some of them with overlapping and often conflicting responsibilities. The institutions include urban and rural/district councils of both Zambia and Zimbabwe, the national parks and wildlife departments of both countries, the ZRA, the SADC Lake Fisheries Project, and the University of Zimbabwe Lake Kariba Research Station. The last three institutions are involved in various areas of research, including water pollution and fisheries.

The establishment of ZAMCOM should help break down some of these sectoral boundaries. However, the fact that discussions on the establishment of ZAMCOM seem to be dragging on, may be indicative of the complexity of "de-sectoralising" initiatives, let alone de-emphasising national interests for the benefit of regional cooperation.

Regional cooperation in the Zambezi Basin cannot be considered in isolation from other socio-economic, environmental and energy policies. The basin states have, individually and collectively, various policies on economic development and growth, and environmental management. Since the mid–1980s, the SADC member states have either adopted National Conservation Strategies (NCS) and/or National Environmental Action Plans (NEAPs). It is important to acknowledge that the objectives of the various policies are different and may even lead to conflict between them. For example, the SADC Energy Protocol promotes the use of coal whose gaseous emissions contribute to global warming. The contradiction lies in the fact that SADC countries have committed themselves to reducing the emission of greenhouse gases (GHGs) under the UN Framework Convention on Climate Change.

At the regional level, they have also adopted the 1996 *SADC Policy and Strategy on Environment and Sustainable Development* and other instruments aimed at promoting sustainable development and economic integration. The framework is already in place to strengthen the policies. At the regional level, the SADC countries have also committed themselves to effective management of natural resources. In the 1992 Declaration: Towards the Southern African Development Community, the heads of state and government noted that:

- The exploitation and utilisation of natural resources, especially land, water and minerals will contribute to human welfare and development. However, such exploitation requires good management and conservation, to ensure that development does not reduce or impair the diversity and richness of the region's natural resource base and environment.[15]

In this context, policy measures should be taken, and mechanisms instituted to protect the environment, and manage natural resource utilisation with a view to achieving optimum sustainable benefits for the present and future generations of southern Africa.[16]

SADC countries should adopt various strategies to strengthen national and regional policies on water and environmental management. The first step should be information gathering and the comprehensive review of policies that have a bearing on transboundary resource management regionally. Following the review, they should introduce new policies, which take into account the following:

- Legislation
- Harmonisation of laws/regulations
- Economic policies
- Population policies
- Regional and international conventions
- Institutions
- Sources of pollution
- Integrated management — impacts on climate, natural resources such as land and water
- Data collection and analyses
- Modelling
- Health
- Monitoring and enforcement
- Emission standards
- Scientific research and development
- Technology use
- Polluter Pays Principle
- Environmental Impact Assessment (EIA)
- Strategic Environmental Assessment (SEA)
- Exchange of information
- Establishment of waste processing plants
- Public awareness and education
- Stakeholder participation
- Alternative energy sources and efficient energy use.

In reviewing and updating relevant policies, SADC countries can adapt initiatives that have been introduced in other parts of the world both at regional and international levels. For example, Agenda 21, which SADC countries have adopted as a blueprint for sustainable development, provides a framework for governments in the basin to cooperate in transboundary water management. Agenda 21 has noted that "transboundary water resources and their use are of great importance to riparian states. In this connection, cooperation among those states may be desirable in conformity with existing agreements and/or other relevant arrangements, taking into account the interests of all riparian states concerned."[17]

Ultimately, the success of regional cooperation within the basin depends on involvement of NGOs, the private sector and civil society in governance and cooperation issues, and political commitment. But it also depends on whether or not governments are willing to match their good intentions by investing both funds and human resources into the management of the Zambezi Basin. Funding support from international partners is critical in the short-term, but sustainability can only be guaranteed with resources from the region.

LINKAGES TO OTHER CHAPTERS

Chapter 1: Regional Overview: People and Environment
For regional cooperation to be successful a multi-sectoral approach is necessary, taking into account not only environmental management, but also social and economic aspects such as poverty and economic development.

Chapter 2: Physical Features and Climate
A regional understanding of present-day distribution of natural resources in the Zambezi Basin needs to be based on knowledge of changes in the climate, hydrology and physical features which have occurred, as well as the modifications to the environment caused by human activity.

Chapter 3: Water and Wetland Resources
The quality and quantity of water and wetland resources depend on the SADC countries cooperating in the use and management of those resources.

Chapter 4: Biological Resources and Diversity
Different ecosystems and different species transcend national and political boundaries and often do occur in two or more countries and are subject to different and sometimes opposing management practices. This is one of many arguments justifying regional cooperation.

Chapter 5: Agriculture
Land is an important resource in the Zambezi Basin on which most of the basin governments are pinning hopes for poverty alleviation. However, such hopes are being undermined by unfavourable land tenure arrangements that reduce food security.

Chapter 6: Industry
Industrialisation is related to increasing discharge of both solid and liquid waste as well as transboundary air pollution. Economic development in the form of industrialisation, therefore, provides an opportunity for regional cooperation.

Chapter 7: Energy
At present there are several energy sharing projects that are either ongoing or planned within the SADC region to meet the demands of a growing population and industry.

Chapter 8: Tourism
If well managed, tourism may foster sustainable regional growth in the basin countries. Achievement of this cooperation within the tourism industry, and between the industry, governments and tourists themselves, is critical.

Chapter 9: Pollution
Water and air pollution knows no boundaries and both have negative impacts on sustainable development, influencing not only ecological aspects, but also social and economic development.

Chapter 10: Poverty
Poverty forces people to overexploit the environment for survival, worsening environmental degradation and reducing the ability to derive a living from the same environment. Poverty alleviation is, therefore, a major catalyst for regional cooperation.

Chapter 11: Gender and Women's Roles
Both women and men play important roles in environmental management. But this fact still has to be recognised and gender inequalities have to be addressed. Women still have to be recognised as major players in environmental management and this has to be translated into project and programme development.

Chapter 13: Trends and Scenarios
The environmental quality of the resources of the region in future depends on what is being done at present. As these resources are affected by human activities, the future state of the environment cannot be achieved through fragmented programmes.

ENDNOTES

1 ZRA, "Kariba Dam's Operation Noah re-launched", Harare/Lusaka, 1999
2 UNDP/SADC/SAPES Trust, SADC *Regional Human Development Report 1998*, SAPES Books, Harare, 1998, p 53
3 Mugwara, personal communication, 1999
4 ANZ, "Angolan relief operation threatened", The *Daily News,* 9 June, 1999, p. 8
5 SADC, "Food, Agriculture and Natural Resources Report", 1-4 February, Johannesburg, 1996
6 SADC, *Declaration Treaty and Protocol of the Southern African Development Community*, Gaborone, 1993
7 SADC, *SADC Policy and Strategy for Environment and Sustainable Development*, ELMS, Maseru, 1996
8 ibid
9 op. cit 7
10 op. cit 3
11 Fakir, S., IUCN South Africa, personal comm., 1999
12 UNECA, *African Common Position on the African Environment and Development Agenda*: *Annex IV – Programme Briefs*, Addis Ababa, 1992, pp 6-7
13 ibid
14 Ministry of Environment and Tourism: Zimbabwe, *10th Conference of the Parties to CITES*, Department of National Parks and Wildlife, Harare, 1997
15 SADC, *Declaration Treaty and Protocol of the Southern African Development Community*, Gaborone, 1993
16 ibid
17 UN, *Earth Summit Agenda 21: The United Nations Programme of Action From Rio*, New York, 1992

REFERENCES

Butchart, D., *Wild about the Okavango*, Southern Book Publishers, 1995

Chikwenhere, G. P., "Floating aquatic weeds in Zimbabwe". In Matiza, T. and S. Crafter: *Wetlands ecology and priorities for conservation in Zimbabwe*, IUCN, Harare, 1994

Hiscock, E., "Zambezi Basin Wetlands Conservation and Resource Utilisation Project: An Overview". In Jansen, R. and B. Kamweneshe (eds.), *Transboundary approaches to the wetlands conservation and utilisation of the Chobe-Caprivi wetlands*, IUCN, 1999

Jansen, R.and B. Kamweneshe (eds.), *Transboundary approaches to the wetlands conservation and utilisation of the Chobe- Caprivi wetlands*, IUCN, 1999

Jeffrey, R. C. V., Chabwela, H. N. Howard G. and P. J. Dugan (eds.), *Managing the Wetlands of Kafue Flats and the Bangweulu Basin*, IUCN, 1992.

Kamukala, G. L. and S. A. Crafter (eds.), *Wetlands of Tanzania*, IUCN, 1993

Masundire, H. M., Eyeson, K. N. and S. F. Mpuchane (eds.), *Wetlands Management in Botswana*, WCC., 1995

SADC, *Environmental Effects of Mining in the SADC Region*, Mining Sector Coordinating Unit, Lusaka, 1992

SADC, *Regional Strategic Action Plan for Integrated Water Resources Development and Management in the SADC Countries (1999-2004):* Summary Report, 1998

SADC Secretariat, SADC Handbook, (www.sadc-usa.net), Gaborone, 1997

SADC Secretariat, "SADC Industry and Trade Report", Gaborone, 1999

SADC Secretariat, "SADC Energy Sector Reports", 10-12 February, Lusaka, 1999

SADC, *Water*, 10-12 February, Lusaka, 1999

SADC, *SADC Policy and Strategy for Environment and Sustainable Development*, ELMS, Lesotho, 1996

UNEP, *Register of International Treaties and other Agreements in the Field of Environment*, Nairobi, 1996

World Bank, *Monitoring Environmental Progress: A Report on Work in Progress*, Washington, D.C., 1995

ZRA, *Zambezi River Authority: An Overview*, 1997

13

TRENDS AND SCENARIOS

"People like to speculate about the future, to tell stories about alternative futures, about what tomorrow might look like. Perhaps the only certainty for anyone looking ahead is that the future is uncertain, unpredictable, and complex." – Peter H. Gleick[1]

The past and the present are the umbilical cord that ties the eight Zambezi Basin countries to the future. Many strands woven into their human and natural resource capital, their history, their geographical location and relationships with other parts of the world will, as in the past, continue to influence the future of the Zambezi Basin as governments and people strive to attain a state of sustainability (SOS). While the acronym is synonymous with an urgent call for help, achieving a state of sustainability is also an urgent priority to reverse the current unsustainable path. State of sustainability, which has its roots in sustainable development, is the point at which the health

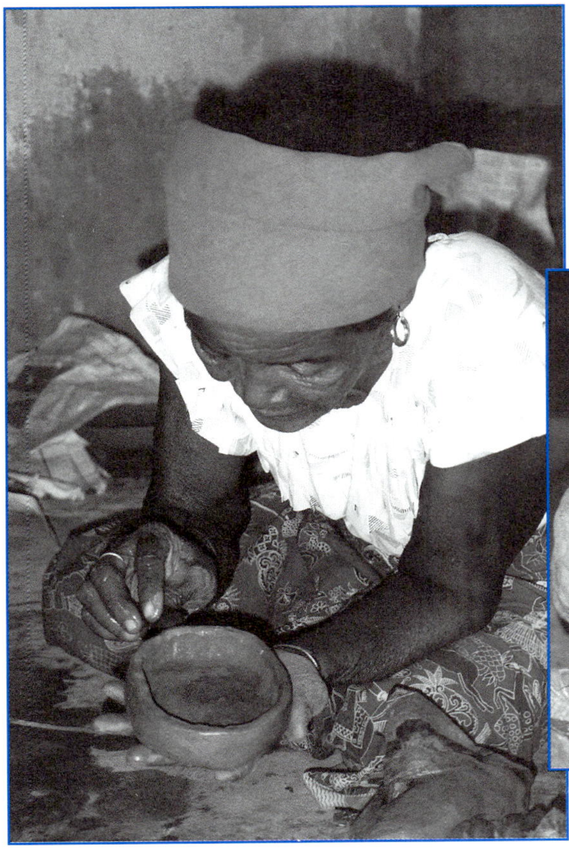

Photos: far left IUCN Mozambique, left M Chenje

The history of the people is the prologue of their future, and the current trends will have a bearing on their development.

of both people and the environment in the Zambezi Basin and the rest of the Southern African Development Community (SADC) region is achieved.

Barring possible but unlikely earth movements which could completely alter the landmass that is the Zambezi Basin, the boundaries of the Zambezi Basin states are the only factor that will remain constant in the next 25 years and beyond. All other factors, from climate and soils composition, population growth and distribution, species and ecosystems biodiversity, to science and technological innovations, economic growth and development, will certainly change over time. History and scientific research has shown that people and the environment on which they depend are constantly in a state of flux, changing and adapting to both natural and human-caused phenomena. Change will have a bearing on the health of the people and environment in the Zambezi Basin. Depending on the level of change and whether it is positive or negative, countries in the region will either continue with the unsustainable trends, further impoverishing the people and the environment; or will implement critical interventions necessary to reverse such unsustainable trends. It is also likely that a combination of both sustainable and unsustainable trends may prevail in the region. The state of sustainability can be achieved by trying to understand the complex issues surrounding human ingenuity and development together with ecosystems processes.

Perhaps the greatest impediment to achieving both human and environmental health is war. Fortunately, war is to a large extent now a footnote in the current history of most of the basin states, with the exception of Angola, although nothing can be guaranteed given the natural resource base that is fast diminishing. Peace hopes in Angola have been dashed time and again with fierce fighting taking place between government forces and the Unita rebel movement. More than 1.5 million people have been internally displaced, including from areas in Angola's part of the basin.

Virtually all the eight countries of the Zambezi Basin have in the past suffered many years of armed conflict, which saw the destruction of environmental management structures in some countries and massive injection of scarce financial resources into military hardware. For example, at the height of the struggle against apartheid South Africa, the SADC countries lost US$60 billion between 1980–88 "through lost exports and investments, forced high defence expenditures and infrastructure damage as a result of economic destabilisation and sabotage. The countries lost US$10 billion in 1988 alone."[2] Some of the money could have been invested in environmental management. Even though most of the countries are enjoying relative peace, several are host to hundreds of thousands of refugees fleeing war. Both Tanzania and Zambia have since 1998 been hosting refugees fleeing war from the Democratic Republic of Congo. While the refugees in Tanzania are outside the Zambezi Basin, the reverse is true in Zambia. The impact of the refugees on the environment in Zambia has not been fully studied, but research elsewhere shows that large refugee populations exert pressure on the environment, often exacerbating land degradation.

Table 13.1: Military expenditure in the basin states 1985-96

Zambezi Basin States	US$ millions (1995 prices)		As a % of GDP		Per capita (US$ 1995 prices)		Military expenditure (as a % of combined education and health expenditure)		Total armed forces Thousands	Index (1985=100)
Country	1985	1996	1985	1996	1985	1996	1960	1990-91	1996	1996
Angola	883	441	15.1	6.4	101	40	-	208	97	196
Botswana	51	224	1.1	6.7	47	147	-	22	75	188
Malawi	29	23	1.0	1.2	4	2	-	24	9.8	185
Mozambique	326	61	8.5	3.7	24	3	-	121	11.0	70
Namibia	-	71	-	-	3.0	-	42	23	8.1	-
Tanzania	191	83	4.4	2.5	9	3	4	77	34.6	86
Zambia	55	58	1.1	1.8	8	6	42	63	21.6	133
Zimbabwe	232	232	3.1	3.9	28	20	-	66	43.0	105
	5,877	3,966	3.4	2.2	41.8	21.5		41	402.2	85

Source: UNDP/SADC/SAPES, SADC Regional Human Development Report 1998, UNDP/SADC/SAPES, Harare, 1998, pg. 163

Figure 13.1: Military expenditure as percentage of education and health expenditure, 1990-91

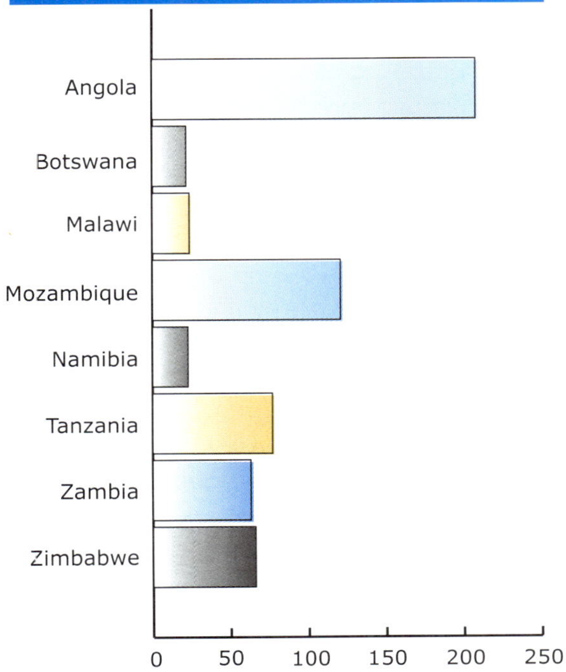

Source: UNDP/SADC/SAPES, SADC Regional Human Development Report 1998, UNDP/SADC/SAPES, Harare, 1998, pg. 32

For most of the basin states, removal of colonial/settler regimes has generally helped people in the different countries to focus on development and other relevant governance issues. This trend is expected to continue consolidating programmes aimed at improving both human and ecological welfare. However, such programmes face a much stronger enemy: poverty.

The trend in the Zambezi Basin states shows worsening poverty with millions of rural people in the basin eking a living from subsistence agriculture, which is dependent on rainfall, a resource that is fast becoming scarce in the basin. Reduced agricultural production due to drought, particularly since the 1990s, has seen more and more people overexploiting their natural resources for survival. At one point, more than 200,000 people invaded about 4,600 sq km of some of Zimbabwe's major rivers, including the Zambezi River tributaries, to pan for gold.[3] Environmental damage was massive in these areas. The magnitude of the ecological disaster was lost on those whose lives depended on the mineral for survival.

The following quotes from two gold panners in Zimbabwe illustrate just how poverty can alienate people from working together with others to conserve the environment: "I am in this business because I don't want my family to starve," said one panner.[4] Another was more blunt: "I can't help it if the future generations find the rivers silted. People are more important than the environment."[5] Poverty and subsistence livelihoods are a threat to sustainable development.

CURRENT TRENDS

Trends are about current positive or negative impacts on sustainable development. Water, environment and economic development are key determinants of current trends within the basin and will continue to be so in future. In these three areas, human interventions and policies (or lack of them) have contributed to the current state of the environment in the Zambezi Basin. Significant strides have been made towards economic development and raising the standard of living of many people. But this has not been without costs, both in terms of human and ecosystems welfare. Millions of people, both in urban and rural areas of the Zambezi Basin are living in poverty with no hope of their situation getting better. The costs have also been felt in terms of ecological welfare with many areas of the basin experiencing land degradation, habitat conversion and loss of biodiversity.

Photo: IUCN

Millions of people in both urban and rural areas are living in poverty.

While the status of human welfare is generally known as a result of systematic data collection and interpretation, data on the overall status of the basin's ecosystems is not conclusive. This reflects a general weakness in science's ability to understand the complex interactions of nature.

Institutional collaboration among SADC sector coordination units is a current trend, especially with the Forestry and Wildlife Sector Coordination Unit in Lilongwe, Malawi. Perhaps the greatest challenges facing the SADC sector coordination units are human capacity and adequate financial resources to discharge their duties. Many of the sectors have limited human resources and are unable to cope with an ever-increasing regional mandate and demands. It may be necessary for national governments to ensure that sector contact departments are strengthened to help the coordination units. The following have been identified as the crosscutting themes on current trends:

- Climate variability
- Population
- Gender
- Poverty and inequality
- Landuse and productivity of natural resources
- Urbanisation and industrialisation
- Research, monitoring and capacity building (of individuals and institutions)
- Integrated management (cooperation)
- Information, communication and education (including awareness)
- Policy and strategy.

Climate

The Zambezi Basin falls within three climatic zones: semi-arid, dry sub-humid and moist sub-humid. Climate in the Zambezi Basin is variable and so is rainfall, making planning a challenge. As early as the mid-19th century, a scientist working in southern Africa hypothesised about climate change. The 1862 drought in southern Africa,[6] provided further credibility in scientific circles to theories of "widespread climatic change". Research in Africa and elsewhere contributed to the debate about climate change, which had "become international in reference and relevance by the mid–1860s".[7] It was recognised then that unrestrained human activities posed an ecological threat.

Drought is an overriding factor in the degradation of cultivated lands and rangelands in many parts of the basin, impacting on plant cover. Drought increases soil degradation problems, while soil degradation magnifies the effects of drought. As a result of drought during the 1994–95 season, cereal harvests in southern Africa declined by 35 percent compared to harvests of 1993–94, with maize harvests alone falling by 42 percent.[8] As table 13.2 shows, drought was equally devastating during the 1991–92 drought, one of the worst in living memory.

With about 80 percent of the people in the Zambezi Basin dependent on agriculture and the region's food security worsening, the basin states are always facing a serious challenge to achieve food self-sufficiency. Since some parts of the basin are arid or semi-arid, drought will continue to be a factor, limiting economic performance and exerting pressure on the environment. Because of the variability of rainfall throughout the basin, the impact of drought varies from country to country with some being affected more than others.

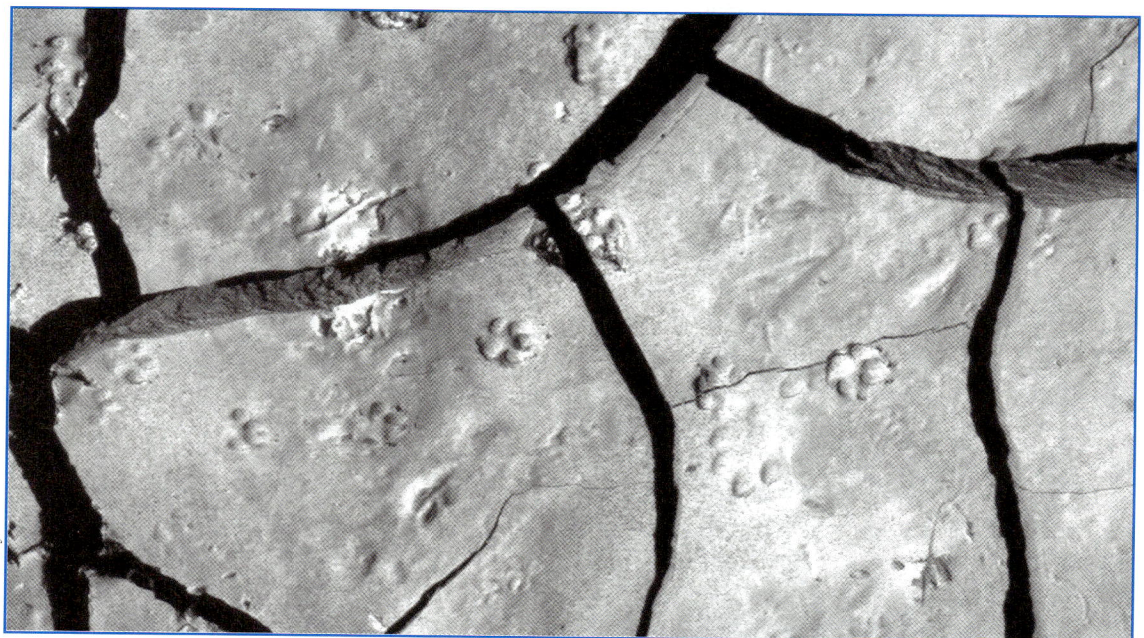

Photo: D Minty

Drought is an overriding factor in the degradation of cultivated lands and rangelands in many parts of the basin.

Table 13.2: Impact of drought on cereal production in the basin states in 1992

Country	Population (millions)		Cereal production ('000 t)	
	Total	Affected	Normal years	1992
Angola	10.3	1.4	316	71
Botswana	1.3	0.3	60	21
Malawi	8.8	6.1	1,474	678
Mozambique	16.1	3.9	551	226
Namibia	1.8	0.2	91	32
Tanzania	25.3	0.6	3,820	3,400
Zambia	8.4	1.7	1,536	603
Zimbabwe	10.1	5.6	2,295	505
SADC total	82.1	19.8	10,143	5,536

Source: Lone, S., Laishely, R., and Bentsi-Enchill, N.K., 'Africa Recovery', Briefing Paper, No.9, United Nations Department of Public Information, New York, 1993

Photo: M Chenje

Pollution of water bodies has become one of the pressing ecological issues in many parts of the basin.

"The human victims of ecological degradation are typically the rural poor — the land-poor and landless masses for whom ecological questions are a matter of life and death — and the unemployed and underemployed in the cities...".[9]

While the advent of drought is always met with trepidation, the countries in the basin have over the years been cooperating in trying to minimise its impact both on people and wild animals. The SADC Drought Monitoring Centre was established to serve as an early warning unit. The SADC Food Security sector also monitors food reserves, ensuring that countries have adequate grain supplies to meet their needs. The trend is that such efforts will continue to be strengthened as the countries move towards regional integration under the framework of SADC.

Population

Trends show that the population in the basin states will continue to grow, with the majority continuing to reside in rural areas. However, many of the basin states will see urbanisation grow. In Malawi, for example, the population is getting more urbanised with projections showing that by 2012, 3.8 million Malawians will be living in urban areas, a 400-percent jump from less than one million in 1988.[10]

There is generally more consumption of resources in urban areas, placing heavy demands on the environment. Large concentrations of people also generate huge volumes of waste, resulting in pollution. For example in Harare, one of the basin's large cities with a population of more than 1.4 million people, urbanisation has been associated with solid and liquid waste, increasing pressure on natural resources, for example water.

Air and water pollution, solid waste and siltation due to urban agriculture have become major environmental issues in the cities of Harare, Lilongwe, Lusaka and other smaller urban areas. Generally, the main sources of pollution are found in urban areas and major developments such as mines, and some agricultural estates. Water contaminated with toxic substances from agricultural estates, mines and industries runs through some areas where it may be used for drinking and washing or watering livestock and gardens.

Industrial and consumer waste is also increasing at alarming rates in the basin, particularly due to rapid economic and population growth in these areas. For example, the Zambian capital, Lusaka which has a population of 1.32 million inhabitants,[11] produces 1,400 tonnes of solid waste daily of which only 10 percent is collected by the city council.[12] Ninety percent of the solid waste is not collected due to a number of factors, including lack of human, financial and material resources. Poor collection of waste is also evident in Harare, with some areas going without waste collection for weeks.

Cholera and other water-borne diseases are often found in water contaminated with untreated human waste and sewage. For example in 1994, a total of 61,960 cases of cholera were reported in five SADC countries —

Angola, the Democratic Republic of Congo, Malawi, Mozambique, and Tanzania, resulting in 4,389 deaths.[13] More than 171,000 cases of dysentery were reported for Malawi, Mozambique and Zimbabwe, with at least 600 lives lost.[14] These cases are examples of negative urban environmental management trends.

Population growth per se does not necessarily result in overexploitation of resources. It is, however, the accessibility and distribution of those resources that may lead to overexploitation. Where population density is high, resources limited and livelihood options have been foreclosed, people overexploit natural resources. It is also important to acknowledge that where the economy is industrialised and diversified, a large consumer base helps to drive economic growth. The issue of population growth in the basin states is not just restricted to controlling the birth rate, but is more complex, requiring strategic planning and innovative approaches.

The significance of population growth is that there will be increased pressure to expand agriculture and grow more food, expand housing, and industrial production, and build more infrastructure. To meet the food needs of people, among other things, the Zambezi Basin states do not only have to bring forested or grazing land under cultivation but to intensify production.

With an average population growth rate of about 2.9 percent, more than 960,000 people are added to the basin population annually. However life expectancy in the basin states will decline due to the HIV/AIDS pandemic. Between 1990–95, life expectancy decreased by 12 years in Zambia, eight years in Botswana and seven years in Malawi due to the AIDS pandemic. Trends show that life expectancy will continue to decline, dropping from a projected average of 58 years between 1995–2000 to 48 years in the same period.[15] Botswana, Malawi, Mozambique, Zambia and Zimbabwe would be the most affected. HIV/AIDS is now the leading cause of death in Africa.[16] In 2015, Zimbabwe's population is expected to be 19 percent lower than it would have been without the impact of AIDS.[17]

Another trend is that the economically productive age group of 18–45 years within basin populations has been the most affected by HIV/AIDS. While the economic impact has not been fully analysed, this situation has had a strong negative impact on agriculture, poverty and general economic development.

Poverty

Related to high population growth is mass poverty, which in the past has been described as one of the root causes of environmental degradation in the region. Poverty and environmental degradation are linked in a vicious circle. A degraded environment produces less, so people directly dependent on it become more vulnerable to disasters. The less it produces, the more it is exploited — for survival. Little regard is paid about the future, let alone about sustainable development. More than half of the population of the region has no access to clean water, sanitation and health services, exposing the majority of the people to water-borne diseases.

"There is a close and growing link between poverty and vulnerability to recurring emergencies, particularly those caused by drought. If the food security of the most vulnerable people could be improved at the household and community level through development projects, the continued need for emergency assistance could be considerably reduced."[18]

Table 13.3: Status of and trends in urbanisation in the basin states

Country	Urban population (as a % of total)		Year	Period	Annual Urban Population Growth rate (%)	City	Largest City Population (thousands)
Year	1970	1995	2015	1970-1995	1995-2015		
Angola	15	31	44	5,7	4,9	Luanda	2,018
Botswana	9	60	89	11,8	4,1	Gaborone	165
Malawi	6	13	23	6,4	5,3	Blantyre	430
Mozambique	6	34	52	10,0	4,7	Maputo	2,212
Namibia	19	36	53	5,4	4,4	Windhoek	187
Tanzania	7	24	38	8,6	5,0	Dar-es-Salaam	1,747
Zambia	30	43	52	4,1	3,4	Lusaka	1,317
Zimbabwe	17	32	46	5,7	3,9	Harare	1,410

Source: UNDP/SADC/SAPES, SADC Regional Human Development Report 1998, UNDP/SADC/SAPES, Harare, 1998, pg. 142

Trends show that poverty will prevail in the basin states into the future, forcing the poor to continue to overexploit their natural resource capital.

Debt

Linked to poverty and environmental degradation is the issue of debt servicing for many of the countries in the Zambezi Basin. Foreign debt continues to rise, at least doubling in most countries in SADC between 1980–1993. In some cases, the debt burden increased by three or even six times during that period. In 1995, Mozambique's total external debt was US$5.8 billion, representing 444 percent of the gross national product (GNP). Its debt service ratio was 35 percent.[19] However, the international community has agreed to cancel some of Mozambique's foreign debt following massive flooding, in February–March 2000, which affected about a million people, particularly in the south of the country.

Some countries in the basin have "severe external indebtedness and extreme aid dependence. Poor external debt management and inadequate debt relief means that for some of the countries, indebtedness will continue to be a problem for macroeconomic stability".[20] Former Commonwealth secretary general, Sir Shridath Ramphal, said "debt repayment is very much at the heart of poverty alleviation" in southern Africa. Poor economic performance is also related to low levels of investment and debt servicing.

Structural adjustment programmes, although widely touted as the answer to most problems of the region, have contributed to the increased vulnerability of both the urban and rural poor. Thousands of breadwinners who in the past could afford to maintain both urban and rural homes are now unemployed, increasing household vulnerability.

The World Bank, one of the institutions which has prescribed economic structural adjustment programmes (ESAP) for many developing countries, including some basin states, recently acknowledged that such programmes exacerbate poverty. Reviewing the impact of ESAP on Zimbabwe, the World Bank resident representative Thomas W. Allen, says the "standard of living has declined for many, particularly urban households, with the percentage of households classified as poor rising from 40 percent in 1991 to over 60 percent in 1995".[21] He says the programme was flawed because of the under-estimation of its social consequences. "The massive retrenchments which resulted in a large number losing their jobs in an economy where employment growth was slowing down cannot be ignored."[22]

Demand for water

Demand for water in the basin is sectoral: urban, rural, agriculture, mining, energy, transport, recreation, and environment. The sectors' demand for water will grow and exert pressure on water and other ecological resources. The increase in water demand is due to general population growth. Such demand is coupled with the reduction of water quality due to waste disposal, siltation, agrochemicals/fertilisers, gold panning, mining effluent and industrial waste. While water pollution in the basin is monitored, there is need to strengthen such measures, covering many locations of the basin to get a clear picture of its extent and magnitude.

Photo: M Chenje

Demand for water has become one of the critical issues facing basin states.

Groundwater

Groundwater is the main source of water for many rural people, catering for about 80 percent of the human and animal populations in Botswana[23] and at least 40 percent in Namibia.[24]

Despite the dependence on groundwater for the majority of people in the Zambezi Basin, information on the occurrence of groundwater is scanty. Little is known on the rates of aquifer recharge. However, some observations of high iron content, sulphur, salinity and nitrates have been made.

Wetlands

The Zambezi Basin wetlands are under threat from both natural and human factors such as drought, pollution, agriculture and grazing. This is likely to continue in future as human populations increase and their associated activities expand. Already some wetlands are under pressure from droughts, livestock and habitat conversion. Satellite imagery indicates increasing incidences of fire threats to wetlands in the Zambezi Floodplains of Zambia.

Land degradation

Land degradation in the Zambezi Basin is increasing due to poor land husbandry and overexploitation of natural resources. Reports from different countries show that there is widespread deforestation due to agricultural expansion and fuelwood collection. In Zimbabwe, for example, 80 percent of the basin area is under commercial and communal farming, each taking 40 percent of the land.[25] In Malawi, land under agriculture increased from 21,000 sq km in 1965 to more than 45,400 sq km by the beginning of the 1990s.[26] Tobacco curing in Malawi consumes 84,826 cubic metres of wood annually.[27] On average, the country loses about 20 tonnes of soil per hectare annually. With growing population and expansion of settlements, ecologically fragile land is being opened up for agriculture and livestock production, often with disastrous impacts on the environment.

The situation in the basin is representative of what is happening on the continent. Africa accounts for over 60 percent of the world-wide deforestation. Between 1980–85, about 1.3 million hectares of closed, broadleafed forests were cleared in Africa and another 2.3 million hectares of open woodlands were chopped down.[28] The annual rate of deforestation was about 0.7 percent between 1990-95, with the highest rate of one percent in the west moist parts of the continent, and the lowest — 0.2 percent in non-tropical southern Africa.[29]

Soil erosion is probably the most important factor in the decline in agricultural productivity, degrading about 15 percent of the SADC region's lands. About two percent of southern Africa's soils are damaged by physical degradation, which includes changes in sealing and crusting of topsoil, loss of water-holding capacity, compaction of topsoil, waterlogging and acidification.

Some of the main causes of land degradation and soil erosion include poverty, skewed land tenure systems, population pressure, and increasing resource demand. Land degradation is also a factor in increased habitat loss in the basin.

Air pollution and waste management

Economic growth in the basin states has resulted in problems of waste production and management, air and water pollution, among others. It has also led to heavy energy demands to satisfy both a growing population and industry. With limited natural locations to generate electricity, some of the countries have turned to thermal power generation, resulting in serious air pollution. Growing industrialisation, transportation and an increase in the vehicle population in the basin states are also contributing to pollution. Fuel consumption has increased virtually in all states over the past decade. For example, fuel consumption in Zambia increased from 84 percent of total consumption in 1980 to 91 percent in 1995.[30]

In addition to atmospheric pollution from thermal power generation, copper smelters, ferro-alloy works, steel works, foundries, fertiliser and cement plants, and pulp-and-paper mills are other pollution sources within the basin. The mining industry contributes significantly to air pollution.[31] For example, less than 10 base metal smelters emit about one million tonnes of sulphur dioxide annually.[32] Such emissions are not only a source of concern, particularly to people with respiratory problems but also represent an economic loss. The disposal of coal fines in slurry dams is a major environmental problem, contributing to acid drainage in some parts of the region.

While virtually all countries in the basin have laws to control air pollution, the state of air pollution in the basin is not well known. In many of the countries, air pollution control is treated as a health rather than an environmental issue such that monitoring and enforcement are a problem due to different priorities, capacity and resources between health and environment departments. Fuzzy responsibilities, where different government departments lack clarity on which of them has the line function, for example, air pollution

Photo: M Chenje

Ecologically fragile land is being opened up for agriculture due to growing population.

control or environmental education, can weaken implementation, monitoring and enforcement. The result is poor environmental management. Table 13.4 shows just how much air pollution is a low priority in the basin in terms of data collection and analysis.

Table 13.4: Environmental Risks to Human Health

COUNTRY	Air	Water	Nutrition	Average Rank
Angola	x	High	High	High
Botswana	x	Moderate	Moderate	Moderate
Malawi	x	High	High	High
Mozambique	x	Moderate	High	Moderate
Namibia	x	High	High	High
Tanzania	x	High	High	High
Zambia	x	High	High	High
Zimbabwe	x	Moderate	Moderate	Moderate

Note: x under the column on air denotes lack of information.
Source: World Resources Institute, 1998

Global initiatives on the impact of greenhouse gases have raised the issue of atmospheric pollution, placing it on the regional agenda. The trend is that atmospheric pollution will continue to attract attention, but financial and human resources capacity may limit the impact of monitoring and enforcement initiatives.

Biodiversity

The people of southern Africa have always had a special bond with the natural environment, and biological diversity has always been part of their livelihood. Traditional societies in the basin developed strategies, which were deliberately aimed at conserving resources for the benefit of the present and future generations, and were deeply enshrined in the traditional values and environmental management practices of these societies.

The emergence of cash economies, the opening up of land for agriculture, mining and settlements not only contributed to the overexploitation of both animal and plant species but also increased biodiversity loss. The need to arrest the loss of biodiversity is especially important in the basin where people depend on biological resources to a greater extent than elsewhere.

Biodiversity loss in southern Africa has been a consequence of human development, as species-rich woodlands and forests have been converted to relatively species-poor farmlands and plantations. However, the number of threatened species could be higher as the full extent of the region's species diversity is unknown. There is a serious lack of species inventories and other baseline data that are fundamental for monitoring biodiversity trends.

Other parts of the region where biodiversity is under threat, include Angola where the wildlife populations of all national parks and reserves have been reduced to about 10 percent of their 1975 levels.[33] The civil war has made protected area management difficult.

Photo: M Chenje

A warthog with a wire trap around its neck is a reminder of the need to tackle rural poverty as part of conservation efforts.

Despite various problems such as habitat loss and overexploitation, the basin is rich in biodiversity, offering a wide spectrum of habitats and ecosystems. Woodlands are the most extensive ecosystem in Africa and prevalent in the Zambezi Basin. They have a high incidence of indigenous plants and animals, and the world's greatest concentration of large mammals. Wetlands are found in every country in the basin and the largest ones include the Okavango Delta and floodplains of the Zambezi River.

Climate change may become a major threat to species composition in the next century. It has already been identified as a cause of the decline in certain amphibian populations due to the drastic reduction in the volume of water bodies as a result of persistent dry weather conditions and human activities.

Alien or exotic species have led to biodiversity loss in some parts of the region. Most of the introduced species invade natural habitats and outcompete the native vegetation. The root causes of biodiversity losses lie in the growing human population, the way in which people have progressively broadened their ecological niche, unsustainable consumption of natural resources, economic systems that fail to set a proper value on the environment, and weaknesses in legal and institutional systems.

While overexploitation of biological resources is acknowledged in the basin, trends show that there is a

Photo: M Chenje

The water hyacinth has become a serious threat in many of the basin's water bodies.

greater commitment, from the highest to the lowest levels of decision-making, to conserve wildlife, particularly large mammals. The introduction of community-based natural resource management programmes in many of the basin states has seen communities playing a role in biodiversity management. Such trends are certain to be strengthened as the programmes expand their activities from large mammals to other resources such as forests and fish.

Table 13.5: Economic growth rates in basin states, 1991-97

COUNTRY	1991	1992	1993	1994	1995	1996	1997
Angola	-2.5	6.9	-24.7	2.5	11.2	12.1	5.9
Botswana	8.9	6.3	-0.1	4.1	3.1	7.0	6.9
Malawi	7.8	-7.9	10.8	-11.6	13.4	12.0	5.3
Mozambique	4.9	-0.8	19.3	4.4	1.4	6.4	7.9
Namibia	7.4	8.2	-1.9	6.5	4.1	3.0	3.0
Tanzania	4.3	1.8	0.4	1.4	3.6	4.2	4.0
Zambia	-0.4	-1.8	6.8	-8.6	-2.3	6.4	3.5
Zimbabwe	5.6	-6.8	4.2	5.3	-3.2	7.2	3.7

Source: UNDP/SADC/SAPES Trust, SADC Regional Human Development Report 1998, Harare, 1998, pg. 29

Technology and economic development

Trends in the basin show that technology and human capacity to deal with various environmental problems are limited. Virtually all the countries have not had significant investment in their economies in the past two decades. According to the SADC Regional Human Development Report 1998, countries are experiencing a "foreign exchange crisis", forcing them to rely on official development assistance (ODA). In Mozambique and Zambia, for example, net ODA rose from US$169 million in 1980 to US$1,103 million in 1995, and from US$318 million to US$2,035 million respectively. This represents a jump from US$14 ODA per capita in 1980 to US$66 in 1995 for Mozambique and from US$55 to US$221 in Zambia.[34] In Zimbabwe, a serious shortage of foreign currency from the end of 1999 led to widespread shortages of fuel, negatively impacting the transport industry for most of the beginning of 2000. Limited foreign currency reserves impact on the basin states' ability to introduce cleaner production by investing in much needed capital equipment to upgrade outdated technology. For example, significant steps have been made in some countries to phase out products, which use ozone-depleting substances. Human resource capacity to develop appropriate technologies in the basin states is also limited.

Ideally, countries in the basin should be investing in research and development, but limited investment by both government and private sector in this area also limits significant changes to the current situation. Trends show that the basin countries will continue to be producers of raw materials, limiting not only the diversification of their economies but also their ability to expand their foreign currency earnings base. Virtually all countries in the Zambezi Basin depend on agriculture, mining and tourism for the bulk of their foreign currency earnings.

Policy and strategy

At the beginning of the 1990s, the basin states, under the auspices of SADC, candidly admitted to the world in the document presented at the UN Conference on

Environment and Development that they have inadequate environmental monitoring, research and planning capabilities. They also said they had "inadequate institutional arrangements, legal framework and enforcement measures for environmental protection and improvement."[55] They also admitted that the then policies on development were weak. In less than 10 years since UNCED, the SADC countries have made great strides in developing policies and strategies aimed at ensuring that future development is both economically and ecologically sustainable. For example, in the water sector, the SADC countries have established a Water Sector Coordination Unit (WSCU) and through it, developed an integrated water resources development and management plan. In 1996, SADC countries adopted the Policy and Strategy for Environment and Sustainable Development.

The trend in SADC has been the drafting and ratification of sectoral policies dealing with diverse issues such as trade, mining, shared watercourse systems and transport, meteorology and communications. However, there is no other cross-sectoral agreement that provides for an integrated approach to environmental management other than the 1992 SADC Treaty. Another weakness is that SADC-developed policies are not being quickly translated into national policies, leaving a gap which results in fragmented approaches in different member states.

At the national level, governments have also been reviewing policies and strategies aimed at enhancing natural resource and environmental management programmes. The trend is the strengthening of such activities, but the challenge is ensuring that these are effectively applied on the ground. Obstacles include lack of human resources and financial capacity. Many of the natural resource and environmental programmes are donor funded, with many countries in the region now depending on overseas development assistance.

SCENARIOS

While the trends discussed above are not exhaustive, they are signposts to the future, providing a snapshot of what life is likely to be in the basin. The scenarios inform society of where it is likely to be if it makes certain interventions today. Two possible scenarios emerge from the history of development in the Zambezi Basin; the beaten track on which society has been travelling all along, and the sustainability highway. The first scenario would be more familiar to travel along as people already know it. The second scenario, however, exists more in the mind than in reality, making it difficult to appreciate that it actually leads anywhere. There are so many uncertainties with the second scenario that any success cannot be measured by the present generation but rather by the next generations.

THE FIRST SCENARIO ROUTE: TOWARDS UNSUSTAINABILITY

The familiar road of environmental management in the Zambezi Basin is clear: it is defined along national boundaries, administrative and sectoral tracks along, which deviation is limited. Tackling issues in the basin using the ecosystem approach is a new concept fraught with many uncertainties and risks national governments may not want to take. It is therefore, more comfortable and convenient to use such an approach while tinkering at the edges with new innovations. This, unfortunately limits the many brave steps that have already been taken to raise the profile of the Zambezi Basin as a critical ecosystem of the region in terms of human development, economic growth and environmental health.

Population growth

Population growth per se is not an issue as such, as growth in a diversified economy can help drive economic growth. Population growth is also an emotive issue with the Cassandras predicting doom if it is not controlled, and Pollyanas taking a more optimist perspective. In the Zambezi Basin, the addition of about one million more people every year, the majority of whom have to depend on the natural resource capital for survival, is unsustainable. With a total population of more than 38.4 million people, whose number will double in about 25 years, the basin will continue to experience demands on water for domestic, industrial and irrigation use, drawing upon water required for environment regeneration.

In terms of the environment, a high population growth rate impacts on the rate at which resources are exploited. Left uncontrolled, overexploitation can lead to the extinction of some species. The expansion of settlements due to population growth not only contributes to the utilisation of ecologically sensitive areas but also exacerbates human-animal conflicts. Reports of such conflicts are on the increase in the basin.

Uneven distribution of population and resources

Population in the basin and resources are unevenly distributed, causing overpopulation in some areas. The end result is that resources in some areas, such as those in parts of

Box 13.1 The Scenario-building exercise

The Zambezi Basin provides the most wanted freshwater for human consumption and economic development, pasture for livestock and wildlife, fertile soils for agriculture, yield a major harvest of fish protein and support some of the largest contiguous wildlife populations and habitats on the African continent. At the same time, the Zambezi Basin provides for the majority of the region's present power generation, represents a key asset in the region's tourism and recreation industry, and supports the subsistence economies of some of the poorest and unique local communities of southern Africa.

Sustainable living in the basin depends on the people's knowledge of ecosystems, climate and the natural resource, and how well they understand and account for the human cultural, social and economic forces interacting with these parameters and how they use the knowledge to guide their behaviour.

These are the issues that were debated by more than 30 experts who gathered in Harare in August 1999 to look at the trends and scenarios in the basin. The deliberations form the basis of this chapter. Scenarios are projections or forecasts about the future that have logical plots regarding the manner in which events unfolds — hypothetical sequence of events constructed for the purpose of focusing attention on causal processes and decision points. They:

- Direct attention to the unfolding of alternatives and to branching points at which actions can significantly affect the future.
- Clarify different views and values, challenge conventional thinking, encourage debate, and provide a common framework within which different stakeholders can address critical concerns and identify alternatives.

Some scenarios may be more analytical (focusing on the unfolding of basic processes), while others may be more impressionistic (introducing specific vivid events for illustration). All scenarios need to be constructed with rigour, detail, and creativity, and they need to be evaluated for plausibility, consistency, and sustainability. Well-articulated scenarios include quantitative insight from scientific models, such as accounting. In scenario formulation, projections of social, economic, and demographic trends become unreliable as time horizons expand from months and years to decades and generations. It is important to note that any future outcomes will depend on human choices that have yet to be made. It is also important to note that future scenarios are unfortunately based on society's current values and biases, factors which will certainly change with time. Therefore, the future only becomes real when it is present.

Underpinning issues affecting environmental and natural resources management in the Zambezi Basin that were identified ahead of the workshop include the following:

- An increasing population heavily dependent on natural resources for survival.
- Increasing poverty across the basin.
- General degradation of the natural resources (deforestation, pollution of water resources, weed infestation).
- Inequities in distribution and access to natural resources.
- Natural calamities such as drought and climate change.
- Poor governance and national policies/strategies.
- Poor macro-economic policies that fail to account for the depletion and degradation of natural resources.

These issues are operating in a situation where there is:

- Increasing population densities per square kilometre.
- Increasing economic dependency on the environmental capital. Throughout the basin, more and more people are facing a bleak future where alternatives for survival (other than the exploitation of natural resources) are quickly vanishing.
- Worsening poverty both in the rural and urban areas of the basin. Scenarios from many parts of the developing world show that those who live in poverty are often too preoccupied with immediate survival to be concerned with environmental issues.
- An increasing trend in rural-urban drift that is resulting in increasing urbanisation and mushrooming of squatter settlements and the informal sector industry. This in turn has created problems of sanitation and poor health as the urban authorities fail to cope with increasing numbers of people.
- Increasing incidence of the HIV/AIDS pandemic. Although the population growth rates are high, signs are that the HIV/AIDS problem will have an impact on the growth in many countries of the basin.
- A worsening unemployment crisis. The employment chances are bleaker due to the current economic reform programmes (that have resulted in massive job losses in many sectors of the economy) and recurrent droughts.

- Increasing economic development demands on the basin's natural resources. The demands on the Zambezi Basin's resources to meet the needs of economic development are increasing. There is increasing demand on the basin's natural resources against a background of dwindling resources.
- Increasing occurrence of drought and the incidence of climate change. Past trends in the region and basin show that temperature will increase by 0.5°C per decade. A steady increase in the temperature is expected to continue in the 21st century. Past trends in rainfall (1900–2000) show a reduction in rainfall. Based on the climate change models for southern Africa, three climate change scenarios are possible, i.e. core scenario with modest drying, dry scenario with a marked reduction in rainfall and a wet scenario in which most of the region (and the basin) will be wetter. These three rainfall scenarios will have implications on the basin's water supply, agriculture, energy generation and flora and fauna. There could be migration of natural vegetation zones in response to climate change.
- Progressive environmental pollution and degradation caused by unsustainable patterns of resource consumption and productive processes, and increasing wildlife populations. Increasing quantities and changing composition of urban waste imposes a burden on urban authorities unable to afford clean disposal technologies. The major problems currently experienced relate to toxic and hazardous substances, effluent, industrial and other chemical wastes. The dumping of toxic wastes that are not biodegradable is gradually becoming a major problem in large cities.
- Water pollution and weed infestation, mainly caused by industrial waste from manufacturing companies in cities, towns and mining areas. Water pollution is becoming a serious threat to the limited available water resources.
- Growing interest and efforts to harmonise environmental policies and legislation. The ZACPLAN programme, SADC protocols (in Water, Trade, Forestry, etc.) energy power pool, are examples of the progress and future plans towards this effort.
- Progressive creation of basin institutions and joint planning in areas of water, environment and general natural resources management.
- Increasing awareness and realisation that the management of the resources in the basin should be tackled using the ecosystem approach.
- Increasing political tensions as the wave of new democracies come into play. Conflicts over water are increasing in the basin as more and more areas are being opened for economic development.
- Availability of and limitations in technological advancement.

The future state of the environment in the Zambezi Basin is/and will be determined by the nature of human activities. This means that human activities and development are the main drivers for possible scenarios for the state of environment in the Zambezi Basin. Possible priority environmental scenario drivers that the workshop used included:

- Changes in economies; i.e. economic growth, shift to market economies, globalisation, growth in tourism, etc.
- Changes in population; i.e. growth, consumption patterns, poverty levels, etc.
- Climate change.
- Changes in technology; availability and non-availability.
- Changes in information and communications.
- Social performance, i.e. values, human security, etc.
- Governance and institutions, i.e. devolution, environmental rights and responsibility, decentralisation, privatisation, commercialisation and changing relationships among institutions.

The workshop participants were divided into the following groups and addressed questions as specified:

- Water: What are the current trends of water quantity and quality, and availability? What are the likely future trends in water availability, water development, etc. and what will be likely impacts (both positive and negative) on the environment? What are/will be the main drivers for change in the state of water in the basin? What are the possible actions that the basin states can take in the short, medium and long-term?
- Environment: What are the current trends in the state of environment, i.e. climate, biological resources, land resources, etc. What are the likely future trends and what will be likely impacts (both positive and negative)? What are/will be the main drivers for the change in the state of environment in the basin? What are the possible actions that the basin states can take?
- Economic development: What are the current trends in the state of economic development. What are the likely future trends and what will be likely impacts (both positive and negative)? What are/will be the main drivers for the change in the state of economic development in the basin? What are the possible actions that the basin states can take?

Photo: M Chenje

The growing population influences the rate at which resources are exploited for immediate survival.

Malawi, are overexploited. The harvesting of forests in such areas, for example, has contributed to serious deforestation. Rural-urban migration is unsustainable if countries in the basin do not take steps to "bring the bright lights" and jobs to the rural areas. Overcrowding and unplanned settlements will continue to cause water supply and sanitation problems. The management of waste as well as uncontrolled pollution due to industrialisation will continue to be environmental challenges for local authorities.

Poverty

The failure to effectively deal with rampant poverty in the basin will continue to pose a serious threat to both human and ecosystems health. With the majority of the people lacking human dignity due to poverty, the challenge for the basin states is to pursue economic development to raise the standard of living of the majority and not just facilitate the amassing of wealth into a few hands. This is generally the case in many of the countries, and is a time bomb if the majority continue to be alienated from accessing even the basic of resources. With no security of tenure and access to resources, the "tragedy of the commons" as postulated by Garrett Hardin in 1968 will set back any sustainable development initiatives as the poor overexploit common resources. The tragedy of the commons is not only linked to poverty but also weak or lack of clearly defined property rights. The "business-as-usual" approach or following this "beaten track" is the way to ruin, and the widening of the circle of poverty.

HIV/AIDS

Southern Africa is the region worst hit by HIV/AIDS with about 19 million infected in 1997.[36] HIV/AIDS, which has been described by SADC as "one of the greatest social problems facing Africa", will continue to wreak havoc in the basin with many of the countries experiencing a decline in life expectancy. Despite the recognition that AIDS is a pandemic, regional responses to date have not been very effective. In fact, some countries tried to sweep the problem under the carpet so as not to jeopardise investment and tourism. HIV/AIDS has also placed heavy demands on the health sector, sending service costs spiralling and increasing the costs of importing drugs. In fact, many thousands of people have no ready access to essential drugs and where such drugs are available, the costs are prohibitive. HIV/AIDS is killing many of the basin's people still in their productive years, contributing to low human capacity in many areas. The true cost of this pandemic is yet to be fully understood.

The emerging scenario is that governments recognise the need to tackle the HIV/AIDS problem but resources for implementation are diminishing and hence economies will suffer and populations will be more youthful.

Gender

One of the major developments of the 1990s is the recognition of gender issues in the Zambezi Basin states. While gender awareness is increasing, application of gender considerations on the ground is still weak. One of the contributory factors is lack of reliable data, disaggregated along gender lines. The continued weak collection of such data, as is the case at the moment, will hamper efforts to mainstream gender in all aspects of human development and environmental management. Without investment in data collection and analyses, the magnitude of women's role in natural resource management will continue to be anecdotal, minimising their input towards sustainable development. Another threat to appreciating the role of women is the apparent chauvinism displayed by some who describe the fight for women's rights as an urban phenomenon with no rural support.

Basin states have shown a commitment to gender issues through the SADC Declaration on Gender and Development signed by Heads of States and Governments in 1997. However, trends show that women in the basin are still a vulnerable group that is yet to enjoy equality in status and access to services and resources with their male counterparts.

Although SADC ELMS has a gender strategy, a lot still needs to be done to mainstream gender into environmental policies and programmes. Environmental programmes acknowledge women as repositories of indigenous knowledge on natural resources management, yet men continue to be targeted for environmental management programmes. Gender focal points with all the necessary institutional and material support should be established in ministries of environment and other relevant institutions to address the problem.

Water

Growing demands on water resources in the basin can provide greater opportunities for regional cooperation or may cause conflict. Ineffective planning, which does not take into account the ever-increasing needs of people and industry, is unsustainable and will result in unnecessary competition within basin states and among themselves. Already some of the countries have experienced water-use conflicts, which have pitted one community against another. Water conflicts have been experienced in other parts of Africa, for example, Democratic Republic of Congo, Rwanda, Uganda and Nigeria.[37] According to projections, Malawi will experience absolute water scarcity by 2025 while Mozambique, Tanzania and Zimbabwe will be water stressed. The remaining basin states will have water quality and dry season problems.[38] Such projections could end up being a disaster if planning for the next 25 years does not start now.

Often, little attention is paid to groundwater resources even though the majority of the rural people and some in urban areas depend on such resources. While the contribution of groundwater is now recognised, progress in research and mapping the resources is slow. The danger is that groundwater will always be seen as an "add-on" rather than a critical part of the whole integrated water resources management debate. In some areas, where groundwater is acknowledged as an asset, there have been reports of "mining" the resources, resulting in the collapse of aquifers. Also, not much is known on the rates of aquifer recharge. There are also reports of waste dumping in mineshafts, putting groundwater resources at the risk of pollution.

Continued use of antiquated irrigation systems, which lose between 40–60 percent of

Box 13.2 Water a growing scarce resource

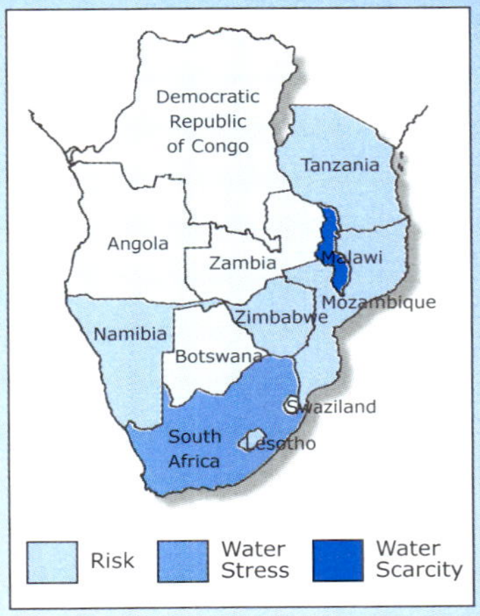

The water resources in the basin are not evenly distributed within national boundaries and across countries, with some countries having abundant resources while others are parched. For example, Malawi was by 1990 water "scarced", meaning that less than 1,000 cubic metres (cu m) were available to each person in that country in that year compared to Zambia whose people had 11,779 cu m each in that period.

Climate, particularly rainfall, is a critical factor in water distribution and availability. Rainfall is extremely variable and unreliable in the region, with droughts and floods being a common occurrence in any one part of the region at any one time. For example, the 1991–92 and 1994–95 droughts devastated most parts of the region, leaving millions of people dependent on food aid. In the 1999–2000 rainy season, hundreds of human lives were lost and about a million people displaced by the worst floods in living memory in the region, particularly in Mozambique. The cost in terms of infrastructural damage runs into hundreds of millions of dollars. The floods set back development in Mozambique by many years.

Water variability in time and space, and increasing pollution of this finite resource are combining to make environmental management for water resources sustainability a challenge as the human population rapidly grows. As the number of people in the basin increases annually, the demand for water is also rising. Unfortunately, the volume of freshwater resources available for use does not increase. It is in fact, decreasing due to increasing pollution from settlements, agriculture and industry. The result is that the people in the basin are forced to share the same cup of water whose contents are being reduced annually. For example in Zimbabwe, a total of 7,062 cu m of water was available to each person in 1955. This amount had declined to 2,323 cu m by 1990, putting the country on the road towards water stress. The water available is projected to be only 1,083 cu m by 2025, placing Zimbabwe on the precipice of water scarcity.

By 2025, three countries in the SADC region (Lesotho, Malawi and South Africa) will be water scarce and four others (Mauritius, Mozambique, Tanzania and Zimbabwe) water stressed. Water scarcity concerns are increasing in southern Africa due to various factors, which include the following:

- Climate variability
- Rapid population growth
- Economic growth
- Inefficient use of available water resources
- Pollution leading to water degradation
- Ineffective management
- Access and user rights.

The challenge facing the countries in the region is to effectively manage the available water resources to meet the needs of a growing population and the environment on which life itself depends. The sustainability of the water environment is critical to development in southern Africa.

Source: Population Action International, "Sustaining Water: An Update," PAI, Washington D.C., 1995
World Bank, African Water Resources: Challenges and Opportunities for Sustainable Development, Washington D.C., 1996

Figure 13.2: Population (in thousands) and annual water availability 1900–2025

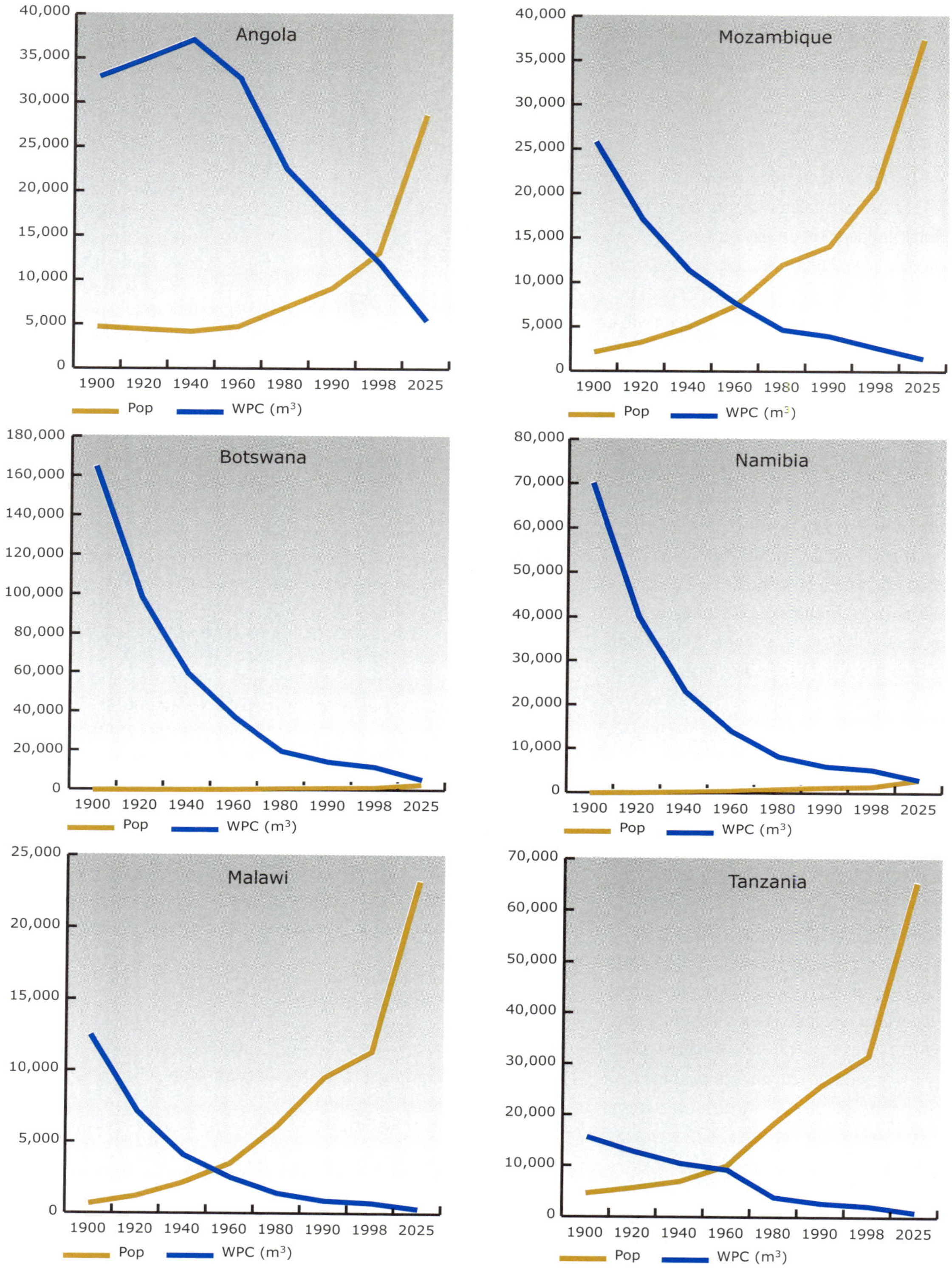

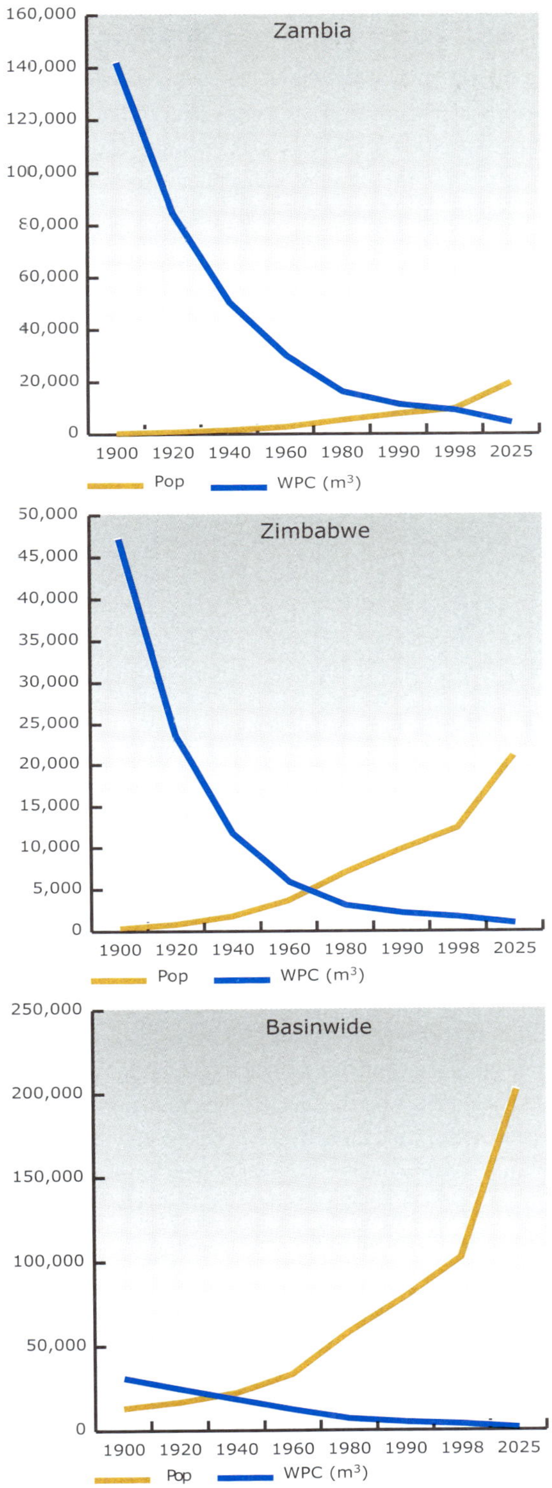

Pop – population (Population figures for 2025 are UN high projections)

WPC (m^3) – water availability per capita

- Scarcity is defined as when water available per capita is less than 1,000 cu m annually and stress is when water available per person is less than 1,700 cu m annually.
- Statistics on water available per person annually calculated from population figures from the following sources:

Klein Goldewijk, C.G.M. and J. J. Battjes, *A Hundred Year (1890-1990) Database for Integrated Environmental Assessments (HYDE, version 1.1)*, National Institute of Public Health and the Environment, Bilthoven, February 1997

Denconsult, "ZACPRO 6 Sector Study 3 – Final Report", Lusaka, 1998

Population Action International, "Sustaining Water: An Update", PAI, Washington D.C., 1995

the water through seepage and evaporation, will exert more pressure on water resources. Intensive agriculture, to feed the growing population, will also see increased use of chemical fertilisers whose runoff can result in the pollution of rivers.

Wetlands

The basin's network of wetlands covering about 66,000 sq km is threatened with human activities, including agricultural expansion. The value of these wetlands extends beyond the immediate communities, which depend on them for livelihood. They provide goods and services, which contribute to making the basin one of the richest wildlife areas in Africa. For the immediate wetlands communities, the net annual income from livestock production in the Zambezi Floodplains and Chobe-Caprivi is US$100 and US$1,200 per household respectively.[39] Net annual income from cropping yields is US$30/hectare in Barotseland, US$118/ha in Chobe-Caprivi, US$16/ha in the Lower Shire (maize only) and US$77 in the Zambezi Delta.[40] The Zambezi wetlands play a significant role in raising the standard of living for communities living in these areas. Such wealth could, however, be lost and more rural people impoverished if settlements and other human activities are allowed to expand in the wetlands. Such expansion could also lead to the decline in tourism activities. The destruction of wetlands is a reality. In South Africa, for example, 50 percent of the wetlands have been lost due to overexploitation and agricultural expansion.[41] In Mauritius, tourism development sites such as hotels and lodges are replacing most wetlands.

Biodiversity

Perhaps the international fight over ivory trade best illustrates the costs that can arise if species become endangered. While the international community reluctantly agreed in 1997 to allow three of the basin states: Botswana, Namibia and Zimbabwe to sell their stockpiled ivory under strict international supervision, many more non-charismatic species in the basin are endangered right under the nose of the same international community. The bias towards one species or a number of species at the expense of others provides cover for exploitation to extinction of those species. Selectivity in terms of biodiversity conservation is shortsighted and unsustainable. Perhaps one of the non-charismatic species which faces extinction is Zimbabwe's most popular medicinal plant, *Warburgia salutaris*.[42] The plant, which is used to treat various ailments, has been overexploited. Threats to biodiversity also extend to habitats, many of which are being destroyed in the name of development.

Policy and strategy

Virtually all countries in the basin have one or more policies relating to environment and development. Unfortunately, some of these policies are not in sync, often resulting in competing sectoral interests. Such an approach to environmental management, which has characterised most of the 20th century, can only exacerbate an unsustainable situation. National policies, which also are not in sync with regional trends, can only serve to enhance unsustainable activities. As has been observed by Maurice Strong (1997), former UNEP Executive Director and UNCED chief: "Far too many governments, companies, institutions, communities and citizens have yet to make the choices and changes necessary to advance the mutually reinforcing goals of sustainable development."[43]

At the regional level, SADC has been described as "one of the most successful regional organisations ever to operate in sub-Saharan Africa, perhaps even the entire developing world". The danger to such cooperation is sovereignty.[44] It has been argued that countries should relinquish some degree of decision-making to SADC for regional cooperation to produce significant benefits. For example, for a free trade area to work in the region, it requires "major institutional steps and loss of sovereignty".[45] This would also be true of cooperation in all other sectors, particularly in the Zambezi Basin. This is unlikely in the short-term, and may hamper efforts being made to look at the basin as one unit. The result would be that national interests would continue to dominate, contributing to unsustainable policies and strategies. Political and institutional will has to be nurtured.

THE SECOND SCENARIO ROUTE: TOWARDS A STATE OF SUSTAINABILITY

Taking an unfamiliar road has its own shortcomings, but accepting such a challenge has brought about many inventions and innovations over the centuries. Scientific breakthroughs have improved our understanding of ecosystems, enabling society to adapt and introduce programmes aimed at better management of the environment. However, some innovations have speeded up the exploitation of nature. For example, industrialisation is believed to be a major factor in the warming of the earth due to excessive emissions of greenhouse gases. The challenge is striking a balance, ensuring both human and ecological health and well being.

Photo: M Chenje

Global warming may see more parts of the basin becoming drier, making rivers run dry and exacerbating water availability.

Population growth

Little agreement exists on the exact population of the Zambezi Basin states, particularly in the various research and policy documents that have been produced over the years. However, unanimity exists in terms of rapid population growth, the fact that the majority of the people live in the rural areas, women outnumber men and that there is greater urbanisation now than ever before. The majority of the people will continue to be rural about 10 years into the 21st century but this should be the complete reverse from about 2020 when most of the people would be in urban areas.

Environmental and water issues are, however, certain to change with increased densities in the urban areas. Urban environmental issues, which may not be as significant today, would assume more prominence and present a serious challenge to urban local authorities and governments. Other actors are likely to emerge and take over the previous role and responsibilities of local authorities in the form of private interests and community groups. These should, however, be easier to deal with as countries in the region would have acquired more experience and appropriate technology in the preceding years.

Poverty

Poverty will continue to be a challenge well beyond 2025. Attention is, however, certain to shift from rural to urban areas since the majority of the people would be urbanised. However, with the advent of SAPs, more people could move to rural areas as they face retrenchments and unemployment in urban areas. With diversified economies and economic growth, governments in the region would be better placed to reduce poverty in the basin. Aware that failure to alleviate problems would set them back in trying to reach the state of sustainability, governments will introduce various measures and social security nets to improve the standard of living for the majority of the people.

One of the critical issues in terms of HIV/AIDS in the basin today is that the drugs, which are imported, are too expensive for the majority, particularly rural residents. Many are condemned to die without treatment. With economic growth and reduction in poverty, countries will be able to make drugs readily available and cheaper than they are today. However, the capacity of basin states to counter HIV/AIDS will diminish under globalisation and so other strategies and options are necessary.

Gender

Awareness of gender issues has been relatively achieved throughout the basin. Applied gender-sensitive programmes will be common over the next five years. This will be evident in data collection and analyses, and acknowledgement of the different roles men and women play in environment and development. In line with Agenda 21, governments in the basin will have introduced and implemented fully strategies to "eliminate constitutional, legal, administrative, cultural, behavioural, social and economic obstacles to women's full participation in sustainable development and in public life".[47] The Zambezi Basin states could learn from South Africa which in 1999 established a strong gender unit with an annual budget of about US$1 million in the Department of Communi-cations. Among its other objectives, the department will "assist in providing gender training programmes and monitor production of gender disaggregated data".[48] Such activities will be common by 2025 as SADC countries strengthen the harmonisation of their programmes.

Table 13.6: Child survival and development

Basin States	Infant mortality rate (per 1 000 live births)			Under-five Mortality Rate (per 1 000 children under 5 years)	Births Attended by trained Health Personnel (%)	Maternal Mortality Rate (per 100 000 live births)
Country	1960	1989	1996	1996	1990-96	1995
Angola	208	173	170	292	15	1,500a
Botswana	116	64	40	50	78	220
Malawi	206	147	137	217	55	620
Mozambique	190	173	133	214	25	1,512
Namibia	146	103	60	77	68	225
Tanzania	147	103	93	144	53	748
Zambia	135	78	112	202	51	940a
Zimbabwe	109	63	49	73	69	570a

Source: UNDP/SADC/SAPES, SADC Regional Human Development Report 1998, UNDP/SADC/SAPES, Harare, 1998, pg. 147

HIV/AIDS

By 2025, regional efforts to tackle the HIV/AIDS pandemic should start being felt as AIDS deaths start to decline. Since poverty and poor education, among other factors, have contributed to massive deaths, particularly since the beginning of the 1990s, both should be under control. Increased awareness about the disease will help reduce infections and ultimately deaths, which are currently running into tens of thousands annually.

In the United States, for example, the number of AIDS deaths has gone down significantly. In August 1999, the U.S. Centres for Disease Control and Prevention (CDC) reported that AIDS deaths in the country had declined by 20 percent between 1997–98. The number of new cases was down 11 percent during the same period.[46] The U.S. experience provides hope for people in the basin.

Employment

With the population growing at an average 2.9 percent in many of the basin states, the workforce is also increasing at a fast rate. Given the slow economic growth rate, hundreds of thousands of people are finding themselves unemployed. In Zimbabwe, unemployment is one of the reasons that have been given for the increase in gold pan-

Table 13.7: Women's participation in politics in the basin states

Country	Economically active population 15 years and above Estimated economic activity rate (%) 1990	1990	Female administrators and managers (%) 1992-96	Female administrators and managers (as % of male) 1990	Female professional and technical workers (%) 1992-96	Female professional and technical workers (as % of male) 1990	Female clerical and sales workers (%) 1990
Angola	52.0	87.0	-	-	-	-	-
Botswana	42.0	85.0	36.0	57.0	61.0	159.0	60.0
Malawi	57.0	89.0	5.0	5.0	35.0	53.0	33.0
Mozambique	79.0	91.0	11.0	13.0	20.0	26.0	-
Namibia	83.0	24.0	21.0	26.0	41.0	69.0	-
Tanzania	77.0	89.0	-	-	-	-	
Zambia	33.0	87.0	6.0	7.0	32.0	47.0	58.0
Zimbabwe	44.0	88.0	15.0	18.0	40.0	67.0	34.0

Source: UNDP/SADC/SAPES, SADC Regional Human Development Report 1998, UNDP/SADC/SAPES, Harare, 1998, pg. 144

ning. Unemployment has also seen growth in the informal sector. While this is now an important part of the economy, some informal sector activities such as the production of curio pieces, has led to the overexploitation of forests to support the carving industry.

Economic restructuring has resulted in the gradual decline in formal employment. For example, Zambia had an unemployment rate of 13 percent in 1986, and this rose to 19.6 percent in 1996, representing a total of 730,000 jobless people.[49] The implementation of the economic structural adjustment programme has contributed to the decline in formal employment. For example in Zambia, unemployed people are settling in forested areas in the Kafue River watershed area, causing massive deforestation and leading to watershed degradation as well as affecting water quality and quantity. Dependence on resource sectors will continue to be a risk for the basin states, increasing the unemployment figures and indirectly contributing to the "mining" of natural resources and resultant degradation of the environment.

Investment

The basin states as well as the rest of the SADC region lag behind the rest of the world in terms of competitiveness.[50] Of the eight basin states, only Botswana (3) and Namibia (4) were in 1998 ranked within the first 10 African countries listed in the Africa Competitiveness Report in terms of openness, governance, finance, labour, infrastructure and institutions. Out of a list of 24, the other basin states were ranked as follows: Zambia (12), Tanzania (16), Mozambique (18), Zimbabwe (20), Malawi (21) and Angola (23). However, all the eight basin states have been

Table 13.8: Population per aridity zone by sub-region and country for Africa (Population in '000)

	Hyper-arid Pop.	Hyper-arid %	Arid Pop.	Arid %	Semi-Arid Pop.	Semi-Arid %	Dry sub-humid Pop.	Dry sub-humid %	Moist sub-humid Pop.	Moist sub-humid %	Humid Pop.	Humid %	Total Pop.
Angola	6	0	55	0	2,666	23	899	8	7,035	61	957	8	11,618
Botswana	0	0	20	1	1,483	99	0	0	0	0	0	0	1,503
Malawi	0	0	0	0	298	3	5,623	54	3,980	38	552	5	10,453
Mozambique	0	0	0	0	4,231	25	4,910	29	6,752	40	959	6	16,852
Namibia	35	2	308	20	1,179	77	0	0	0	0	0	0	1,522
Tanzania	0	0	0	0	6,516	23	10,724	38	10,065	36	831	3	28,135
Zambia	0	0	0	0	1,746	20	2,895	33	4,128	47	30	0	8,798
Zimbabwe	0	0	17	0	7,515	68	3,264	29	283	3	35	0	11,113
Total	41	0	400	1	25,634	28	28,315	31	32,243	36	3,364	4	89,994
Africa Total	58,764	9	51,617	8	172,964	26	88,182	13	160,351	24	126,287	19	658,165

Source: UNDP/UNSO, An Assessment of Population Levels in the World's Drylands, New York, 1997, pg. 14

receiving direct foreign investment over the years, ranging from US$58 million for Zambia in 1996 to between US$250 million – US$300 million for Namibia in 1998.[51]

Despite this investment, it has not been significant enough to reduce the ever-increasing number of the unemployed. For many of the countries, the investment has been mainly in the resource sectors such as mining. Generally, mining operations have been regarded as some of the most environmentally unfriendly activities although this is likely to be arrested by the growing environmental impact assessment policy regime. Mining operations influence settlement patterns in Zambia and Zimbabwe. Most major settlements, including cities are located in mining areas which influenced the development of infrastructure such as railways, roads, schools and power.

The basin boasts some of the world's major minerals such as copper, cobalt, coal, gold and chrome which contribute substantially to the economies of the basin states. Investment in the mining sector is viewed as one of the likely areas to bring about economic recovery of the region. The mining industry is making efforts to promote environmentally sound operations by establishing environmental management programmes at major operations.

Table 13.9: Population in productive land (Plpop) and productive land vulnerable to desertification (Plvdpop)

Basin states	Total Pop.	PL Pop.	PLVD Pop.	PLVDPOP/ PLPOP (%)	PLVDPOP/ TOTALPOP
Angola	11,618	11,612	3,620	31	31
Botswana	1,503	1,503	1,503	100	100
Malawi	10,453	10,453	5,921	57	57
Mozambique	16,852	16,852	9,141	54	54
Namibia	1,522	1,487	1,487	100	98
Tanzania	28,135	28,135	17,240	61	61
Zambia	8,798	8,798	4,641	53	53
Zimbabwe	11,113	11,113	10,796	97	97

Note: Population numbers are in thousands.
Source: UNDP/UNSO, An Assessment of Population Levels in the World's Drylands, New York, 1997, pg. 19

Photo: M Chenje

Economic activities, including tourism, will contribute to increasing demand of water and services.

Water

Armed with the March 2000 World Water Vision for the 21st Century and other policies and strategies at the national and regional levels, the Zambezi Basin states would have succeeded in making their people appreciate both the economic and social value of water. Regional issues such as legal and regulatory framework, institutional strengthening, sustainable development policies, information acquisition, management and dissemination; awareness building, education and training, public participation, and infrastructural development would have been tackled to ensure effective management of water resources for both consumptive and non-consumptive use.[52]

Currently, the Zambezi Basin's legal and regulatory framework is weak. National water laws in most of the countries are weak. Some of the countries, such as Zimbabwe, are introducing innovative laws, which promote catchment committees. Water pollution management lacks a strong mechanism for inter-sectoral planning and coordination. Economic activities such as agriculture, tourism, mining and settlements will increase demand for water as they grow, and hence a holistic approach for integrated planning becomes necessary.

The basin still lacks socio-economic instruments, which encourage cost effectiveness, increase investments in water infrastructure, and give incentives for efficient use of water as well as pollution control. Top-down approaches, which are common in the basin in terms of water supply also limit public participation, promoting unsustainable water consumption patterns. Such unsustainable activities will be of the past by 2025. In addition, the current unsustainable competition between water users such as domestic, industrial, irrigation and the environment will be resolved through integrated water resources management.

The SADC water sector has already noted that poverty alleviation in the basin and the rest of SADC depends, among other things, on reliable access to water. "Providing sufficient water supplies can only be accomplished by, amongst other measures, expanding irrigation systems, improving water harvesting techniques and implementing water and soil conservation practices."[53]

Wetlands

The social, economic and ecological value of the Zambezi Basin's 66,000 sq km of wetlands has already been recognised at various levels, national, regional and global. Despite such recognition, the wetlands remain under threat from human activities, particularly from agriculture and expansion of settlements. Aware that alleviation of poverty in the rural areas cannot be successful without wetlands, countries in the basin will recognise wetlands as conservation areas, slightly different from the conventional national parks. They will be managed almost like biosphere resources where human activities are permissible within some buffer zones. The wetlands will become centres of community-based natural resource management programmes, expanding such programmes from the current focus of large mammals to include resources such as fish.

Just like national parks whose value extends beyond their immediate environs in that they generate revenue from tourism, wetlands can do the same. Development projects that impact on wetlands include hydroelectricity generation plants and reservoirs. Countries in the region will stringently enforce environmental impact assessment regulations, which are already in place or being introduced, to ensure minimum disruption of the natural function of wetlands.

Biodiversity

Different species have over the past decades been either overexploited to extinction or are now endangered, for example the rhino. Overexploitation has also impacted on other species, including plants. The clearance of forests is one of the major threats to biodiversity through the destruction of habitats. In the basin states for example, a total of about 1.15 million hectares of forest were converted to other uses every year between 1990–95. The basin's total forest cover declined from about 147.1 million ha to 141.3 million ha in the same period.[54] By 2025 biodiversity conservation will be more holistic, covering many species and focusing on the whole basin rather than selected or more popular areas, an approach which will also help provide understanding of the complex interactions that make up ecosystems. New species, which had hitherto been ignored or unknown will get attention, unlocking some of the biodiversity treasures, which exist in the region.

Policy and strategy

Policy and strategy initiatives will be very responsive to scientific developments. Even though developed on a sectoral or issue basis, policies and strategies will take a more integrated approach, reflecting the complexity of both human development and ecological functions. The SADC sector coordination units as well as the national contact points will have greater coordination both at national and regional levels. As the policies and strategies become more cohesive, the SADC region will be able to make stronger contributions at various international fora convened to discuss environment and development issues.

In the Zambezi Basin, one of the challenges facing people and their governments is translating integrated water resources management into reality. Delays in establishing the Zambezi Basin Commission (ZAMCOM) can only hamper regional efforts to foster closer links in managing the basin. The proposed ZAMCOM should chart a new course in terms of river basin institutions.

Photo: M Chenje

Policy and strategy initiatives should be responsive to regional needs.

Box 13.3: Key opportunities for improving regional water resources management

1 A political environment and level of awareness conducive to promoting and implementing regional opportunities for integrated water resource management.
2 Political stability in the region, a precondition for cooperation and development.
3 SADC Treaty which provides a framework for regional economic integration and cooperation, and identifies water as an area of cooperation.
4 Reforms in line with the international consensus on integrated water resource management, already started in a number of countries.
5 SADC Protocol, which forms a basis for the management of, shared watercourse systems and the establishment of the SADC Water Sector.
6 SADC water resources programme proposal for the promotion of integrated water resources management.
7 Precedents such as ZACPLAN and the Lesotho Highlands Water Project, which illustrate that cooperation, can work to benefit the Zambezi river basin riparian states, and Lesotho and South Africa.
8 Past, ongoing and proposed involvement in the region by numerous external support agencies.

Source: SADC, *Regional Strategic Action Plan for Integrated Water Resources Development and Management in the SADC Countries (1999-2004): Summary Report*, WSCU, Maseru, 1998

Historically, these institutions have tended to comprise of only officials from water departments of the involved countries. ZAMCOM should be multi-disciplinary, bringing in officials from various ministries as well as other non-government stakeholders.

The challenge is to ensure that ZAMCOM does not become an unwieldy bureaucracy. The danger is that each of the eight countries could assign a total of about six officials (from departments such as water, environment, development planning and finance, transport and energy, local government, justice and agriculture), making the commission large, difficult to manage and very costly. Perhaps the solution lies in having national-level structures, for example, committees which could then feed into the line department such as of water.

The establishment of sub-basin catchment committees could also facilitate further stakeholder participation at the country level. Such catchment committees in all the basin states could facilitate stronger public and community participation in various issues related to the management of the Zambezi Basin. Some experts have even suggested that sub-basin committees could in future become national-level administrative units instead of the present rural or urban councils some of whose current boundaries defy logic. While catchments form natural ecological boundaries, some environmental groups may prefer other methods of defining such boundaries, for example, ecological zones (ecozones).

Vegetation is a good criterion for defining ecozones because it embodies all the conditions in which it grows. Soils provide nutrients and store water used by plants. Ecozones are affected by water, soils and nutrients, fire, animals, climate, including rainfall, as well as human activities. While the Zambezi Basin has 13 sub-basins, they fall into four ecozones; moist savanna woodlands, dry savanna woodlands, afromontane/temperate forest and grasslands and lowland/coastal forest, woodlands and grasslands.

The moist savanna woodlands ecozone covers the largest part of the Zambezi Basin, particularly the following sub-basins: Upper Zambezi, Lungue Bungo, Luanginga River, Cuando/Chobe River, Kabompo River, Kafue River, Barotse, Shire River/Lake Malawi, parts of Kariba, Tete, Mupata and Luangwa River. These sub-basins are found in Angola, Namibia, most of Malawi, Mozambique, Zambia and Zimbabwe and a small part of Botswana and Tanzania.

The dry savanna woodlands ecozone covers parts of the Kariba, Mupata, Luangwa River, Tete and Shire River/Lake Malawi sub-basins including the Zambezi Valley; parts of southern and northeastern Zambia, most of northern Mozambique in the Tete province, and a small part of the basin in Botswana and Tanzania.

The Afromontane/Temperate forest and grasslands ecozone lies in part of the Tete, and Shire River/Lake Malawi sub-basins in some parts of Malawi, a small part in Tanzania and in northeastern Zimbabwe bordering Mozambique. Finally, the Coastal/Lowland Tropical forests, woodlands and grasslands ecozone is found in the Zambezi Delta sub-basin at the mouth of the Zambezi River in Mozambique.

All these considerations should give way to the most appropriate and effective way of implementing integrated management measures in the basin. Measures that ensure both human and ecological health and not sectoral or professional interests can help the basin states attain a state of sustainability.

The sun should never set on the people's efforts to chart a course that will see them achieve a state of sustainability.

Within the integrated basin management umbrella are issues such as:

- Providing water to meet various user needs, including the environment.
- Managing the demand both in times of scarcity and surplus.
- Monitoring and enforcing regulations aimed at recognising both the economic and social good of water.
- Minimising water pollution to levels that enable water to break down the pollutants.
- Effective environmental management.
- Implementing both national laws and regional protocols aimed at effective environmental and resource management.
- Promoting sound landuse activities.
- Promoting the use of indigenous knowledge systems.
- Planning for a growing population, which continues to place heavy demands on natural resources.
- Reversing growing poverty and ultimately alleviating poverty.
- Providing sustainable energy alternatives in terms of fuelwood use and electricity generation.
- Promoting industrial activities that use cleaner production techniques.
- Raising awareness among the people of the basin of the environment and development issues, and need to play an effective role in natural resource management.
- Promoting the strengthening of community-based natural resource management programmes in all basin states where there is potential.
- Ensuring that peace and security are maintained in the region.

Perhaps the second scenario is more optimistic than the many lessons history provides. History shows that since the Industrial Revolution in the 18th century, the human race has continued to provide pressure on the environment despite mounting evidence that this was unsustainable. While many measures have been introduced to deal with the resultant environmental problems, they have not been very effective due to lack of commitment. It is this lack of commitment that will ultimately decide at basin level, SADC and global levels whether or not the state of sustainability will be a reality. The Zambezi Basin states will know in 2025.

LINKAGES TO OTHER CHAPTERS

CHAPTER 1: Regional Overview: People and Environment
A state of sustainability involves costs and benefits, which cut across regional and global developments, national and regional interests, local and national issues, sectoral and integrated considerations, economic and social interests, and ultimately the popular participation of people in all areas.

CHAPTER 2: Physical Features and Climate
The physical features of the basin pre-determine what the basin states can or cannot do. The limiting factors to what can be done are the fact that land in the basin cannot be expanded and that the countries are in an arid to semi-arid region, which forecloses some options. Climate is an important factor on whether or not the Zambezi Basin records significant progress on environment and development.

CHAPTER 3: Water and Wetland Resources
Sustainable development in the basin depends on effective management of water resources and conservation of wetlands. Both resources are critical to environment and development in the Zambezi Basin.

CHAPTER 4: Biological Resources and Diversity
The Zambezi Basin is rich in biological resources, which support economies and raise the standard of living of basin inhabitants. Destruction of habitats and overexploitation of biodiversity will impoverish the basin, short-circuiting all efforts towards sustainability.

CHAPTER 5: Agriculture
Population is expected to rise in the Zambezi Basin states and as this occurs, more and more mouths will need to be fed. It is also expected that agriculture will expand especially in the form of irrigation and this will result in increased pressures on the natural resources within the basin. Planning for such scenarios is therefore, an urgent matter.

CHAPTER 6: Industry
To satisfy demand for manufactured goods, industry is expected to grow together with the human population. The growth of industry could be seen as a positive development as it could wean some of the people from the overdependence on natural resources. However, such growth must take into account the negative aspects of pollution in all its forms and necessary legislative mechanisms must be seen to be effective in mitigating environmental disasters.

CHAPTER 7: Energy
Both traditional and scientific approaches to energy use exert pressure on ecosystems, exacerbating land clearance and altering natural functions of ecosystems due to the construction of dams for hydroelectricity generation.

CHAPTER 8: Tourism
Tourism is the golden goose in the Zambezi Basin. Over-development of this sector will kill this golden goose by making it less attractive for tourists who, ultimately, will be forced to visit other sites in the world. It is, therefore, important that tourism, just like other sectors, moves towards a state of sustainability.

CHAPTER 9: Pollution
Air and water pollution reduce options available to the people of the basin. Both are likely to worsen as urbanisation of the majority of the people becomes a reality. Cleaner technology is one of the solutions to reduce air and water pollution.

CHAPTER 10: Poverty
The greatest impediment towards a state of sustainability is poverty. As long as the majority of the people remain poor and are forced to overexploit natural resources, a state of sustainability will remain a mirage.

CHAPTER 11: Gender and Women's Roles
If gender is to be mainstreamed in all environment and development issues the motto should be 'from awareness to action'. One of the critical steps is ensuring the collection and analysis of gender-disaggregated data.

CHAPTER 12: Environmental Management and Regional Cooperation
Sovereignty is giving way to regional cooperation. This will become more evident as national governments give SADC powers to make decisions on their behalf and which they will be expected to implement.

ENDNOTES

1 Gleick, P.H., *The World's Water: The Biennial Report on Freshwater Resources 1998-99*, Island Press, Washington DC, 1998, p183
2 Johnson, P. and Martin, D., Apartheid Terrorism - A Report for the Commonwealth Committee of Foreign Ministers on Southern Africa, Commonwealth Secretariat, London, 1989
3 Zimbabwe Inter-Africa News Agency, "Call to involve gold panners in legislation", *Ziana*, Harare, 31 Oct, 1994
4 Zimbabwe Inter-Africa News Agency, "300 Penhalonga families resort to gold panning", *Ziana*, as reprinted in *The Daily Gazette*, Harare, 8 Sept, 1994
5 op. cit 3
6 Grove, R., "Threatened Islands, Threatened Earth: Early Professional Science and the Historical Origins of Global Environmental Concerns", 1990
7 ibid
8 SADC, "Food, Agriculture & Natural Resources", *SADC Annual Consultative Conference*, Johannesburg, 1996
9 O'Connor, J., "Uneven and Combined Development and Ecological Crisis", *Race and Class*, 1989, pp 1-11
10 Environmental Affairs Department, *State of the Environment Report for Malawi 1998*, Malawi Government, Lilongwe, 1998, p93
11 UNDP, *Human Development Report 1997,* Oxford University Press, New York, 1997
12 Agyemang, O., et al, *An Environmental Profile of the Greater Lusaka Area: Managing the Sustainable Growth and Development of Lusaka*, Lusaka City Council/Government of Zambia, Lusaka, 1997
13 WHO, Cholera-1994: Situation in the African Region, Harare, 1995
14 Holloway, Ailsa, "Challenges for Long-term Disaster Reduction: Elements of Telemedicine", Harare, unpublished, 1995
15 SARIPS/SAPES Trust, *SADC Regional Human Development Report 1998*, SARIPS/SAPES Trust, Harare, 1998, p41
16 UNFPA, *The State of the World Population 1999: 6 Billion - A Time for Choices*, United Nations Population Fund, New York, Sept 1999, p23
17 ibid
18 Shaw, D.J. and Cranshaw, B., "Disaster Mitigation and Rehabilitation in Sub-Saharan Africa: A World Food Programme Perspective", The University of Arizona, *Arid Lands Newsletter*, vol. 34, 1993, pp 6-7
19 ibid
20 Southern African Development Community, "Finance and Investment Annual Report", *SADC Annual Consultative Conference*, Johannesburg, 1-4 Feb., 1996, p. 4
21 Allen, T.W., "Structural Adjustment in Zimbabwe: The World Bank Perspective", Paper Presented at Belvedere Teachers' College, Harare, Sept 1999
22 ibid
23 Government of Botswana, "Botswana Country Paper", in *Proceedings of the Workshop on Water Resources Management in Southern Africa*, Victoria Falls, 5-9 July, 1993
24 Heyns, P., "Water Management in Namibia", in *Proceedings of the Workshop on Water Resources Management in Southern Africa*, Victoria Falls, 1993
25 ZRA, "ZACPRO 6 Sector Study 3: Patterns of Land Use and Conservation Practices", ZRA, Lusaka, 1998, pp42-43
26 Environmental Affairs Department, *State of the Environment Report for Malawi 1998*, Malawi Government, Lilongwe, 1998, p29
27 ibid
28 Cheru, F., *The Silent Revolution in Africa*, ZED Books/Anvil Press, London/Harare, 1989
29 FAO, *State of the World's Forests 1997*, Words and Publications, Oxford, 1997
30 op. cit 15
31 SADC, *Environmental Effects of Mining in the SADC Region*, Mining Sector Coordinating Unit, Lusaka, 1992
32 ibid
33 Huntley, B.J. and Matos, E., *Biodiversity: Angolan Environmental Status Quo Assessment Report*, IUCN Regional Office for Southern Africa, Harare, 1992
34 op. cit 15
35 SADCC, *Sustaining Our Common Future*, ELMS, Maseru, Oct 1991, p10
36 SADC Secretariat, *1997/98 SADC Annual Report*, Gaborone, 1999, p7
37 Department of Environmental Affairs and Tourism, http://fred.csir.co.za/www/deat/, July 1999
38 Fruhling, P., *A Liquid More Valuable Than Gold*, Sida, Stockholm, 1996, p33
39 Turpie, J., et al, *Economic Value of the Zambezi Basin Wetlands*, IUCN-ROSA, Harare, Oct 1998, p vi
40 Turpie, J., et al, *Economic Value of the Zambezi Basin Wetlands*, IUCN-ROSA, Harare, Oct 1998, p vi
41 op. cit 38
42 The Standard, "Local plant faces extinction", 5-9 Sept, 1999, p 17
43 ICSW, 'Rio + 5', *Social Development Review*, March 1997, vol 1, No. 3; International Council on Social Welfare, Montreal
44 Rowlands, I.H., *Climate Change Cooperation in Southern Africa*, Earthscan/Unep, London, 1998, p49
45 Mandaza, I., "SADC: The way forward", *The Zimbabwe Mirror*, SAPPHO, Harare, Sept 1999, p6
46 Cooper, M., "US sees slowing decline in AIDS deaths", 1999, http://dailynews.yahoo.com/h/nm/19990831/ts/aids_deaths_3.html
47 UN, *Earth Summit: Agenda 21 - The United Nations Programme of Action from Rio*, UN, New York, 1992, p221
48 Streek, B., "R6m for gender monitor", *Mail & Guardian*, M & G Media Ltd., Johannesburg, Sept 3-9, 1999, p6
49 SADC, *Employment and Labour*, Lusaka, Feb 1999, p27
50 SADC, *Finance and Investment*, Lusaka, Feb 1999, p2
51 ibid
52 SADC, *Regional Strategic Action Plan for Integrated Water Resources Development and Management in the SADC Countries (1999-2004): Summary Report*, WSCU, Maseru, 1998, pp vi-viii
53 ibid
54 op. cit 30

REFERENCES

Agyemang, O., et al, *An Environmental Profile of the Greater Lusaka Area: Managing the Sustainable Growth and Development of Lusaka*, Lusaka City Council/Government of Zambia, Lusaka, 1997

Environmental Affairs Department, *State of the Environment Report for Malawi 1998*, Malawi Government, Lilongwe, 1998

FAO, *State of the World's Forests 1997*, Food and Agriculture Organisation, Rome, 1997

Gleick, P.H., *The World's Water: The Biennial Report on Freshwater Resources 1998-99*, Island Press, Washington DC, 1998

Grove, R., "Threatened Islands, Threatened Earth: Early Professional Science and the Historical Origins of Global Environmental Concerns", 1990

Heyns, P., "Water Management in Namibia", *Water Resources Management in Southern Africa Workshop*, Victoria Falls, 1993

Holloway, A., "Challenges for Long-term Disaster Reduction: Elements of Telemedicine", Harare, unpublished, 1995

Huntley, B.J. and Matos, E., *Biodiversity: Angolan Environmental Status Quo Assessment Report*, IUCN Regional Office for Southern Africa, Harare, 1992

O'Connor, J., "Uneven aknd Combined Development and Ecological Crisis", *Race and Class*, 1989

Rowlands, I.H., *Climate Change Cooperation in Southern Africa*, Earthscan/UNEP, London, 1998

SADC, "Food, Agriculture & Natural Resources", *SADC Annual Consultative Conference,* Johannesburg, 1996

SADC, *Regional Strategic Action Plan for Integrated Water Resources Development and Management in the SADC Countries (1999-2004): Summary Report*, WSCU, Maseru, 1998

SADCC, *Sustaining Our Common Future*, ELMS, Maseru, October 1991

SADC/UNDP/SAPES Trust, *SADC Regional Human Development Report 1998*, SAPES Trust, Harare, 1998

Shaw, D.J. and Cranshaw, B., "Disaster Mitigation and Rehabilitation in Sub-Saharan Africa: A World Food Programme Perspective", The University of Arizona, *Arid Lands Newsletter*, vol. 34, 1993

Turpie, J., et al, *Economic Value of the Zambezi Basin Wetlands*, IUCN-ROSA, Harare, October 1998

UN, *Earth Summit: Agenda 21 – The United Nations Programme of Action from Rio*, UN, New York, 1992

UNDP, *Human Development Report 1997,* Oxford University Press, New York, 1997

UNFPA, *The State of the World Population 1999: 6 Billion – A Time for Choices*, United Nations Population Fund, New York, September 1999

Zambezi River Authority, "ZACPRO 6 Sector Study 3: Patterns of Land Use and Conservation Practices", ZRA, Lusaka, 1998

1A GLOSSARY

Aa

abiotic – physical elements of an ecosystem, including, for example, climate, soil, water, geology, ice, and non-living organic matter like peat

acid mine drainage – low-pH drainage water from mines, usually caused by the oxidation of sulphates to sulphuric acid. Mine drainage can also contain high concentrations of metal ions

acid rain – industrial pollutants such as sulphur dioxide and nitrogen oxide gases from burning fossil fuel pollute the air, which during rain mixes with water droplets to form acid rain. It damages buildings, destroys vegetation and acidifies water bodies

adsorption – surface retention of solid, liquid, or gas molecules, atoms, or ions by a solid or liquid. Adsorption onto suspended or bottom sediments is an important process in the removal of pesticides and other organic compounds from the water column of rivers, lakes, and oceans

aestivation – dormancy that occurs in some animals during a period when conditions are hot and dry

afforestation – planting trees in an area where there are no trees

agrarian – of agricultural land and its cultivation

agrochemicals – chemical fertilisers, herbicides and pesticides used in agriculture

agroforestry – planting trees and crops together

algae – plants, which contain chlorophyll and have no stems, roots or leaves and live in water or in moist conditions

alien – also known as exotic, introduced, or non-native: any organism that enters an ecosystem beyond its normal range through deliberate or inadvertent introduction by humans. Includes genetically modified organisms that result from biotechnology

alloy – metal made from a compound of two or more metallic elements

alluvial deposits – material transported by a stream or river and deposited on the river flood plains

alluvial soils – these are soils, which are grey to dark brown in colour formed from alluvium

anoxic – absence of oxygen

anthropogenic – influences that originate from humans

anticyclone – an area of high atmospheric pressure around which air circulates in a clockwise direction in the northern hemisphere, or anticlockwise direction in the southern hemisphere

antiquity – ancient times

aquatic – marine and freshwater ecosystems

aquifer – porous layers of sand, gravel, or bed rock able to store groundwater

arable land – land that can be cultivated

arid – a dry area which receives less than 250 mm rainfall every year

aridity – condition in which evaporation always exceeds precipitation

artisan – a skilled workperson in industry or trade

asbestosis – a lung disease caused by inhaling asbestos particles

axiom – an accepted general truth or principle

Bb

basin – a large, low-lying area drained by a stream or river system
bio- – Prefix referring to living organisms
bio-accumulation – a process by which a chemical substance is ingested and retained by organisms, either directly from the environment or through consumption of food containing chemicals
biodegradable – capable of being broken down by living organisms into inorganic compounds
biological diversity – also known as biodiversity: the variability within species, between species and of ecosystems
biomass – quantity of living matter in a defined area expressed in units of living or dead mass. In the energy field, any form of organic matter (from both plants and animals) from which energy can be derived through burning or bioconversion such as fermentation
biome – large ecological region characterised by similar vegetation and climate (such as forest, woodland, grassland) and all the living organisms in it
biosphere – that part of the earth's environment in which living organisms are found, and with which they interact to produce a steady state system, effectively a whole planet system
biota – collectively, the living organisms in a given ecosystem, including bacteria and other micro-organisms, plants, and animals
biotic – living elements of an ecosystem, such as plants and animals
borehole – hole, drilled vertically or at an inclination into the ground and usually fitted with a mechanical or motorised pump to draw water from the ground
briquettes – a block of compressed coal dust or other vegetal matter such as saw dust, grass, etc
bubonic plague – a contagious disease that usually causes inflammation in the groin or armpits
by-products – a substance produced in the course of making something else

Cc

caches – a hiding place for treasure or store
cash crop – crop grown for sale as opposed to consumption by the grower
carbon dioxide (CO_2) – one of the greenhouse gases, which is released to the atmosphere by both natural and human activities
carbon monoxide (CO) – a colourless, odourless, and tasteless gas released primarily by incomplete combustion of fossil fuels (especially by automobiles)
carcinogenic – cancer-producing substance
chiperoni – weather condition experienced in Malawi whereby moist easterly trade winds from the Mozambique channel rise along the slopes of the eastern highlands and reach south-east Zimbabwe and south Malawi as a cold and nearly saturated air mass accompanied by a heavy deck of low clouds with light rain or drizzle, with very little change of temperature during day and night
cholera – an infectious and often fatal disease causing severe diarrhoea
chronic – affecting a person for a long time or constantly recurring
climate – the sum total of the weather experienced in a region or area in the course of the year and over the years
climate change – a change of climate, which is attributed directly or indirectly to human activity and alters the composition of the global atmosphere in addition to natural climate variability observed over comparable time periods
climate system – the totality of the atmosphere, hydrosphere, biosphere and geosphere and their interactions
coalface – the exposed surface of coal in a mine
coastline – shoreline
colluvial – weathered rock debris that has moved down a hill slope either by creep or by surface wash
coliform bacteria – group of bacteria predominantly inhabiting the intestinal tracts of humans and other warm-blooded animals, but also occasionally found elsewhere
commercial farming – growing crops or keeping animals, usually on a large-scale, for sale
common pool – resources (commons) commodities such as land, air, water and forestry to which all people have equal right of access and use, and no one has the right to spoil
condensation – transition from the vapour to the liquid state
conservation – sustainable use of natural resources for the benefit of present and future generations
consumptive use – use of natural resources which involves changing them from their natural state through various forms of harvesting
contamination – the introduction of any undesirable substance – physical, chemical or biological – into an ecosystem
convention – agreement made by nations over partic-

ular issues that are for the benefit of all

culvert – a drain that crosses under a road or railway

Dd

dambo – a shallow, seasonally or permanently waterlogged, grass-covered depression

DDT – dichloro-diphenyl-trichloroethane. Toxic synthetic insecticide introduced for widespread use just after World War II to control malaria. Mostly banned around the world

decimate – to destroy a large proportion of something

deforestation – indiscriminate removal of trees from a forested area without adequate replanting

desertification – land degradation processes occurring in dry sub-humid areas as a result of various factors, including climatic variations and human activities. It is a process of conversion of non-desert land to desert, usually as a result of overgrazing, depletion of organic matter in the soil, overuse of groundwater, changing patterns in precipitation, etc

desiccation – a long-term loss of water that is associated with regional climatic change or the drying out of an organism that is exposed to air

dialectical – vocalisations among a population of animals that differ from those of another population of the same species

diarrhoea – an illness in which the bowels are emptied too often and in too liquid a form

downstream – the direction in which a stream or river flows towards the mouth

dredging – scooping debris from the bottom of a river or sea

drought – continuous absence of precipitation during the period when it is normally expected to occur. All other definitions of drought are related to the effect or impact of below average rainfall on agriculture, water resources, social and economic activities, hence the terms *agricultural drought*, *hydrological drought*, etc

dysentery – a disease with inflammation of the intestines, causing severe diarrhoea

Ee

ecology – study of natural environment and of the relations of organisms to each other and to their surroundings

ecosystem – all living and non-living components of the environment that interact and influence one another as a functional unit

ecotourism – purposeful travel to natural areas to understand the culture and natural history of the environment, ensuring not to alter the integrity of the ecosystem, while producing economic opportunities that make the conservation of natural resources beneficial to local people

effluent – liquid waste material that is a by-product of human activities such as liquid industrial discharge or sewage

El Niño – the warming of eastern and central Pacific Ocean waters which bring about changes in global weather patterns, by affecting air and ocean temperatures and is associated with below average rainfall in southern Africa

encephalitis – inflammation of the brain

encroachment – illegal occupation of an area set aside by law for use other than settlement and its associated activities, conflicting with the designated use of the land

endemism – uniqueness of a species to a particular geographic region

enunciate – to state clearly

environment – components of the earth, including air, land, and water, all layers of the atmosphere, all organic and inorganic matter and living organisms, and the interacting systems that include all these components

environmental degradation – depletion or destruction of a potentially renewable resource such as soil, grassland, forest, or wildlife by using it at a faster rate than it is naturally replenished

equity – fairness, impartiality

eradicate – to get rid of or to remove all traces of

estuary – part of the river where it meets the sea, characterised by a mixture of sea water and fresh water

euphemism – a mild or roundabout expression substituted for one considered improper or too harsh or blunt

eutrophication – also known as nutrient enrichment. The process of over-fertilisation of a body of water by nutrients that produce more organic matter than the self-purification reactions can overcome

evaporation – the process by which a solid or liquid turns into vapour; loss of moisture as vapour

evapotranspiration – the loss of water from the soil to the atmosphere by evaporation and plant transpiration

extinction – the total loss of a species from the earth due to natural or unnatural causes

Ff

fauna – animal life

fecundity – ability to produce many offspring

ferrallitic – these are soils derived from various parent rocks, including granite, gneiss, sandstone, and schist

ferralsols – also known as tropical red soils; these are acidic, leached out tropical soils of low fertility

ferro- – Prefix referring to or containing iron

ferruginous soils – these are among the most fertile soils, generally red in colour with sandy clay loam topsoil over a sandy clay or clay subsoil

feudal systems – a method of holding land by giving one's service to the owner

floodplain – area beside a river or lake, which is seasonally flooded when water levels rise due to high rainfall

flora – plant life

food chain – food relationship in an ecosystem in which energy and nutrients are transferred through a series of organisms by each stage feeding on the preceding one and providing food for the succeeding stage

food security – assurance that food will be available when needed through growing, storing or importing it

fossil fuels – oil, gas, coal, and other fuels that were formed under the Earth's surface from the fossilized remains of plants and tiny animals that lived millions of years ago

freehold – complete ownership of property (land in this context) for an unlimited period

frost – the condition that exists when the temperature of the earth's surface falls below freezing point i.e. 0°C or 32°F

fuelwood – wood used to produce energy for heating, cooking and lighting

Fynbos – a South African name for the sclerophyllous vegetation, physiognomically similar to chaparral, which occurs in the Cape Province

Gg

game ranch – a farm specialising in wildlife used in licensed hunting for food, sport and photography

gastro-enteritis – inflammation of the stomach and intestines

gender – refers to differences between boys and girls; and women and men; that are socially constructed, changeable over time, and that have wide variations within and between cultures. While sex refers to the permanent, given characteristic of individuals, gender refers to what it means to be born male or female, and it varies with time, place, class, culture and ethnicity

gender analysis – studies the roles played by women and men, including such key issues as gender-based division of labour, productive and reproductive activities, access to and control over resources and benefits, and socio-economic and environmental factors that influence women and men. Gender analysis also refers to the systematic investigation of the differential impacts of development on women and men. Gender analysis takes into account factors such as race, class, ethnicity, sexual preferences, and age. A gender analysis assists in understanding the full picture and provides a project with a better chance of meeting its objectives than one based on incomplete information

gender awareness – means a commitment to placing women's and men's needs and priorities at the centre of development planning. It includes an awareness of different needs of women and men, and awareness that programmes and activities impact differently on women and men

gender blindness – is the inability to consider the differences between women's and men's needs, benefits, access to resources, access to power, or socio-political status; it is the absence of gender analysis from thought or practice

gender division of labour – means an overall societal pattern of women being allocated one set of gender roles, and men to another. In practice, discrimination against women in the gender division of labour means that women are mostly confined to the more menial and lower paid jobs; whereas men have the jobs with the higher status and higher pay

gender equality – a state where there is no discrimination on grounds of a person's sex in the allocation of resources or benefits, or in access to services. Gender equality may be measured in terms of whether there is no discrimination – equality of opportunity; or whether there are no gender gaps – equality of results. The UN Convention of the Elimination of All Forms of Discrimination Against Women (CEDAW) can be understood as a statement of what the principle of gender equality of opportunity should mean in practice, for all aspects of life, and all sectors of the economy

gender gap – a measure of gender inequality. It is used to measure actual results arising from differences between females as a group and males as a group, in the material well-being, their allocation of resources, or their utilisation of services

gender issue – arises where an instance of gender inequality is recognised as undesirable, or unjust

gender roles – are roles which are classified by sex, but where this classification is social, and not biological

gender sensitivity – is the ability to recognise gender issues, and especially, the ability to recognise women's different perceptions and interest arising from their different social location and different gender roles

gender disaggregated data – is broken down by gender, where gender is one of the independent variables and where the information also distinguishes among the different activities, aspirations, needs and interests of women and men as well as their access to resources

genera – group of closely related species, the members of which are more closely related to each other than they are to members of other species

geology – study of the earth's surface and its underlying strata

geomorphology – study of the physical features of the earth's surface, their development and how they are related to the core beneath

granary – a storehouse for grain

grasslands – a type of vegetation dominated by grass with some trees, but differing from savannah woodland by being generally cooler and drier

greenhouse gases – gaseous constituents of the atmosphere, both natural and anthropogenic, that absorb and re-emit infrared radiation

greenhouse effect – preservation of warmth in the earth-atmosphere system caused by the presence in the atmosphere of gases such as carbon dioxide that transmit short wave energy (visible and ultraviolet radiation) to the earth's surface and temporarily absorb and retain heat radiating from the loss of energy to space

gross domestic product (GDP) – unduplicated value of production originating within the boundaries of a country, regardless of the ownership of the factors of production

goitre – an enlarged thyroid gland, often showing as a swelling in the neck

gorge – narrow valley with steep sides

gross reproduction rate – average number of female births that a woman would have under a certain schedule of childbearing represented by the current age specific fertility rate, a measure indicating the degree to which women are reproducing themselves

groundwater recharge – replacement of water, usually through rainwater percolating into the ground, to replenish water lost from the groundwater through abstraction, evaporation or transpiration

gullies – small narrow valley cut especially by heavy rain and usually growing in years of above average rainfall, particularly after a period of drought when the plant cover has died

Hh

habitat – area used by an animal or plant as a natural home

hazardous waste – waste that poses a risk to human or ecological health and requires special disposal techniques to make it harmless or less dangerous

heavy metal – metal such as lead, cadmium and zinc, which has a high atomic number and which, if present in soil, has a stunting effect on plants

hepatitis – inflammation of the liver

herbaceous – plant without perennial stems above the ground

horticulture – gardening, growing of plants for food or decoration

humidity – mass of water contained in a unit volume of air

hydroelectric power – electricity generated by waterpower. This is usually achieved by damming rivers and running the water under pressure through turbines

hydrology – science dealing with the properties of water and its occurrence in space and time

hydrological cycle – *See* water cycle

hydromorphic – these are clays with strongly developed, very coarse blocky to prismatic structure and are seasonally waterlogged

Ii

indigenous – people, plants or animals, which originate naturally in a region

ingenuity – inventiveness and cleverness

insidious – proceeding inconspicuously but harmfully

inundated – flooded

indigenous knowledge systems (IKS) – the development over many generations, through a traditional way of life of an in-depth knowledge of an ecosystem or ecosystems by indigenous people in which they live

impediment – hindrance or obstruction

infant mortality rate – the number of infant deaths per 1,000 live births

insolation – incoming solar radiation; solar radiation received at the earth's surface

interfluves – the elevated part of the landscape that extends between two adjacent valleys

Intertropical Convergence Zone – a zone of intense rain cloud development oscillating between the tropics created when the southeast trade winds meet with the northeast monsoon

invertebrates – organisms, which do not have the bony framework of their bodies such as the small water beetles and shrimps

Jj

Jurassic – one of the three Mesozoic periods, which followed the Triassic and preceded the Cretaceous periods

Ll

lacustrine – of lake
lagoon – a salt-water lake separated from the sea by a sand bank or coral reef or a small freshwater lake near a large lake or river
landfill – tipping of wastes into holes in the ground. A relatively cheap method used for both domestic waste and more hazardous chemicals
land tenure – the type of land ownership system
leaching – a process by which a substance is washed out of the soil by water passing through it. Leachate from landfills may contain organic pollutants such as heavy metals including lead, and high salt levels, which all are a potential source of groundwater pollution
leasehold – holding of property, land in this context, by contract in which the owner allows another person to use it for a specific period
lethal – causing or able to cause death
liquidation – closing down a business and dividing its assets between its creditors; also, to completely wipe out
lithodial soils – poorly developed, coarse-grained stony soils that occur in areas of transition between humid and semi-arid zones
lithosols – soils that are usually shallow located on escarpments and have not developed beyond their present depth either because of climatic conditions or other environmental factors
littoral – pertaining to the shore

Mm

macrophyte – large water plant
malnutrition – a dietary condition resulting from the absence of some foods or essential elements necessary for health and growth
mangrove – tropical plants which grow in salt-water swamps
marginal lands – lands with poor soils not suitable for cultivation
marine – of the sea or ocean
marsh – area of permanently wetland and the plants growing on it
masonry – stonework
massif – mountain heights forming a compact group
matriarch – a woman who is head of the family
measles – an infectious disease producing small red spots on the body
microcosm – a small scale, simplified, experimental ecosystem, laboratory or field based, which may be derived directly from nature or built up from cultures until the required conditions of organisms and environment are achieved
mitigation – reduction of impact or effect, for example, of natural disasters
mixed cropping – growing more than one species of plant on the same piece of land at the same time
molasses – uncrystallised syrup drained from raw sugar
monocropping – the planting of a single crop over a large area, a system vulnerable to soil erosion because the soil is left bare and exposed after harvesting
monsoon – seasonal winds usually associated with rainfall

Nn

native – local inhabitant or an indigenous plant or animal
net reproduction rate – average numbers of female births that a woman would have in her lifetime taking into account her survival prospects and a current schedule of childbearing
nexus – a connected group or series
niche – comfortable or apt position in life
non-consumptive use – not allowing for complete withdrawal
non-point source – source of pollution in which pollutants originate from over a wide area or from a number of small inputs rather than from distinct, identifiable sources
non-renewable resource – natural resource that cannot be replaced, regenerated, or brought back to its original state once it has been extracted
nutrients – essential elements such as carbon, nitrogen, calcium, phosphorus absorbed by plants and animals from food that enable growth and protection from diseases
nutrient depletion – type of chemical degradation of soil in which soil nutrients are lost through too-frequent fires, topsoil loss and/or leaching
nutrient enrichment – *See* eutrophication

Oo

ore – solid rock or mineral, found in the earth's crust, from which metal or other useful or valuable substances can be extracted

organic compounds – compounds based on carbon and usually containing hydrogen, with or without oxygen, nitrogen, or other elements. Organic originally meant "of plant or animal origin", and is still sometimes used in this way

organic pollutants – organic compounds that pollute

organochlorine compounds – chlorine-containing organic compounds, which may contain oxygen and other elements, such as phosphorus. The term includes many pesticides and industrial chemicals, such as dioxins, furans, and dieldrin

orthodox – conventional or currently accepted beliefs

orthoferralitic – types of ferralitic soils, which result from deep weathering of underlying rock followed by intense leaching of bases

Oshanas – local name for the system of interconnected drainage channels that flow through central Owambo in Namibia

overgrazing – destruction of vegetation when too many grazing animals feed too long and exceed the carrying capacity of a rangeland area

Pp

palaeo- – Prefix, meaning ancient or prehistoric

palustrine – of marsh, swamp, pond, pan, lagoon, dambo, spring

pan – very shallow seasonal body of water, usually salty

panacea – a remedy for all kinds of diseases or troubles

paradox – a statement that seems to contradict itself or to conflict with common sense but which contains a truth

paraferrallitic – types of ferrallitic soils, which are derived *in-situ* from granitic parent material

pathogen – an agent, such as bacteria or virus that causes a disease

patriarch – the male head of a family

pelagic – in marine ecology, applied to the organisms that inhabit open water

perennial – lasting throughout the year

permeable – capable of being penetrated through

pesticide – chemical used to kill unwanted plants and animals

piscivorous – fish eating

pneumatic – filled with or operated by compressed air

point source – a source that is distinct and identifiable, for example, smokestacks and outfall pipes from industrial plants and municipal sewage treatment plants

policy – a set of government or corporate objectives and guidelines deliberately chosen to influence future decisions

pollution – the poisoning of land, air or water with anything that reduces its ability to support life

polyps – sedentary aquatic organisms with a tube-shaped body

pontoon – a kind of flat bottomed boat

population density – the average number of plant and animal inhabitants living per unit area

population growth rate – annual growth rate of population calculated from mid-year

potable – drinkable

precipitation – water, which falls from clouds to the ground as rain, snow, dew, fog, frost, hail, mist, sleet, etc., or to cause a substance to be deposited in solid form from a solution in which it is present

pristine – in an ancient and unspoilt state

prologue – introductory event

promulgate – make known to the public or disseminate

protected area – an area set aside by law for the preservation of given aspects of cultural and natural heritage

prospect – to explore in search of something

Rr

radar – a system for detecting the presence, position, or movement of objects by sending out short radio waves which they reflect

radiation – spreading out in all directions from a central point

rainfall variability – pattern of rainfall where the amount of rain and where it falls differ widely from year to year, The changes from year to year in rainfall from its long-term average

rain-fed agriculture – the practice of growing crops or rearing animals that is entirely dependent on rainfall

Ramsar – a city in Iran where the United Nations Convention on Wetlands of international importance especially as habitat for wildfowl was concluded in 1975. Wetlands under this convention are known as Ramsar sites

recession – a temporary decline in economic activity or prosperity or water levels

recycle – processes for reclaiming material that would otherwise be disposed of as waste

refugee – a person who has left their country sheltering from war, persecution or natural disaster

repository – harbour of information

reservoir – a large natural or artificial collection of water forming a small lake especially used as a source of water supply, also a component or components of the climate system where a greenhouse gas or a precursor of a greenhouse gas is stored
reticulated – divided into a network
riparian – of or on a riverbank; sharing basin
riverine – of river, floodplain or swamp

Ss

sabotage – deliberate damage to productive capacity especially as a political act
saturation deficit – the difference between saturation vapour pressure and actual vapour pressure
seafaring – working or travelling on the sea
sedges – a grass like plant growing in marshes or near water
seine – a large fishing-net that hangs vertically with floats at the top and weights at the bottom, the ends being dawn together to enclose fish as it is hauled ashore
semi-arid – areas where the mean annual rainfall is between about 250 and 600 mm, usually seasonal and variable, and where potential evaporation is high
shifting cultivation – also known as slash-and-burn cultivation. A traditional form of agriculture in which an area of forest is cleared by burning, cultivated for a few years and then abandoned as soil fertility and crop yields fall
silicosis – an abnormal condition of the lungs caused by inhaling dust that contains silica
siltation – the deposition of sediments by water in a river channel or reservoir
sink – process, activity or mechanism, which removes a greenhouse gas, an aerosol or a precursor of a greenhouse gas from the atmosphere
slash-and burn-agriculture – *See* shifting cultivation
smallholder – a person who farms a small agricultural unit under 20 hectares in area, usually run as a small family concern
smith – a person who makes things in metal
smelt – to heat and melt (ore) so as to obtain the metal it contains
soil degradation – declining productivity of soils resulting from a combination of physical factors such as drought, management factors such as cultivation of marginal land, and historical and socio-economic factors such as inequitable distribution of land
soil erosion – detachment and movement of soil by the action of wind (wind erosion) and moving water (water erosion)
soluble – able to be dissolved in liquid
spawn – produce as eggs or young
speciation – process of forming a new species
species – group of related individuals with common hereditary morphology, chromosome number and structure, physiological, and way of life, separated from neighbouring groups by a barrier that is generally sexual in nature
sport hunting – an example of consumptive use of wildlife that usually constitutes a big and lucrative industry where it is permitted
subsidence – sinking (sometimes of land) motion of air characteristic of anti-cyclones
subsistence – a situation in which people provide for all their own needs from their immediate environment, rather than earning wages to pay for goods and services
suffruticose – perennial plant, which has a wooden base to its stem and does not die down to ground level in winter
sustainability – the ability to maintain ecological and other processes and functions over long time
sustainable development – term coined by the World Commission on Environment and Development to denote development which meets the needs of the present without compromising the ability of future generations to meet their own needs, development that does not require a continuous input from outside to sustain itself
swamp – area of waterlogged ground and the plants that grow on it

Tt

tamarind – a tropical tree bearing fruit with acid pulp
taxonomy – the scientific classification of organisms
telemetry – the process of recording the readings of an instrument at a distance, usually by means of a radio
tectonics – the scientific study of the earth's structural features as a whole
topography – a detailed description of the physical features of a region
total fertility rate – average number of children that a woman would have if she bore children according to a certain specified schedule of childbearing represented by the age specific fertility rates
transitory – existing for a time but not lasting
trepidation – fear or anxiety
Triassic – the earliest of the three periods of the

Mesozoic Era, which lasted from 248 million years to 213 million years ago

trophic – processes of energy and nutrient transfer from one or more organisms to other in an ecosystem

tropical cyclone – an intense tropical storm with high winds and heavy rainfall, revolving (clockwise direction) round a low pressure area

trough – an area or zone of relatively low pressure with cold air in it, often extending outwards from a depression

trypanosomiasis – sleeping sickness, a disease attacking both livestock and humans, spread by the tsetse fly

turbid – water that is cloudy or murky as a result of suspended sediment

turbidity – state of being muddy or unclear

typhoid – a serious infectious disease that attacks the intestines, caused by bacteria taken into the body in food or drink

Uu

unpalatable – unpleasant to the taste or to the mind

upstream – the direction which is opposite to the flow of a river, towards the source; the upper part of a catchment basin

usufruct rights – rights to something you do not own

Vv

vagary – changes or fluctuations

volatile organic compounds (VOCs) – a chemically diverse group of compounds that have at least one carbon atom and are highly volatile

volcanism – movement of magma or molten rock onto or towards the Earth's surface

Ww

water cycle – a natural and continuous process in which water changes from one state of matter to another and redistributes itself replenishing itself in the process. Such changes include evaporation to clouds, rain, stream or river flow

water logged – state of water saturation in the soil

water hyacinth – a plant with fragrant bell-shaped flowers, growing from a bulb regarded as an indicator of water pollution in a water body

wetlands – areas of marsh, fern, peat land or water, whether natural or artificial, permanent or temporary, with water that is static or flowing, fresh, brackish or salty, including areas of marine water, the depth of which at low tide does not exceed six metres

wildlife – animals and plants that are not domestic

2A

TRENDS AND SCENARIOS WORKSHOP

Harare, 28 August, 1999

Chairperson
Prof. C. Magadza, University of Zimbabwe

Facilitator
Tabeth Matiza-Chiuta, IUCN-ROSA

Baltazar, T.	Mozambique High Commission in Zimbabwe
Bolaane, M. M. M.	University of Botswana – Botswana
Chatta-Chipepa, L.	SARDC-WIDSAA
Chenje, J.	IUCN-ROSA – Zimbabwe
Chenje, M.	SARDC-IMERCSA
Chief Mukuni, S.	Lumpasa Palace, Livingstone, Zambia
Garanganga, B. J.	Drought Monitoring Centre – Zimbabwe
Guveya, E.	IUCN-ROSA – Zimbabwe
Kalaote Kalaote	Department of Water Affairs – Botswana
Kaluwa, P.W.R.	Ministry of Water Development – Malawi
Kullberg, M.	SARDC-IMERCSA
Laisi, E.	SARDC-IMERCSA
Mafuta, C.	SARDC-IMERCSA
Marongwe, D.	Ministry of Mines, Environment and Tourism – Zimbabwe
Muganda, A. A.	Tanzania High Commissioner to Zimbabwe
Murty, A.S.	Zambezi River Authority
Musarira, P.	SARDC-IMERCSA
Mutambirwa, C.	SARDC-IMERCSA
Obol, C.	SADC ELMS – Lesotho
Ramoeli, P.	SADC WSCU – Lesotho
Sola, L.	SARDC-IMERCSA
Timberlake, J.	Biodiversity Foundation for Africa, Bulawayo – Zimbabwe
Zulu, E.	SADC MSCU – Zambia

INDEX

Nn

Oo

Pp

Rr

Ss

Zz

SOEPROZ PROGRAMME PARTNERS

SADC – Southern African Development Community
SADC is made up of 14 members states; Angola, Botswana, the Democratic Republic of Congo, Lesotho, Malawi, Mauritius, Mozambique, Namibia, South Africa, Swaziland, Seychelles, Tanzania, Zambia and Zimbabwe. Being an intergovernmental organisation, SADC is structured into development sectors whose respective coordination is entrusted to the governments of specific member states. The government of Lesotho hosts the SADC Environment and Land Management Sector (SADC-ELMS) responsible for overall coordination of environmental issues, and the SADC Water Sector Coordination Unit (SADC WSCU) whose vision is to attain sustainable, integrated planning, development, utilisation and management of water resources in the SADC region. The WSCU has the overall responsibility of coordinating water resources programmes and projects of the SADC water sector, including the Zambezi River Action Plan (ZACPLAN).

IUCN – The World Conservation Union
The World Conservation Union is the world's largest conservation organisation bringing together 76 states, 104 government agencies, 720 non-governmental organisations, 35 affiliates, and some 10,000 scientists and experts from 181 countries in a unique worldwide partnership. For more than half a century, IUCN has endeavoured to shape a just world that values and conserves nature. The IUCN Regional Office for Southern Africa (ROSA) based in Zimbabwe and with country offices in Botswana, Mozambique, South Africa and Zambia, coordinates regional programmes that are implemented through partnerships with members and other key organisations in all the SADC countries except the Democratic Republic of Congo, Seychelles and Tanzania. The IUCN mission in southern Africa is to facilitate and strengthen an integrated approach for the sustainable and equitable use of natural resources and the conservation of biological diversity.

ZRA – Zambezi River Authority
The Zambezi River Authority (ZRA) is a corporate body that is jointly owned by the Governments of Zambia and Zimbabwe and primarily mandated to operate, monitor and maintain the Kariba Dam, reservoir and any other installations owned by the Authority. In partnership with other organisations, the ZRA has been commissioned to undertake research, collect and process hydrological and environmental data of the Zambezi River and basin for the better performance of its functions, and ensure effective and efficient use of the waters and other resources of the Zambezi. Of particular interest is the Zambezi River Action Plan's Zambezi River Action Projects (ZACPRO) whose research findings are widely referred to and provided a good information base for this report.

SARDC – Southern African Research and Documentation Centre
SARDC is an independent institution whose primary mandate is to research, collect, analyse and disseminate information about the southern African region, with offices in Harare and Maputo. Under the Communicating the Environment Programme (CEP), SARDC established the Musokotwane Environment Resource Centre for Southern Africa (IMERCSA) whose objective is to bridge the environmental information gap by producing state-of-the-environment reports and providing researchers with up-to-date information on southern Africa's environment among other activities. IMERCSA coordinated the preparation of this report.

Sida – Swedish International Development Cooperation Agency
Sweden has, through Sida, been engaged for many years in the water sector, especially in rural water supply, sanitation and in the field of hydropower in developing countries. In southern Africa, Sida initiated in 1996 a three-year pilot programme with an overall objective to support integrated management of shared water resources in the SADC region. Programmes and projects, which receive either financial or technical support from Sida are of a regional nature, involving at least two SADC countries. All activities supported promote regional integration and cooperation. The programme is coordinated by the Swedish Embassy in Harare, Zimbabwe, on behalf of the Department for Natural Resources and the Environment at Sida Stockholm.